Contents

Preface

The book is written to provide a first point of reference for those interested in, working in or studying subjects related to the countryside. We have put at the end of each chapter a section on references, further reading and websites that we hope will be useful to you.

The countryside is a place of industry, a major resource for leisure and a haven for wildlife. *The Countryside Notebook* started life as a collection of ideas formed during the 15 years of editing *The Agricultural Notebook*. With each edition of *The Agricultural Notebook* the debate ran as to which subjects could or should remain. Despite the book increasing from 500 to 720 pages in length, the pressure to include more on existing topics meant that some subjects had to go. I am delighted that with this new title, *The Countryside Notebook*, contemporary subjects such as social capital are included, along with some more traditional areas such as game, aquaculture and equine.

We do hope you enjoy reading *The Countryside Notebook*, not I suspect from cover to cover but as an essential reference book on the shelf. We would welcome feedback from readers on aspects of the book that we might amend in subsequent editions.

We hope this book will be welcomed by the many, many people who live, work or simply enjoy non-urban areas.

Acknowledgements

Particular thanks to all the contributors who have found time in their busy schedules to write for this first edition of *The Countryside Notebook*. We hope this may be the first of many editions.

I should also mention Primrose McConnell who started the idea of an *Agricultural Notebook* back in 1883. Twenty editions later the book is selling better than ever, although to a different audience. His *Agricultural Notebook* has always been part of the countryside.

Grateful thanks also to the team at Blackwell Publishing, particularly Nigel Balmforth as publisher, who helped nurture the idea into reality, and also Josie Severn, Laura Price and Shahzia Chaudhri. Many thanks to Julie Musk for her work as copyeditor on the script – her attention to detail as ever is so impressive.

Finally, thank you to my family, Nicci, Will, Libby, Loretta and Charlotte, for their support.

Richard Soffe

Contributors

Richard Soffe, MPhil, MCIM, M Inst M, ARAgS

Present appointment: Director of Professional Development, Faculty of Social Science and Business, University of Plymouth

Richard's previous appointments have included a lectureship at Sparsholt College, Winchester, and assistant farm manager of a large estate in Hampshire. Research interests include rural leadership, management and marketing. Richard is Course Director of the innovative and well-respected Challenge of Rural Leadership course at the University of Plymouth. He was co-editor of the eighteenth and editor of the nineteenth and twentieth editions of *The Agricultural Notebook*.

Sue Blackburn, BSc

Present appointment: Senior Lecturer in Planning and Environmental Management, University of Plymouth

Sue is a professional planner and spent time in practice as a consultant and in the public sector before pursuing an academic career. She has research interests in environmental transitions, community participation and governance, gender and environment, and representation of countryside in children's literature.

Paul Brassley, BSc (Hons), BLitt, PhD

Present appointment: Senior Lecturer in Rural History and Policy, School of Geography, University of Plymouth

After reading agriculture and agricultural economics at the University of Newcastle-upon-Tyne, Paul went on to research in agricultural history at Oxford. Since his appointment to his present post at the University of Plymouth, he has taught agricultural economics and policy and researched agricultural history from the seventeenth to the twentieth centuries.

Paul R. Brunt, BSc, PhD, MILT

Present appointment: Head of Marketing, Tourism and Hospitality, Plymouth Business School, University of Plymouth

Paul started work at Seale-Hayne Faculty of Agriculture, Food and Land Use in 1989 as a lecturer in tourism, following PhD research in the Department of Tourism at Bournemouth University. He led the development of the tourism management degree programme in 1992, and has been its programme manager since its inception. He has also assisted in the development of numerous other undergraduate and postgraduate programmes within the University and throughout its partner college network. More recently, curriculum development work and other professional activities include external examination, chair of programme approvals, QAA Subject Specialist Reviewer and executive committee member of the Association for Tourism in Higher Education. Paul's teaching and research interests include tourism and crime and tourism research methods (Subject Specialist for Tourism and Hospitality Research Methods by the LTSN). Paul has published widely the results of research in the tourism and crime area including media representations of terrorist acts against tourists. He is the author of *Market Research in Travel and Tourism* (Butterworth Heinemann) and co-author of *Tourism: A Modern Synthesis* (Thomson International Publishing). He has recently embarked on research investigating social exclusion and crime in English seaside resorts.

Adam Carter, BSc (Hons), MSc, MBBS, MRCPCH (Part 2)

Previously: Lecturer in Forestry, Faculty of Land, Food and Leisure, University of Plymouth

Adam graduated in rural environment studies from Wye College, University of London, before completing an MSc in forestry at the Oxford Forestry Institute, University of Oxford. His previous experience includes British silviculture and forestry extension work in Sudan with Voluntary Services Overseas (VSO). Current research interests are in forest ecology, silviculture and medicinal uses of forest products.

Peter Child, MA (Cantab), Dip Conservation Studies, IHBC

Peter has recently retired from Devon County Council where he was Historic Buildings Adviser, a post he held for 14 years, following other jobs in building conservation and planning. The study and conservation of farm buildings was a significant part of his work and he has published on this subject. He is a committee member of the Historic Farm Buildings Group.

R.A. Cooper, CDA (Hons), NDA, MSc, PhD

Previously: Principal Lecturer, Animal Production, University of Plymouth

Dr Cooper taught animal production at the Shropshire Farm Institute before moving to the University of Malawi as lecturer in animal husbandry and assistant farms director. He joined Seale-Hayne in 1974 following an MSc at Reading University and in 1982 completed a PhD on interactions between growth promoters and reproductive physiology in ewe lambs.

Tim Felton, LLb (Hons) of the Middle Temple, Barrister at Law

Present appointment: Senior Lecturer, Bar Vocational Course, BILP, University of West of England

After reading law at Leeds University, Tim was called to the Bar of the Middle Temple. Following a period of legal practice he took up a career in practical farming and obtained a Diploma in Farm Management from the University of Plymouth. Prior to taking his present position he was share-farming a mixed dairy, arable and beef farm of 185 ha at Tiverton, Devon. Particular interests include access to the countryside and employment law issues in the rural environment.

Patrick Haughton, MSc, BSc (Hons)

Present appointment: Senior Lecturer, Sparsholt College, Winchester

Pat joined the fishery section at Sparsholt College in 1983. He lectures in sport fishery and fish farming business management and finance. He jointly owns and manages a coarse fish farming business, Hampshire Carp Hatcheries.

Dan Horsely, BSc (Hons)

Present appointment: Sparsholt College, Winchester

Dan completed his degree in aquaculture and fisheries management at Sparsholt College in 2003. He now works for the Environment Agency performing fisheries monitoring work, and is looking to take on a fisheries-related PhD.

Jeremy Houghton Brown, FRAgS, BPhil Ed. (Hons), BHSSM, PrinDip ABRS

Jeremy is a practical and experienced horseman who was manager of the British National Equestrian Centre and the British Racing School prior to becoming Head of the Horse Department at Warwickshire College, where he led the development of equine education in Britain. He is also the author of several equine text books.

Nicolas H. Lampkin, BSc (Hons), PhD

Present appointment: Senior Lecturer, Agricultural Economics, Institute of Rural Sciences, University of Wales, Aberystwyth

Nicolas graduated in agricultural economics from the University of Wales, Aberystwyth, and then undertook PhD research on the economics of conversion to organic farming. Subsequent research has focused on the financial performance of organic farming and the role of organic farming in the development of the Common Agricultural Policy. He is currently co-ordinator of organic agriculture teaching and research at IRS and director of the Organic Centre Wales. He is author/editor of *Organic Farming* (Farming Press) and the *Organic Farm Management Handbook* (UWA/EFRC).

Matt Lobley, BA (Hons), PG Cert, MSc, PhD

Present appointment: Assistant Director and Senior Research Fellow, Centre for Rural Research, School of Geography, Archaeology and Earth Resources, University of Exeter

Current research interests lie in the areas of CAP reform and the transition to European rural policy; environmental and social impacts of agricultural restructuring; farm household processes including lifecycle and succession issues; and the design, delivery and response to agri-environmental policy.

Matt Reed, BSocSc (Hons) MSc, PhD

Present appointment: Research Fellow, Centre for Rural Research, School of Geography, Archaeology and Earth Resources, University of Exeter

Matt graduated in social science, specialised in politics and for several years worked in the catering industry in a variety of roles. On returning to university he spent several years at the University of the West of England, as a Research Student and Visiting Lecturer, before taking up the post of Research Fellow at the University of Plymouth. He has recently concluded a doctoral thesis on the development of the Soil Association and is engaged in a DEFRA study of the contribution of organic farming to the rural economy. Other research interests included rural protests, new technologies in rural areas and rural social exclusion.

Alison Samuel, BSc (Hons)

Present appointment: Senior Lecturer, School of Biological Science, University of Plymouth

Alison's current research interests lie in the fields of organic farming, crop protection and potato quality, with teaching interests in crop production and protection and agricultural systems.

Mark A.H. Stone, BA (Hons), MSc (Econ), MCIPD

Present appointment: University of Plymouth Teaching Fellow and Principal Lecturer in People Management and Electronically Supported Open and Distance Learning

After reading economics and politics at the University of Central Lancashire, Mark went on to a masters degree in employment studies at University College, Cardiff. He spent five years working in industry, during which time he achieved membership status within the Chartered Institute of Personnel and Development. He is involved in people management and information technology teaching. Mark recently led a national project into student progression and transfer and is now Director of The Higher Education Learning Partnerships Centre for Excellence.

Stephen Tapper, BSc, PhD

Present appointment: Director of Policy and Public Affairs, The Game Conservancy Trust, Fordingbridge

Stephen has written and edited numerous papers on game conservation.

Paul Tyler, BSc (Hons), DipTP, MRTPI

Present appointment: Strategic Implementation Officer, Devon County Council

Paul joined Devon County Council in 1982 as a strategic planner specialising in retail and rural policy issues. He is currently involved with the development of two new towns in Devon, particularly advising on service delivery and sustainability. He also lectures part time in town planning at the University of Plymouth.

Pauline Warner, BSc (Hons), DipTP, MRTPI

Present appointment: Strategic Planner, Devon County Council

Pauline is currently employed as a Strategic Planner working on regional spatial strategy and rural planning issues.

Roger J. Wilkins, BSc, PhD, DSc, FIBiol, CBiol, FRAgS

Present appointment: Visiting Professor, University of Plymouth, Visiting Professor, Department of Agriculture, University of Reading, Research Associate, Institute of Grassland and Environmental Research, Academician, Russian Academy of Agricultural Science

After completing a PhD at the University of New England, Australia, he was appointed to the Grassland Research Institute, Hurley, in 1966. He held a series of posts in that Institute and its successor, the Institute of Grassland and Environmental Research, including Institute Deputy Director, Head of the North Wyke Research Station, Devon, and Head of the Soils and Agroecology Department, until retirement in 2000. Research interests include the production and utilization of grassland and the environmental implications of grassland farming. He was recipient of the RASE Research Medal and the British Grassland Society Award, and President of the British Grassland Society, 1986–87 and 1995–96.

Eirene Williams, BSc, PhD, MIEEM

Present appointment: Principal Lecturer, University of Plymouth

Eirene's current research interests lie in the fields of plant ecology and mycology, with recent work on rare species on the Culm grasslands and multi-access computerized keys to plants and fungi.

Ian Willoughby, BSc (Hons), MBA, MIC (For)

Present appointment: Silviculturist, Forestry Commission Research Agency, Alice Holt Lodge, Farnham, Surrey

After joining the Forestry Commission as a forest manager, Ian moved into research and has published extensively in scientific journals and technical manuals in the field of forest vegetation management. His current research interests include integrated forest vegetation management, reducing pesticide use, and direct seeding.

Michael Winter, BSc, PhD

Present appointment: Head of the School of Geography, Archaeology and Earth Resources, University of Exeter, Professor of Rural Policy and Co-Director of the Centre for Rural Research, University of Exeter

Current research interests lie in the fields of rural policy, sustainable agriculture and food systems, rural social change, and animals in rural society. Michael has held visiting academic positions at the Universities of Bath and Wageningen. He is currently a Research Associate of the Institute of Grassland and Environmental Research and a Visiting Professor at the University of Plymouth. He is Chair of the South West Rural Affairs Forum, a member of the Chamber for Rural Enterprise, Chair of the Hatherleigh Area Project and Chair of the Devon Rural Network.

Part 1

RURAL DEVELOPMENT

1

One thousand years of rural life

P. Brassley

Introduction

One thousand years is a purely arbitrary period, and it is important to remember that people have been living in what we now call the English countryside for many thousands of years. However, from the tenth and eleventh centuries A.D. we begin to have written evidence such as Anglo-Saxon estate records and, most famously, in 1086, the Domesday Book. Upon this small base (and with much assistance from the work of archaeologists) we can begin to write a history of rural England. This chapter is far too short even to summarize all that historians know about the subject, but that is not its purpose; rather, it is an attempt to remind the reader that the English countryside discussed in succeeding chapters did not just emerge, fully fledged, in the last few years, but that it developed over many hundreds of years, and sometimes it is only possible to understand what is happening now by knowing something about events long past.

Rural population, society and government

Perhaps the most obvious question concerns the number of people living in rural England. How many were there, and were they a big or a small proportion of the total English population? The first historical evidence comes from the Domesday Book, but it only provides an idea of the number of households in the places it mentions. There were other places, including much of the north of England, that were not mentioned at all, and there is much argument over the number of people found in each household. Given these uncertainties, it is not surprising to find that estimates of eleventh-century population vary enormously, from as few as 1.2

million to as many as 2.4 million (Bartlett, 1993, p. 108). Of these, one estimate suggests that about 10% of the population lived in urban areas. After the Norman conquest in 1066 the population grew more or less steadily for the next 250 years, although again the precise figures are widely variable and hotly disputed, with some claiming as many as 7.2 million English people in 1300, and others only 4.5 million. The lower figure now seems more likely. Both town and countryside were dramatically affected by the depredations of the Black Death in 1348–49. In some ways this was the culmination of a series of years of high mortality which had affected the country since 1300. Neither was it the only visitation of the plague. There were a further series of outbreaks in the late fourteenth and early fifteenth centuries (Hatcher, 1977, pp. 68–71; Campbell, 1995). In consequence the population was reduced to around 2.5 million for the whole period up to about 1500. Thereafter there are more secure population estimates which demonstrate the growth in population and the transformation from a rural to an urban society (see Table 1.1)

The relative rise in the urban population is clear from the sixteenth century onwards. It increased rapidly with the onset of industrialization, in the eighteenth century, but it was not until 1851 that more than half of all English people lived in towns. The rural population peaked 10 years later, and from then onwards into the interwar years the rural population fell, partly as a response to falling agricultural employment, and partly due to decreasing employment in the rural manufacturing and service trades. However, from the early 1950s the recovery in the rural population that was such a feature of the second half of the twentieth century is apparent.

In what kind of society did these people live? If the question is difficult enough to answer concisely for our

Table 1.1 Changing urban and rural populations (numbers in millions).

Year	Total population	Of which	
		Rural	Urban
1520	2.4	2.27	0.13
1750	5.8	4.6	1.2
1801	8.7	6.3	2.4
1851	17.9	8.9	9.0
1861	20.1	9.1	11.0
1871	22.7	8.7	14.0
1901	32.5	7.5	25.0
1951	43.7	8.4	35.3
1991	47.0	9.2	37.8

Sources: Saville, 1957, pp. 2 and 61; Wrigley, 1987, p. 170; www.odpm.gov.uk (*Urban and Rural Area Definitions: A User Guide*, Section 1, p. 2).

own times, it is still more so for times past. Perhaps the most obvious difference from today is that those living in the eleventh-century countryside appeared to live in a much more static society. In terms of social status, Domesday book names about 180 tenants in chief (i.e. tenants of the king), who were the great landowners, about 1400 lesser landowners, and roughly 6000 small landowners. Occupying and working the land for this group were about 270 000 freemen, villeins, cottagers and slaves, and their families (Wood, 1986, p. 159; Harvey, 1988, pp. 47–8). To rise from the status of villein to landowner, in one lifetime, was virtually impossible. Over several generations members of a family might manage to move into trade or the professions (effectively meaning the church) and so eventually be in a position to purchase land, but it would take a very long time. Geographical mobility was equally limited: few people would work land in more than one manor. On the other hand, the idea that everybody stayed within the narrow confines of their own fields from one year's end to the next is not supported by the evidence. They would at least go to local market centres, and many of them would be called to move around the country at some point in their lives, perhaps as members of armies, or following their trade. When King Edward I was having castles built in North Wales in 1282–83, masons and carpenters from as far away as Northumberland, Norfolk, Kent and Somerset were impressed to help with the work. In 1355 Richard of Winchcombe, master mason, was sent with 48 other masons to work on the fortifications of Calais; the same

thing happened in 1373, when 20 carpenters and 10 masons were sent (Brown *et al.*, 1963, pp. 183 and 427).

By this time the social rigidities of the early medieval period were beginning to break down under the pressures of economic and demographic change. Labour shortages after the Black Death meant that tenants could obtain better terms from landlords, so that serfdom gradually disappeared and a peasant land market began to develop. In consequence, the more successful farmers began to acquire more land, and the range of farm sizes began to increase. In 1484 at Cheshunt in Hertfordshire one third of the tenants held over two thirds of the land (Wrightson, 2002, p. 101). Nonetheless, an account of English society written at the beginning of the sixteenth century still described the three 'estates' of society that medieval people would have recognized (Dyer, 2003, p. 363): the clergy, who prayed for all; the nobility and gentry, with their duties of government and defence; and the common people. However, it was now possible to recognize significant social status differences among this last group, from the wealthy merchants and substantial farmers, through the tradesmen and craftsmen and small farmers, down to the labourers and household servants (Wrightson, 2002, p. 28).

This pattern proved remarkably stable, lasting in many rural districts down to the beginning of the twentieth century. What changed it was the decline of the landowning class (Howkins, 2003, pp. 11–14). In 1876 about a million people in England and Wales owned land, but over 70% of them owned less than an acre. At the opposite extreme, about 7000 families owned 80% of the land, most of which they rented out to the farmers. Then farm prices fell, tax laws changed, and more and more landowners put their estates on the market. Although the extent, speed and timing of these sales have been the subject of much argument, there can be little doubt about the eventual outcome: at the end of the nineteenth century the majority of the agricultural land was tenanted; at the end of the twentieth the majority was owner-occupied. (Hill and Ray, 1987; Howkins, 2003, pp. 55–60). It was probably the middle-sized estates – those between 3000 and 10 000 acres – that changed most, but almost all were affected to some degree. Furthermore, with the change in landownership went changes in political and social influence. At the end of the nineteenth century landowners still influenced national politics to a degree that is remarkable in an industrial economy. Similarly, their social influence was felt at a local level: even in 1930 the squire's views on anything from religious observance to game preservation might affect the behaviour of villagers. However, by the 1950s it was clear that the farmers, operating through the National Farmers' Union, were dominating national politics on rural issues; indeed, the 1947 Agriculture Act required that the Union be consulted at the annual round of price negotiations. At a local level, the farmers took over the social position vacated by the landowners, occupying leading positions on county and parish councils and village institutions from the church to the flower show (Newby, 1980, p. 184). It did not last. While they had been climbing to the top of the rural social pile, those at the bottom of it had been drifting away. There were 891 000 farm labourers in 1947 and only 251 000 in 1995, less than half of whom worked full time (Marks and Britton, 1989, p. 138; Brassley, 1997, p. 77). It is more difficult to quantify the leakage of rural craftsmen and tradesmen, but it was no less apparent. At the beginning of the twentieth century many villages had not only their blacksmith and carpenter, but also their wheelwright, baker, bootmaker and tailor, all of whom had largely disappeared by the middle of the century (Saville, 1957, p. 212). However, their houses remained, occupied now by the commuting or second-home-owning middle class. Farmers and landowners bemoaned the disappearance of the rural labour force and deplored the rise of the one-class village in the late twentieth century; whether such communities worked better or worse than their predecessors as social units is more a question for the sociologist than the historian (see Bell, 1994).

These developments in rural society were reflected in the government of rural communities. It is often thought that the manor, the village and the parish were more or less synonymous, and although this was certainly true in some cases, there were many parishes containing more than one settlement or village, and more than one manor. Equally, some large manors might extend over more than one parish. The manor was initially the principal unit of administration, through the manor court. In theory, there might be two kinds: the court baron, dealing with administrative issues, such as the transfer of land, and the court leet, with responsibility for minor misdemeanours or crimes.

In practice, the records often suggest that both were dealt with in the same court. Then as manorial control of farming and land tenure began to break down in the fifteenth and sixteenth centuries, so the parish took over as the principal unit of local administration. Each parish was given responsibility for its highways in 1555, and shortly after for the relief of the poor and the punishment of vagrants. These duties were carried out by annually elected constables, surveyors of highways and overseers of the poor. Since these individuals were often the more substantial tenant farmers, and the farmers paid the bulk of the rates that financed their work, they had some influence on government at the parish level, notwithstanding the dominant position of the gentry. This remained so after 1834 when poor law administration was transferred to Boards of Guardians, and after rural police forces were formed in 1839. The system therefore remained essentially unchanged until the establishment of parish councils and rural district councils in 1894 (Edwards, 1993, pp. 9–10; Digby, 2000, pp. 1425–6). The other part of the system, the county councils, were created by the Local Government Act of 1888, and took over much of the local administrative work that had been done by the sheriff in the Middle Ages and the Justices of the Peace from the sixteenth century onwards. Finally in 1974 local government reorganization established the present pattern of parish, district and county councils.

The rural economy

Until recently, government policy has assumed that the rural economy is synonymous with the agricultural industry. This may have been true between about 1935 and 1985, but it was not always so. According to Professor Wrigley's calculations, 80% of the population was involved in agriculture to some extent in 1520, but of these many may have had part-time employment in handicraft industries of some kind. By 1801, he calculates, only half of the rural population was employed in agriculture (Wrigley, 1987, p. 171).

The industries of the countryside fell into two categories: those that were widespread, and catered for a local market, and those that were geographically concentrated, but catered for regional or national markets. Examples of the former included the blacksmiths, carpenters, shoemakers and millers that might be found in most villages. Domesday Book records about 5000 mills, which would have been almost all watermills, since the windmill was not introduced until the late twelfth century (Hurst, 1988, pp. 928–30). Of the second category, the big five in the medieval countryside were saltmaking, stone quarrying, metalliferous mining, fishing and the cloth industry. The last of these became especially important in the pastoral areas in the fourteenth and fifteenth centuries. Places that are now small country towns, such as Lavenham in Suffolk, Castle Coombe in Wiltshire or Totnes in Devon can trace their expansion to the rise of the rural textile industry in this period. It was not just the presence of the right raw materials that explained rural industrialization. The right sort of landholding and inheritance customs, those allowing the proliferation of small farms in pastoral areas, were perhaps more important. In such areas there was not the heavy demand for field labour as in the arable districts, but there was often a need to supplement the small family income from farming. In Staffordshire, for example, the pottery industry was already growing in the seventeenth century, and the range of trades at that time included lead, ironstone and coal miners, scythesmiths and other edge tool makers, other metal workers who produced harness fittings, nails, door and window handles and locks, wood turners and tanners, leather workers, and hemp and flax weavers. The growth of these trades sometimes led to subsequent industrialization, as in the Potteries, but it was not inevitable. Needwood and Kinver Forests, which were involved in this early Staffordshire metal working, are clearly rural areas today: deindustrialization is not just a recent phenomenon. It affected the Weald of Kent and Sussex, where the iron and cloth industries were declining by the middle of the seventeenth century, and the charcoal and textile trades in the Lake District at about the same time. The woollen industries in East Anglia and the West Country declined in the eighteenth and nineteenth centuries. The concentration of manufacturing that occurred in what is usually called the Industrial Revolution clearly affected the rural districts, but still it did not mean that they were entirely dependent upon farming. Quarrying, ore extraction, brickmaking, saltworking, as well as the local service trades continued to produce a demand for rural labour, and new activities, such as the development of reservoirs and sewage treatment works, emerged in the nineteenth century. Opencast mining

and electricity generation and distribution were added to the list in the twentieth century (Thirsk, 1973; Glasscock, 1992; Palmer and Neaverson, 1994).

By the end of the twentieth century the most important industry in many parts of the countryside, in terms of numbers of people employed, was tourism. It is tempting to think of it as a relatively late development, and indeed, in comparison with some of the activities discussed earlier, so it is. However, it is possible to detect its beginnings as early as the eighteenth century. Visitors were employing local people to show them round the caves at Castleton in the Peak District in the 1750s, a guidebook listed country houses worth visiting in 1760, and at the beginning of the nineteenth century Stonehenge featured as one of the major attractions in another guidebook, *The Beauties of Wiltshire*. (Ousby, 1990, pp. 65, 96, 133). At this time rural tourism remained largely an activity for the elite; transport developments opened it to a mass market. First the railways in the mid-nineteenth century, then the bicycle and finally the motor vehicle in the twentieth century. By the inter-war years public transport operators advertised the pleasures of the countryside for the walker and the cyclist as the petrol companies advertised them to the rather smaller number of motorists. The invitation was enthusiastically accepted, and walking and cycling organizations and youth hostels proliferated. The demand for access to the countryside was not always eagerly supplied at that point, as the mass trespass on Kinder Scout in the 1930s demonstrates, but there could be little doubt even then that the desire to escape from the town would be matched by the need to find new ways of making a living in the countryside. By the 1980s there were more people working in tourism than in agriculture in Devon, and some areas could even be identified as the heritage landscapes of popular culture: Yorkshire had Emmerdale Farm country and Herriot country, Tarka country, with its Tarka Trail, was in Devon, and Cornwall had Daphne du Maurier country (Atkins *et al.*, 1998, p. 243). The rural landscape itself had become a commodity.

Whether or not the same could be said of its wildlife is debatable, but it is certainly true that a small proportion of the rural labour force still makes its living from field sports. The principal field sports have been hunting, hawking, shooting and fishing, although it should not be forgotten that within these broad categories is an enormous range of variations. In addition to fox hunting, there was also deer and boar hunting, and hunting with beagles, harriers (for hares) and otter hounds. Partridge and pheasant shooting is still accompanied by grouse shooting, wildfowling, pigeon and other rough shooting, and clay pigeon shooting. Fishing includes sea fishing, coarse fishing and fly fishing, and the conventional differences between them in Britain do not always apply in other countries. All of these have their own histories, and have in the past conflicted with each other, with poachers and with protestors. Other activities might be included within the category of traditional country sports, such as bear, bull and badger baiting, bull running, dog fighting, goose riding, cock throwing, cock fighting, rat catching and, at the other end of the acceptability continuum, steeplechasing (Malcolmson, 1973, pp. 45–51, 66–7; Holt, 1989, pp. 16–18; Billett, 1994, pp. 14–15).

In centuries past hunting with dogs was more widespread than today (indeed in 2004 it became illegal). Predominant in the early medieval period was deer hunting, which involved a slow ride through wooded country accompanied by hunt servants on foot. By the late seventeenth century wild deer were becoming rare, and hunters turned to the hare and fox. The hare was regarded as morally superior to the fox, and a supreme test of the hunter's skill, but by the late eighteenth century fox hunting had emerged as the fashionable sport. It was partly a matter of faster horses, of improved hounds and especially the influence of Hugo Meynell, who developed a new fast and exciting style of hunting during his mastership of the Quorn Hunt between 1753 and 1800. Paradoxically, the numbers of participants rose most quickly after the development of the railways. By the mid 1840s it was possible to leave Euston station in London at 6.30 a.m., hunt with the Quorn, and be back in London by the late evening: the rich no longer had to choose between country and town sports (Carr, 1976).

By the end of the nineteenth century, hunting, with its demand for hunt servants and other workers to take care of the horses, hounds and hunters, was a significant employer in some parts of the countryside such as Leicestershire. Its main rival was shooting, which in a way developed from hawking, in the sense that falcons were flown at the same prey – plover, snipe, duck, grouse, pheasant – that later became targets for guns. Sixteenth-century guns took a long time to fire, used small quantities of irregular shot and enveloped the shooter in smoke, so the original practice was to get close to the birds on the ground and shoot as many as possible with one blast. As gun design improved it became possible to shoot flying birds, but it was not until the middle of the nineteenth century that technology permitted the modern form of shooting flying birds driven over the guns. From then on the shooting business developed rapidly. Prince Albert took to grouse and deer shooting, and by 1900 there were thousands of let grouse moors in Scotland, and numerous shoots in England, all requiring their keepers, gun and game dealers, and beaters, to the extent that a report on timber production in the early twentieth century complained that woods were being managed for the benefit of pheasants more than trees (Vandervell and Coles, 1980; James, 1981).

The conflict between production and pleasure is thus an enduring theme in the history of the countryside. It can be seen from the early medieval period, when farmers complained that forest laws in Devon made agricultural expansion difficult (Stanes, 1986, p. 49) to the present day, with conflicts over access to the countryside. In general, the producers have had the better of it, but not always and everywhere. The same story of continuity applies to farming: roughly the same number of farmers are farming about the same amount of land now as farmed it at the time of Domesday book, nearly a thousand years ago (Brassley, 2000, p. 15).

At the time of Domesday Book, farmland in England was going through the process of being reallocated so that it could be farmed in common, in the two or three great open fields that we commonly associate with the medieval village. It was a process that took place between the ninth and thirteenth centuries, and in different ways in different regions, but in the end it affected most of the country in one way or another. At the same time, as the population increased, more land was being taken into cultivation, and farming for the market, as opposed to simply providing for the needs of the family and the local labour force, became more common. Parts of northern England began to concentrate on producing sheep and cattle, and farmers in southern and eastern England specialized in grain production. By 1300 the London grain market was served from as far as 20 miles away by land carriage and 60 miles by water, so that the coastal areas of East Anglia were drawn into the trade (Campbell et al., 1993, p. 173). In the 250 years after the Norman conquest the population almost certainly doubled, and its food and drink (apart from a few luxury products such as wine) came from home production, so agriculture was clearly thriving and expanding.

The fourteenth and fifteenth centuries were different. The population no longer increased, the grain market was no longer so buoyant, and the social certainties of the feudal system began to be eroded. Historians continue to debate the reasons for these changes: some argue that population and food production had expanded to the limits of the technology then available, so that poorly fed people had lowered resistance to disease; others blame the wars of the early fourteenth century, with their associated taxation and disruption of trade (Postan, 1975, pp. 42–3; Campbell, 1995, pp. 95–6). Whatever the reason, it culminated in the crisis of 1348–49, when the Black Death killed between one third and one half of the population. For the following 150 years it remained at the new lower level. In consequence, the demand for labour exceeded supply, while the demand for corn was correspondingly reduced. Arable was converted to pasture, farmers turned to wool and dairy products, farm sizes and layouts began to change, and a land market began to develop. Despite these changes, it would be a mistake to regard agriculture in 1500 as similar to today's industry, but with lower levels of technology. A majority of people remained responsible for producing a high proportion of their own food, some as substantial yeomen who regularly hired labour, some as small farmers, some as cottagers with a few acres who regularly worked for others, down to those with little more than a garden who were almost entirely dependent upon what they could earn from others. There was little rural land that had no agricultural use, even if it was only extensive grazing, so sixteenth-century farmers, given that the woodland area was a little larger than it is now, and the urban area much less, were probably using much the same total area of land as we use today, only less intensively. In some areas, such as the midlands, most of the land was in arable, meadow or regulated common pasture. The only lightly grazed waste in Leicestershire was in Charwood Forest (Thirsk, 1958, p. 14). Conversely, the fens remained as marshy pasture, yet to be converted into arable.

From about the middle of the sixteenth century to the middle of the nineteenth, technical and organizational changes transformed agriculture in England. New food crops were introduced, most notably the potato. New fodder crops appeared too: turnips, swedes, mangel wurzels, ryegrass and legumes such as clover, trefoil and sainfoin. They fed new breeds of cattle and sheep that had been bred to grow faster and provide a greater proportion of their carcases as meat. Water meadows were created to provide an earlier bite of grass in spring. The fens were drained by dykes and pumps in the seventeenth century, and the invention of a machine for making drainage tiles meant that thousands of acres of farmland could be underdrained between 1850 and 1870. From the early eighteenth century farmers learned the value of lime to control soil acidity, and farmyard manure was supplemented, from the 1830s, by imported fertilizers (Brassley, 2000). Additionally, the open fields disappeared. It was a process that began slowly in the sixteenth century, speeded up in the seventeenth and was concluded in the eighteenth and early nineteenth centuries with the aid of numerous Acts of Parliament (Turner, 1980; Wordie, 1983).

In 1751 English farmers produced almost all of the food for about six million people. By 1851 they were producing 84% of the food for a population of nearly 17 million, so their output had more than doubled, which is why several historians have described this century as a period of agricultural revolution. Whether or not it was the most revolutionary century is a matter of debate. In 1951 English farmers produced slightly less than half of the food eaten by 41 million people, whereas in 2000 their output accounted for two thirds of the food consumed by 50 million. In effect, they fed another 15 million people (Brassley, 2000). Which change is the greater depends on whether we value relative change more highly than absolute or vice versa. Clearly both were enormously important. The difference was that in the late twentieth century the change was accompanied by environmental consequences that attracted more and more attention from the consuming population.

The rural environment: wildlife and landscape

It is often assumed that a pleasant landscape benefits wildlife and vice versa. While this may be true in some cases, it is not always so. Not everyone likes to see overgrown unkempt hedges and boggy meadows, yet these often produce ideal habitats for some species of plants and animals. The following brief history of the rural environment therefore separates the two, insofar as it is possible to do so, and concentrates on wildlife and wildlife habitats since landscape changes are dealt with at greater length in Chapter 7.

The expansion and intensification of economic activity described above was the biggest influence upon the development of the rural environment. Since its population was so small, it is tempting to think of the England of Domesday Book in 1086 as an ideal wildlife habitat. In fact, it was already extensively affected by human activity, and had been so for thousands of years. The Domesday survey suggests that about a third of the country was under cultivation, roughly 15% in meadows, gardens and built-up land, probably the same quantity in woodland (which is roughly double the present-day figure), and thus about a third remained in pasture, heath, marshes, moors and waste (Rackham, 1980, pp. 126–7). Even this final third was by no means untouched by human activity: most of it would have been grazed to some extent by domesticated animals, it was a source of gorse or furze, which had all sorts of uses from animal fodder to firewood, and some people hunted over it. Neither was it uniformly distributed, for there was more of it in highland than in lowland England. Over much of midland and south-east England, therefore, the Domesday landscape was one of islands of uncultivated land in a sea of farmland, rather than the reverse.

Over the next 200 years many of the moors and wastes, as well as some woodland, were preserved by being made subject to forest law, which means that hunting and its requirements took precedence over other activities, not that these areas were devoted to timber production. In fact, the twelfth and thirteenth centuries were a time of rapid expansion of the arable area at the expense of woodland, a process known as 'assarting'. The woods that remained were often carefully managed to produce charcoal, tan bark and other coppice products as well as timber. Marshlands too were increasingly exploited in this period. In Somerset, at Sowy island, where the rivers Tone, Cary and Parrett meet, the records of Glastonbury Abbey show that nearly 400 ha of marshland had been converted into meadow by the middle of the thirteenth century (Silvester, 1999, p. 130), and similar changes were happening at the same time in the fens around the Wash and the Norfolk Broads. This process of wetland reclamation went on sporadically for hundreds of years. The great fenland schemes of the seventeenth century are well known, but farmland underdrainage in the late nineteenth century and coastal reclamation in the late twentieth also converted what had been marshland into farmland.

To put it crudely, the main impact of human activity was to control the amount of wildlife habitat, and thus to determine the numbers of plants and animals that could exist. However, it was more complex than this. Many plants and animals can co-exist more or less happily with agriculture or woodland management. Periodic coppicing, for example, increases the light level on the woodland floor and produces a flush of bluebells; hares make their forms in grassland or cereal crops; cinnabar moth caterpillars live on ragwort, a pasture weed. Many species are well adapted to traditional agriculture, and in only a few cases can the extinction of a species in Britain be attributed to the expansion of human settlement. Among these would probably be the disappearance of the brown bear, which had already gone from Britain by the tenth century, the beaver, gone by the thirteenth century, and the wolf, the last recorded sighting of which in England was made in 1486 (Ponting, 1991, p. 163). On the other hand, wildlife was sometime actively persecuted. The impact of nineteenth-century gamekeepers on stoats, weasels and raptor species is well known; what is less appreciated is that a late seventeenth-century Act of Parliament empowered churchwardens to pay a bounty for the destruction of many species of birds and mammals. At Tenterden in Kent, for example, in 1688–89, they paid for numerous raven, magpies and crows, and even a few kites. At Great Budworth, Cheshire, in 1678, Ralph Shawcross was paid 1s. 4d. (approx. 7p) for killing 43 bullfinches (Jones, 1972, p. 113). Human influence also accounts for the introduction of some species: the grey squirrel, brought in at the end of the nineteenth century, is perhaps the best known, but it should not be forgotten that the rabbit was introduced, probably from France, shortly before 1100 A.D. (Sheail, 1971, p. 17). Later, between 1825 and 1850, Japanese knotweed (*Fallopia japonica*) was introduced from Japan and Himalayan balsam (*Impatiens glandulifera*) from India. They began as garden plants, but were both found growing wild by 1900 (Mabey, 1996, pp. 106–7 and 274–5). Most recently, the biggest impact on the environment has come from what might be termed molecular introductions: first fertilizers, then pesticides, and most recently genetically modified organisms.

Conclusions

The English countryside is now much more heavily populated than it was a thousand years ago. The number of farmers has not changed much. They now produce much more than their forebears, with the aid of more technology rather than more people. The big increase has been in the number of people who work outside agriculture. Until the end of the nineteenth century they were largely occupied in rural crafts and trades; now the majority of them commute to jobs in the town. Timber production has probably declined in importance, and the underwood trades have virtually vanished. The social elite left some evidence that they valued the landscape from the eighteenth century onwards, and their preference has now become more widely distributed among society. A thousand years ago wildlife appears to have been valued to the extent that it could be hunted; now it is valued for its rarity and as an indicator of the health of the environment. Whether there are now any social or cultural differences between people who live in the town and those who live in the country is increasingly debated.

References

Atkins, P., Simmonds, I. and Roberts, B.K. (1998) *People, Land and Time.* Arnold, London.

Bartlett, R. (1993) *The Making of Europe.* Penguin, Harmondsworth.

Bell, M.M. (1994) *Childerley: Nature and Morality in a Country Village.* University of Chicago Press, Chicago.

Billett, M. (1994) *History of English Country Sports.* Robert Hale, London.

Brassley, P. (1997) *Agricultural Economics and the CAP: An Introduction.* Blackwell Science, Oxford.

Brassley, P. (2000) One thousand years of English agriculture. *Journal of the Royal Agricultural Society of England,* Millennium edition, pp. 14–24.

Brown, R.A., Colvin, H.M. and Taylor, A.J. (1963) *The History of the King's Works.* HMSO, London.

Campbell, B.M.S. (1995) Ecology versus economics in late thirteenth- and early fourteenth-century English agricul-ture. In: *Agriculture in the Middle Ages* (ed. D. Sweeney). University of Pennsylvania Press, Philadelphia.

Campbell, B.M.S., Galloway, J.A., Keene, D. and Murphy, M. (1993) *A Medieval Capital and its Grain Supply: Agrarian Production and Distribution in the London Region c. 1300.* Historical Geography Research series no. 30. Institute of British Geographers, University of Pennsylvania Press, Philadelphia.

Carr, R. (1976) *English Fox Hunting: A History.* Weidenfeld and Nicolson, London.

Digby, A. (2000) The local state. In: *The Agrarian History of England and Wales* (ed. E.J.T. Collins), vol. 7, pp. 1850–1914. Cambridge University Press, Cambridge.

Dyer, C. (2003) *Making a Living in the Middle Ages: The People of Britain 850–1520.* Penguin, Harmondsworth.

Edwards, P. (1993) *Rural Life.* Batsford, London.

Glasscock, R. (ed.) (1992) *Historic Landscapes of Britain from the Air.* Cambridge University Press, Cambridge.

Harvey, S. (1988) Domesday England. In: *The Agrarian History of England and Wales* (ed. H.E. Hallam), vol. 2, pp. 1042–350. Cambridge University Press, Cambridge.

Hatcher, J. (1977) *Plague, Population, and the English Medieval Economy 1348–1530.* Macmillan, London.

Hill, B. and Ray, D. (1987) *Economics for Agriculture: Food, Farming and the Rural Economy.* Macmillan, London.

Holt, R. (1989) *Sport and the British. A Modern History.* Oxford University Press, Oxford.

Howkins, A. (2003) *The Death of Rural England: A Social History of the Countryside Since 1900.* Routledge, London.

Hurst, J.G. (1988) Rural building in England and Wales: England. In: *The Agrarian History of England and Wales* (ed. H.E. Hallam), vol. 2, pp. 1042–350. Cambridge University Press, Cambridge.

James, N.D.G. (1981) *A History of English Forestry.* Basil Blackwell, Oxford.

Jones, E.L. (1972) The bird pests of British agriculture in recent centuries. *Agricultural History Review,* **20** (2), 107–25.

Mabey, R. (1996) *Flora Britannica.* Sinclair Stevenson, London.

Malcolmson, R.W. (1973) *Popular Recreations in English Society, 1700–1850.* Cambridge University Press, Cambridge.

Marks, H.F. and Britton, D.K. (1989) *A Hundred Years of British Food and Farming: A Statistical Survey.* Taylor and Francis, London.

Newby, H. (1980) *Green and Pleasant Land? Social Change in Rural England.* Penguin, Harmondsworth.

Ousby, I. (1990) *The Englishman's England.* Cambridge University Press, Cambridge.

Palmer, M. and Neaverson, P. (1994) *Industry in the Landscape, 1700–1900.* Routledge, London.

Ponting, C. (1991) *A Green History of the World.* Sinclair-Stevenson, London.

Postan, M.M. (1975) *The Medieval Economy and Society.* Penguin, Harmondsworth.

Rackham, O. (1980) *Ancient Woodland: Its History, Vegetation and Uses in England.* Arnold, London.

Saville, J. (1957) *Rural Depopulation in England and Wales 1851–1951.* Routledge and Kegan Paul, London.

Sheail, J. (1971) *Rabbits and Their History.* David and Charles, Newton Abbot.

Silvester, R. (1999) Medieval reclamation of marsh and fen. In: *Water Management in the English Landscape* (eds H. Cook and T. Williamson). Edinburgh University Press, Edinburgh.

Stanes, R. (1986) *A History of Devon.* Phillimore, Chichester.

Thirsk, J. (1958) *Tudor Enclosures.* Historical Association Pamphlets, General Series no. 41. Macmillan, London.

Thirsk, J. (1973) The roots of industrial England. In: *Man Made the Landscape* (eds A.R.H. Baker and J.B. Harley). David and Charles, Newton Abbot.

Turner, M. (1980) *English Parliamentary Enclosure.* Dawson, Folkestone.

Vandervell, A. and Coles, C. (1980) *Game and the English Landscape: The Influence of the Chase on Sporting Art and Scenery.* Debrett's Peerage, London.

Wood, M. (1986) *Domesday: A Search for the Roots of England.* BBC Books, London.

Wordie, J.R. (1983) The chronology of English enclosure, 1500–1914. *Economic History Review*, **36**, 483–505.

Wrightson, K. (2002) *Earthly Necessities: Economic Lives in Early Modern Britain, 1470–1750.* Penguin, Harmondsworth.

Wrigley, E.A. (1987) *People, Cities and Wealth: The Transformation of Traditional Society.* Blackwell, Oxford.

Useful website

www.odpm.gov.uk/stellent/groups/odpm_planning/documents/page/odpm_plan (accessed 25 Nov 2003).

Part 2

RURAL SOCIETY AND GOVERNMENT

2

An introduction to contemporary rural economies

M. Winter and M. Lobley

Introduction

The rural economy has been the focus of considerable attention in recent years – from the media, policy makers and academics. The 2001 foot-and-mouth disease outbreak highlighted the dependency of many rural businesses on the physical fabric of the country-side (i.e. the rural environment), the interdependency between different rural business sectors (e.g. tourism and agriculture) and the diminishing role of farming in the rural economy. At the same time, economic policy for rural areas, which at one time was primarily delivered through a narrowly agricultural sectoral approach, is increasingly shifting to a territorial focus with a growing emphasis on rural development. Against this background, this chapter provides a brief overview of contemporary rural economies. Drawing on a recently completed literature review (commissioned by DEFRA) on the English rural economy (Winter and Rushbrook, 2003) the chapter explores some of the key characteristics of rural economies and rural firms. In doing so it also illustrates the changing nature of research into rural economies which, in recognizing that economic behaviour is essentially social behaviour, is increasingly turning toward an understanding of the way in which economic actors (i.e. people) interact as a means of understanding the processes behind some of the key characteristics of rural economies.

Characteristics of rural economies

The decline of agriculture as the economic driving force in rural areas has been well documented (Performance Innovation Unit, 1999). It is several decades since agriculture provided the majority of jobs in rural areas and its contribution to GDP (gross domestic product) is now as little as 1.3% in the UK as a whole. That said, there are some geographical areas where agriculture continues to account for a high proportion of jobs. For example, the Government's Performance Innovation Unit (PIU) report on *Rural Economies* (1999) identified several districts in the South West, West Midlands, East Anglia and the North where agriculture accounts for more than 10% of employment. In addition, agriculture continues to have knock-on effects in upstream and downstream sectors. Nevertheless, it has been argued (Winter and Rushbrook, 2003) that possibly the greatest local economic significance of agriculture is its role in underpinning the environmental quality of rural areas, which, in turn, acts as a magnet for new rural residents and new economic activity (see below).

As the economic prominence of agriculture has declined, the occupational profile of rural areas has become increasingly diverse, with large numbers of jobs in the service and manufacturing sectors. Indeed, other than the presence of agriculture, fisheries and forestry, there are few significant differences between rural and urban areas in terms of business types (Countryside Agency, 2002). There are, however, some significant differences beyond the broad business type classifications of services, manufacturing, etc. The last four decades have seen a well-documented shift of manufacturing capacity from urban locations to rural areas to take advantage of greater space and beneficial differences in operating costs (Keeble and Taylor, 1995) and rural areas now contain more SMEs (small- and medium-sized enterprises) than large towns (Keeble, 1998). Rural economies also display a number of other distinctive characteristics (PIU, 1999; Countryside Agency, 2002):

- a high proportion of micro businesses, with over 90% of rural firms employing fewer than ten people and a smaller proportion of large firms;
- a higher rate of new small firm formation;
- a higher proportion of self-employment (9% of people of working age in rural areas compared to 6.5% in urban areas);
- low wages and a higher rate of unemployment and underemployment in remoter rural areas (such as Devon and Cornwall);
- declining levels of service provision (e.g. village shops and post offices).

Rural businesses

A small but growing body of literature points to some distinctive rural business characteristics, including a greater degree of specialization in niche markets. Research undertaken in the early 1990s suggests that firms in remote rural areas tend to specialize in niche markets serving retailing, tourism and agricultural sectors, whereas in accessible rural areas, firms specialize in markets created by increasing business and technological complexity (Keeble et al., 1992). As we have seen above, rural businesses tend to be small, often a reflection of high rates of new firm formation (NFF), and many are operated by managers who cite a range of advantages stemming from a rural location, such as attractive living conditions, good labour relations, lower wage and premises costs and greater space for expansion. Indeed, evidence suggests that the rural environment itself is a significant driver of rural economic development. Keeble et al. (1992) point to environmentally induced migration to rural areas and link this to high rates of rural NFF.

In contrast to urban areas, most rural entrepreneurs are in-migrants attracted by the residential and business opportunities afforded by rural areas. Keeble and Taylor (1995) argue that high rates of NFF and growth result from the dynamic effects of environmentally induced migration fuelled by rising real household incomes and a proliferation of market niches for specialized products and services. For example, in a survey of 949 manufacturing SMEs in north Devon and south Warwickshire, 58% of the Devon respondents and 46%

of those in south Warwickshire felt that their rural location offered them business advantages, and close to one-third claimed that the perceived attractiveness of their operating environment had helped to initiate visits from potential customers (Jarvis et al., 2002). However, it is important not to overemphasize the role of the rural environment in business development, as some respondents to the same survey cited negative factors such as the imagery of Devon as a county lacking in dynamism. Moreover, some authors have argued that there are still fundamental contradictions between economic development and environmental protection (Sneddon, 2000), with protective designations and restrictive planning control limiting the availability of suitable land and property and shortages of affordable housing (see Chapter 3) adversely affecting the availability of skilled labour.

Evidence on issues relating to labour in rural businesses is somewhat confusing. While remoter firms frequently cite labour shortages, especially of managers, as a constraint to expansion, some authors point to the relatively more labour-intensive development paths of rural businesses, particularly those in remote rural areas (North and Smallbone, 1996). Smallbone et al. (2002) explain this in terms of rural SMEs' adaptability to local labour market conditions. Compared to their urban counterparts, rural SMEs have fewer opportunities for subcontracting out, which, in turn, contributes to them being more self-reliant in production terms. According to Smallbone et al., the impact of this is that for a given increase in sales the increase in employment in rural firms is greater than that in urban firms. In turn, this can help explain the higher growth performance of many rural firms compared with urban firms in the 1980s and early 1990s. Since then, however, there is evidence that rural SMEs have been outperformed by their urban counterparts (Smallbone et al., 2002), although there are notable differences between different types of rural business and the results are also sensitive to the definitions employed of rural and urban areas.

It is one thing to describe differences between rural and urban firms but quite another to explain the processes behind any observed differences (although the latter is of greater importance from a policy perspective). Most attempts at explanation focus on the characteristics of rural areas compared to urban areas or the impact of rural area characteristics on firm and entrepreneur behaviour. There is some consensus that the geographical constraints of distance are not usually considered a problem by the owners and managers of rural businesses in the UK. For example, it is thought that the majority of rural firms do not experience problems with regard to access to business services, finance capital or key infrastructure services such as communications (Keeble and Taylor, 1995), although others report that most rural firms do not consider that they suffer a competitive disadvantage as long as they are located within one hour of a motorway (North and Smallbone, 1996). For those in more remote and 'hostile' business environments, the environment itself seems to stimulate business innovation to overcome local constraints such as increased transport costs and labour shortages. Greater attention to marketing, training provision and/or enhanced wages are all cited as examples of business innovation and adaptability in the face of otherwise remote and hostile business environments (Vaesson and Keeble, 1995; North and Smallbone, 1996).

More recently, researchers have begun to examine the social processes associated with business success and failure. Terluin and Post (2000), for example, examined 'leading' and 'lagging' regions across nine EU members states and argue that successful economies are characterized by strong internal networks (enhanced by the attitudes of local actors, solidarity, easy communication and strong local leaders), whereas lagging regions were associated with weak local networks. In their study of paired rural areas in four countries, Bryden and Hart (2001) also examined the role of networks and point to a more ambiguous role, suggesting that strong networks can be a cause of poor economic performance through the exclusion of others and retention of information. On the other hand, networks that were effective in promoting access to information and co-operation between entrepreneurs and institutions were associated with good economic performance. In explaining the differential economic performance of rural areas, Bryden and Hart argue that, while at one level the explanatory variables are unique to the individual place, it is the interaction between a range of tangible and less tangible factors that accounts for differential economic performance:

- cultural traditions and social conditions;
- infrastructure and peripherality;

- governance, institutions and investment;
- entrepreneurship;
- economic structure and organization;
- human resources and demography.

As the authors point out, these six themes do not represent discrete categories and are closely related to each other. The important implication for policy is that measures that are not integrated or 'joined-up' in terms of design, delivery and implementation will be unable to successfully influence rural economic development (Bryden and Hart, 2001).

Local economies

The shift in rural policy towards more of a territorial focus and the growing policy emphasis on regional and local sustainable economic development are associated with several recent research projects addressing interactions within 'local' economies. For example, Courtney and Errington (2000) focus on local economic linkages as 'a network of transactions of varying nature which either contribute to the income generation within, or leakage from, the "local economy"'. Reducing economic leakages from particular localities, and therefore enhancing local economic multipliers, can provide a means of promoting endogenous economic development. The renewed focus on the local economy, however, extends beyond traditional concerns with economic multipliers, and in recent research on the economy of rural areas there has been a resurgence of interest in the importance of clusters, networks and innovation (Winter and Rushbrook, 2003). In summary:

- Frequently, the strengthening of local ties is seen as being a prerequisite for the formation of a stronger rural economy with the benefits of local enterprise cascading into the rest of the rural economy.
- In turn, this takes the study of endogenous development beyond the consideration of economic multipliers alone to consider the importance of a whole range of interactions and transactions which may strengthen the local economy (Courtney and Errington, 2000).
- Finally, this explicitly links rural development with the concerns of social capital and the concept of embeddedness (see below).

Errington and colleagues (Courtney and Errington, 2000; Errington and Courtney, 2000) have explored local economic linkages through a series of national and international studies. In examining transactions between firms and also between firms and households, these authors have sought to trace the 'economic footprint' of small market towns, exploring the extent to which they are integrated into their local economy. This has important implications for policy makers seeking to promote prosperous rural economies through the development of small 'market' towns. For instance, Courtney and Errington (2000) compare two small towns: Kingsbridge in a relatively remote part of Devon and Olney in a relatively accessible area of the English Midlands, close to the major urban centres of Milton Keynes, Northampton and Bedford.

- The results indicate that Kingsbridge is considerably more integrated into its locality than Olney. In Kingsbridge, over 38% of all sales revenue and 14% of all input purchases went to suppliers located within either the town or its hinterland, compared to 5% and less than 1% respectively for Olney.
- In both cases the local economy accounted for a higher proportion of sales revenue than supply expenditure, suggesting that contemporary small towns are more important as a market for sales than as a source of inputs for local firms.
- However, further analysis revealed some important differences according to the type of firm considered. For instance, small and independent firms (i.e. those that are not branches of larger national or international businesses) were more likely to be closely integrated into the local economy both in terms of their purchases and their sales. In addition, service sector, consumer service and non-agricultural firms were more closely tied to their local economy than manufacturing, producer services and agricultural firms.

The 'market towns' research both confirms and contradicts some earlier research findings (e.g. Gripaios *et al.*, 1989; Blackburn and Curran, 1993) and has raised some, as yet, unanswered questions. As Courtney and Errington (2000) acknowledge, differences in local economic integration are in part a function of economic structure within locations (i.e. the relative proportion of different types of firm) as well as a function of locational characteristics (e.g. remoteness and accessibil-ity), although it is unclear which (if either) is the most powerful determinant of local economic integration. In addition, it raises questions about the degree of interaction and linkage between economic actors – firms, entrepreneurs and household members.

Embeddedness, networks and social capital

In attempting to understand the development and functioning of local economies the concept of embeddedness has prompted growing attention to the wider social and cultural significance of economic relations. The seminal work on embeddedness is widely regarded to have been undertaken by Granovetter (1985), who stressed the role of social relations in generating the trust necessary for economic transactions to take place. Thus, research on embeddedness has concentrated on the social components of economic behaviour, particularly the role and value of networks of association and exchange. Given that economic transactions are ultimately socially based, differentiation in economic behaviour and action is also social, and the notions of social capital and social networks are receiving considerable attention as a potential explanatory factor in differential economic performance and rural development. Social capital forms from repeated social interactions which develop trust, reciprocity and social norms. Importantly, social capital can provide access to information, and key actors, can reduce transaction costs and facilitate innovation and development:

- According to Arnason and Lee (2003), 'the amount of social capital built is seen to depend on the quality and quantity of interactions. Social capital represents the ability of actors to secure benefits through membership of social networks and other social structures'.
- Maskell *et al.* (1998) state, 'some geographical environments are endowed with a structure as well as a culture which seem to be well suited for dynamic and economically sound development of knowledge, while other environments can function as a barrier to entrepreneurship and change' – an argument echoed in Bryden and Hart's comments on networks reported earlier.

- A review of the longer-term impacts of an experiment in integrated rural development in the Peak District National Park suggested that part of the success of the scheme was attributable to its ability to capitalize on exiting social networks in the development and implementation of rural development proposals (Errington *et al.*, 2000).

However, despite the growing interest in embeddedness, networks and social capital, there is limited empirical research within the non-farming rural business community, and while many commentators argue that the three interrelated concepts are important in rural development, the process is not well understood.

Governance

Another recent theme in rural economic research is that of governance. Deployed with increasing frequency in recent times, the notion of governance helps us to characterize 'the development of governing styles in which boundaries between and within public and private sectors have become blurred' (Stoker, 1998). In addition to this blurring of boundaries, Stoker (1998) identifies the significance of autonomous self-governing networks of actors and government playing a role of steering and guiding as well as, or in addition to, legislative provision (see also Jessop, 1997; Majone, 1997). These developments are crucial to the way in which economic development is promoted in policy terms. No discussion of rural economic development is now complete without considering the complex array of stakeholders and actors involved. To date, however, there have been few empirical studies addressing issues surrounding rural governance, leading Winter and Rushbrook (2003) to call for a programme of research exploring:

- how power and influence in rural areas affect the drivers of rural development as well as the constraints;
- in-depth studies of policy implementation within the new governance context; and
- an examination of partnerships emerging in rural regeneration policy.

Conclusions

As this chapter has demonstrated, while rural economies are often very similar to urban economies, they display some distinctive characteristics, such as a preponderance of small firms and high rates of new firm formation. The latter is also often linked to a defining characteristic of rural areas – the rural environment itself. Many entrepreneurial in-migrants are attracted to the rural environment and then proceed to establish a new business or relocate an existing business. While the precise nature and role of the 'rural environment' driver in rural economic development has yet to be fully understood, other new areas of interest are the focus of increasing attention. In seeking to understand and explain economic processes, policy makers, academic researchers and other commentators are increasingly turning to social factors, particularly the role of social capital and various networks of association. Again, while evidence remains limited, there are indications that social capital can provide an important 'lubricant' in economic processes, creating conditions of trust and reciprocity and reducing transaction costs.

Acknowledgements

This chapter is based on a literature review of the English rural economy (Winter and Rushbrook, 2003) funded by DEFRA. The authors are grateful to Liz Rushbrook for her contributions to the original report.

References

Arnason, A. and Lee, J. (2003) *Crofting Diversification: Networks and Rural Development in Skye and Lochalsh, Scotland*. The Arkleton Centre, University of Aberdeen.

Blackburn, R. and Curran, J. (1993) In search of spatial difference: evidence from a study of small service sector enterprises. In: *Small Firms in Urban and Rural Locations* (eds J. Curran and D. Storey). Routledge, London.

Bryden, J. and Hart, K. (2001) *Dynamics of Rural Areas (DORA): The International Comparison*. The Arkleton Centre, University of Aberdeen.

Countryside Agency (2002) *The State of the Countryside 2002*. The Countryside Agency, Cheltenham.

Courtney, P. and Errington, A. (2000) The role of small towns in the local economy and some implications for development policy. *Local Economy*, **15**, 280–301.

Errington, A., Blackburn, S., Lobley, M. *et al.* (2000) *Review of the Peak District IRD, eleven years on.* Final Report to the Countryside Agency. University of Plymouth.

Granovetter, M. (1985) Economic action and social structure: the problem of embeddedness. *American Journal of Sociology*, **91**, 481–510.

Gripaios, P., Bishop, P., Gripaios, R., *et al.* (1989) High technology industry in a peripheral area: the case of Plymouth. *Regional Studies*, **23** (2) 151–7.

Jarvis, D, *et al.* (2002) Rural industrialisation, 'quality' and service: some findings from south Warwickshire and north Devon. *Area*, **34**, 59–69.

Jessop, B. (1997) Capitalism and its future: remarks on regulation, government and governance. *Review of International Political Economy*, **4**, 561–81.

Keeble, D. (1998) North–south and urban–rural variations in SME growth, innovation and networking in the 1990s. In: *Enterprise Britain: Growth, Innovation and Public Policy in the Small and Medium Sized Enterprise Sector 1994–1997* (eds A. Cosh and A. Hughes). ESRC Centre for Business Research, University of Cambridge, pp. 99–113.

Keeble, D. and Taylor, P. (1995) Enterprising behaviour and the urban–rural shift. *Urban Studies*, **32**, 975–97.

Keeble, D., *et al.* (1992) *Business Success in the Countryside: The Performance of Rural Enterprise.* HMSO, London.

Majone, G. (1997) From the positive to the regulatory state: causes and consequences of changes in the mode of governance. *Journal of Public Policy*, **17** (2), 139–67.

Maskell, P., *et al.* (1998) *Competitiveness, Localised Learning and Regional Development: Specialisation and Prosperity in Small Open Economies.* Routledge, London.

North, D. and Smallbone, D. (1996) Small business development in remote rural areas. *Journal of Rural Studies*, **12**, 151–67.

Performance Innovation Unit (1999) *Rural Economies.* Cabinet Office, London.

Smallbone, D., *et al.* (2002) *Encouraging and Supporting Business in Rural Areas.* Report by the University of Middlesex to the Small Business Service.

Sneddon, C. (2000) 'Sustainability' in ecological economics, ecology and livelihoods: a review. *Progress in Human Geography*, **24**, 521–49.

Stoker, G. (1998) Governance as theory: five propositions. *International Social Science Journal*, **155**, 17–28.

Terluin, I. and Post, J. (2000) *Employment dynamics in leading and lagging rural regions of the EU: some key messages.* Paper presented at the European Rural Policy at the Crossroads Conference, The Arkleton Centre, Aberdeen.

Vaesson, P. and Keeble, D. (1995) Growth-oriented SMEs in unfavourable regional environments. *Regional Studies*, **29**, 489–505.

Winter, M. and Rushbrook, L. (2003) *Literature Review of the English Rural Economy.* Final report to DEFRA, University of Exeter.

More than the picturesque: An introduction to contemporary rural society

M. Reed and M. Lobley

How wide the limits stand
Between a splendid and a happy land

Oliver Goldsmith, *A Deserted Village* (1770)

Introduction

Rural life and, more importantly, rural living continues to fascinate many people. A swift perusal of the TV schedules quickly reveals an increasing number of programmes devoted to various aspects of country living and the rural idyll. Many people have a deep-seated desire to 'escape to the country'. Indeed, approximately 80 000 people a year migrate to rural areas of England and with a few exceptions most rural areas are gaining in population. However, the popular portrayal of rural areas in the media, frequently one of affluence in terms of material wealth, community well-being and a rich environment, is at odds with the lived experience of many people in rural areas. According to one commentator, combining the words 'English', 'rural' and 'community' produces an effect 'like a chemical chain reaction which grows and glows, suffusing everything in a good, green light – but an ideological light which can obscure as well as ornament the object of analysis' (Short, 1991). Significant numbers of rural dwellers find that the myth of the rural idyll is just that, and that the reality of rural living involves many problems and often considerable hardship.

The relatively recent focus of policy attention on rural development and capacity building in rural areas has helped expose some of the problems faced by members of rural communities. In reiterating the overall goal of establishing 'sustainable rural communities', the UK government has recognized the need to tackle the issue of social exclusion in rural areas which is leading to sharp differentiation both between and within rural areas (Beckett, 2003). Against this background, rather than attempt a fully comprehensive review of rural society (which would require a book in itself), this chapter focuses on the nature of social exclusion in rural areas and the role of *social capital* in both creating that exclusion and as a potential route out. In doing so we aim not to ignore questions of material poverty, far from it, but in focusing on social exclusion we aim to consider the full complexity of what makes people vulnerable to material need and how some of the processes that lead to this have become hardwired into contemporary rural society. This chapter considers many of the routes that lead to social exclusion in rural areas – lack of diverse employment, low educational attainment and the distance from major markets – but focuses in particular on one emerging problem – that of access to affordable housing. The price of private housing and the relative lack of social housing are leading to a situation where many rural areas are becoming hostile to any social group other than a wealthy middle class. Indeed, according to the Joseph Rowntree Foundation (JRF), 'planning for, and resourcing, affordable housing provision is fundamental to the economic, social and cultural sustainability of rural communities and to the life chances of many people' (JRF, 2000).

In considering any complex social problem a multitude of views emerge regarding causes and solutions.

Consequently, this chapter provides a brief overview of some of the different standpoints, the conceptual tools that are used to examine them and details about some of the problems. In this chapter we consider what is different about poverty in rural areas and the nature of poverty and social exclusion. This is followed by a discussion of social capital both as a way of explaining social exclusion and as a possible solution. We then review some of the evidence of social exclusion in rural areas and the processes that contribute to it as well as those that can be harnessed to create more inclusive rural areas. Before considering any of these issues, the chapter begins with a consideration of the nature of the powerful and persistent myth of the rural idyll and its role in obscuring important aspects of contemporary rural life.

The rural idyll

Myths about the virtues and value of rural living appear to have a particular potency and can be traced back to at least 300 BC and the writings of the *Idylls* by Theocritus. Although the modern usage of the word idyll to denote an idealized view of rural life does not apply to the work of Theocritus, he did employ the idea of countryside as a symbol for loss (Short, 1991). The modern myth of the rural idyll projects the countryside and country life as more wholesome, possessing greater moral affluence, a more kindly, civilized and peaceful place, and, in particular, the last resting place of traditional values (see, for example, Newby, 1985; Short, 1991; Bunce, 1994). As such, the rural idyll is a profoundly complex and emotive cultural symbol which serves to complicate and mask rural social exclusion. This often links closely to the argument that the proximity to nature and the support of a rural community compensates for a lack of material resources. Although there is widespread agreement in British cultural life about the beauty and splendour of rural areas, there is little discussion of the social processes that underpin these aesthetic creations. As many cultural geographers have indicated it is important to consider whose idyll the current shape of the countryside represents (Matless, 1998). Although seemingly a very abstract argument, it has very practical consequences. Preserving rural areas as a seeming idyll creates contradictions and conflicts. For example, in preserving a village,

restrictions on its development may be put in place that limit the building of new houses, forcing poor or younger people from the village. Alternatively, a hamlet may be very picturesque, making it desirable to those seeking holiday homes, once again forcing the less affluent from the village and lessening the community life of the area. The splendour of rural areas may have to be secured by limiting and channelling the opportunities of those who live there.

The second aspect of this common conception of the rural idyll is that rural identity takes the place of all other differences, creating a form of equality as all share in the rural life. North American anthropologists studying rural English villages in the past decade have arrived at widely divergent views about the importance of this in the social life of rural areas. Mayerfeld-Bell, in his study of a village in Hampshire, found that discussions of rural life were often ways of trying to cover-over class differences in the community (Mayerfeld-Bell, 1994). Rather than allowing the obvious financial and material differences in the village to become part of the discussion, residents would often assert a common rural identity. In doing so the poorer and more excluded in the village were denied opportunity to discuss their needs. Stephens, in his study of a village in Gloucestershire, having read Mayerfeld-Bell, found that there was a widely held idea by the residents that everyone in the village was reasonably well off and it was their rural identity not their financial situation that was important (Stephens, 2000). However, as Stephens noted he was only able to talk with middle class villagers and in doing so confirmed Mayerfeld-Bell's argument that discussion of material differences was downplayed. Instead of a single rural idyll or rural identity this suggests that there are many countrysides and many ways of living life in rural areas. Any attempt to assert one of these above all others will always exclude the voice of someone.

Poverty and rural life

Before moving to any detailed discussion of what we mean by social exclusion and social capital, we need to acknowledge that poverty exists in rural Britain. As suggested above, the first response of many people is to be surprised that there is hardship in rural areas; it runs counter to the dominant images of rural life on the

TV and in most of the media. All too often the image of rural life, shared by those who live in the country-side as much as by those who dwell in towns and cities, is that it is a wealthy one. Certainly it is true that very few large concentrations of poverty exist in rural areas, but by definition they could not. Rural hardship is more widely distributed. Poor people in rural areas often live next door to very wealthy people. There are no very large 'sink' estates where poverty is concentrated. As it is so dispersed and occurs in often scenic areas, the impact and importance of this hardship is frequently discounted. Those experiencing social exclusion in rural areas are socially invisible and often are encouraged to believe their lot is less hard because of their surroundings.

Poverty has often been broadly defined as being either absolute or relative. Absolute poverty is not having the wherewithal to keep oneself fed, clothed and sheltered appropriately. Although for many years this was thought to be absent from Western societies, there are certainly those who are struggling to make ends meet. Relative poverty, on the other hand, refers to the situation when individuals are poor in comparison to their peers. So, as incomes have risen some groups are either excluded from earning enough money and so are reliant on state benefits, or are working but do not earn enough to enjoy the standard of living of the majority. As is apparent in these definitions, they capture only a narrow aspect of social hardship (Alcock, 1993). For example, in considering absolute poverty there can be discussions about what is absolutely necessary. Is a car absolutely necessary? Possibly so if it is the only means of getting to a shop or place of employment in a remote rural area. Similar questions could be raised about relative poverty: should the level of poverty be set at 50% of the national average income or should it be based on the average income of the immediate community? Each of these approaches to poverty focuses on money and income rather than the experience of being poor, which is perhaps more complex than money alone. The lack of money, whether absolute or relative, is an outcome of a range of processes, which exclude individuals, families and other groups from what most would think of as an ordinary lifestyle. Consequently, as our understanding of the causes and impacts of poverty have become more sophisticated, so the more simple measurements of poverty have become less useful.

Social exclusion

Social science research has allowed us to understand that poverty causes problems with children's development, has impacts on people's physical and mental health and leads to them participating less in civic society. Indeed, there are many ways in which people are excluded from the benefits of modern society. In shifting the focus away from poverty alone, the social exclusion perspective involves a more multifaceted approach to understanding the processes that lead to hardship, poverty and a lack of social integration. That said, there is no agreed definition of social exclusion. The government's Social Exclusion Unit (SEU) defines social exclusion as 'a shorthand term for what can happen when people or areas suffer from a combination of linked problems such as unemployment, poor skills, low incomes, poor housing, high crime, bad health and family breakdown' (SEU, 2001). Taking a different approach, Shucksmith argues that there are three competing ways of thinking about social exclusion, which he pithily describes as 'no work' (integrationist), 'no money' (poverty) and 'no morals' (underclass) (Shucksmith, 2000). Working back through these competing schools of thought reveals more about the understanding of social exclusion. Those who subscribe to the underclass view of social exclusion emphasize the role of a 'culture of poverty' or 'dependency culture' which leads to individuals being poor. In this view those who are poor are generally morally or culturally deviant and through their actions bring poverty on themselves and their children. Those who argue for the poverty view emphasize the lack of material resources alone, the redistribution of which in itself would lead to a reversal of the effects of poverty. As might have been guessed from the paragraph above we have adopted a form of the modified integrationist approach.

The modified integrationist approach emphasizes the role of work, important not only for the money it provides, but also for what it brings individuals and their families socially. Those with work have a social identity, a role that they fulfil and a purpose in broader society. This not only fuels feelings of self-worth and confidence but also provides a way of integrating with wider society. Work brings colleagues with whom one can share information, form friendships and sometimes create new social and work opportunities. In focusing on the role of work in integrating people into society, the importance of relationships is brought to the fore. These are not just relationships between people but also between society and the individual. Social exclusion in this view is not about the failing of victims but rather the systems which fail individuals and push them out of the main flow of society. This is not to deny individuals a role in their own destiny, but it is to ensure an emphasis on the context within which they are trying to make their way (which is largely beyond the control of the individual).

The role of social networks and social capital

'Bonding social capital constitutes a kind of sociological superglue, whereas bridging social capital provides sociological WD-40.' (Putnam, 2001, p. 23)

One of the ways of trying to discuss the degree of connectedness of an individual or family within society is to consider their social capital. As with social exclusion, the term social capital is used in different ways by different authors. Social capital can be thought of in terms of knowledge and skills and of knowledge of, and association with, other people. Not all social capital is the same. Some is only useful in the dense networks of a specific locality, giving the holder special access or status – the insider status of a 'local person', while other forms of social capital, such as educational qualifications, may be more formalized and have credence in much wider formal networks. A consideration of social capital and its linkages to the concept of social networks can help provide further insight into the multifaceted processes of social exclusion and can also point to routes towards greater inclusion. However, as the popularity of the concepts of social capital grows, particularly its application in rural development, it should also be noted that developing and sustaining social capital is not a panacea and that in some circumstances social capital can hinder desirable change (see below).

According to Putnam, social capital is 'the theory that social networks have value' and that from the networks of daily life a number of civic virtues can emerge (Putnam, 2001). In this sense social capital is simultaneously a private and a public good, held by individuals but frequently only useful or present in their interactions with others. It emerges in the networks that people are involved in; these networks must be more than contacts in that it requires norms of reciprocity. On occasions, these mutual obligations are specific, and in others they are generalized. This means that within the network, there must be trust; indeed, Putnam illustrates this point with the aphorism 'Trustworthiness lubricates social life' (Putnam, 2001).

Social capital can be highly variable: the loose connection of an internet chat-group, the familial ties of blood in an extended family or the tight bonds of a professional group. Not all social capital is the same and not all of its uses are benign. Similarly, not all social networks are the same. Some networks are repeated, intimate and informal; groups of friends or workmates who meet on the weekends. Others are formal and often infrequent; parents' meetings or annual memorials. In understanding these networks Putnam argues that the most important distinction is between *bonding* and *bridging* social capital. Bonding social capital excludes others, creating a group around a common identity. It offers specific reciprocity and solidarities, providing members of the network with easily recognized boundaries. Examples may be ethnic groups, church groups or expensive clubs. Such networks provide psychological support for their members, as well as frequent opportunities that might not otherwise be available such as work or important information. Alongside this exclusivity can run prejudice, enforced homogeneity and even violence toward others. Bridging social capital looks out to other networks; it is based on weaker ties, but offers potentially greater opportunities. It can bring to individuals a wider range of information and opportunities. These are not 'either/or' forms of capital but possessed by individuals and within networks to lesser or greater extents. Some networks may regularly interact with certain other networks but not with any others, whilst others may be in constant fleeting contact with a diversity of different groups.

When faced with hardship, bonding capital, embedded in dense exclusive networks, may allow people to hunker down to wait until times change. Some people may find succour in the unquestioning acceptance of a close network. However, people previously engaged in a dense successful network, who lose their job, may feel the weight of now unfulfillable obligations bearing down on them and so withdraw. By exiting from the network they increase the strain on themselves, stripping away the succour and identity of the bonding capital. At the same time, existing within a dense local network can inhibit individuals who may feel socially stigmatized by being seen or known to be applying for state welfare payments.

Beyond the motorway network: the experience of social exclusion

'Somewhere, it is believed, at the far end of the M4 or A12, there are "real" county folk living in the midst of "real" English countryside in – that most elusive of all rustic utopias – "real" communities'. (Newby, 1985)

So far we have considered social exclusion, social capital and social networks from a largely conceptual position. The discussion now turns to describing and understanding the experience of social exclusion in rural areas. Most of the literature on social exclusion emphasizes the importance of locality (see, for example, King, 1999; Shucksmith, 2000), or at least the importance of understanding the local context in which larger societal forces and system failures are played out. Whilst accepting the locally contextualized nature of social exclusion (and the fact that different groups within the same locality can experience social exclusion in very different ways, e.g. older people and younger people in rural areas), a number of shared factors to rural social exclusion can be identified:

- Socially excluded households will be geographically scattered, often living alongside the far more affluent.
- Distance, isolation and poor access to jobs and services are a general feature of rural exclusion, placing importance on the ability to use private means of transport.
- Problems with housing tend not to be focused so much on the quality of housing, as it might be in urban areas, but on the availability of any affordable housing at all.
- Absolute lack of employment is not as much of a problem as underempolyment, temporary, low wage and seasonal employment.
- The myth of the rural idyll leads to misunderstandings about rural hardship, making it harder for those

living in rural areas to have their voices heard, by both those who live in urban areas and those who live in the countryside.
- Traditional rural attitudes about self-reliance may lead to individuals not making claims about their needs but also families and communities supporting others.

One of the defining characteristics of rural areas is that people and their homes are more geographically spread out. For those who make their living directly from working the land this provides them with an immediate site for their livelihood, while for others it means having to seek employment across a wider area. This has a direct impact on the demands made on people's incomes and social opportunities. With public transport severely limited in many rural areas, ownership of a car, motorbike or scooter is essential to get to work or to go shopping. Although some areas of rural Britain have been described as ex-urban – so close to cities and major towns with the major amenities these offer that the impacts of rural isolation are not important – at the other extreme some remote rural areas (such as the far north of Scotland, with a population density of 2.2 persons/km^2) are among the most sparsely populated in Europe. The impacts of this isolation are various, sometimes boosting an active tourist trade and in other cases undermining the local economy to the disadvantage of many local residents.

In terms of income level, Shucksmith reports research indicating that there are proportionately fewer people with low incomes in rural areas. However, he also goes on to point out that a third of people in rural areas have experienced a period where their income has fallen below half of the average income and that 'the relative prosperity of rural areas is not so much the result of strong rural economies but reflects the movement of wealthy people into rural areas'. Living beside the wealthy and prosperous but often 'hidden', the main groups experiencing rural poverty are single older people, low-paid manual workers and their households, self-employed people and those who for a variety of reasons have become detached from the labour market (Shucksmith, 2000).

By considering in turn the two extreme ends of the UK, the far north of Scotland and the far South West of England, some of the impacts of geographical isolation can be illustrated. The county of Cornwall is the second largest English county and has the longest coastline. Lacking any major urban centre, two-thirds of its half-million residents live outside of a large town. As a county, Cornwall has the lowest gross domestic product (GDP) per capita of any English county, only 69% of the EU average. This is matched by a low wage and low skill economy. Gross average weekly earnings for men were £362.30 in 2001, compared to £388 in Devon, £412.10 in Dorset and £451.80 for the region as a whole. For women the average was £287.50, compared to £308 in Devon, £330.60 in Dorset and £333.50 for the region as a whole[1]. Moreover, to earn this lower wage it would appear that people living in Cornwall are working longer, with the average working week being 42.5 hours. Unemployment in the county is complicated by the large number of people dependent on the tourist trade for their income, which is characterized by its seasonality and informality of employment.

In contrast, a recent study by Lindsay and colleagues focused on the long-term unemployed in remote rural areas of Scotland (Lindsay *et al.*, 2003). Their study centred on Wick and Sutherland. Wick is 162 km (260 miles) north of Edinburgh and over 62 km (100 miles) from Inverness, the only city in the Highlands. They found that those who were experiencing long-term unemployment were more likely to be men (80%) and a slightly higher proportion (43%) were over 45. The long-term unemployed were more likely to have a low level of formal education, live in public-sector rented accommodation and have gaps in their work experience. When searching for work, the long-term unemployed were also less likely to have access to the social networks that are regularly used by people to find work. In the terms used in the previous section, their social capital, which may have already been low, had become depleted. None of this is particular to rural areas, but what changed the experience of long-term unemployment was the impact of the remote rural location. Because of the extreme remoteness of the area, even those with private transport were challenged in commuting to work in urban centres. At the same time the formal systems such as the Employment Service were unable to provide the level of services that they would in urban areas. Employment in rural areas is frequently

found by word of mouth and many of the better jobs are filled through informal networks and not formally registered with the Employment Service. Consequently, those who are well connected and perceived to be 'good workers' have an advantage over those moving from another area, those who have broken locally accepted social norms or have otherwise become detached from key local networks.

From these examples it is evident that a proportion of the rural population is characterized by having a low level of work experience, fewer qualifications and that these people are often older. The study from the Highlands also highlights that men may be disproportionately represented amongst the long-term unemployed. This raises the question of what the young and women do to remain economically active and where those with educational qualifications work. To understand this requires a fuller understanding of the role of migration in rural areas and the dynamics of the domestic property market. In many areas the young, particularly women, have left in search of work, and at the same time older, wealthier people have moved in search of a better quality of life.

'Getting on your bike'

One of the most important exports from rural areas is educated young people. Almost all degree-level education is delivered in cities and further education is often only available in larger urban centres. Young people need to move to become educated and then they may find that there are few job opportunities in the rural areas that they have left. Finding work in a rural area does not resolve the forces that push young people to move from rural areas. A recent study in rural North Yorkshire identified the tensions in remaining in the area, which largely revolved around achieving meaningful independence (Jones and Rugg, 1999). They focused on the struggle of young people in achieving any measure of autonomy from their family home. Apart from those who were attending college or university away from the area, young people found themselves in a number of vicious circles. Often they found themselves in a 'catch-22' situation, needing a car to get to a job but unable to afford a car without first having employment. The result was frequently extended parental financial support and the need to

---

[1] Figures compiled by the South West Regional Development Agency and the South West Economy Centre, 2003.

remain in the family home to be able to afford the costs of running a car. This was particularly the case for young men who were in low-paid jobs with few prospects. Rental property was in very short supply and owner-occupation, without considerable parental assistance, was not possible for single people and even young couples found it hard to buy a house.

Those unable to secure accommodation in rural areas can face some stark choices. Homelessness, like poverty, is one of the secrets of rural life. Those who acknowledge homelessness in rural areas often point out that it actually fuels the problem of urban homelessness. Many rural areas do not have any emergency housing or night-shelters for those who find themselves roofless, and the ever-tightening squeeze on public housing leaves urban areas as the only resort for many. Homelessness in rural areas, as recent work by Paul Cloke and colleagues has illuminated, is far more complex than simple rural–urban migration (Cloke *et al.*, 2003). As in other areas, homeless people in rural areas often 'sofa-surf', moving in a complex circuit between family and friends trying to find a combination of work or accommodation that will allow them to make a home. The hidden homelessness of 'sofa-surfing' is thought to be more prevalent in rural than urban areas, although, as Cloke *et al.* acknowledge, the paucity of research in this area makes such claims difficult to verify. Others experience rural homelessness in a different manner, following patterns of seasonal work in either agriculture or the tourism industry, moving within and between rural areas often on a seasonal basis. For some these patterns may reflect their ethnic background, lifestyle choice or an accident of personal history, but the denial of their role in rural life is also part of the emphasis placed on boundaries and property in rural areas. The final group who find themselves homeless in rural areas are those who were homeless in urban areas, who move to or through rural areas as a respite from the problems of urban areas. The appeal of the rural idyll is not just confined to those who are looking for a cottage with roses over the porch.

In contrast to the well-kept secret of rural homelessness, for many years a house in the country has represented the British ideal of success and the recent property boom has not seen this desire diminish. The rise in house prices has had obvious benefits for those who already owned property, as their assets have appreciated. In no small way, this has saved many farm busi-

Table 3.1 Property affordability by English Government Office region (The Countryside Agency, 2002).

Region	Mortgage index[1]		Cost as % of income		
	Rural	Urban	Rural	Urban	Difference
North East	2.49	3.74	40	27	13
North West	3.03	3.59	33	28	5
Yorkshire and Humberside	2.51	3.75	40	27	13
West Midlands	2.18	2.88	46	35	11
East Midlands	2.11	2.13	47	47	0
South East	1.59	1.79	63	56	7
South West	1.75	2.17	57	46	11
Total	**2.08**	**2.64**	**38**	**48**	**10**

[1] The mortgage index is a hypothetical one based on using average house prices, average earnings and mortgage lending guidelines.

nesses from the worst of the agricultural recession (anecdotal evidence suggests that many banks have been lending against the asset value of farms rather than business viability). For those who did not own property before the house price rise, their opportunity to do so now or in the future has dramatically diminished. For many young people in rural areas or those who were previously tenants, house price inflation has forced them out of the communities they had previously lived in. It has also had many pronounced consequences for rural communities.

Looking at the figures in Table 3.1, some of the unevenness of these processes becomes evident. The mortgage index is a way of indicating the affordability of housing against average incomes: the lower the number the less affordable the houses are. Unsurprisingly the price of houses in the South East in both rural and urban areas is very high, in part reflecting the concentration of wealth and opportunity in the metropolis. The South West of England is both the most rural English region and the most expensive for housing outside of the South East. For those who earn below-average income the percentage of income taken up in servicing their mortgage alone will be higher than these figures suggest, making them highly vulnerable to changes in either the economy or personal circumstances. Given the preponderance of lower-paid employment in rural areas, those who live and work in rural areas are increasingly vulnerable and disadvantaged.

The next process illustrated in Table 3.1 is that patterns of advantage or disadvantage are not evenly spread. Whilst property in parts of rural Devon may be in high demand, property in the remoter parts of Northumberland, for example, may not be under such pressure. In part this will reflect the opportunity presented in each area and is a warning of thinking of the countryside as being undifferentiated. Nevertheless, it is estimated that nationally 40% of newly formed households in rural areas are unable to afford to purchase a home (Shucksmith, 2000). An alternative is to seek rented accommodation, but this is also problematic. The private rented sector in rural areas is slightly larger than in urban areas (Cloke *et al.*, 2003) but is often prohibitively expensive, particularly in the more attractive and accessible rural areas. Social housing, on the other hand, offers an obvious route to tackling housing issues but is in short supply in rural areas. In 1999 social housing provision accounted for only 15% of total rural housing stock compared to an average of 23% for England (Cabinet Office, 2000). Social housing, where it is available, is important not only because it can provide affordable accommodation but also because it can allow people to remain within their community and networks of kinship and association, which, in turn, provide support (for example, friends and family helping with childcare) and are a vital source of information about job opportunities (Shucksmith, 2000).

Finally, Table 3.1 demonstrates starkly, by implication, that those in the South East or several other urban areas who owned property before the rise in prices and earn above the average will have considerable purchasing power. For many people a small property in a rural area will have become affordable as never before, either as a holiday home or as a place to which they can retire (Countryside Agency, 2002). Holiday homes in rural areas are not new and they have certainly been a feature of many rural communities in Wales in particular for many years. Certain villages near beauty spots have felt the rise in the number of holiday homes, which have had a profound effect on the existing communities. The worst hit area is the Lake District where in some parts up to 60% of houses are used as holiday flats and local people are not able to compete with the rise in house prices. Until recently second homeowners only paid 25% of the normal level of council tax, but councils have now been given the opportunity to levy up to 90% of the council tax in the future (al Yafai, 2003; Mathiason, 2003). Although the impact of second homes may not be altogether negative on the rural economy, as they bring in people looking to spend money on leisure pursuits and services, these effects may not be immediately of benefit to those who may feel as if they have been displaced.

Not all of those who buy rural property intend to use it only for holidays; many bring their families or new businesses to a rural area. Others come to retire, to take up the higher quality of life that rural areas offer to many. Although often they start their new life with resources, as they grow older they find rural life increasingly harder to negotiate. Older people are very vulnerable to becoming isolated and excluded in rural areas, whether they have lived there all of their lives or are recent migrants. Older people in rural areas fall foul of many of the factors that those on lower incomes or without capital assets find difficult in rural areas – transportation, access to services and isolation. Yet they also face problems peculiar to their circumstances in that often they are living on fixed or modest incomes that may not have the scope to allow for their changing needs. The costs of maintaining private transport or of being able to use taxis on a frequent basis can become prohibitive but nevertheless essential in order to access the services that are needed. Without the support of close family living locally older people can become increasing isolated, and those who have lived in an area for a longer period may find that their family are unable to afford to live near them. Moving to a country cottage on retirement remains a dream for many people but one that once realized may turn sour.

Conclusions

As Goldsmith observed (Goldsmith, 1770), there is a great deal of difference between splendour and happiness, and few would disagree that the contemporary British countryside is splendid, but the question of whether it is happy is far more contentious. This chapter has considered the question of social exclusion through the lens of considering the assets of individuals and families. Some of these assets are socially held, embedded in communities and only built up over a long time; these can be the friendships generated by an individual or his or her educational achievements. Whilst these do not guarantee that people will be safe from material hardship, they represent the resources that would allow people the opportunity to escape their situation. The combination of who you know and what you know often allows people to change their own circumstances with relatively little help from the state. We have argued that these abilities in rural areas are being undermined by the unequal distribution of financial assets mostly in the form of residential property. The demand for rural property has created a situation whereby social capital and educational attainment, even when coupled with the most positive of attitudes, are always trumped by property assets accumulated elsewhere.

The creative force behind the movement of people towards rural areas has been idyllic images of rural life that are very deeply engrained in our culture, and the differences in economic wealth in the regions of the UK mean that some are able to achieve their dreams at the expense of others less fortunate. Whilst this may be morally unjust, it is also very highly socially and economically damaging. It creates the impression that there are no substantial social problems in rural areas, it forces younger and poorer people out of rural areas and stifles economic dynamism. We are not suggesting that rural life is unendingly bleak or that is it not worthwhile, but it needs to be viewed in a balanced way and that balance needs to be realized in rural communities. Access to affordable housing is one of the key measures that will not only lessen social exclusion in rural areas but also contribute to the renaissance of community life that so many people hope for in rural life.

Acknowledgement

The authors would like to acknowledge the valuable assistance of Dawn Wakefield in the preparation of this chapter.

References

Alcock, P. (1993) *Understanding Poverty*. Macmillan, London.

al Yafai, F. (2003) From rural idyll to holiday park. *The Guardian*, 22 April.

Beckett, M. (2003) England's rural future. *Keynote speech to stakeholders*, Institute of Civil Engineers, London, 4 Nov.

Bunce, M. (1994) *The Countryside Ideal: Anglo-American Images of Landscape*. Routledge, London.

Cabinet Office (2000) *Rural Economies*. The Stationery Office, London.

Cloke, P., Milbourne, P. and Widdowfield, R. (2003) The complex mobilities of homeless people in rural England. *Geoforum*, **34**, 21–35.

Countryside Agency (2002) *The State of the Countryside*. Countryside Agency, London.

Goldsmith, O. (1770) *A Deserted Village*. Goldsmith Press Ltd.

Jones, A. and Rugg, J. (1999) *Housing and Employment Problems for Young People in the Countryside – Summary*. Joseph Rowntree Association, York.

Joseph Rowntree Foundation (2000) *Exclusive Countryside?* Joseph Rowntree Foundation, York.

King, E. (1999) *Defining and Measuring Social Exclusion*. Discussion Paper. Office for Public Management, London.

Lindsay, C., McCracken, M. and McQuiad, R.W. (2003) Unemployment duration and employability in remote rural labour markets. *Journal of Rural Studies*, **19**, 187–200.

Mathiason, N. (2003) Labour plans tax on rural building. *The Observer*, 28 Sept.

Matless, D. (1998) *Landscape and Englishness*. Reaktion Books, London.

Mayerfeld-Bell, M. (1994) *Childerley – Nature and Morality in a Country Village*. University of Chicago, Chicago and London.

Newby, H. (1985) *Green and Pleasant Land? Social Change in Rural England*, revised edition. Wildwood House, Aldershot.

Putnam, D. (2001) *Bowling Alone*. Simon and Schuster, London.

Short, B. (1991) *The English Rural Community: Image and Analysis*. Cambridge University Press, Cambridge.

Shucksmith, M. (2000) *Exclusive Countryside? Social Inclusion and Regeneration in Rural Areas*. Joseph Rowntree Trust, York.

Social Exclusion Unit (2001) *Preventing Social Exclusion*. SEU, London.

Stephens, W. (2000) *Civility in an English Village*. Severn Books, Tallahassee.

4

Countryside law

T. Felton

Introduction

Unlike established classifications of subject areas such as contract law and criminal law whose basic content is well defined, countryside law as a discipline in its own right is an emerging area. Its subject matter is taken from a number of existing disciplines such as land law, criminal law, environmental law, European law and public law areas such as town and country planning and private rights to be found in the law of tort. As such, countryside law remains in its infancy and is ill-defined as to its scope and function. This chapter seeks to provide an introduction, for those involved in the study of countryside and rural land management, to three key areas that underpin the subject matter of countryside law, namely ownership of the countryside, protection of the countryside and access to the countryside. That is to say:

- how the land that makes up the countryside is owned and the legal rights attached to that ownership;
- the law used to protect and enhance the countryside;
- the extent and means by which the public may gain access to the countryside for leisure and recreation purposes.

The chapter does not seek to cover all areas that may fall under the title countryside law and the reader is encouraged to research other chapters in this text for relevant material such as that on town and country planning, the equine industry and the Common Agricultural Policy of the European Union.

In addition, other textbooks, journal articles and internet sources will give valuable assistance on matters of countryside law, and a list of useful internet reference sites may be found at the end of this chapter.

Amongst others, the following bodies also provide much information related to countryside law: the Royal Institution of Chartered Surveyors, Country Land Owners and Business Association, National Farmers' Union, Ramblers' Association, Environment Agency and Agricultural Law Association.

The English legal system

Any greater understanding of the substantive law that seeks to regulate ownership, protect and enhance the countryside and allow the public access to land must commence with a review of the system that creates and administers that law. English law shares with the countryside, as we know it today, a long history. Our law remains different in substance and procedure from continental European practice – although this is not to say that ideas of justice differ – and it should be noted that in the rulings of the Court of Justice of the European Union there now exists a unifying factor of particular importance in relation to environmental law which has been identified above as a key ingredient of countryside law. Countryside law is not a separate part of English law; the same general principles apply here as in other areas.

Law consists of those rules of conduct that the courts will enforce. In this connection we include in the term 'court' all those bodies recognized by the judges as having an obligation to act judicially. So in addition to the civil and criminal courts we recognize various tribunals and other bodies set up by Act of Parliament.

Civil courts decide disputes between fellow citizens where one, the claimant, alleges that he or she has suffered injury or loss by the unlawful act or omission of another, the defendant. If the defendant is adjudged legally responsible for the injury, he or she will be required to compensate the claimant, normally by a money payment called damages. When a state agency causes such an injury, an action brought against it will also be a matter of civil law.

The criminal courts decide cases where the state is involved as prosecutor against a citizen accused of committing a criminal offence. Accused persons, if found guilty, are punished by fine or imprisonment or both.

The two sets of courts are kept separate; magistrates' courts and the crown court hold criminal trials while county courts and the High Court hear ordinary civil cases. Certain matters are, however, reserved for specialized tribunals of which the agricultural land tribunal and the lands tribunal are particularly relevant to rural land use. At the top of the hierarchy of courts, the Court of Appeal and the House of Lords, sitting as a court, provide an appeal structure with the possibility of reference to the European court on questions of European Union law. It should be noted that in 2003 the Government signalled its intention to replace the judicial role of the House of Lords with an entirely distinct Supreme Court.

Most hearings in court are concerned with matters of fact: in criminal trials the prosecution will have to prove 'beyond reasonable doubt' that the defendant has committed the crime of which he or she is accused; in civil trials the court will decide on the balance of probabilities, whether the events concerned took place as alleged by the claimant or not. However, in some cases the decision depends on a question of law; for example, in the case of *Mirvahedy* v. *Henley* (2003) the House of Lords was called upon to interpret the law contained in the Animals Act 1971 and in particular s2(2)(b). This case concerned the liability of owners whose horses had strayed onto the public highway. Three horses had become terrified by an unexplained incident, forced their way out of their field through some dense vegetation and run up a track onto the main road where a road accident then occurred. Mr Mirvahedy who was seriously injured in the accident sued the owners of the horses. Section 2(2) Animals Act 1971 provides:

'Where damage is caused by an animal which does not belong to a dangerous species, the keeper of the animal is liable for the damage, except as otherwise provided by this Act, if:

(a) the damage is of a kind which the animal unless restrained, was likely to cause or which if caused by the animal was likely to be severe; and
(b) the likelihood of the damage or of its being severe was due to the characteristics of the animal which are not normally found in animals of the same species or are not normally so found except at particular times or in particular circumstances; and
(c) those characteristics were known to that keeper or were known to a person who at that time had charge of that animal.'

The words that caused the difficulty in this case were from s2(2)(b) 'or are not normally so found except at particular times or in particular circumstances'.

The House of Lords found that s2(2)(b) makes the keeper of an animal liable for the damage it has caused when, in causing the damage, it displays behaviour that has arisen as a result of circumstances in which the animal has found itself. The result of this is that the keeper will be responsible for an animal's behaviour except where a pure accident occurs which is not related to behaviour, say an injury caused to a person by an accidental stumble.

There are two principal sources of law: decided cases and statute. Case law has been built up into a system termed the common law because when a superior court makes a decision turning on a point of law, that decision becomes a precedent binding on judges dealing with subsequent cases involving the principle. Statute consists of Acts of Parliament and Statutory Instruments, i.e. orders and regulations made under the authority of Acts of Parliament by duly authorized ministers. EC treaty provisions are incorporated into English law by virtue of the statutory force to regulations passed by the Council of the European Union. European Union directives normally require legislative action by the UK parliament before implementation, but they are capable in certain circumstances of having direct effect. Decisions of the European court also create precedents that UK courts must follow. For example, in *Von Menges (Klaws)* v. *Land Nordrhein-Westfalen* No. 109/84 1986 CMLR 309, the claimant had kept a herd of dairy cows until 1980 when he obtained a premium for going out of milk production. He undertook, as required by Regulation 1078/77 EEC, not to market milk or milk products for 5 years. In 1981 he let his farm to another farmer who planned to use it for milk-sheep. He asked in the German courts for a declaration that the marketing of sheep milk would not be contrary to the undertaking given. The local administrative court referred to the European Court, under Article 177 EEC, the question of the meaning of 'milk and milk production' in the Regulation. The court held that the words concerned ewes milk and ewes milk products as well as cows milk and cows milk products, since otherwise the premium payments would only encourage the replacement of dairy cows by milk-sheep, leading to new surpluses. This decision is binding upon courts throughout the European Union which of course includes UK courts, who will in future follow this interpretation of 'milk and milk products' in this context.

A statutory provision overrules any common law precedents that directly conflict with it. Nowadays the definition of criminal offences and the powers of courts to penalize offenders depend almost exclusively upon statute, and a vast range of legislation covers fiscal and commercial matters together with the social activities of government, e.g. housing, employment, public health and social security. Even in the field of property law, contract and tort (see below), formerly the domain of common law, parliament has codified or amended many of the rules developed in the courts.

Thus, to find the law relevant to a particular topic it is necessary to know if statutory rules apply. For instance, in respect of security of full-time farm workers occupying service cottages, the position is governed by one of two statutory codes: the Rent (Agriculture) Act 1976 for occupancies entered into prior to 15 January 1989, and the Housing Act 1988 Part 1 Chapter 3, which relates to occupancies that commenced on or after that date. Questions concerning the rights of the farmer and worker when employment comes to an end can only be resolved by reference to these acts.

When the topic is one that is covered by common law, the decisions of relevant cases can be found in the Law Reports, where decisions of significance are recorded. In practice, a legal practitioner will depend upon books of reference and texts dealing with specific areas of the law to enable him or her to find the statutes and cases that are relevant. The rural land manager or conservationist must know about changes in the law relating to his or her business or area of expertise. Details of many new statutory rules are publicized by

government agencies, e.g. on environmental issues by the Environment Agency, but to keep up to date all those interested in countryside law should read the professional journals.

Criminal offenders are liable to punishment and those committing civil wrongs are liable to pay damages or to have their activities stopped by an order termed an injunction made by a court. These are the means by which the law is enforced. By the standards of other legal systems, enforcement of civil judgments in England is reasonably effective. However, before embarking on expensive litigation all would do well to seek advice on available methods of alternative dispute resolution, such as mediation or expert determination, which may lead to a satisfactory conclusion far more expeditiously and economically.

The Human Rights Act 1998

No commentary on the English legal system at the start of the twenty-first century would be complete without mention of the Human Rights Act 1998 (HRA) which came into force on 2 October 2000. It was heralded as bringing about the most important constitutional changes for over 300 years. It is legislation that by its very nature will affect every aspect of law in this country and the way in which public authorities carry out their duties. As such it will affect land managers and conservationists. In recent cases the noise created by aeroplanes at night, the compensation provided to mink farmers on the banning of mink fur farming in England and Wales, the designation of land as a Site of Special Scientific Interest and planning applications within the Dartmoor National Park have all been subject to arguments relating to human rights law.

The HRA incorporates the European Convention on Human Rights (ECHR) into our domestic legal system. That means that the ECHR is enforceable in our own courts. The ECHR to which the UK has been a signatory since 1950 guarantees certain fundamental freedoms. These include:

- the right to life (Article 2);
- the right to respect for private and family life (Article 8);
- freedom of expression (Article 10);
- the right to a fair and public trial by an impartial independent tribunal (Article 6);

- the right to peaceful enjoyment of possessions (Protocol 1 Article 1).

The HRA operates in three key ways. First, it requires all legislation to be interpreted and given effect as far as possible compatibly with ECHR rights. Second, it makes it unlawful for a public authority to act in a manner incompatible with ECHR rights. Third, UK courts and tribunals must take account of ECHR rights in any case they hear, thus providing the opportunity for the common law to be developed. In addition, a minister in charge of a bill must confirm that the bill is compatible, or ask the House to proceed although he or she cannot say that it is.

Remedies under the HRA

Essentially a court or tribunal can provide any remedy that is within its powers and is just and appropriate; for example, award damages, quash unlawful decisions, make an order preventing a public authority from making a decision that would otherwise be unlawful.

Although the ECHR does not explicitly mention the environment, the Spanish case of *Lopez* v. *Spain* [1995] 7 ELM 49 provides a graphic example of how human rights and environmental issues are related. In 1988 a new waste treatment plant was opened. The plant was allegedly the cause of both health and nuisance problems, as a result of which nearby residents were evacuated. Despite being ordered to cease part of its operations environmental degradation continued. It was claimed that this state of affairs amounted to a breach of Article 8 of the ECHR, the right to respect for one's family and private life. The court held that 'severe environmental pollution may affect an individual's well-being and prevent them from enjoying their homes in such a way as to affect their private and family life adversely, without, however, seriously endangering their health'.

Legal aspects of ownership, possession and occupation of the countryside

Most of the countryside is owned and managed by owner-occupiers or tenants of agricultural holdings or farm business tenancies. However, increasingly conservation charities are establishing themselves as substantial landowners in their own right, for example the Royal Society for the Protection of Birds owns and manages nature reserves covering 121 082 ha, the Wildlife Trusts own nature reserves of 80 988 ha and the National Trust owns in excess of 248 000 ha. In many cases land will be owned by single individuals or partnerships, and the persons involved will be legally responsible for any obligations that arise. Where the enterprise is run by a limited company, the company is recognized as a legal person with rights and duties separate from those of its members. The directors of an enterprise organized in this way must, however, recognize their obligations under the Companies Acts which contain provisions designed to prevent them from using the advantages of corporate personality to defraud creditors, members and employees of the company.

English law recognizes only two ways by which land may be held: freehold and leasehold. If it is desired to tie up freehold land in the ownership of succeeding generations within a family, this can only be done by the creation of a trust set up in accordance with certain rules – known as equity – developed in the Court of Chancery, now a part of the High Court. This definitely requires professional expertise.

A freeholder has the largest possible freedom to decide how to use his or her land that the law recognizes. It has never been possible for freeholders to do whatever they pleased to the detriment of neighbours, but, in addition to the limits already imposed by the law of nuisance have been added the constraints of town and country planning and the compulsory acquisition of land. Public and private rights of way may also diminish a freeholder's privacy, and land can be lost by adverse possession (see below). A freeholder's rights may be reduced by restrictive covenants and his or her obligations increased by a mortgage.

Although ownership of land normally includes rights over what lies in and beneath the soil, mineral rights may be held separately from the freehold, and coal and oil deposits belong to the state. Water may be taken from surface streams by riparian owners for normal agricultural purposes such as watering stock (not for aerial spraying), but it is generally the case that for other purposes it may only be taken under a licence from the Environment Agency and the same applies to most water taken from underground sources.

The law of tort includes trespass, negligence and nuisance which may be used by individual landowners to protect the environment, though it cannot be argued that this is achieved in a cohesive or structured pattern. So, for example, the League Against Cruel Sports has used the law of trespass to prevent the Devon and Somerset Stag Hounds from hunting on its land (*League Against Cruel Sports* v. *Devon and Somerset Stag Hounds* [1986] 2QB 240) and the National Trust has been held responsible for the collapse of land onto its neighbouring landowner (*Leakey* v. *National Trust* [1980] QB 240 CA).

As a result of the leading case of *Rylands* v. *Fletcher* (1868) LR 3 HL 330, an occupier is strictly liable for the escape of any potentially dangerous thing kept on his or her land if it escapes and injures neighbouring property or people. Defences to such a claim are very limited and the occupier is liable even if the escape was caused by a contractor. The liability is, however, limited to non-natural uses of land. Both the doctrine of *Rylands* v. *Fletcher* and the tort of nuisance were extensively reviewed by the House of Lords in the 1993 case of the *Cambridge Water Company* v. *Eastern Counties Leather plc*, 'the Cambridge Water Case'. In this case Eastern Counties used a toxic solvent in its tanning process, a considerable quantity of which was spilt onto the floor. Over a period of years the solvent leaked underneath the property into an aquifer and thence into a borehole owned by Cambridge Water, rendering the water undrinkable. Two important points to note from the judgement are:

- Foreseeability of harm of the type suffered is a requirement of liability under the rule in *Rylands* v. *Fletcher*.
- The fact alone that usage of a particular item is common to an industry does not bring that use within the definition of 'natural' for the purposes of the rule in *Rylands* v. *Fletcher*.

In the Cambridge Water Case, Eastern Counties was not liable as the pollution was unforeseeable.

The Land Registration Act 2002

A review of property law in England and Wales would not be complete without reference to the Land Registration Act 2002 (LRA 2002) which came into force in October 2003. It has been claimed this that act intro-

duces the most fundamental reform of the law relating to property since the raft of property legislation created in 1925. The Act is largely concerned with the registration of land in England and Wales and providing the way for the introduction of electronic transfer of property from one party to another, or, as it is popularly called, e-conveyancing. The law for the most part is of a highly technical nature and thus professional advice should be sought from a solicitor.

However, for those concerned with the management of rural land it should be noted that the law relating to adverse possession or 'squatters' rights' has been altered fundamentally by the LRA 2002. There is now a new system in place for *registered land*. Under the new rules:

- A squatter may apply to the registrar at the Land Registry to be registered as the proprietor if he or she has been in adverse possession for a period of 10 years.
- The registrar will then inform persons affected by the application and they may give notice to the registrar that they intend to oppose the application.
- If no opposition is registered then the squatter is entitled to be registered as proprietor.
- If the application is opposed the squatter will only succeed in limited circumstances laid down by the Act. For example, in all the circumstances it would be unfair for the paper owner to dispossess the squatter.
- Finally, if the squatter's application is rejected yet he or she remains in possession for the following 2 years, he or she may make a second application, an application that does not have to be referred to the paper owner.

It should be noted that transitional provisions will be in force relating to claims that pre-date the coming into force of the LRA 2002 and that the old rules will apply to land that is not registered (which should act as a further incentive for land owners to register their land). It is therefore important that professional advice should be taken regarding this area of law.

Common land

The term common land often leads to a misunderstanding as to the nature of ownership and the rights of the public over such land. Common land is not land that is in public ownership; in fact, of the 585 000 ha of common land in England and Wales the vast majority is in private ownership. Common land may be defined in two ways: (1) land registered under the Commons Registration Act 1965 and (2) land to which rights of common are attached.

Rights of common are exercised by commoners who are identified, for example, through their ownership of certain property or by living in a particular locality. The rights that may be exercised will vary from area to area but may include the right to graze cattle or ponies, the right to collect fallen wood, the right to dig turf, the right to take fish, or extract gravel or sand.

Common land has taken on increased relevance in terms of recreational access to the countryside with the passing of the Countryside and Rights of Way (CROW) Act 2000. By virtue of s2 of the Act, once the Act has been fully implemented, access on foot for recreational purposes will be available, subject to any restrictions put in place under the terms of the Act, to registered common land. This new situation highlights the potential conflicts in the use of the countryside as much common land is made up of land of high conservation value and as such is designated as a Site of Special Scientific Interest (SSSI); 180 000 ha in England and 66 000 ha in Wales are designated as SSSIs. The complexities of registration and law relating to the use and management of commons have been recognized by government. Following an extensive consultation period, the New Labour government produced the policy document *Common Land Policy Statement* which provides a comprehensive review for updating this area of law in order that commons are protected, continue to provide increased opportunities for access to land and are managed appropriately for agricultural purposes. The Commons Registration Act 1965 also required the registration of village greens; all traditional greens that were not registered between 1967 and 1970 ceased to be greens. However, since the case of *Regina v. Oxfordshire County Council ex parte Sunningwell Parish Council* (1999) 3 WLR 160, it has been possible to register land as a green if it can be shown that there has been open, unchallenged use for at least 20 years for recreational purposes by a significant number of local inhabitants.

Farm business tenancies

By the final decade of the twentieth century the freedom of landlords and tenants to negotiate the terms under which property was let had been greatly circumscribed by statute – in no area more so than agriculture. As a result there was a fall-off in let land from some 90% in 1910 to 36% of the total agricultural land available in 1991. In 1991 the government produced a consultation paper entitled *Agricultural Tenancy Law Proposals for Reform*, the objectives of the reform being:

(1) to deregulate and simplify;
(2) to encourage the letting of land;
(3) to provide an enduring framework which can accommodate change.

Having undertaken a full consultation process, the government introduced the Agricultural Tenancies Bill to parliament in November 1994; the new legislation took effect on 1 September 1995. From this date all new tenancies created are entitled 'farm business tenancies'. The terms of such tenancies are freely negotiable between the landlord and tenant within a much simplified legal framework designed primarily to prevent disputes arising. The legislation:

(1) defines farm business tenancies;
(2) requires the service of notices to quit;
(3) prescribes arrangements for compensation;
(4) prescribes fall-back procedures for disputes;
(5) provides fall-back procedures on rent reviews.

To fall within the definition of a farm business tenancy the Act stipulates a number of conditions, as follows.

Business condition

The first is the business condition. Section 1(2) requires that during the life of the tenancy all or part of the land comprised in the tenancy is 'farmed for the purposes of a trade or business'. Thus it is not possible to diversify completely out of agriculture. The condition also precludes hobby or recreational farming from falling under the 1995 legislation.

Agriculture condition

Once the business condition has been satisfied it is also necessary to demonstrate that the character of the tenancy is primarily or wholly agricultural [s1(3)]. In determining whether the use of the land is agricultural a definition of agriculture is provided at s38(1) and the court is required to consider the terms of the tenancy, any commercial activities undertaken on the land and other relevant circumstances.

Notice condition

Section 1(4) creates the notice condition. This allows the landlord and tenant to identify that the tenancy being created is to be a farm business tenancy by exchanging appropriate notices prior to the commencement of the tenancy. For a successful exchange of notices the written notices must:

- be exchanged before 'the earlier of the beginning of the tenancy and the date of any written agreement creating the tenancy';
- identify the land to be let in the proposed tenancy; and
- state that the person giving the notice intends the proposed tenancy to be and to remain a farm business tenancy.

Notice to quit

Where a tenancy is for a term of less than 2 years the agreement will come to an end on the term date. If a periodic tenancy has been created the common law principles of notice will apply. For example, a monthly periodic tenancy will require 1 month's notice.

Any farm business tenancy for a term of 2 years or more can only be terminated by a notice to quit of at least 12 months and less than 24 months. Should the landlord fail to serve notice the tenancy will continue as a tenancy from year to year until it is properly terminated.

Compensation

Compensation is available for any tenant's improvement that is a physical improvement made wholly or partly at the tenant's expense. However, to be eligible the tenant must have obtained the landlord's consent to the improvement. A tenant may also seek compensation for 'intangible advantages', for example milk quota.

Resolution of disputes

The Act contains provision for the resolution of disputes by arbitration or alternative dispute resolution.

Rent reviews

The Act provides freedom for the parties to agree rent review dates and periods. If the agreement is silent as to rent review the Act provides for rent to be reviewed every 3 years to an open market rent. If they so wish the parties may agree a rent formula outside the statutory framework.

Although there is general satisfaction with the working of the Agricultural Tenancies Act 1995, at the time of writing the Act is under review. The report of the Tenancy Reform Group (TRIG) with its recommendations for change is available through the Agricultural Law Association website at www.ala.org.uk.

Access to the countryside and public rights of way

The latter half of the nineteenth century and the entire twentieth century witnessed increasing pressure for public rights of access to the countryside over and above those provided by public rights of way. This pressure was ultimately successful when the Countryside and Rights of Way (CROW) Act received the Royal Assent in November 2000. CROW provides a further illustration of the development of the countryside from agricultural production to public resource. From a legal standpoint it is now necessary to consider public rights of access to the countryside, first, under those rights and obligations created under Part 1 of CROW and popularly but inaccurately known as the 'right to roam', and second, under those statutory and common law rights relating to public rights of way as amended by CROW.

The Countryside and Rights of Way Act 2000

It is important to note that the rights and obligations created by CROW will be introduced over a number of years. The consequences of this are twofold. First, it allows all interested parties whether landowners or users the opportunity to take part in and influence the manner in which the Act is implemented; second, it requires landowners and users to be responsive to the gradual introduction of the legislation. Influence over the way in which the new rights are managed may be gained by lobbying either directly or through pressure groups the national and local access forums that CROW requires to be created to oversee implementation of the legislation.

The Act itself is made up of five parts each subdivided into chapters. These deal with:

(1) Access to the countryside;
(2) Public rights of way and road traffic;
(3) Nature conservation and wildlife protection;
(4) Areas of Outstanding Natural Beauty;
(5) Miscellaneous and supplementary.

Access to the countryside

Part I introduces a new statutory right of access for open-air recreation to 'access land' defined as 'open country' which is 'land which is wholly or predominantly mountain, moor, heath, down' and registered common land. The Act also includes a power to extend this definition to coastal land and for landowners voluntarily to dedicate irrevocably any land to public access. Land will be identified as 'access land' on a map to be issued by the countryside bodies. Following extensive consultation both locally and nationally the countryside bodies will be responsible for determining the extent of any mountain, moor, heath and down. Any land over 600 m and registered common land immediately qualifies as access land. It is imperative that those who own or occupy land that is or may be subject to the new regime become actively involved in the consultation process so that maps accurately reflect the extent of 'access land'. As indicated above, one opportunity for this will be through representation on a local access forum. A right of appeal will lie with the secretary of state (or National Assembly for Wales) against the inclusion of land on provisional maps as access land. By mid-2003 the Countryside Agency was able to provide appeals figures for the first two areas to be mapped, that is to say area 1, the South East, and area 2, the Lower North West. In area 1 there were 160 appeals and in area 2 650 appeals; these figures represented approximately 10% of the comment forms entered at the draft map stage. Details of all appeals entered may be found on the planning inspectorate

website at www.planning-inspectorate.gov.uk. Schedule 1 of the Act lists land excepted from access; this includes 'Land on which the soil is being, or has at any time within the previous 12 months been disturbed by any ploughing or drilling undertaken for the purposes of sowing crops or trees'.

Rights and liabilities of landowners

During the debate leading to the passing of the CROW Act much time was given to the issue of the liabilities of persons with an interest in land towards those seeking access. Section 12 of the Act provides that the right of access does not increase the liability of a person interested in the land in respect of the state of the land or things done on it. The position now is as follows: a person exercising his or her rights of access under s2(1) of the CROW Act will not be a visitor within the terms of the Occupier's Liability Act 1957 and so the land owner will not owe him/her a duty of care under that Act. Section 13 of the CROW Act amends the Occupier's Liability Act of 1984. The duty under this Act is owed but in a form modified so as not to include risks arising from any natural features or landscapes and this includes all rivers, streams, ditches, ponds, plants, shrubs or trees. The result is that 'access' land under the CROW Act is subject to a less rigorous regime than is owed to either visitors to land or trespassers on land. However, it should be noted that those who seek to deter access by the production of misleading information will be penalized under clause 14 which creates an offence of displaying a notice containing false or misleading information. The use of any path or area of land in exercise of the statutory right of access will not provide the basis for a claim for the existence of a right of way or of a town or village green.

Exclusion or restriction of access

Land may be closed as of right for a period of 28 days. However, the days on when this right is available are in themselves limited. For example, it must not include more than four Saturdays or Sundays and these Saturdays or Sundays are limited to certain periods of the year to avoid restrictions at the height of the summer. Above the 28 days, further closures may be applied for on a number of grounds including land management; nature conservation and heritage preservation; defence or national security; and exclusion or restriction of access in the case of emergency. Landowners and occupiers should confirm, prior to an application to restrict access, that the days they are seeking to achieve this on are permitted days. The ability to exercise dogs is also restricted (Schedule 2) in that they must be on a lead both in the vicinity of livestock or during the period beginning 1 March and ending 31 July each year. These restrictions may be relaxed by direction of the relevant authority.

Access forums

Sections 94 and 95 of the CROW Act place a duty on appointing authorities (highway authorities and national park authorities) to establish an advisory body to be known as a local access forum. Local access forums are statutory advisory bodies only and have no executive powers.

Local forums have the primary purpose of giving advice to their appointing authorities on how to make the countryside more accessible and enjoyable for open-air recreation, in ways that address social, economic and environmental interests.

Although there are statutory functions laid down for local forums and, for example, regulations as to how members should be selected, it is anticipated that local forums may well adopt different methods of working which best meet the needs of the public they serve. However, the areas on which they will be advising may be summarized as follows: the implementation, management and review of the statutory right of access to the countryside, including:

- byelaws made by the appointing authority;
- appointment of wardens;
- directions to restrict or exclude access to land;
- the Rights of Way Improvement Plan.

CROW 2000 – conclusion

There can be little doubt that the introduction of CROW 2000 will come to represent a significant development in the public's perception of their relationship with the countryside. However, despite the inevitable teething problems and individual difficulties that many will have to face, the likely reality of CROW is that a highly managed right of access will emerge for areas gener-

ally well defined by map. It is important that all involved in the ownership, management and use of the countryside play their part in that management process. Following the foot-and-mouth disease outbreak in February 2001 and the response of government to provide powers for local authorities to close public rights of way and impose fines of up to £5000, the immediate response of many such authorities and organizations representing ramblers suggests that there is a growing awareness of the necessity for a truly responsible approach to access in the countryside.

Public rights of way

The legal provisions that relate to public rights of way occur as the result of both common law and statute, are numerous and can be complicated. However, this should not deter all those using and managing rights of way from understanding their rights and responsibilities. Below is an outline of the main responsibilities of farmers and landowners with regard to public rights of way over their land as amended by CROW 2000. Note: the provisions of CROW will be introduced over a period of time. Where reference is made to this legislation those seeking to rely on it should confirm that the amendment has taken effect.

Redesignation of roads used as public paths

Most members of the public are aware of the designations footpath and bridleway. To these has been added the restricted byway (CROW 2000). Section 43 repeals s54 of the Wildlife and Countryside Act 1981. It redesignates all roads used as public paths (RUPPS) as restricted byways; these are defined as including a right of way on foot, on horseback or leading a horse, and a right of way for vehicles other than mechanically propelled vehicles.

(1) Obstructions

No person may wilfully obstruct the free passage along a highway (highway includes both footpaths and bridleways and restricted byways) (s137 Highways Act 1980). This means that if a person 'without lawful authority or excuse, intentionally as opposed to accidentally, that is, by an exercise of their own free will, does something or omits to do something which will cause an obstruction, he or she is guilty of an offence' [Parker LCJ *Arrowsmith* v. *Jenkins* (1963)]. Any person may serve the highway authority with notice requesting the authority to secure removal of an obstruction (CROW 2000). The highway authority must, within 1 month of the date of service, inform the complainant of the intended action. If the complainant is not satisfied that the obstruction has been removed an application may be made to the magistrates' court.

(2)   Gates and stiles

The maintenance of a gate or stile that crosses a footpath or bridleway is the responsibility of the landowner (s146 Highways Act 1980). It must be maintained in a safe condition so that there is no unreasonable interference with the public's rights of access.

(3)   Overhanging vegetation

Where any overgrowth interferes with a public right of access, the highway authority or district council may serve a notice under s154 of the Highways Act 1980 requiring the owner or occupier to cut back the growth. Section 154 has been amended by CROW to include overhanging vegetation that endangers or obstructs horseriders.

(4)   Bulls: Wildlife and Countryside Act 1981 s59

It is an offence for an occupier of a field crossed by a public right of way to keep a bull in that field unless one of the following exceptions are met: (1) the bull is less than 10 months old or (2) it is not of a recognized dairy breed and is running with cows or heifers. Dairy breeds include Ayrshire, British Friesian, British Holstein, Dairy Shorthorn, Guernsey, Jersey and Kerry.

This means that any bull over the age of 10 months is prohibited on its own and any bull over 10 months that is of a recognized dairy breed is prohibited whether or not it is accompanied by cows or heifers.

(5)   Misleading notices

Section 57 of the National Parks and Access to the Countryside Act 1949 makes it an offence for any person to place or maintain, on or near any way shown on the definitive map, a notice that contains false or misleading statements that are calculated to deter the public from using the way.

(6)   Barbed wire: Highways Act 1980 s164

Where barbed wire is placed on land adjoining a highway in a position in which it is likely to be injurious to persons or animals using the highway, the highway authority may serve a notice on the occupier of the land requiring him or her to remove the nuisance.

(7)   Chief obligations imposed by the 1990 Rights of Way Act

Under the Act farmers may plough or disturb the surface of a crossfield footpath or way as long as it is not convenient to avoid it. Farmers may not disturb field edge paths.

When the surface of a public right of way has been 'disturbed' (disturbance including all necessary operations for cultivation):

(a)   The period allowed for restoration of the surface is 24 hours; however, where there is a sequence of operations leading up to sowing, a period of up to 14 days is allowed.

(b)   When restoring the path or way the farmer is required to indicate the line of the route on the ground.

(c)   A specific duty is imposed on farmers to prevent crops from encroaching onto paths by:
–   growing through the surface;
–   overgrowing onto them from the sides.

(d)   Paths and ways that have been disturbed by cultivations must be restored to the following minimum widths:
–   crossfield path 1 m;
–   crossfield bridleway 2.5 m.
The minimum for a field edge path is 1.5 m.

(e)   Offences committed under the Act are punishable by fine which may be increased when an offence is repeated.

(f)   If a farmer fails in his or her obligations under the Act, the local authority may act in default and seek reimbursement from the farmer.

(g)   Excavations and engineering activities that disturb the surface of a public right of way may only be carried out if you first get written permission from the highway authority.

As indicated, the above is merely an outline of the obligations placed upon those who have public rights of way over their land. Farmers and landowners should also be aware that other activities that interfere with the public's rights of access may be a nuisance at common law or under statute. In recent years the Environmental Protection Act 1990 Part 3 s79 has added to the list of statutory nuisances that a farmer might breach when hindering the public's rights of access. Likely future pressures suggest that it is the wise farmer who has a complete knowledge of the law as it affects his or her particular circumstances.

Extinguishment of rights of way

A further important addition to public rights of way law is the extinguishments of unrecorded rights of way. A cutoff date of 1 January 2026 has been introduced. This means that rights of way predating 1949 and not recorded on the definitive map will be extinguished. Unrecorded higher rights will also be extinguished. This in effect creates a 25-year deadline for the investigation into alleged rights of way, the production of evidence to prove their existence and orders seeking to make appropriate amendments to the definitive map, on which all public rights of way are recorded.

Discovering lost ways

In order to offset the effect of the 2026 cutoff date the Countryside Agency is co-ordinating the Discovering Lost Ways Project. The first phase of this project in 2002 reported on, amongst others, the following issues:

- the state of the definitive map in terms of how up to date the map was on a national basis;
- estimating the extent of unrecorded rights of way;
- the nature of archive and source material available for researching rights of way.

The research showed that at the date of the report (2002), only one authority, the Isle of Wight, reported that its definitive map and statement were up to date, and that there were in the region of 20 000 unrecorded rights of way in England with an estimated length of 16 000 km and 2170 in Wales with an estimated length of 1600 km. This represents an increase of 9% to the existing rights of way. However, to unearth these rights of way would take 54 000 days of research and cost £5.5 million. The researchers recommended that to take this work forward an 'Archive Research Unit' should be created which would be responsible for the retrieval of information alongside a 'Rights of Way Claims Trust' which would take responsibility for the conversions of researched material into applications for modifications to the definitive map. The board of the Countryside Agency has now agreed to the above process to make extensive research into historic rights of way by establishing: (1) an arm's-length Archive Research Unit to identify and record lost ways; and (2) an independent Rights of Way Claims Trust to convert the unit's findings into formal applications to modify the definitive map to standards agreed with appropriate professional bodies.

For countryside managers, highway authorities, researchers and lawyers the roll out of this project is likely to prove both a fascinating and frustrating process.

Rights of Way Improvement Plans

Section 60 of the CROW Act requires access authorities to prepare Rights of Way Improvement Plans. The function of such plans is to review the existing network, comment upon its adequacy and propose actions to secure an improved network for all users. Although there is a duty on authorities to prepare such plans, no funding has yet been promised by central government to see through their implementation. However, if appropriate resources are found this could prove to be one of the most significant outcomes of the CROW Act in seeking to achieve enhanced public access to the countryside.

Nature conservation

The Wildlife and Countryside Acts 1981 and 1985 provide the cornerstone for the protection of birds and wild animals. Apart from the destruction of common pest birds it is an offence to kill wild birds, destroy their nests or disturb them near their nests. The Secretary of State for the Environment, working with English Nature and the Countryside Commission for Wales, has powers to make orders to protect Sites of Special Scientific Interest, and either the minister or a local authority may enter into a management agreement with owners and occupiers of land to preserve and enhance the natural beauty of the countryside. The land involved will be subject to restrictions as to the use for which compensation is payable, which will be binding not only upon the owner or occupier who makes the management agreement, but also upon successors in title. The Act is also used as the mechanism (by using statutory instruments) for incorporating into our law European directives related to nature conservation such as the directive on the Conservation of Wild Birds (79/409/EEC) and the directive on Natural Habitats and Wild Fauna and Flora (92/43/EEC).

The Wildlife and Countryside Act is made up of four parts:

(1) protection of wildlife;
(2) countryside and national parks including the designation of protected areas;
(3) public rights of way;
(4) miscellaneous.

Under Part 1 of the Act not only are birds protected as identified above but also animals and plants. Subject to certain exceptions, for example to nurse or humanely destroy an animal, the Act prohibits the intentional killing, injuring or taking, the possession of or trade in wild animals listed in Schedule 5 attached to the Act. Plants are protected by making it illegal to intentionally uproot wild plant species without authorization and the Act forbids the picking, uprooting or destruction of plants listed in Schedule 8 attached to the Act. As these Schedules are subject to addition it is very important that legislation is regularly referred to.

Nature conservation designations

Appropriate management and protection of wildlife habitat is sought through designations of vulnerable habitats. The following designations may be made:

- national nature reserves;
- Sites of Special Scientific Interest;
- special nature conservation orders;
- limestone pavements;
- UK biosphere reserves;
- marine nature reserves.

In the European context the EC Council Directive on the Conservation of Natural Habitats and Wild Fauna and Flora (Directive 92/43/EEC) – the Habitats Direc-

tive – came into force in June 1994. The following designations are provided for:

- Special Areas of Conservation (SACs);
- Special Protection Areas (SPAs).

Sites designated as such will be subject to restrictions when a proposed development is likely to have a significant effect on the area. The local planning authority will be required to assess the impact of a development and respond accordingly. PPG Note 9 has been issued to advise on the planning implications of regulations that implement the EC Habitats Directive.

CROW 2000 and nature conservation

Part III of the Countryside and Rights of Way Act 2000 (CROW) seeks to enhance nature conservation and wildlife protection. The following issues have been addressed.

Biological diversity

A new duty has been placed on government departments and the Welsh National Assembly to have regard to biodiversity conservation and maintain lists of species and habitats for which conservation steps should be taken or promoted.

Sites of Special Scientific Interest (SSSI)

CROW creates new procedures designed to enhance the protection of SSSIs including:

- the provision of powers to conservation agencies to refuse consent for damaging activities and to encourage positive land management;
- a statutory duty for public bodies to further the conservation and enhancement of SSSIs;
- increased penalties for damage to SSSIs by owner-occupiers and other parties.

Wildlife protection

CROW amends and updates the Wildlife and Countryside Act 1981 to increase legal protection for threatened species by:

- providing increased search and seizure powers to the police and making certain offences 'arrestable';

- creating a new offence of reckless disturbance;
- allowing courts to impose heavier fines and prison sentences for most wildlife offences.

Areas of Outstanding Natural Beauty (AONBs)

The management of AONBs will be improved by the production of management plans by local authorities for AONBs in their area, creating the mechanism for the creation of conservation boards where there is local support for such a scheme. The role of a conservation board would be to take over the role of the local authority in the production of a management plan and other management matters concerning an AONB.

Agri-environmental schemes

Increasingly wildlife and habitat protection is being achieved by the limitation of agricultural production methods. This is intimately related to the Common Agricultural Policy (see Chapter 9), in particular the Single Payment Scheme (SPS) which decouples production and subsidies, and links payments to environmental goods. The introduction of milk quotas by EEC Council regulations in 1984 introduced a new and complicated area of law to the rural scene. This scenario has had further complexities added in the form of the Suckler Cow Premium Scheme and Sheep Annual Premium Scheme. It should be noted that they differ from milk quota in two basic respects. First, they are owned by the producer, and, second, they are a right to receive premium each year and not to produce. With the continuously shifting sands and perilous nature of this area of law, those involved in land and/or quota transfers are respectfully referred to take expert advice.

Product limitation coupled with conservation is to be found in the agri-environmental schemes, for example, the *Environmentally Sensitive Areas Scheme* (introduced by the Agriculture Act 1986) of which there are now 22 in England and Wales. Payments are made to farmers in such areas for farming in accordance with traditional methods, thus maintaining established landscape patterns and improving the environmentally beneficial aspects of farmland. This 'carrot' approach to the protection and enhancement of the countryside has also manifested itself in schemes such as 'Countryside Stewardship'. Such initiatives are complemented by

further schemes that are part of the Rural Development Programme, the response to the European Rural Development Regulation; they include the following:

- Land-based schemes
  - energy crops scheme
  - farm woodland premium scheme
  - hill farm allowance scheme
  - organic farming scheme
  - woodland grant scheme
- Project-based schemes
  - processing and marketing scheme
  - rural enterprise scheme
  - vocational training.

Pesticides and pollution

The law relating to the use, supply and storage of pesticides is controlled by the Control of Pesticides Regulations 1986 (SI 1986 No. 1510) of the Food and Environment Protection Act 1985, and the Control of Substances Hazardous to Health Regulations 1988 (1988 No. 1657), more familiarly known as COSHH, of the Health and Safety at Work Act 1974. The term pesticide includes herbicides, insecticides and fungicides but does not include those substances applied directly to livestock (e.g. sheep dips which are regulated by separate legislation, namely the Medicines Act 1968). The legislation seeks to protect both human health and the environment by requiring ministry approval for all pesticides and prescribing the training necessary for the use of such products. Those working in the countryside seeking to use pesticides who were born after 31 December 1964 must obtain a certificate of competence before using an approved pesticide.

All users should be aware of the two statutory approved Codes of Practice that accompany the legislation. These are the Code of Practice for Suppliers of Pesticides to Agriculture, Horticulture and Forestry, and the Code of Practice for the Safe Use of Pesticides on Farms and Holdings (the 'COSHH Combined Code').

Those users whose actions cause a deterioration in water or air quality or cause other statutory nuisances may find themselves held legally responsible under the Water Resources Act 1991 or the Environmental Protection Act 1990. Under the former it is a criminal offence to cause or knowingly permit any poisonous, noxious or polluting matter, or any solid waste matter, to enter any controlled waters. In the Control of Pollution (Silage, Slurry, and Agricultural Fuel Oil) Regulations 1991 (SI 1991 No. 324), minimum standards are set for the construction of silage, slurry and fuel oil installations. These regulations apply to constructions built after 1 March 1991. However, it should be appreciated that the Environment Agency may give notice to a user to achieve the statutory standards on an installation built before 1 March 1991 if its condition presents a 'significant' risk of pollution.

Users should always seek to follow the Codes of Good Agricultural Practice for the Protection of Water, Soil and Air. These codes have statutory approval but do not in themselves create any criminal liability. The Water Code does not furnish a defence for charges brought under the 1991 Act. This is a major change from the position under the previous legislation (Control of Pollution Act 1974) where a farmer could state by way of defence that he or she had acted in accordance with the Code.

As a result of regulations made under the Environmental Protection Act, air quality has sought to be protected by the prohibition of straw and stubble burning as from 1993. Further emissions of odours or smoke from farms may create statutory nuisances under the Act as defined by s79(1). If a local authority is satisfied of the existence of such a nuisance, it must serve an abatement notice. Failure to comply with such a notice, without reasonable excuse, constitutes a criminal offence.

Conclusions

As identified at the outset of this chapter it may well be difficult to identify countryside law as a distinct discipline. What cannot be denied though is that a knowledge of the law as it affects the countryside, the mechanisms by which law is created and the direction in which it is progressing is now a necessity for all those who own, manage, conserve and utilize the countryside. This is particularly true if we are to maintain those qualities of the countryside that we admire as individuals, respect the qualities admired by others and protect and enhance the rights of those who seek to access and enjoy the countryside.

Further reading

Sydenham, A. (2003) *Public Rights of Way and Access to Land*, 2nd edition. Jordan, Bristol.

Useful websites

Gateways and portals to legal information on the web:
http://library.ukc.ac.uk/library/lawlinks/
http://www.venables.co.uk/
http://www.bailii.org/
http://www.justask.org.uk/

UK Government and the Administration of Justice
http://www.open.gov.uk/lcd/lcdhome.htm
http://www.courtservice.gov.uk/
http://www.defra.gov.uk
http://www.dca.gov.uk/ – Department for Constitutional Affairs

Acts of Parliament
For example, the Countryside and Rights of Way Act 2000, Human Rights Act 1998, Land Registration Act 2002
www.legislation.hmso.gov.uk/acts/acts2000/20000037.htm
http://www.environment-agency.gov.uk
http://www.planning-inspectorate.gov.uk/access/
http://www.countryside.gov.uk
http://www.lcd.gov.uk/hract/studyguide/index.htm

Europe
http://europa.eu.int/
http://www.ecnc.nl/doc/europe/legislat/convglob.html
http://www.eel.nl/

Access to the countryside
http://www.ramblers.org.uk

Countryside law
http://naturenet.net/
www.cla.org.uk – Country Land and Business Association
www.ala.org.uk – Agricultural Law Association

5

Rural planning

P. Tyler and P. Warner

Introduction

Rural planning in England operates within the context of the Town and Country Planning legislation. This chapter falls into two parts. First, to assist the reader in understanding the planning system the preliminary sections describe the nature and operation of the planning system. The second part of the chapter looks in more detail at the way the system interacts with typical rural issues.

This chapter quotes a number of legislative and other references. These are not intended to be comprehensive but will 'signpost' the reader to other relevant information. The planning system is currently undergoing the reforms outlined in the Planning and Compulsory Purchase Bill. The text does flag the main changes that are likely to occur, but the reader is encouraged to monitor the progress of the legislation.

1 THE NATURE AND OPERATION OF THE PLANNING SYSTEM

The aims of the planning system

The aims of the planning system can be summarized as follows:

* to secure the most efficient and effective use of land in the public interest;
* to do so within a framework that seeks to promote sustainable development, balancing social, economic and environmental objectives in a way that ensures that the quality of life of future generations is not prejudiced by decisions taken today. Ideally, decisions to bring forward essential development to meet the need for new homes, jobs and infrastruc-

ture should also benefit the environment. Assessing the impact of development proposals on the environment is a key aspect of decision making in the planning system.

The operation of the planning system is rooted in the use of land, rather than its management (see Chapters 3 and 4). For example, agricultural land is a 'land use' whereas the type of agriculture that is pursued (arable, stock rearing, mixed farming, etc.) represents the management of land within the agricultural land use category. Planning views land as a finite resource which is lost to other potential future uses once it has been developed. Developed land very rarely returns to agriculture or its 'green field' state. This is mainly due to the high value that developable land commands, which in turn is a result of restrictions on the release of land for development by the planning system. As a consequence the prudent use of land is a high priority.

In order to achieve efficient and effective use of land the rights of private interests to develop their land as they choose have been restricted for the public good. This may result in significant loss of potential value for a land owner. However, the planning system normally provides no compensation for such notional losses (there are some circumstances in which compensation is made). Conversely, there are no means to benefit the public or the public bodies for the significant uplift in value that can occur when planning permission for development is granted. Nevertheless the developer can be made to pay for necessary infrastructure, services and other needs of the local community which have been generated by that development. This is done through planning conditions and a 'Section 106' legal agreement under the Town and Country Planning Act (TCPA) of 1990. These can require off-site highway works or the provision of other appropriate infrastructure where the need is directly attributable to the development.

A 'Key concept in planning for the built environment is that planners try to guide and shape development by positive guidance backed up by negative control'. (Gilg, 1996; see Further reading, under 'General'). Positive guidance will take the form of policy advice which provides a framework for decision making, while the development control system which regulates the development of land is the negative control.

The definition of development is (TCPA, 1990).

'the carrying out of building, engineering, mining or other operations in, on, over or under land, or the making of any material change in the use of any buildings or other land'.

Processes and players

Having considered the aims of the system we can now move on to the process itself and the players involved in the system. These players or 'development partnerships' are increasingly important to deliver public goods.

The legislative and policy context

The operation of the planning system at the local level is guided by legislation and policy emanating from European and national governments.

The European Union has influenced much recent planning policy development by publishing the European Spatial Development Perspective (European Commission, May 1999). This is not binding on member states, but the Government has taken account of it in the advice it issues to plan makers. There are also several European directives that have an impact on planning, for example, the Directive on Strategic Environmental Assessment (42/EC/2001). The Government introduced legislation in July 2004 to require the strategic environmental assessment of certain types of plans and programmes. This will help to deliver 'sustainable development'.

The national legislation underpinning the operation of the planning system is complex. More detailed information can be found on the Office of the Deputy Prime Minister's website (www.odpm.gov.uk). The general framework (Figure 5.1) can be broken down into:

- Primary legislation: successive Town and Country Planning Acts, the most recent of which is the Town and Country Planning Act 1990.
- Secondary legislation: instruments such as the Use Classes Order; the General Development Order; the Permitted Development Order – which govern the more detailed day-to-day operation of decision making in development control. Regulations also proscribe the way in which primary legislation is implemented.

Figure 5.1 also outlines the relationships between the legislative framework and the policy documents that interpret the system. The 'top-down' system is plan led, and comprises:

- legislation and other statutory instruments;
- Planning Policy Guidance notes (PPGs), currently being reviewed by Government and reissued as Planning Policy Statements (PPSs);
- Regional Planning Guidance; Spatial Strategy for London;
- Structure Plans;
- District-wide Local Plans and Minerals and Waste Local Plans.

This represents an important change to earlier systems and for the countryside a more positive control system.

The Structure Plan and associated Local Plans comprise the Development Plan. Section 54A of the Town and Country Planning Act 1990 requires that applications for planning permission should be decided in accordance with the Development Plan, unless material considerations indicate otherwise. This is to ensure there is rational and consistent decision making and to minimize abuse of the planning system. The role and content of the Development Plans is dealt with more fully below.

In London and the metropolitan areas, and in a few non-metropolitan unitary areas, councils produce Unitary Development Plans (UDPs) which combine the functions of Structure Plans and Local Plans.

Planning authorities also prepare other forms of guidance which may be taken into account in decision making. Supplementary Planning Guidance amplifies policies in Development Plans. Typical examples would be policies for provision of affordable housing, or site development briefs setting out in more detail how the planning authority wishes to see a particular

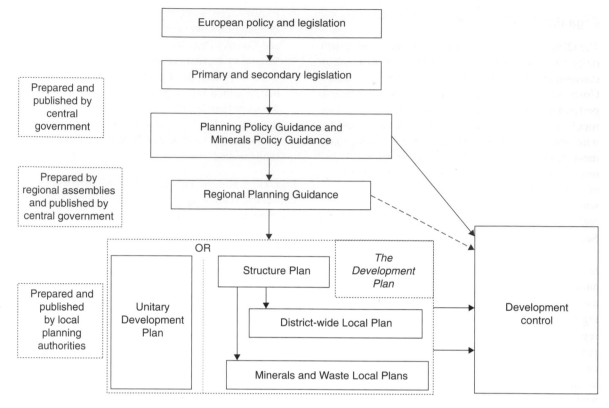

Figure 5.1   Framework of national legislation.

arca of land developed. These documents can be used to provide positive guidance for developers.

The plan-led system operates in the form of cascading guidance. Each 'tier' in the system must prepare the guidance it is responsible for in such a way that it takes account of and interprets the guidance sitting above it in the system.

Forthcoming changes to planning legislation

At the time of writing, the Government is steering new planning legislation through Parliament (as set out in the Planning and Compulsory Purchase Bill, 2002). The main changes to the system aim to speed up and simplify the plan-making process and to acknowledge the growing role of regional assemblies in developing regionally relevant policy. The aims and objectives of the planning system remain essentially the same. In terms of plan making the main changes proposed can be summarized as:

- A statutory role for regional assemblies, who will prepare Regional Spatial Strategies;
- the revision of PPGs and their replacement by more streamlined PPSs and complementary good practice guidance;
- the abolition of Structure and Local Plans and their replacement by Local Development Frameworks;
- greater recognition of the role of community plans in setting the 'vision' for the future of an area.

Details about the proposed operation of the new system can be found on the ODPM website (www.odpm.gov.uk/planning). Royal assent for the new system is expected in the near future, but Government is advising local planning authorities to work to the new processes and guidelines in advance of finalized legislation.

Organizational context

The Office of the Deputy Prime Minister has responsibility for ensuring that the planning system operates consistently and fairly, and in a way that reflects Government's objectives. Day-to-day scrutiny of the performance of local planning authorities (unitary, metropolitan, county, district and National Park authorities with responsibility for preparing plans and development control decision making) rests with the Government offices for the regions (an example of the range of work undertaken by regional offices can be seen at www.gosw.gov.uk). The Government has the power to 'call in' plans and decisions to be determined by the Secretary of State if it is considered necessary.

Regional assemblies exist in all eight English regions. Most of the membership is drawn from local authorities in the region, but one-third of the members are social and economic partners. As well as scrutinizing the work of the regional development agencies, regional assemblies must prepare regional planning guidance for the Government. The Government is currently consulting on an enhanced role for regional assemblies. Further details can be found at www.odpm.gov.uk.

Local planning authorities can be divided into:

- Those that prepare Strategic Planning guidance (such as Structure Plans which provide a planning framework for more detailed interpretation in Local Plans). These are county councils (in two-tier local government areas), unitary or metropolitan councils or London boroughs. National Park authorities also have a role in preparing Strategic Plans, often in partnership with shire counties. Strategic planning authorities are also responsible for preparing Minerals and Waste Local Plans.
- Those that prepare Local Plans. These are district councils (district-wide Local Plans), metropolitan or unitary councils (sometimes as part 2 of a Unitary Development Plan), London boroughs and National Park authorities.

Mention has already been made of the crucial role the Development Plan has in guiding the decisions taken in response to planning applications.

The Development Plan can consist of a number of documents (see Figure 5.1) and all relevant policies must be considered when reaching decisions. The Structure Plan usually plans for a 15-year period. The Local Plan, which interprets the Structure Plan in more detail, usually has a shorter (often 10-year) time horizon. Both plans are generally reviewed and updated on a 5-yearly basis. Plans are also monitored so that their performance can be assessed and policies and proposals revised at any time if necessary. This helps a local authority adapt for changing local circumstances, but does mean that there are always plans evolving as well as the adopted plan.

Government guidance advises on the form and content of the Development Plan. A key requirement is to reflect the principles of sustainable development, balancing economic, environmental and social issues. To ensure that this occurs, all development plans must undergo appraisal to make sure the effects of the policies and proposals they contain are fully understood.

The Structure Plan (or in some areas, part 1 of a Unitary Development Plan)

This document will:

- reflect national planning guidance and the spatial strategy, policies and proposals of regional planning guidance;
- set out a strategy, policies and proposals for further refinement by Local Plans;
- illustrate its strategic approach through diagrams (rather than maps);
- set out broad policies on housing; green belts; conservation and improvement of the physical and natural environment; the economy; transport and its integration with land use; mineral working and protection of mineral resources; waste treatment and disposal, land reclamation and re-use; tourism, sport and recreation; and energy generation, including renewable energy;
- give estimates of the future need for housing and employment land;
- set out guidance for the distribution of development.

Structure Plans do not contain detailed policies or site-specific proposals (which are matters for Local Plans). Further information about the role and content of Structure Plans is set out in PPG12.

District-wide Local Plans (or in some areas, part 2 of a Unitary Development Plan)

District-wide Local Plans:

- must reflect national and regional planning advice, and reflect and interpret the Structure Plan;
- offer detailed guidance on land use;
- consist of policies and a proposals map, prepared on an OS map base;
- allocate sites for essential development (and in sufficient quantity to conform to the estimates of need set out in the Structure Plan);
- reflect local issues and priorities and give certainty to developers about where development is likely to be acceptable;

Minerals and Waste Local Plans are prepared and presented in a similar way, although their content is circumscribed by their topic-based nature. The Government publishes specific advice for minerals and waste planning authorities.

Involving the public in the plan-making process

The process for preparing plans allows for public consultation and participation at key stages. Most local planning authorities involve partners, interest groups and the public in early discussion about the issues the plan should address. All plans must be made available in draft form for an 'official' period during which formal objections can be made to its contents (the 'deposit' period). The local planning authority is obliged to consider the objections that are made. Many objections will be resolved by changes to the plan that suit all parties. However, where objections cannot be overcome by negotiation provision is made for resolving them in the following way:

- Regional Planning Guidance and Structure Plans: objections are considered as part of an Examination in Public (EiP) before an independent panel. Participation at an EiP is at the invitation of the panel, who decide what further information they need to reach recommendations about how the plan should be changed. An EiP is discursive in nature, focusing on broad strategic issues and the conceptual framework of the plan.

- Local Plans: objections are considered as part of a Public Local Inquiry at which individual objectors have the right to appear. As well as considering the overall approach of the plan, a Public Local Inquiry will also hear representations about the use of individual sites. This form of Inquiry is held by an independent planning inspector and is a more formal process, often involving the use of legal representation.

Following an Examination in Public or Public Local Inquiry, the local planning authority will receive a report recommending how the plan should be amended. There are further opportunities for the public to object to the way the local planning authority chooses to amend the plan.

Once the plan is finalized the local planning authority formally adopts the contents as policy guidance.

Further information on procedure is set out in PPG12 and in free literature available from the ODPM (see website www.odpm.gov.uk/planning).

The development control process

A developer or individual wanting to develop land (see the earlier definition of 'development') must apply to the local planning authority for planning permission. Planning officers will recommend approval or refusal of the application, with the final decision usually resting with elected councillors (usually a planning or development committee). The route to final decision involves consultation with other interested parties and there are also opportunities for public comment on the proposal. Figure 5.2 shows how the various parties in the process interact. The decision-making process attempts to build consensus and adjudicate between competing interests. In reaching a decision there will be a number of key tests:

- Is the proposed development sustainable (does it meet social, economic and environmental objectives as set out in Government guidance and Development Plans)?
- Does the proposed development conform to the development strategy set out in the Regional Planning Guidance, Structure Plan and Local Plan?
- Is the proposal in line with detailed development policies in the Local Plan?

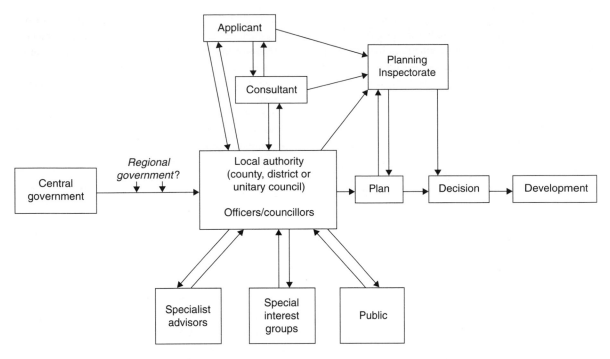

Figure 5.2   The development control process.

What are the immediate impacts on neighbouring land uses and land users?

Can any adverse impacts be offset by mitigating measures?

What can be done to ensure this happens (e.g. use of planning conditions)?

The progress of a planning application

The Government asks local planning authorities to process planning applications within 8 weeks wherever possible. The time taken will vary according to the complexity of the issues raised. A typical route to final decision on an application involves the following steps:

- Pre-application discussions, where the applicant discusses ideas with the authority and receives early advice about the acceptability of the proposal.
- Planning application is submitted, with payment of the application fee.
- The planning application is validated – checks are made to ensure the relevant land owners and neighbours have been notified.
- The application is allocated to a planning officer.

- The planning application is advertised.
- The planning officer considers the planning history of the site and will visit the site. The conformity of the proposal with relevant Development Plan strategies, policies and proposals will be considered. Further information may be requested from the applicant.
- A number of consultations are undertaken, for example:
  - parish council;
  - neighbour interests;
  - other relevant expert advisers to give views on, for example, the archaeology, ecology, waste, education, transport, etc. implications of the proposals. Typically, consultees will include the highway authority, the Environment Agency and other organizations where relevant (for example, the police, Ministry of Defence, airport authority, Regional Development Agency).

The case officer will consider all the consultation responses and letters of objection and any additional information that may have been requested. The case officer then prepares a report for the council's planning committee recommending whether the proposal should be approved or refused. The councillors then formally approve or refuse the application. All refusals must be accompanied by clear reasons which can be challenged through the planning appeal process.

Planning appeals

Legislation (see s78 of the Town and Country Planning Act, 1990) gives applicants the right to appeal against planning decisions. The circumstances when this might occur are:

- if a planning application is refused;
- if a decision is delayed;
- if conditions are imposed on a planning permission that the applicant considers are inappropriate.

An independent planning inspector will be appointed to deal with an appeal. The form this takes may vary. Some appeals are dealt with in writing, while informal hearings are also common. Appeals may also take a quasi-legal adversarial form (a Public Inquiry), usually with legal representation of the parties involved.

Third parties

Under the current system there is no form of appeal for third parties who feel aggrieved by the decision. The planning system will take account of third party interests in the following way:

- Those directly affected by a planning application are notified.
- Local planning authorities make publicly available lists of planning applications received.
- Third parties may write letters of representation for the case officer to consider and report to the planning committee.
- Councillors can be lobbied.
- Many councils make provision for the public to ask questions or speak at planning committee meetings.

Outcomes

The planning system is:

- discretionary, which means that, unlike the legal system, the authorities involved have great latitude in reaching decisions on the acceptability of development in any given location, i.e. approvals and refusals may result from exactly the same set of circumstances at different locations or times;
- consultative, which means that decisions are based not only on the proposal itself but also upon the concerns/support of statutory and professional bodies, neighbours, the public at large and pressure groups. Ultimately the decision is made at a local level by democratically elected laymen who, after hearing professional advice, employ their individual opinions and perspectives to conclude on the approval or refusal of development

The development control system has been criticized as bureaucratic and time consuming, which can be attributed to the factors outlined above. However, the current approach has safeguarded large areas of countryside from development and redirected development into more suitable locations. Conversely, the system can be criticized for aggregating land uses into single types (for example, large housing estates) and, in the past, paying insufficient attention to the relationships between development and transport.

Where do I get advice?

The local planning authority will offer advice freely. Early discussions about ideas can save both time and money. Free booklets and advice leaflets can often be obtained from local planning authority offices. Planning consultants and agents can help with professional representation and complex matters, although this can often be expensive and unnecessary for straightforward issues. The Royal Town Planning Institute maintains a list of professional consultants. The Institute also administers a planning aid service, which may make professional help available (on a voluntary basis) to help individuals and groups with particular planning problems.

2 RURAL PLANNING ISSUES

The strategic context

The planning process is about the management and control of development. Most attention is therefore directed to areas where the majority of development occurs – the towns and cities. In many cases Regional Planning Guidance and Structure Plans deal with rural areas by thematic or generic policies which are sometimes difficult to apply to local circumstances. The strategy diagrams that accompany these plans appear to treat rural areas as 'left over' after the major areas of change have been dealt with.

Strategic rural planning issues tend to focus on:

* the maintenance and enhancement of environmental assets and the need for essential development, particularly where it supports the rural economy. An example of the case for maintaining and enhancing the environment as an economic asset is set out in *The Environmental Prospectus for South West England* (March 1999; report available from www.swenvo.org.uk), prepared by regional partners to inform the work of the Regional Development Agency.
* the integration of planning with other rural strategies (for example, in those areas benefitting from European structural funds; areas with rural strategies);
* sustainable patterns of development, the role and function of settlements in the countryside, the nature and scale of development appropriate to them and their role in service delivery;
* linkages between urban and rural areas – the use of the urban fringe for public access and the role of green belts (see PPG2);
* quality in countryside planning.

The maintenance and enhancement of environmental assets

The degree to which the open countryside is protected by planning policy will vary, with landscapes and habitats of international and national importance being accorded the greatest protection. However, even within areas such as Areas of Outstanding Natural Beauty and National Parks, some development is necessary to support social and economic need. Planning tools that are often used to help assess the sensitivity of areas to development include:

* constraints mapping – identifying the most important environmental assets and other constraints, such as land liable to flood;

* landscape character assessment and the countryside capital approach (both promoted by the Countryside Agency);
* Strategic environmental assessment, sustainability appraisal and environmental impact assessment.

Integrating planning with other rural strategies

Some rural areas have a declining agricultural sector and a poor economy. Targeted funding for economic diversification and regeneration is often dependent on a strategy and action plan-led approach to problem solving. Funding from European structural funds and regional development agencies is often of this type. Typical examples may include:

* European Objective 1 and Objective 2 programming documents – which give guidance on the issues for which funding is applicable;
* regional rural strategies, e.g. the rural strategy for Cumbria;
* plans prepared in response to specific issues, e.g. the foot-and-mouth disease recovery plan prepared for Devon;
* plans prepared by subregional partnerships such as local strategic partnerships, LEADER+ partnerships, etc.;
* town and village plans prepared by regeneration groups, often aided by market and coastal town initiatives run by regional development agencies.

Strategies and plans of this type may contain proposals that need inclusion in development plans if conflicts are to be avoided. An example might be an objective to provide for more employment land in rural areas.

Sustainable development and settlements in the countryside: service provision

A sustainable pattern of development is one that generally provides for the economic and social needs of residents in a way that protects the environment and minimizes the need to travel. This presents particular challenges in rural areas.

Regional Planning Guidance and Structure Plans will contain general policy guidance (often criteria led) regarding the nature of sustainable development. They also contain policies setting out the role and function of settlements, usually in the form of a 'settlement hierarchy'. Market towns and larger villages are often identified as the most suitable locations for development and services to meet the needs of surrounding rural areas. The *Rural White Paper* (November 2000) envisaged market towns as the 'rural hubs' for homes, jobs, shops, facilities and transport. In remote rural areas, market towns may be the only place where there is sufficient critical mass to maintain services and where homes, jobs and transport can be successfully integrated. Many market towns are also in need of regeneration and have problems associated with declining town centre functions. Since the publication of the *Rural White Paper*, rural service standards have been developed for key service areas and these are monitored (see DEFRA, 2003). The Government, in PPG6, sets out policies for retailing which discourage further out-of-town shopping and seek to maintain the important role of town centres.

Linkages between urban and rural areas

Some examples of linkages between urban and rural areas are:

- 'city region' relationships where people choose to live in a high quality rural environment and travel to urban areas to work, shop and socialize. Similar areas of 'accessible countryside' exist around most towns of any size. Where large cities have 'green belts' (see PPG2), development pressure may have led to the expansion of nearby market towns;
- the use of the countryside for leisure and recreation. The quality of the countryside supports the tourist industry and gives opportunities for leisure. This can create its own problems at popular visitor locations and close to urban areas (the so-called urban fringe);
- supply chain linkages, where goods and services produced in rural areas support other business ventures.

Quality in countryside planning

The Countryside Agency, the Government's champion for rural issues, has led the debate about the quality of planning for the countryside (Countryside Agency, *Planning Tomorrow's Countryside*, 2000). There are three main areas for debate:

(1) how to ensure the countryside develops in a sustainable way that supports rural communities and where development benefits the environment;

(2) how to involve rural communities in policy making and decision making. Regeneration strategies and town and village plans give opportunities for this to happen. The formation of Local Strategic Partnerships to prepare community plans (Local Government Act 2000) is another opportunity for local opinion to shape policy and delivery;

(3) the quality of development itself, the integration of development with the surrounding environment and the design and materials used. The Countryside Agency has promoted Village Design Statements, which encourage local people to identify the distinctive elements of their area and develop policies to ensure they are taken into account in proposals for new development. A range of advice is also available regarding the use of renewable energy and energy efficiency in new buildings, sustainable drainage systems and sustainable construction (www.futurefoundations.com).

Local rural planning issues

This part of the chapter examines in more detail some topic-based issues affecting rural areas.

Housing

Estimating how much housing land should be provided in plans depends on:

- demographic and economic forecasts, including the relationship between births and deaths, the degree to which migration occurs and the rate at which new households are formed;
- understanding the social and policy context;
- identifying the potential for 'recycling' land that is already developed;
- assessing the degree to which meeting housing need will require the release of additional land.

The national planning context

Key documents include the following:

- The OPDM *Housing Green Paper Quality and Choice: A Decent Home for All* (April 2000; see further reading under 'Housing') sets out the Government's aim that everyone should have the opportunity of a decent home. This objective is reiterated in PPG3, *Housing* (March 2000).
- The OPDM *Rural White Paper Our Countryside: The Future* (November 2000; see Further reading under 'General'). This contains the core objective of 'a living countryside' and supports measures to enable people to live in the communities in which they grew up.
- *Sustainable Communities, Building for the Future* (ODPM, 2003; see Further reading under 'Housing') reiterated the Government's intention to provide more resources for housing. The Sustainable Communities action programme contains an aspiration to provide over 5000 affordable homes in villages. Following the Sustainable Communities statement, regional housing boards for each English region have been set up, to work alongside other regional agencies in the coordination and prioritization of funding for housing provision in their areas.
- PPG3, *Housing* (March 2000). This focuses on the role of the planning system in housing provision. It promotes sustainable patterns of development, focusing additional housing on existing towns and cities, making better use of previously developed land and improving design and the quality of living environment. The guidance promotes a 'plan, monitor and manage' approach to housing. The Government has subsequently published a number of good practice guides to inform implementation of the PPG: *Better Places to Live by Design – A Companion Guide to PPG3* (CABE and DETR, 2000); *Tapping the Potential – Assessing Urban Housing Capacity: Towards Better Practice* (DETR, 2000); *Planning to Deliver: The Managed Release of Housing Sites: Towards Better Practice* (DETR, 2000); *Monitoring the Provision of Housing Through the Planning System: Towards Better Practice* (DETR, 2000).
- Circular 6/98 *Planning and Affordable Housing* offers advice to local authorities on negotiating and facilitating affordable housing as part of development proposals.

A full list of publications can be found at www.odpm.gov.uk/planning under 'Planning Policy'.

At the time of writing, the Government is considering consultation responses to proposed amendments to PPG3. The objective of these amendments is to improve the supply of housing.

Housing and development plans

Regional Planning Guidance will assess the broad requirement for housing in a region, and set out a spatial strategy that guides its distribution. Regional housing boards also set out their priorities for regional investment in housing. The regional housing requirement is refined in Structure Plans, and subsequently translated into specific site proposals in Local Plans. However, the planning process cannot foresee all the sites that will come on to the market. Many sites are 'windfalls'. The capacity of towns and villages to re-use previously developed land, and the likelihood of 'windfall' sites arising is taken into account before new greenfield sites are allocated in plans.

Most Local Plan proposals maps will show details of proposals for specific towns and villages. They may show a development boundary surrounding a settlement. The development boundary is a line that effectively divides the 'settlement' from the 'open countryside'. Within the development boundary new proposals are often acceptable in principle; on small infill plots in villages, for example.

Affordable housing

Not everyone can afford to buy their own home in today's housing market. The Countryside Agency's report *The State of the Countryside 2003* says that roughly half of all rural residents (compared to one-third of urban residents) would have to commit more than 50% of their income to mortgage payments to buy an average-priced house in their area. The gap between housing and income is greatest in the South West and South East of England, where studies show that house prices can be up to nine times higher than income (see Wilcox, 2003; see Further reading under 'Housing'). National planning guidance recognizes the need for affordable housing to be a material planning consideration. 'Affordable' housing can take many forms, from subsidized rented housing, through shared ownership, to 'cheap' housing.

The current planning system does not allow local planning authorities to allocate sites specifically for affordable housing. Provision is through negotiation, based on Local Plan policies and guidance set out in PPG3 and Circular 6/98. This means that most Local Plans will set out:

- what is meant by 'affordable' in terms of the relationship between housing costs and local incomes;
- the minimum site size on which the local planning authority will seek to negotiate for affordable housing;
- the allocated housing sites where it will be seeking to negotiate affordable housing;
- guidance on the percentage of affordable housing it will seek from particular sites;
- policy guidance on the 'exceptional' circumstances in which land will be released for affordable housing in small villages.

The details of the council's negotiating approach are sometimes set out in supplementary planning guidance.

When negotiating for affordable housing, planning officers work closely with housing colleagues. The type of affordable housing best suited to an area will depend on detailed housing need assessment [*Local Housing Needs Assessments: A Guide to Good Practice* (see Further reading under 'Housing').] Some areas benefit from the work of Rural Housing Enablers whose role is to help communities identify and bring forward housing to meet need (see www.devonrcc.org.uk).

Land for affordable housing in rural areas can be brought forward in a number of ways. Some examples are set out in the Countryside Agency and Housing Corporation publication *Affordable Rural Housing: An Opportunity for Business* (June 2003; see Further reading under 'Housing').

In small rural communities land is not always allocated for development and the opportunity to influence supply through negotiation is more restricted. Where an identified need for affordable housing exists, land that would not otherwise be considered for housing may be released. These sites are known as 'exceptions sites'. Land provided in this way is often 'gifted' to the community, or sold at less than commercial rates.

This makes it easier to build houses at a lower cost. The exceptional nature of this type of development means it will nearly always be subject to legal agreement ensuring the housing remains available to meet local need. The Government is currently consulting on whether local authorities should be able to allocate sites for affordable housing in rural communities.

Agricultural workers' dwellings

Building new houses in the open countryside is strictly controlled by planning policy. However, farmers often require workers to live on or near the farm in order to look after stock or perform other essential duties. Before a house for an agricultural worker can be built, the proposal must undergo a series of tests as set out fully in Annex C of PPG7. These examine the way the farm unit operates and the consequent need for a full-time worker on site (the 'functional' test) and the financial viability of the enterprise (the 'financial' test). The financial test is particularly important when new farming enterprises are being established. If the future of a new farm is uncertain, local authorities may grant permission for temporary accommodation on site, to enable the viability of the business to be tested.

The special treatment given to agricultural workers' dwellings by the planning system is open to abuse. To guard against this, local authorities will use planning conditions and, in some circumstances, legal agreements to restrict the occupancy of the house. Steps may also be taken to tie other dwellings on the holding to the land, to prevent their disposal for open market housing. Removing a planning condition restricting occupancy to an agricultural worker is complex. Although there may no longer be a need for a farm worker on the holding itself, it is necessary to show that there is no need for agricultural workers' housing in the locality.

Other supervisory dwellings

Other businesses may also have a need for supervisory staff to be on site 24 hours a day. The Government is currently consulting on proposals (see consultation draft of PPS7) that would apply the same approach to supervisory dwellings as is used for agricultural workers' dwellings. The suitability of proposals will vary according to the business concerned and the circumstances of its operation. The conversion of existing buildings on the site to provide workers' accommodation will often be preferred to building a new house.

Agricultural diversification

The national planning and farming context

*   *A New Direction for Agriculture* (MAFF, December 1999; available from www.defra.gov.uk/farm/agendatwo/agendatwo.htm) announced more help for farm business diversification, including clearer planning guidance and advice.
*   *The Action Plan for Farming* (MAFF, March 2000; available from www.defra.gov.uk/farm/agendatwo/agendatwo.htm).
*   *The England Rural Development Programme* (MAFF, 2000; available from www.defra.gov.uk/erdp/default.htm) offers funding opportunities to help deliver social, economic and environmental objectives in rural areas.
*   The *Rural White Paper* (November 2000) supports the objective of a working countryside, with a diverse economy giving high and stable levels of employment.
*   PPG7, *The Countryside – Environmental Quality and Economic and Social Development* (1997, amended March 2001) gives planning advice on rural and agricultural diversification.
*   The definition of agriculture, as set out in s336 of the Town and Country Planning Act 1990, is:
    *   horticulture, fruit growing, seed growing, dairy farming;
    *   the breeding and keeping of livestock (including any creature kept for the production of food, wool, skins or fur, or for the purpose of its use in the farming of land);
    *   the use of land as grazing land, meadow land, osier land, market gardens or nursery grounds; and
    *   the use of land for woodlands where that use is ancillary to the farming of land for other agricultural purposes).

Farm enterprises

Modern farm enterprises often depend on income from sources that are not directly related to agriculture. A recent benchmarking study for DEFRA (Centre for Rural Research, 2003; see Further reading under 'Farm diversification') showed that almost 60% of farms surveyed were engaged in some form of diversified activity. Activities identified include:

*   agricultural services, e.g. contract machinery services;
*   trading enterprises, e.g. selling products to the public via farm shops, for example;
*   accommodation and catering;
*   equine enterprises;
*   recreation and leisure services, e.g. golf courses; community forests; fishing;
*   unconventional crops or crop-based processing activity, e.g. energy crops;
*   unconventional livestock and livestock processing, e.g. rare breeds; exotic animals.

Government planning policy encourages farmers to diversify in order to retain jobs in rural areas and help maintain economically viable farm units. Changes to the farming regime itself (new crops or grazing regimes) will not require planning permission and some development associated with farming can occur under agricultural permitted development rights (see Annex E of PPG7). Where buildings are erected using permitted development rights, the local authority may require prior notification of details of the siting, design and appearance. Certain types of development may also require an environmental impact assessment.

The main aim of the planning system is to try to integrate proposals for diversification in such a way that they benefit not just the local economy, but the environment as well. The acceptability of proposals will depend on the location and scale of the development, and the impact of the use and its associated servicing (by transport, for example) on the surrounding countryside and other nearby land uses. Useful background references include:

*   the ODPM publication *A Farmer's Guide to the Planning System* (2003; see Further reading under 'General') contains useful advice about the need for planning permission and the issues to be taken into account when thinking about a new enterprise.
*   the Countryside Agency-commissioned study *Traffic and Transport Issues in Farm Diversification – An Investigation into the Treatment of Traffic Issues in the Determination of Planning Applica-*

tions for Farm Diversification Proposals; see Further reading under 'Farm diversification' (2002).

A key message of advice is to engage the local authority in discussions about proposals at an early stage.

Certain proposals raise particular issues, as discussed below.

Farm shops

Shops that sell principally produce grown or raised on the farm will not require planning permission. However, if a significant proportion of the stock is produced elsewhere then a planning application will be necessary. The local planning authority will assess the likely impact of the shop on other nearby retail businesses and also take account of servicing and traffic movements. Further advice on retailing can be found in PPG6, *Town Centres and Retail Development* (June 1996).

Product processing and packaging

Many farms operate food-processing businesses that add value to primary produce from the farm. Planning policies encourage this type of activity for the contribution it can make to the rural economy (cf. Annex C PPG7 paragraph C18). The scale of the proposal, the likely transport impacts and the impact of development on the landscape are likely to be the main planning issues. Local authorities may use planning conditions to ensure that the size of the business remains appropriate to its location.

Golf courses and other leisure development

Land used for golf courses and other informal leisure uses can often be returned to farming if required. There may be environmental issues to be resolved if there is heritage or conservation interest in the land – Sites of Special Scientific Interest, for example, or archaeological remains, where ground disturbance may have adverse impacts. Most of the concerns will relate to the acceptability of ancillary support facilities such as clubhouses and accommodation. Farm ponds dredged for agricultural purposes will not generally require planning permission, but if used for recreation (fishing, boating, etc.) they will.

Forestry and energy crops

The *Energy White Paper* (February 2003) sets out the Government's commitment to developing renewable energy. Management of existing woodlands and the introduction of new crops such as short-rotation coppice and miscanthus present new opportunities for farmers to diversify. Farm slurry and other waste can also be used for energy generation. Most new crops will not require planning permission, although generating plant and other processing facilities will. Opportunities also arise where wind energy is viable (see section on 'Renewable energy').

'Horsiculture'

PPG7 states that 'A planning application is normally required for the use of land for keeping horses and for equestrian activities, unless they are kept as "livestock" or the land is used for "grazing"' (paragraph F3). Generally, loose boxes, jumps, menages and riding school operations, etc. will all need planning permission. Further advice can be found in Annex F of PPG7.

Farming and other regulations

Many on-farm activities may not require planning permission but may require licences or permissions from other regulatory authorities such as the Environment Agency. Examples include the spreading of waste, tipping of hard core to improve tracks and hard standings, the disposal of slurry and operations affecting water courses.

Economic development (land and buildings)

The national planning context

- The *White Paper Our Competitive Future: Building the Knowledge Driven Economy* (1998) made clear the Government's commitment to supporting and growing the economy.
- The *Rural White Paper* (November 2000) gave regional development agencies a specific remit for rural areas, to ensure a dynamic local economy and vibrant communities, able to respond to changes in traditional industries. Regional development agencies are key consultees on significant proposals for economic development. Economic development policies for rural areas are set out in Regional Development Agency (RDA) Economic Strategies.
- PPG4, *Industrial and Commercial Development and Small Firms* (January 1988). Revision of this guidance is long overdue and revised guidance is expected shortly.

- PPG12 sets out the scope and content of plans in relation to the economy and making provision for a range of economic development.

Development for employment purposes

The Government's central economic objective (as set out in Planning Policy Guidance) is to achieve high and stable levels of growth and employment, whilst ensuring that the benefits of economic growth can be shared by everyone and so deliver a better quality of life.

The planning system is expected to make provision for an adequate supply and mix of development land to support a healthy economy. The requirements of particular sectors, the need for strategic employment sites to meet regional needs and the locational requirements of certain technologies to locate near each other (clusters) have to be taken into account (see DETR 2000 in Further reading under 'General'). Authorities work closely with the RDA to ensure that these factors are taken into account in allocating employment land and in developing criteria-led policies that guide the location of development.

Areas of land allocated for economic development are shown on Local Plan proposals maps. Employment land will often be in locations that are easily serviced and accessed, close to labour markets and with good public transport connections. Topography may also be a limiting factor in the case of larger buildings or groups of buildings, which usually require relatively flat land. Allocated areas of land may be subject to policies that restrict the types of employment uses considered suitable for the site; for example, if it is close to a residential area. Occasionally, necessary 'bad neighbour' uses may require land well away from existing settlements. Factors to be taken into account as part of the development proposals will include design and landscaping and measures to minimize any adverse impacts.

In rural areas there may be scope for the re-use of redundant rural buildings for economic uses (see below) or for development of a suitable scale close to market towns and villages. Difficulties sometimes arise when businesses become successful and outgrow their premises, or where the nature of the business changes over time. This can result in unforeseen impacts on neighbouring land users. Occasionally the local planning authority may consider that the level of use has intensified to such a degree that a new planning application is required. More often, however, planning conditions are used to control, for example, the hours of working or number of traffic movements from a site.

Redundant rural buildings

The national planning context

- PPG7 sets out the national policy approach to the re-use of redundant rural buildings. The re-use of such buildings can be a way of assisting farm diversification or facilitating essential rural development without building new premises in the countryside.

The re-use of redundant rural buildings: issues

Currently, PPG7 expresses a preference for new uses for redundant buildings that benefit the local economy, and applicants have to demonstrate they have actively tested the market for such uses. The Government is currently consulting on proposals to remove this requirement (see consultation draft of PPS7 on www.odpm.gov.uk).

Many old farm buildings are listed and have value as historic structures, and their re-use is a way of securing their future. Changes to their structure may, however, impact on the historic integrity of the building, while creation of gardens and other domestic structures can look out of place in a farmstead setting. Occasionally large complexes of farm buildings are converted, effectively creating a small new settlement in the countryside, some distance from facilities.

The state of repair of the building can give rise to problems. A building must be capable of conversion without substantial reconstruction. If there has been no intention to maintain a building over time then it is technically 'derelict' and its re-use is judged against policies for new buildings in the countryside.

Redundant rural buildings suitable for conversion are a diminishing resource. Many local planning authorities are able to offer specialist advice relating to conservation and conversion.

The built heritage

The national planning context

- The Ancient Monuments and Archaeological Areas Act 1979, dealing with ancient monuments and archaeology.

- The Planning (Listed Buildings and Conservation Areas) Act 1990, dealing with historic buildings and areas.
- PPG15, *Planning and the Historic Environment*, 1994.
- PPG16, *Archaeology and Planning*, 1990.

Under proposed changes to the planning system, the Government is proposing to merge and streamline the contents of PPGs 15 and 16.

The legislative framework provides for several categories of protected buildings, monuments or areas: scheduled monuments, listed buildings and conservation areas. Scheduling and listing is the responsibility of the Secretary of State; local authorities are responsible for the designation of conservation areas. English Heritage also maintains a Register of Parks and Gardens of Special Historic Interest. The listings of listed buildings and ancient monuments are available for public inspection and local authorities have lists available for their own areas.

Local authorities are required to take into account the historic environment in their policies and development control decisions. Most development plans will contain proposals and policies relevant to conservation areas, listed buildings, ancient monuments and archaeological remains.

Conservation areas

Local authorities have a legislative duty to designate as conservation areas any areas of special architectural or historic interest, the character or appearance of which it is desirable to preserve or enhance. It is the quality and interest of the areas, rather than that of the individual buildings, that is the main consideration in identifying conservation areas.

Within conservation areas, special care is taken to ensure changes are managed in a way that is sympathetic with the character and qualities that underpin the designation. For example:

- Changes proposed in a conservation area are advertised and care is taken to ensure that new development is sensitively and appropriately designed. Decisions are based on the degree to which proposed change enhances or preserves the appearance or character of the area.

- The demolition or partial demolition of buildings will require consent, and some householder permitted development rights are removed.
- Proposals for development outside of a conservation area will also be judged for their impact if they affect the setting of the conservation area, or views into or out of the area.
- Trees located within a conservation area are also protected. Anyone proposing to cut down, top or lop a tree in a conservation area is required to give 6 weeks' notice to the local planning authority.

Listed buildings

Listed buildings are found within settlements of all sizes and throughout rural areas. Buildings are listed as the result of survey or following proposals by local authorities or other interest groups. In deciding which buildings to include, the Secretary of State takes into account the architectural interest of the building, its historic interest, close historical associations with people or events, and the group value of buildings. Some modern buildings have also been listed as exemplars of particular types of architecture or building processes. As well as the building itself, listing includes objects and structures fixed to the building and objects and structures within the curtilage.

Proposals that affect a listed building (this can include the interior, exterior or setting of the building) will require listed building consent (even if the changes proposed do not fall within the definition of 'development' and require planning permission).

In cases where local authorities (or English Heritage) feel a listed building is not being properly preserved, they may serve a repairs notice on the owner. This must specify the works that are considered necessary for the preservation of the building. If a repairs notice is not acted upon within 2 months, the local authority or English Heritage may begin proceedings to compulsorily acquire the building.

Scheduled monuments and archaeological remains

Under the provisions of the Ancient Monuments and Archaeological Areas Act 1979, unoccupied historic structures and other historic and archaeological remains can be protected as scheduled ancient monuments. However, not all important remains are scheduled.

The Government's approach to the preservation of archaeological remains is set out in PPG16 which states (paragraph 8) that 'Where nationally important archaeological remains, whether scheduled or not, and their settings, are affected by proposed development there should be a presumption in favour of their physical preservation'. If physical preservation in situ is not possible, then provision can be made for excavation and recording of the site as an alternative.

Transport and access

The national planning context

- The *Transport White Paper* (*A New Deal for Transport: Better for Everyone*, DETR 1998). This contained proposals to extend choice in transport and promote transport measures to encourage sustainable development and meet the needs of travellers.
- The proposals contained in the *White Paper* were given a statutory basis in the Transport Act 2000. Prior to this, the Transport Bill required local transport authorities to develop and implement policies for the promotion and encouragement of safe, integrated, efficient and economic transport facilities and services to, from and within their area. These policies are set out in Local Transport Plans, which also provide the mechanism by which local transport authorities make the case to Government for funding for transport measures.
- PPG13, *Transport*, published in March 2001, sets out the Government's approach to integrated planning and transport. The primary objective of Government policy is to minimize the necessity to travel, by co-locating jobs, houses, services and other facilities and integrating them with sustainable transport choices. This means considering the degree to which walking, cycling and public transport can be facilitated in preference to using the private car. Considering movement at the very local level helps to integrate development with nearby land uses, as well as promote sustainable transport choices.

Sustainable transport issues

The issue of service accessibility is closely linked to transport provision. Flexible rural transport measures need to be considered alongside the retention of local services. The recent Social Exclusion Unit's report *Final Report on Transport and Social Exclusion Making the Connections* (2003; see Further reading under 'Access') illustrates the extent to which improved accessibility can help overcome social exclusion.

The *Rural White Paper* and PPG13 both encourage the location of new development in rural areas at key locations such as market towns or groups of villages. It is at such locations that there is the most chance of promoting walking and cycling and achieving the critical thresholds necessary to support public transport measures.

Providing more jobs in the countryside is also a means of reducing long-distance commuting to nearby urban areas. The planning process will consider the traffic that will be generated by employment proposals and the ways in which the transport needs of the business can be met sustainably (see PPG13, paragraph 43). This will involve consideration of both workforce travel and freight transport. Where a business generates significant volumes of traffic it is usually preferable for it to be located close to a sustainable transport corridor.

In determining planning applications, the views of the local transport authority (the county council, unitary or metropolitan authority) regarding the sustainability of the proposals and specific site access issues will be sought. The planning system may also be used to help deliver traffic management measures, particularly where they help influence travel behaviour and promote safety. The management of congestion, parking, traffic calming and improvements to road layouts may be sought as part of development proposals. Land uses that generate significant volumes of traffic may also be encouraged to develop 'green transport plans' to promote sustainable transport choices for staff and visitors.

Service delivery

The national planning context

- The *Rural White Paper* (November 2000) sets out minimum service standards that people living in rural areas should expect.
- PPG7 refers to local planning authorities facilitating and helping to retain existing services in rural areas (paragraph 3.23).

Retention and promotion of services

A good, reliable standard of services in rural areas is important for those who are unable to travel to facilities and can help overcome social exclusion. The planning system can help to secure the conditions that make services viable and safeguard premises such as shops and pubs from harmful changes of use.

The location of new housing close to existing services and facilities will help to sustain the conditions that make services economic to provide. Regional planning guidance for the South West, for example, contains a policy supporting additional rural housing of an appropriate scale where this is the case. However, residential property values often make local business premises more attractive capital assets if converted to houses. Consequently, many Local Plans contain policies protecting premises from conversion, particularly if the building houses the last remaining village facility. Local Plans may also contain other safeguarding policies, support the provision of new community facilities and aim to integrate local public transport with land use.

Information and communications technology

The national planning context

- PPG8, *Telecommunications*, August 2001 sets out the Government's view that 'Modern telecommunications are an essential and beneficial element in the life of the local community and in the national economy'. The *Supporting Guidance Appendix* to PPG8 contains information regarding the requirements of different telecommunications infrastructure and the physical developments associated with them.

Information and communications technology (ICT) issues

Many rural areas suffer from poor communications infrastructure and particularly the absence of broadband technology. This inhibits rural business development and imposes additional costs on small rural enterprises. The planning system is directly involved in decisions regarding the siting of radio and telecommunications masts, and in individual householder applications for satellite dishes. PPG8 encourages Development Plans to set out policies for telecommunications development; the matters to be considered are set out in PPG8 paragraphs 37–42). While the Government expects local authorities to respond positively to proposals for ICT development, there are also safeguards to ensure that the environmental impacts of proposals are adequately considered, particularly where they affect designated landscape areas such as AONBs and National Parks.

Minor telecommunications development may not require planning permission and can be carried out using permitted development rights. In these cases the operator must give prior notice of its proposals to the local planning authority, so that the authority may consider whether the siting and appearance of the development requires approval.

Renewable energy

The national planning context

- In January 2000 the Government announced the aim of generating 10% of UK electricity from renewable sources by 2010. The *Energy White Paper* (February 2003) outlined the Government's intention to double the contribution from renewable energy by 2020. For the future, the Government foresees a greater role for smaller-scale energy generation with greater involvement of the English regions and local communities. The planning system is expected to be more responsive and to weigh the impacts of local renewable energy schemes against the global environmental benefits of moving towards a low carbon economy.
- PPG22, *Renewable Energy* (1993) is currently being revised by the Government. The draft PPS22 is an enabling document that aims to stimulate the renewable energy industry.

Other studies

Further studies have been undertaken to see how national renewable energy targets could be taken forward at the regional level. The Oxera report (*Regional Renewable Energy Assessments*; see Further reading under 'Renewable energy') examined regional findings to see if, in aggregate, they would meet national targets. The study found that indicative regional targets nationwide only satisfy the overall national target when the figures at the high end of the range are used. An example of more detailed regionally led work to establish county-wide renewable energy targets can be seen at http://www.oursouthwest.com/revision2010/.

Funding for renewable energy

The Government has set aside funding for renewable energy projects and has introduced mechanisms to stimulate the market. The first of these support mechanisms was the Non-Fossil Fuel Obligation (NFFO) which provided a premium price for up to 15 years for the electricity generated from renewable sources. This has now been replaced by the Renewables Obligation. This places an obligation on suppliers to acquire a percentage of their supplies from renewable sources. Where they are unable to fulfil their obligation, they may purchase Renewable Obligation Certificates (ROCs) from other suppliers, or buy out their obligation by making a payment to the industry regulator.

Renewable energy technologies and planning issues

Renewable energy can be generated from a number of sources:

- *From the sun*: for example, by using photovoltaics; active solar heating; passive solar design. Planning issues arising from development of this type are most likely to relate to design and local visual impacts.
- *From the wind*: wind turbines, clusters of turbines and wind farms are the forms of development that raise most planning issues relating to landscape and visual impacts; noise; access; grid connection; telecommunications interference; and effects on MoD and aviation interests. To meet short-term targets to 2010 it is likely that a significant contribution will need to be made by onshore wind energy generation. Longer term, offshore wind energy generation may have potential.
- *From water*: hydro power, wave power and tidal energy. Planning powers do not extend offshore, but shoreside facilities will be affected. Proposals are often located in environmentally sensitive areas and

impacts will need to be carefully weighed against the benefits from the scheme.
- *From wood and energy crops*: existing wood resources or energy crops such as miscanthus and short-rotation coppice can be used to generate power, heat or both at a wide range of scales. Energy generation from energy crops can have benefits for the rural economy and provide an alternative crop for farmers. Other agricultural wastes can be used as combustible fuels through anaerobic digestion. Impacts arising from the scale of generating plant and transport of fuel are the two major environmental impacts to be considered.
- *From farm, household, commercial and industrial waste*: a variety of technologies exist to process waste of various kinds. Most raise planning issues relating to public acceptability, the transport of waste to processing points and the environmental impacts of the proposal.

It is the local environmental impacts of renewable energy proposals that raise most planning concerns. Where impacts are significant, an environmental impact assessment may be required.

Further reading

General

DETR (2000) *Planning for Clusters: A Research Report*. Available from www.odpm.gov.uk/stellent/groups/odpm_planning/documents/page/odpm_plan_606006.

Gilg, A. (1996) *Countryside Planning*. Routledge, London.

Office of the Deputy Prime Minister (2000) *The Rural White Paper – Our Countryside: The Future*. Available from www.defra.gov.uk/rural/ruralwp/whitepaper.

Office of the Deputy Prime Minister (2003) *A Farmer's Guide to the Planning System*. Available from www.odpm.gov.uk/planning under 'Planning guidance and advice'.

Rural Development Council (1998) *Rural Development and Land Use Planning*.
This report was published by the RDC before merger. It examined the development system, the policies affecting rural areas in local authority development plans and the planning decisions taken. It covered issues such as sustainable development, the location of development, the rural economy, affordable housing and Green Belts and made a number of recommendations for future planning policy. ISBN: 1-86996-4-70-5; price: free.

TCPA (1990) *Town and Country Planning Act 1990*. HMSO, London.

The Countryside Agency (2000) *Planning Tomorrow's Countryside*. CA 60.

This policy statement is the Countryside Agency's advice to local planning authorities, to the Government and to developers on how the planning system should operate and evolve, to ensure that development is achieved while the countryside remains protected. ISBN: 0-86170-648-X; price: free.

The Countryside Agency (2003) *The State of the Countryside 2003 – Summary of Key Facts*. CA 141.

This leaflet outlines the latest key facts and figures about the English countryside, its people and places.

The Countryside Agency (2003) *Rural Proofing in 2002/03*. CA 146.

In the Rural White Paper the Government made an important new commitment to 'rural-proof' its policies. Rural proofing means that as policy is developed and implemented, policy makers should systematically think about the impacts in rural areas and make adjustments to their initiatives if appropriate. The Countryside Agency was asked to report each year on how Government has performed against this commitment. This is the second year report, detailing the progress made by the Government with its rural proofing commitment. It also contains detailed information on each department in the appendices. Price: free.

Countryside character

Environmental Prospectus (1999) *The Environmental Prospectus for South West England*. Report available from www.swenvo.org.uk.

The Countryside Agency (2002) *Making Sense of Place – Landscape Character Assessment*. CAX 94.

This pamphlet summarizes guidance for England and Scotland and is jointly published with Scottish Natural Heritage. It emphasizes the growing importance of landscape character assessment as a tool for identifying what makes a place distinctive. Price: free

The Countryside Agency (2002) *Is This the Future We Want? Land Management Scenarios in the South West*. CAX 103. Prepared for the Countryside Agency by Land Use Consultants. Price: £2.

Planning and policy appraisal

Department of the Environment, Transport and Regions (2000) *Good Practice Guidance on the Sustainability Appraisal of Regional Planning Guidance*. Stationery Office, London.

Office of the Deputy Prime Minister (2003) *Strategic Environmental Assessment: Guidance for Planning Authorities*. ODPM, London.

Environmental impact assessment

For background information see *Note on the EIA Directive for Local Planning Authorities* available on the ODPM website, www.odpm.gov.uk.

Biodiversity

Association of Local Government Ecologists and the South West Biodiversity Initiative (1998) *A Biodiversity Guide for the Planning and Development Sectors: A South West Regional Perspective*. Available from www.algae.org.uk.

Housing

Department of the Environment, Transport and the Regions (2000) *Local Housing Needs Assessment: A Guide to Good Practice*. Available from www.odpm.gov.uk/housing under 'Housing Research', 'Housing Research Summaries' '2000'.

Office of the Deputy Prime Minister (2000) *The Housing Green Paper – Quality and Choice: A Decent Home for All*. Available from www.defra.gov.uk/housing under 'Consultation Papers'.

Office of the Deputy Prime Minister (2003) *Sustainable Communities: Building for the Future*. Available from www.odpm.gov.uk/communities under 'The Communities Plan'.

The Countryside Agency (2003) *Undertaking Effective Housing Needs Assessments: A Guide to Local Authorities Serving Rural Communities*. The Countryside Agency, Cheltenham. Price: free.

Wilcox (2003) *Can Work – Can't Buy: Local Measures of the Ability of Working Households to Become Home Owners*. Joseph Rowntree Foundation, York.

Access

Office of the Deputy Prime Minister (2003) *Making the Connections. Final Report on Transport and Social Exclusion*. Social Exclusion Unit. Available from www.socialexclusionunit.gov.uk/publications.asp?did=229.

The Countryside Agency (1994) *Managing Public Access: A Guide for Farmers and Landowners*. CCP 450.

A valuable source of reference for all managers of land. This small booklet contains practical advice on managing public access and helps land managers find answers to their questions about legitimate public access to their land. Price: free.

The Countryside Agency (2001) *Great Ways to Go – Good Practice in Rural Transport.* CA 62.

This fully illustrated guide presents 17 case studies, drawn from around the country, that show the varied ways in which local communities have found appropriate solutions to their transport problems. Many different types of project are presented – from social car schemes to young people's scooter projects, community buses and rail schemes. Price: free.

The Countryside Agency (2001) *Countryside Access and the New Rights.* CA 65.

The Countryside and Rights of Way Act 2000 will give people a new right of access to walk over large areas of open countryside and common land. This leaflet describes how these new access arrangements will be put into practice on the ground in England, and explains what they will mean for those involved, especially farmers, landowners and countryside visitors. Price: free.

The Countryside Agency (2001) *Improving Access to the Countryside: Integrated Access Demonstration Projects.* CA 98.

A small two-colour leaflet that briefly describes projects that the Countryside Agency is undertaking to demonstrate how existing access can be improved by linking countryside recreation to wider issues such as public transport and the availability of information. Price: free.

The Countryside Agency (2002) *Traffic and Transport Issues in Planning Applications for Farm Diversification.* CRN 48.

This research note sets out the key findings of research carried out on behalf of the Countryside Agency. The research was undertaken to discover whether there is an unreconciled tension in policy hampering decisions on planning applications for farm diversification, and to investigate the transport impacts of implemented farm diversification development. Price: Free.

The Countryside Agency (2002) *Traffic and Transport Issues in Farm Diversification – An Investigation into the Treatment of Traffic Issues in the Determination of Planning Applications for Farm Diversification Proposals.* CAX 108.

This Working Paper presents research conducted by the University of the West of England on behalf of the Countryside Agency. The research set out to discover whether there is an unreconciled tension in policy hampering decisions on planning applications for farm diversification, and to investigate the transport impacts of implemented farm diversification development. Price: £2.

Recreation/leisure

The Countryside Agency (1995) *The Visitor Welcome Initiative.* CCP 476.

A practical information pack containing everything countryside recreation site managers need to assess the quality of welcome offered by their sites. It concentrates on informal sites to provide a checklist and guidance notes that contain a wealth of helpful recommendations and suggestions. Price: £4.

The Countryside Agency (1996) *Market Research for Countryside Recreation.* CCP 491.

This advisory publication is for countryside managers with limited experience of carrying out visitor surveys. Knowing what visitors really want from countryside recreation sites is essential when planning improvements or new facilities. The basic whys and hows of market research are covered in a practical and highly readable style. Price: £6.

The Countryside Agency (1998) *Site Management Planning: A Guide.* CCP 527.

The guide sets out an approach to management planning that can be tailored to any site. This publication is a toolkit for site managers, helping them to assess a site's needs and make decisions. It has been prepared with the assistance of a range of countryside professionals. Price: £12.50.

Design

The Countryside Agency (1996) *Village Design: Making Local Character Count in New Development – Part 1.* CCP 501.

Detailed advice and background information for local communities on how to prepare Village Design Statements. See also CCP 501, Part 2. This publication is available in pdf format only. Price: free.

The Countryside Agency (2000) *Design of Rural Workplace Buildings: Advice from the Countryside Agency.* CA 36.

This report examines some successful recent workplace buildings for industrial, commercial and public use, and assesses how well they have achieved good design. It is well illustrated with high quality images and should inspire anyone with an interest in sustainable development and in buildings that contribute to long-term social, environmental and economic well-being. Price: £8.

Urban fringe

The Countryside Agency (2002) *Sustainable Development in the Countryside Around Towns – Volume 1: Main Report.* CAX 111.

This is the main report of a study, conducted by the Centre for Urban and Regional Ecology, that aimed 'to determine the current nature of sustainable development in the urban fringe, and to outline the opportunities for the Countryside Agency to influence urban extensions to meet a sustainable vision for the Countryside around towns'. Price: £2.

Renewable energy

Devon County Council. Appeal Decision: Reference APP/W1145/A/02/1105474. Stowford Cross, Bradworthy. A wind cluster development was proposed comprising 3 × 1.0 megawatt wind turbine generators, a substation, the construction of access tracks and ancillary development including a visitor car park. The planning application was initially rejected, and at appeal, planning permission was granted.

Department of Trade and Industry (2002) *Wind Energy and Aviation Interests: Interim Guidelines*. DETR, London.

Oxera Environmental for Department of Trade and Industry (2002) *Regional Renewable Energy Assessments*. Available from www.dti.gov.uk/energy/publications/policy/index.shtml

The Countryside Agency (2002) *The Community Renewables Initiative*. CA 47.
The aim of the Community Renewables Initiative is to help local people and organizations devise and implement renewable energy developments. This leaflet provides information about how the Initiative will operate and lists contact details of local support teams. Price: free.

South West Renewable Energy Agency (2003) *The Appropriate Development of Wind Energy: Guidance for Local Planning Authorities*. RegenSW, Exeter.

Services

Department for Environment, Food and Rural Affairs (2003) *Rural Services Standard 2003*. DEFRA, London.

The Countryside Agency (2002) *Examination of Planning Inspector's Appeal Decisions Concerning Rural Pubs*. CRN 39.
Considers issues relevant to all community facilities in rural areas. Price: free.

Farm diversification

Cabinet Office Performance and Innovation Unit (1999) *Rural Economies*. The Stationery Office, London.

Campaign to Protect Rural England (2000) *Farm Diversification: Planning for Success*. CPRE, London. Price: £3.50.

Centre for Rural Research (2003) *Farm Diversification Activities: Benchmarking Study 2002*. CRR Research Report 4. University of Exeter.

Countryside Agency (2002) *Traffic and Transport Issues in Farm Diversification – An Investigation into the Treatment of Traffic Issues in the Determination of Planning Applications for Farm Diversification Proposals*. A Working Paper. Available from www.countryside.gov.uk/Publications/articles/Publication_tcm2-4410.asp

Shorten, J. and Daniels, I. (2000) *Rural Diversification in Farm Buildings – An Investigation into the Relationship Between the Re-use of Farm Buildings and the Planning System*. Prepared for the Planning Officers' Society by the Centre for Environment and Planning Research, Faculty of the Built Environment, University of the West of England, Bristol.

Other planning resources

Cullingworth, B. and Nadin, V. (2001) *Town and Country Planning in the UK*. Routledge, London.

www.devon.gov.uk/structureplan – Devon Structure Plan (sets out strategic planning policies for development and other land uses).

www.planning.odpm.gov.uk – Government information on planning.

Part 3

THE RURAL ENVIRONMENT

6

Nature conservation

E. Williams

Introduction

The quantity and quality of species and habitats on the planet has become a matter of great public concern. Managers and users of the British countryside are in the front line of local efforts to conserve nature and hence must be equipped with the best possible range of relevant knowledge and skills. Nature, which may also be referred to as wildlife, biodiversity or the environment, is now of equal or greater concern than productive use of countryside resources such as agriculture or forestry, and aesthetic and recreational considerations or development opportunities such as those associated with the tourism industry.

In order to manage natural systems it is necessary to understand and apply a wide range of scientific facts and concepts but also to operate within the prevailing socio-economic context. [A good general reference is Sutherland (2000).] This chapter will outline what needs to be known and indicate where further information can be sought.

Aims and rationale

The conservation of biodiversity sounds an admirable and straightforward aim. The definition of biodiversity itself can be controversial. It is probably most frequently taken to be the number of species at a particular site, or what ecologists call species richness. It is almost impossible to get a complete species list for a site, however, and anyway the delimitation of species is not always rigid. For many economic and other purposes it would be ideal to conserve genetic or even molecular diversity. Conservation below species level, or at population or individual organism or gene level, would ensure adaptability and resources for an uncertain future. It could then be argued that larger populations or communities, which are likely to be more genetically diverse, are more worth conserving even if species-poorer. With this in mind, ecologists often calculate diversity indices, such as Shannon's or Simpson's, which incorporate species richness and abundance into a single index number, at least for sample data or for indicator species. However, these indices need calibrating for particular sites and habitats before they can be used for comparisons or management decisions.

At the other end of the spectrum, habitat, ecosystem and landscape diversity can also be targets of conservation. Within Europe alone, the CORINE (Co-ordination of Information on the Environment) Biotopes database recognizes 2500 different habitats plus several ecosystem groups and biogeographic regions. These are all hard to delimit and should include considerations of structural complexity in both vertical and horizontal planes, abiotic factors such as soils and geology, and successional stages and trophic relationships in the biotic communities present. These latter have very specific ecological meanings.

Succession can be primary or secondary depending on the history of the site and whether the starting point is a bare physical substrate or a developed but disturbed soil. The classical theories of succession explain the inexorable development of climax communities in terms of pioneer 'r'-selected or transient species facilitating and being replaced by competitors or 'K'-selected species over time. The letters r and K refer to parameters in an equation modelling logistical growth of populations. r is the intrinsic rate of natural increase: birth minus death. K is the carrying capacity of the environment: the maximum number of individuals that can survive there. Thus r-selected species have high rates of reproduction and are thought of as 'weedy' in plant terms and do well in disturbed environments.

K-selected species are good competitors with staying power but probably slow growth and reproduction. K-selected species will probably dominate when the community is at its carrying capacity, in a stable environment and/or late in ecological succession. These theories are being superseded by more complex ideas which, for example, account for medium-term stability of some mid-successional communities in terms of inhibition, and help in understanding the effects of perturbing the links within food webs present in a community. These ideas are helping to formulate more effective management.

Previously, much conservation policy and attention has been directed towards 'K' selected species as these are more vulnerable to habitat destruction and over-exploitation due to their typically slow growth, slow reproduction and larger sizes. Conversely, much practical management effort has been aimed at arresting succession and creating mosaics of habitat, and using 'traditional' practices. Now this is not necessarily accepted as the best approach.

Ideally, those involved in nature conservation should always be clear as to the particular purpose of any activity undertaken. The measures of biodiversity or habitat quality chosen in each case need to be defined and agreed by the relevant interested humans on behalf of other species. The WORLDMAP project at the British Natural History Museum attempts to set priorities for *in situ* conservation by developing a single agreed value for biodiversity; distinguishing pro-active processes from re-active attempts to save endangered species; setting the criteria for necessary and sufficient conservation goals using limited resources; and resolving conflicts.

It is clear from sources such as the *Living Planet Report* (Loh, 2002) that recent human impact on biodiversity is detrimental. As measured by the Living Planet Index, which is derived from population data on a few hundred species, the last 30 years have witnessed a fall of 37% in biodiversity. This may be up to 10 000 times the natural extinction rate. However, the true number of species on earth has never been known. It was thought to be about 5 million.

Much nature conservation effort has been rather arbitrarily directed, sometimes by fashion and whim. It is probably better to attempt to use scientific information to decide which species and habitats to conserve actively, given that resources are limited and that some can only be conserved at the expense of others. However, essential relevant information is often lacking. Current criteria for choosing species to which conservation effort should be directed include potential extinction using the *IUCN Red List* criteria (Hilton-Taylor, 2000) and rarity and taxonomic uniqueness including endemics and flagship species. Criteria for habitats or areas include naturalness, species richness, diversity, abundance and size.

The following section should give the impression that nature conservation is high on the agenda at all levels of governance. However, the results of all this concern will need to be judged by future generations in terms of species and habitat survival on the planet.

History

Conservation of natural resources is not a new idea. For example, in previous civilizations and under various monarchs and landowners the rights to hunt and gather in certain areas of countryside were carefully guarded. The reasons were primarily economic or sporting yet usually resulted in conservation of species and habitat.

More deliberate conservation of nature arose mainly in the nineteenth century. Both the Fur and Feather Group, founded in Britain in 1885, and the Audubon Society, founded in 1886 in the USA, were inspired by the slaughter of birds, especially for their plumage for use in hats. The former became the Royal Society for the Protection of Birds in 1904. The Selborne Society (1885), the National Trust for Places of Historic Interest or Natural Beauty (1894) and the subsequent Society for the Promotion of Nature Reserves (1912) and Norfolk Naturalists Trust (1926) were further examples of early British conservation bodies, the modern counterparts of which still operate. These were the initiatives of interested and usually wealthy individuals who recognized the dangers to nature and could take the lead in protecting sites. The British Ecological Society was founded in 1913 and was the first such society in the world. Its focus has remained more scientific and less campaigning. Box 6.1 highlights the example of Wicken Fen in Cambridgeshire, UK. The development of this site reflects the development of nature conservation in Britain.

Impetus was lost during the Second World War, but in 1948 the International Union for the Protection of

Box 6.1   Wicken Fen

Two acres (0.8 ha) of Wicken Fen were purchased in 1899 for the National Trust as Britain's first nature reserve by Verrall and Rothschild. Its value had been highlighted by entomologists associated with Cambridge University including Charles Darwin. Fifty-five conveyances later, it now covers over 800 acres (320 ha), and supports many UK Biodiversity Action Plan priority species, Red Data Book, National Scarce and National Notable species, although a few like the swallowtail butterfly have been lost. The Fen is a National Nature Reserve, a Site of Special Scientific Interest, a Special Area of Conservation and a Ramsar site (see below). It is managed by traditional sedge cutting and peat digging, although this is no longer for economic reasons. Now it is enjoyed by numerous human visitors. (See www.wicken.org.uk.)

Nature (IUPN) was founded, later the International Union for the Conservation of Nature and Natural Resources (IUCN) (www.iucn.org), and also known as the World Conservation Union since 1990. This provided a global framework. Currently six commissions operate within the IUCN:

(1) Species Survival (SSC);
(2) Protected Areas (WCPA);
(3) Environmental Law (CEL);
(4) Education and Communication (CEC);
(5) Environmental, Economic and Social Policy (CEESP); and
(6) Ecosystem Management (CEM).

In Britain, following the work of the Wildlife Conservation Special Committee of 1945 and subsequent Huxley Report in 1947, the Nature Conservancy (NC) was established in 1949 as part of the National Parks and Access to the Countryside Act. Its approach to nature conservation was identification and designation of areas representative of remaining natural and semi-natural habitats to act as reservoirs. The selection criteria included:

- size;
- diversity;
- naturalness;
- rarity;
- fragility;
- typicalness;
- recorded history;
- ecological position;

- potential; and
- intrinsic appeal.

The exact meanings of these terms in the context of nature reserve evaluation are very specific. A good example of their use is given in the Nature Conservancy Council guide to Site Management Plans (Anonymous, 1988).

National Nature Reserves were therefore bought or leased for protection and scientific study of their habitats, but also to some extent for education and enjoyment. Sites of Special Scientific Interest (SSSIs) were defined, on land used for other purposes such as agriculture and forestry. Research stations were also set up. However, neither approach really incorporated earlier ideas of integrating nature conservation into the wider countryside and rural economic activity. It was also assumed that housing and industrial development were the main threat, and that landowners were the best stewards of the countryside despite post-war financial incentives to increase production.

In 1965 the NC became part of the new Natural Environment Research Council (NERC) in the Department of Education and Science but with little change in function. In 1973 the NC became the Nature Conservancy Council (NCC) in the Department of the Environment with new functions: involvement in EEC legislation, financial support to nature conservation NGOs (non-governmental organizations), wider research support and policy advice.

The publication of *Silent Spring* by Rachel Carson (1962) was significant in highlighting threats to habitats and significant losses. In Britain, nature conservation policy was revised in the Wildlife and Countryside Act 1981. This was not altogether different from the 1949 Act but did strengthen species protection and the incentives and regulations governing land-use change especially in SSSIs. Related Section 15 management agreements under the 1968 Countryside Act provide compensation for profits foregone when operations such as drainage are not allowed, or incentives for positive management. Public access was also addressed.

In 1991 the Joint Nature Conservation Committee (JNCC) (www.jncc.gov.uk), established by the Environment Act (1990), became the forum for delivery of nature conservation functions through the four separate parts of the UK.

However, in the 1970s other international conventions were initiated to protect nature and these impinge on local sites and their management. The most important international conventions are:

- 1971, Ramsar, or the Convention on Wetlands of International Importance, especially Waterfowl Habitat. Ramsar sites are agreed internationally and some traditional wildfowling is allowed on these (www.ramsar.org).
- 1972, World Heritage, or the Convention concerning the Protection of the World Cultural and Natural Heritage. Wildlife is not regarded as a heritage *per se*, but World Heritage sites often include and protect habitat (whc.unesco.org/nwhc /pages/home/pages/homepage.htm).
- 1973, CITES, or the Convention on International Trade in Endangered Species. As the name suggests, this regulates trade, most famously in ivory, but also in other plant and animal products (www.cites.org).
- 1979, Bonn, or the Convention on the Conservation of Migratory Species of Wild Animals. Countries through which migratory species range are encouraged to agree to protect them. Cetaceans, birds and bats have been targeted so far (www. wcmc.org.uk/cms).
- 1979, Bern, or the Convention on the Conservation of European Wildlife and Natural Habitats (Anonymous, 1991). This protects European flora and fauna especially vulnerable or migratory species and their habitats (www.nature.coe.int/english/cadres/ bern.htm).
- 1992, Rio, or the UN Convention on Biological Diversity. This put the terms and concepts of biodiversity and sustainability firmly on the agenda of 150 signatory states. Agenda 21 was a manifestation of the intent of this convention (www.biodiv.org).

Within the European Union (EU) there have been further policy and legislation initiatives since the Paris Summit of 1972 and the first Environmental Action Programme. Most details of these are held centrally at the European Centre for Nature Conservation (ECNC) in the Netherlands (www.ecnc.nl).

Important Directives include the following:

- 1979, Birds Directive, or the Directive on the Conservation of Wild Birds. This obliges member states to maintain populations of native wild birds at current levels or better. This obviously requires control of hunting and trade, and also applies to nests, eggs and habitat. Special Protection Areas (SPAs) were also designated.
- 1985, Environmental Impact Assessment (EIA) Directive, or the Directive on the Assessment of the Effects of Certain Public and Private Projects on the Environment. This requires that larger development proposals are subject to scrutiny before approval to describe their likely environmental effects.
- 1992, Habitats Directive, or the Directive on the Conservation of Natural and Semi-Natural Habitats and of Wild Fauna and Flora. This established a coherent network of candidate Special Areas of Conservation (cSACs), which, together with the SPAs, are called Natura 2000. The selection criteria are similar to all previous nature conservation legislation. However, the final list of SACs has yet to be decided at the time of writing.
- 2000, Water Framework Directive, or the Directive establishing a Framework for Community Action in the Field of Water Policy. Biodiversity is a core indicator of good ecological status in this instrument.

Organizations

Global organizations that promote nature conservation measures include the secretariats of the international conventions listed above but also all or part of the following:

- the Food and Agriculture Organization (FAO) (www.fao.org);
- the Global Biodiversity Information Facility (GBIF) (www.gbif.org);
- the Organization for Economic Co-operation and Development (OECD) (www.oecd.org/home);
- the United Nations Environment Programme (UNEP) (www.grid.unep.ch/index.php);
- the United Nations Educational, Scientific and Cultural Organization (UNESCO), especially its Man and the Biosphere Programme (www.unesco.org/mab);
- the World Bank (www.worldbank.org).

Within Europe, various Directorates of the EU in Brussels deal with nature conservation policy and

Box 6.2 Examples of non-governmental organizations (NGOs) involved in nature conservation

General biodiversity NGOs	*Species- or taxon-specific NGOs*
• Biodiversity Action Network (BIONET)	• Bat Conservation International (BCI)
• The Conservation Forum (CF)	• Birdlife International
• DIVERSITAS	• Cat Survival Trust (CST)
• Earthwatch	• International Mire Conservation Group (IMCG)
• Friends of the Earth (FOE)	• Seal Conservation Society (SSC)
• Greenpeace International	• World Owl Trust
• World Wide Fund for Nature	
• (WWF-International)	

legislation, especially the Directorate-General Environment. Most details of these and the following are also held centrally at the European Centre for Nature Conservation (ECNC) in the Netherlands (www.ecnc.nl).

Other global NGOs have arisen, often from national initiatives and subsequent international partnerships and/or concerned with particular species of popular appeal or under great threat. Box 6.2 provides examples of such NGOs.

European NGOs include regional offices of the international NGOs mentioned above, plus some specific to Europe, such as the European Bird Census Council (EBCC) and the European Wolf Network (EWN). Others are specific to regions such as the Mediterranean, the Alps and Eastern Europe, for example the Mediterranean Association to Save the Sea Turtles (MEDASSET).

Within the UK, government authority stems from the Department of Environment, Food and Rural Affairs (DEFRA) (www.defra.gov.uk). Agencies that currently deliver DEFRA policy are:

- the Countryside Agency (www.countryside.gov.uk/index.htm);
- English Nature (EN) (www.english-nature.org.uk);
- the Countryside Council for Wales (CCW) (www.ccw.gov.uk/);
- the Department of the Environment for Northern Ireland (DOENI) (www.doeni.gov.uk);
- Scottish Natural Heritage (SNH) (www.snh.org.uk)

However, a new integrated agency has been proposed in a draft bill published by the Government in February 2005.

The more scientific aspects of research and extension work are dealt with by the following:

- the Joint Nature Conservation Committee (JNCC) (www.jncc.gov.uk);
- the Centre for Ecology and Hydrology (CEH) (www.ceh-nerc.ac.uk);
- the Natural Environment Research Council (NERC) (www.nerc.ac.uk).

Local government is also required to provide Biodiversity Action Plans under the terms of the Rio Convention. Many counties and districts have full-time ecologists on their staff and further involvement in designating local nature reserves, county wildlife sites, etc. In this they will liaise with landowners, NGOs and the general public.

Many NGOS are peculiar to the UK, while others have international counterparts. The former include:

- the British Association of Nature Conservationists (BANC) (www.banc.org.uk);
- the British Trust for Conservation Volunteers (BTCV) (www.btcv.org);
- the British Trust for Ornithology (BTO) (www.bto.org);
- Butterfly Conservation (www.butterfly-conservation. org);
- the Council for the Protection of Rural England (CPRE) (www.cpre.org.uk);
- the Farming and Wildlife Advisory Group (FWAG) (www.fwag.org.uk);
- the National Trust (www.nationaltrust.org.uk/main);
- Plantlife (www.plantlife.org.uk);

- the Royal Society for Nature Conservation (RSNC) (www.rsnc.org/welcome.php);
- the Royal Society for the Protection of Birds (RSPB) (www.rspb.org.uk);
- the Wildfowl and Wetlands Trust (www.wwt.org.uk);
- the Woodland Trust (www.woodlandtrust.org.uk/).

Funds

Within Europe, public funding for nature conservation dates from the 1970s. The ACE (Actions by the Community relating to the Environment) instruments ran through two Regulations. The first (1984–87) was to develop clean technologies, techniques for measuring the quality of the natural environment and to maintain and enhance habitats for species under the Birds Directive. The second (1987–94) expanded the support into habitat restoration including that on contaminated land.

Urgent Action funds were eventually included after 1988 for species other than birds. There were parallel projects for the Mediterranean (MEDSPA) and the northern seaboard (NORSPA). Overall, 38 million ECU was spent on 198 projects.

The Habitats Directive in 1992 paved the way for wider funding requirements. The LIFE (EU Financial Instrument for the Environment) funding Regulations were all-encompassing and designed to implement the EU's environment policy. LIFE I (1992–95) had 400 million Euro and distributed this to 176 out of the 893 proposed projects. LIFE II (1996–99) continued this work and aimed towards favourable conservation status of key habitats and species as well as maintenance of the NATURA 2000 protected areas which had been identified under LIFE I. This involved financing 309 projects with 450 million Euro. LIFE III (2000–2004) works towards the implementation of the Sixth Action Programme for the Environment and has three themes: Nature, Environment and Third Countries. The latter

Box 6.3   Example of a successful proposal for a LIFE project: restoration of the mid-Cornwall Moors for *Euphydryas aurinia*, the marsh fritillary butterfly

Project Partners: Cornwall Wildlife Trust, the Highways Agency, the Environment Agency and Butterfly Conservation.

Total budget:	1 843 502 Euro
Life contribution:	921 750 Euro
Year of finance:	2003
Duration:	17 March 2003 to 31 December 2007
Commission reference:	LIFE03 NAT/UK/000042

The marsh fritillary butterfly has suffered a 20–50% decline in its distribution in Europe over the past 25 years. It is essentially a grassland species found mainly on damp acidic and dry calcareous grassland habitats. It is also totally dependent on its food plant, devil's-bit scabious. The rate of decline in the UK has been particularly severe mainly as a consequence of the loss of traditional livestock grazing, leading to unmanaged pastures. The Breney Common and Goss and Tregoss Moors cSAC in Cornwall, however, remains one of its strongholds. The site is at the centre of a metapopulation which represents 5% of the UK population and approx. 1% of the estimated European population. The lifestyle of the butterfly requires its conservation at the metapopulation level, a collection of interlinked sub-populations, which may frequently die out and re-establish. For the metapopulation to be sustained, a minimum of 70 ha of suitable breeding habitat should be available. The main threats to the species are from habitat change, poor connectivity between suitable sites, reduction in the size of habitat patches at breeding sites and barriers to migration.

The project aims to increase the area, connectivity and quality of suitable breeding habitat across the cSAC and at seven satellite sites. To do this English Nature has brought together a unique partnership of three statutory agencies, two NGOs (the Cornwall Wildlife Trust and Butterfly Conservation) and landowners. The project is strongly linked to the UK Biodiversity Action Plan for the species and it supports the Highways Agency Biodiversity Action Plan. Actions include installing management and grazing infrastructure, scrub removal and reinstating livestock grazing on the site. The Environment Agency will prepare a water-level management plan for part of the cSAC to improve the conditions for the wet grassland, whilst the Highways Agency will help to reduce fragmentation through the downgrading of a trunk road which currently bisects the main site. The actions are focused on achieving a secure and self-maintaining metapopulation. English Nature and the Wildlife Trust will maintain some sites and farmers will be supported through agri-environment payments. The improved understanding of species ecology and metapopulation dynamics generated by the project is expected to have a strong demonstration value and a Europe-wide application for the marsh fritillary and for other species that require a similar landscape-scale approach.

Box 6.4 Example of grants available for hedgerow management in the UK

Ancient and/or species-rich hedges are supported by English Nature under its Biodiversity Grant Scheme which can support hedge planting and hedgelaying. English Nature can provide up to 50% of the cost of approved projects, between £250 and £5000, and anyone can apply. DEFRA supports hedgerow management under its Countryside Stewardship scheme which can assist with hedgelaying and hedge planting. The Countryside Stewardship scheme is geared to farms or applicants with agricultural landholding certificates and is not suitable where there is just a single stretch of hedge to be laid. Small amounts of money may be available within the countryside management section of your local county council where amenity or landscape improvement grant money might be available. To qualify, your hedge is definitely going to have to be visible to the public, e.g. by a road or a footpath.

involves the new accession countries. A fund of 640 million Euro is available to selected proposals. Box 6.3 shows an example of a recent successful application, while Box 6.4 gives examples of grants available for hedgerow management in the UK.

Other sources of grants exist including private charities, banks and commercial sponsors within Europe. Of course admission fees, taxes and linkages with development permissions can also be used. Obtaining funding for nature conservation projects is a time-consuming and increasingly specialist task but usually crucial to the success of any venture.

Internationally, WWF supports an interesting initiative called Debt For Nature Swaps. The swap can be either bilateral or commercial. In the former the creditor government cancels the debt owed by a debtor government in return for the setting up of a counterpart fund in the debtor country for nature conservation projects. This may be facilitated by WWF. In a commercial swap, an NGO such as WWF buys out the debt from the creditor and then sets up the conservation project funding in the debtor country. Several such commercial swaps have occurred since 1988 in the Philippines, Madagascar, Poland and Zambia. A recent example of this is the Tropical Forest Conservation Act in Peru (2002) brokered by WWF, whereby $14.3 million of debt to the USA is to be cancelled in return for $10.6 million to finance sustainable forestry and forest conservation over the next decade.

The limits to the success of nature conservation activities can often be financial, but policies concerning national security, land tenure, energy resources, settlement, trade, transport, etc. may also be major constraints. Policy-makers may not recognize the value of biological resources in financial or any other terms. They may need to be persuaded that continued productivity as a result of biological resources and a supportive local management and investment climate are desirable. Innovative funding mechanisms backed by government policies may overcome many obstacles to the conservation of nature.

Scientific requirements in nature conservation

It is essential to know exactly what individuals, populations and communities of organisms coexist in any area being targeted by countryside management, and to understand, as much as possible, their interactions with each other and the wider environment. The purpose for which the ecological information is collected is usually defined even if this is only in case it is needed for later comparisons. More often management decisions will be made in the light of the data and thus the latter need to be accurate.

In order to be able to propose and carry out a nature conservation project such as that for the marsh fritillary, a number of practical scientific skills are necessary. The next section explains what competences are required by a professional delivering nature conservation objectives in the field.

Identification of species

The most fundamental skill in assessing biodiversity is the ability to recognize species, preferably in the field. For some decades this traditional area of biological expertise has been undervalued as exciting developments in molecular and *in vitro* biology have been understandably fashionable. It is therefore important that project or site managers appreciate the need for taxonomic accuracy and, ideally, possess it for one or more groups of plants or animals. This expertise can be acquired in several ways, as discussed below:

- Formal courses at postgraduate, degree or sub-degree level at universities. Purely taxonomic courses are almost non-existent. The UCAS handbook (www.ucas.com) lists a few courses under botany or zoology, and some other titles such as geography include an element of field work and species identification.

- Short courses such as those run by the British Museum of Natural History (BMNH) (www.nhm.ac.uk/science/rco/idq/), the Institute of Ecology and Environmental Management (IEEM) (www.ieem.org.uk/workshops.htm) and at a more elementary level by the British Trust for Conservation Volunteers (BTCV) (www.btcv.org/dsdetail.html) on particular groups of organisms. Some lead to formal examinations and qualifications such as the IDQ scheme of the BMNH.
- Field days run by specialist groups such as Butterfly Conservation, the British Lichen Society, etc. These can usefully be counted as Continuing Professional Development (CPD), required by many employers and institutes.
- Voluntary or seasonal work with conservation and countryside organizations such as the County Wildlife Trusts, the National Trust, national and country parks. This is often a prerequisite for obtaining paid employment in nature conservation these days anyway.
- Self tuition and practice. This is essential to back up any of the above introductory routes. Acquisition and use of appropriate keys and texts are also a crucial part of the process. Computerized multi-access keys such as those developed at the University of Plymouth for Flora and Fungi can facilitate and speed recognition of species. Experience and appreciation of the range of variation to be found within a species and from year to year can also be relevant in nature conservation.

It is often necessary to engage or refer to a consultant or expert when dealing with particular unfamiliar taxonomic groups. It is not unreasonable to expect this person or company to have qualifications and/or experience recognized by, for example, membership of the IEEM or by holding the relevant IDQ of the BMNH. Increasing professionalism in ecology and taxonomy generally is to be valued and required, even or especially when taxonomic skills are in short supply.

The end result will be the species list for, or species richness of, the site.

Quantification of abundance

Evaluation of the nature conservation status of sites depends on the species list or richness, but also usually on an assessment of the abundance of the populations of these species. Standard techniques for ecological surveying are not complex in theory but require practice and staying power when used in the field.

This expertise can be acquired in several ways similar to those listed above, such as through the following:

- Formal courses at postgraduate, degree or sub-degree level at universities. Detailed questions should be asked about the nature, relevance and extent of field work before embarking on any course as it is often very perfunctory. However, these courses usually deal with the statistical analysis of the results very comprehensively (see below).
- Short courses such as those run by the Institute of Ecology and Environmental Management and the Field Studies Council on particular techniques. These professional bodies also serve a useful function in setting up discussions groups about the standardization of techniques for various taxonomic groups or habitats and for various purposes.
- Field days run by specialist groups such as Butterfly Conservation, the British Lichen Society, etc., although these usually focus on identification of species only. However, they sometimes provide another opportunity to practise, compare and discuss specialist quantitative techniques.
- Voluntary or seasonal work with conservation and countryside organizations such as the county Wildlife Trusts, the National Trust, national and country parks where surveys or monitoring are being carried out. This is often the best way of acquiring more substantial practice alongside more experienced colleagues.
- Self tuition and practice. This is essential to back up any of the above introductory routes.

Techniques for determining plant populations, such as quadrat surveys or transects, are easily learned in principle, but the subtleties of what sizes and how many to use and where to sample the populations require study and experience. Whether quadrats are positioned randomly, systematically or permanently depends on why the survey is being carried. In many cases, standard protocols exist, such as for the National Vegetation Classification (NVC) (Rodwell, 1991–2000) and for the Environmental Change Network (ECN) (Sykes and Lane, 1996), and more can be found in the ecological literature.

Ecologists frequently want to be able to characterize a population or to compare populations. For example, they may want to know the density of beech trees in a forest or the frequency of a rare orchid. In some circumstances it may be possible to deal with complete populations and count every tree in the forest, in which case the total number would be a parameter of the population.

More often there is not enough time, money or energy to deal with the whole population. Instead a fraction of the whole population or sample is examined. The traits that the sample possesses are referred to as statistics. If the sample is representative of the whole population then the parameters can be inferred from the statistics. The sample is representative if it is both unbiased and adequate in size.

Some of the statistics commonly obtained for plant populations are:

- Density: this is the number of a particular species per unit area. It is usually determined by counting the individuals in a quadrat of known area. One drawback is the difficulty of defining the limits of some individual plants such as grasses or sprawling plants.
- Frequency: this is the chance of finding a particular species in any quadrat of given area. It is usually determined by placing a number of quadrats in an area of uniform vegetation and recording whether the species is present or not in each. The frequency is then expressed as a percentage of the quadrats used in which the species occurred. One drawback can be in deciding whether the plant is rooted in the quadrat or not. It is also important to record the size of the quadrat used as this can affect the result a great deal. One advantage of the method is the speed with which an estimate of the vegetation can be obtained compared with the other methods.
- Cover: this is the percentage of the ground occupied by the species, normally expressed as a percentage of the total area sampled. Cover can be measured objectively by using a pin frame or point quadrat, recording each species touched by a large number of pins lowered at random locations in the sward. The percentage cover can then be estimated for each species found. This method is fairly laborious but precise. Cover can also be measured subjectively by eye.
- Yield: this is usually taken to mean the above-ground biomass per unit area of the species, bulked together or individually. It is a destructive and time-consuming measurement involving clipping a sample of the vegetation off to ground level and obtaining its dry weight. It is relevant when food chain or productivity studies are required.
- Energy content: this is the ultimate measurement of ecosystem performance but is rarely used in surveys. It involves destructive sampling of above- and, ideally, below-ground material from a known area and burning it in a bomb calorimeter to obtain the energy content of the living material in the system.

In any of these cases the samples chosen must be unbiased. The usual approach is to sample randomly, when every member of the population or area has an equal and independent chance of being included. Most statistical tests are based upon the assumption that sampling has been random. In some circumstances, however, it is appropriate to sample systematically or in some other way that is more efficient or representative of the vegetation. If this is undertaken by an experienced surveyor with a good ecological eye for avoiding extremes or atypical areas or repeated patterns in the area, bias can still be satisfactorily avoided.

Random sampling does not just mean placing or throwing the quadrats, pinframes or transects haphazardly, avoiding conscious bias. Tests have shown that this rarely succeeds! In a professional survey the samples are positioned using any unbiased methods such as tossing coins or using random numbers from tables to locate a point in the area. The sample space is then orientated on this point in some predetermined way. See the example in Box 6.5.

Quadrats can be any shape or area and made of wood, metal or string, as long as the shape and area are known. Quadrats are used to sample reasonably uniform vegetation by recording the species within them, with perhaps the numbers and/or cover values of each.

Transects can be used to investigate any situation in which a systematic sampling regime is required and/or

Box 6.5   Example of a vegetation protocol for positioning permanent stand-alone 1 × 1 m quadrats for long-term monitoring as part of the Ecological Monitoring and Assessment Network (EMAN) in Canada (Roberts-Pichette and Gillespie, 1999)

Select the community where the monitoring is to take place. To determine the quadrat locations within the community, use a random selection method. One method is to prepare a rough map of the area to be monitored and draw a line along one side (Figure 1a). Randomly select two or more points along this line for setting stakes (base stakes). Randomly select locations along a line perpendicular to the original line starting from the base stake (Figure 1a). Another method is to preselect (using a random selection method) two or more points in the stand. From these points where the base stakes are placed, quadrat locations can be determined by randomly selecting points on a compass star centred on the point (Figure 1b). In selecting the specific locations, ensure that all quadrats are at least 5 m apart.

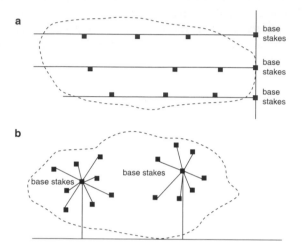

Figure 1   Two plans for placing 1 × 1 m stand-alone quadrats in a grassland community. **a** Three randomly selected lines crossing the community from base stakes, with randomly selected 1 × 1 m stand-alone quadrats. **b** Two randomly selected locations for base stakes from which randomly distributed arms of random lengths radiate, with a stand-alone 1 × 1 m quadrat at its tip.

Using a table of random numbers, or a bag of numbered chips or balls, draw numbers to select points along the line. Each point represents the location of one corner of a quadrat. The decision as to whether the quadrat is to go to the left or right of the line should be predetermined before going into the field. Set at least two stakes at each selected point and drive in flush with the ground (use metal-core plastic or stainless-steel stakes).

A series of lines could be established in this way with a permanent stake (base stake) being placed at the origin of each line or the centre of the compass star arrangement.

if the vegetation is changing in a regular way along an environmental gradient such as across a pathway or up the spatial successional gradient in a saltmarsh. Line transects involve recording the plants that touch a line at all points or at regularly spaced points. However, the data are difficult to present and of limited use. Belt transects involve quadrat sampling of all or regularly spaced areas along a line and lead to frequency or cover data as described above which can be related to any environmental gradient detected.

When carrying out vegetation surveys in the field the following sampling procedures should be adhered to:

- Make sure the area being sampled represents a single plant community, not two or more due to local variation in soil, drainage, etc. If there is any doubt, subdivide the area and sample each division separately.
- Always state clearly the sampling methods, and especially the sizes of any quadrats. Most results are

Box 6.6   Example of a protocol for long-term monitoring of butterflies as part of the Environmental Change Network (ECN) in the UK (Sykes and Lane, 1996)

Butterflies are one of the easier insect groups to identify and monitor and are known to respond rapidly to changes in vegetation abundance and quality (Thomas, 1991). The protocol adopted by the ECN is that already in use for the national Butterfly Monitoring Scheme, operated jointly by the Institute of Terrestrial Ecology and the Joint Nature Conservation Committee and organized from the Biological Records Centre at ITE Monks Wood (Hall, 1981; Pollard *et al.*, 1986). This existing scheme will provide a strong background of information from approximately 15 years of sampling for comparison with ECN data. Analysis of the existing data has already shown interesting changes in the distribution and phenology of individual species (Pollard, 1991) and significant relationships between butterfly population size and climate (Pollard, 1988).

Location
A fixed transect route is set up at each site following the instructions in Hall (1981), and is strictly followed on each sampling occasion. The route is selected so as to be reasonably representative of the ECN site and will often follow existing paths or boundaries and include areas under different management regimes. If necessary, the route should be marked out to ensure that the same route is followed on each occasion. The length of the transect will depend on local conditions but should be capable of being walked at a comfortable, even pace in 30–90 minutes and will therefore usually be 1–2 km. The transect should be divided into a maximum of 15 sections, which may be of different vegetation, structure or management and which are used as sampling strata. The length of each section is recorded on a map, together with information on habitat types and abundant plants, especially butterfly food plants. Management operations in the vicinity of the transect are also recorded. Changes in these characteristics are noted, so as to assist in interpretation of results.

Recording
Recording of the transect takes place weekly between 1 April and 29 September, between 10.45 and 15.45 BST. The temperature should be 13–17°C if sunshine is at least 60%, but if the temperature is above 17°C recording can be carried out in any conditions, providing that it is not raining. At northern, upland sites the appropriate upper temperature is 15°C. The use of more than one recorder would make recording easier at the peak of the season when high numbers of different species occur, or to provide cover for an absent recorder. Transects should be walked by someone with a good knowledge of the British butterfly fauna.

   The transect is walked at an even pace and the number of butterflies that are seen flying within or passing through an imaginary box, 5 m wide, 5 m high and 5 m in front of the observer, are recorded by species for each section of the transect, using the forms provided. Start time is recorded, as are the temperature, percentage sun and wind speed at the completion of the transect. Percentage sunshine is also recorded section by section as the transect progresses.

   Estimated time required: 0.5–1.5 hours/transect (depending on transect length) for 26 weeks each year.

only valid in relation to the particular methods used.
- Species must be correctly identified.
- Bare ground should be recorded as if it were a species, especially in cover analysis.
- Random sampling should ideally be achieved by objective methods as explained above and in Box 6.5, and not by throwing quadrats!

Techniques for animal populations are more varied and dependent on the characteristics and behaviour of the group being surveyed. Certain techniques requiring handling of mammal and bird groups such as Longworth trapping or mist netting can only be carried out by licensed personnel. Observational surveys such as butterfly transects and British Trust for Ornithology breeding bird surveys follow standard protocols and do not require a licence (Box 6.6). Amateurs can and do contribute to these, but once again the ability to identify species in the field is essential. Recognized tech-

niques for many terrestrial and aquatic invertebrates such as pitfall traps and kick sampling exist and should be familiar to nature conservation managers.

Statistical analysis is desirable but not always possible in ecological situations due to lack of symmetry or replication in the data. However, a knowledge of how and when to use parametric and non-parametric statistical techniques and their constraints is important, perhaps more to detect flaws and fallacies in analyses and comparisons carried out and reported by others. Such knowledge can be acquired on a wide range of courses as the basics are not subject-specific and from textbooks such as Eddison (2000).

However obtained, data on the identity and abundance of the natural denizens of an area ought to be used to create a report or record, otherwise such data tend to remain inaccessible and forgotten. Whether or not statistical analysis is possible, information about an area of countryside or site can be summarized, tabulated and graphed so as to be easily comprehended. In the context of nature conservation, key points should be made in the report as to the validity of the data, exactly how, when and where they were collected and any limitations due to access, season, etc. that affected the whole exercise. Comparisons with previous surveys of the same area and with other habitats or areas are likely to be relevant as are realistic judgements on the overall value of the nature conservation interest of the area in question. It is likely that the main thrust of the report will have been dictated by the original aims of the survey work. Frequently the type of survey envisaged above becomes part of a habitat management planning process, or a report to a statutory body, or an application for funding. Biological records centres exist in most counties in England in order to keep information about the distribution and abundance of species in such areas and are always keen to receive and disseminate reputable data and reports on local flora and fauna.

Practical habitat management skills

In addition to scientific skills, involvement in nature conservation often requires physical and chemical intervention at a site, either directly or through supervising contractors. In either case it is advisable to know how to carry out the tasks, how long it should take and what resources are needed as well as the costs of all these. Competent contractors will often belong to a professional body such as the Countryside Management Association (www.countrysidemanagement.org.uk) which will certify their skills.

Management of vegetation is the most common requirement. Scrub clearance by means of manual labour, or using strimmers, clearing saws or chain saws, is frequently necessary. Experience and certification in appropriate skills and particularly in the use of power tools can be obtained through some colleges, through LANTRA (www.lantra.co.uk) and by working with the BTCV (www.btcv.org).

Other skills that are relevant include:

- *Burning and swaling especially of heath and moorland.* Burning is used to manage dwarf shrub vegetation in order to stimulate new growth and create mosaics of different age classes of plants.

- *Hedgelaying*, a traditional skill with interesting regional variations. Hedgelaying is the means of managing hedges throughout large parts of England and Wales. It creates a thick barrier from untidy, gappy and straggling hedges. Hedgelaying is the art of cutting a stem partly through, so that it will bend without breaking and will continue to grow. There are grants available for hedgelaying (see earlier under Funds).
- *Tree planting and woodland management.* Much woodland is now managed by contractors who provide a comprehensive service to landowners and site managers.
- *Transplantation of individual plants or of species-rich turves.*
- *Wall building* in regional styles and including Devon banking. Dry stone walls are found in many parts of Britain and overseas, wherever field, mountain or quarried stone is plentiful. There are regional variations but with the same building principles applying. The skill is in making the best use of what is available.
- *Use of geotextiles.* Geotextile is a term used to cover a wide range of products used in combating erosion, reinforcing and stabilizing soil in, for example, footpaths and embankments. The products are usually biodegradable and designed to produce a natural-looking result as opposed to concrete or other hard engineering solutions.
- *Mulching and brash piling.* Cut scrub can be used in many constructive ways. It can be chipped and used as a mulch in weed control, or left for detritivores and decomposers, which in turn provide nutrients for other members of the food webs.
- *Beetle banks and conservation headlands on farmland.* These are often constructed to encourage biological control of pests.
- *Weedkilling and chemical pest control*, including spraying certification and dealing with notifiable species such as ragwort and Japanese knotweed (for example, see www.cabi.org/bioscience/japanese_knotweed_alliance.htm). *Note:* legislation requires that persons applying chemical pest control have appropriate certification. Testing is organized and validated by the National Proficiency Tests Council (see Useful websites at the end of this chapter).
- *Management with and of animals through con-*

trolled grazing. Many site managers are experimenting with the use of livestock in environmental management. However, this requires familiarity with animal behaviour and welfare as well as the facilities to manipulate their grazing to achieve the desired habitat.

Additional considerations

Finally, management of visitors, interpretation of nature conservation for mutual benefit and regulation of access is another whole area with which the nature conservation professional needs to become involved. This overlaps into the realms of tourism management and is getting beyond the scope of this chapter.

It remains to emphasize that anyone working in nature conservation management in the twenty-first century needs to be equipped with a plethora of knowledge and skills, both academic and practical. He or she should also have the ability to communicate and liaise with specialists in many fields in order to achieve positive results for a site or the planet.

References

Anonymous (1988) *Site Management Plans for Nature Conservation.* Nature Conservancy Council, Peterborough.

Anonymous (1991) *The Bern Convention on Nature Conservation.* Council of Europe, Strasbourg.

Carson, R.L. (1962) *Silent Spring.* Hamish Hamilton, London.

Eddison, J. (2000) *Quantitative Investigations in the Biosciences Using Minitab.* Chapman and Hall, London.

Hall, M.L. (1981) *Butterfly Monitoring Scheme. Instructions for Independent Recorders.* Institute of Terrestrial Ecology, Cambridge.

Hilton-Taylor, C. (ed.) (2000) *2000 IUCN Red List of Threatened Species.* International Union for the Conservation of Nature and Natural Resources, IUCN, Gland, Switzerland.

Loh, J. (ed.) (2002) *Living Planet Report 2002.* Worldwide Fund for Nature, Gland, Switzerland.

Pollard, E. (1988) Temperature, rainfall and butterfly numbers. *Journal of Applied Ecology*, **25**, 819–28.

Pollard, E. (1991) Changes in the flight period of the hedge brown butterfly *Pyronia tithonus* during expansion of its range. *Journal of Animal Ecology*, **60**, 737–48.

Pollard, E., Hall, M.L. and Bibby, T.J. (1986) *Monitoring the Abundance of Butterflies.* Nature Conservancy Council, Peterborough.

Roberts-Pichette, P. and Gillespie, L. (1999) *Terrestrial Vegetation Biodiversity Monitoring Protocols.* Ecological Monitoring and Assessment Network, Burlington, Ontario.

Rodwell, J.S. (ed.) (1991–2000) *British plant communities.* Vols 1–5. Cambridge University Press, Cambridge.

Sutherland, W.J. (2000) *The Conservation Handbook: Research, Management and Policy.* Blackwell Science, Oxford.

Sykes, J.M. and Lane, A.M.J. (eds) (1996) *The United Kingdom Environmental Change Network: Protocols for Standard Measurements at Terrestrial Site.* Stationery Office, London.

Thomas, J.A. (1991) Rare species conservation: case studies of European butterflies. In: *The Scientific Management of Temperate Communities* (eds I.F. Spellerberg, F.B. Goldsmith and M.G. Morris), pp. 149–97. Blackwell Scientific, Oxford.

Further reading

Arnold, H.R. (1993) *Atlas of Mammals in Britain.* HMSO, London.

Gutzwiller, K.J. (ed.) (2002) *Applying landscape ecology in biological conservation.* Springer, New York.

Harrison, P.A., Berry, P. and Dawson, T.P. (eds) (2001) *Climate Change and Nature Conservation in Britain and Ireland: Modelling Natural Resource Responses to Climate Change (the MONARC Project).* UK Climate Impacts Programme, Oxford.

Hendry, G.A.F. and Grime, J.P. (1993) *Methods in Comparative Plant Ecology: A Laboratory Manual.* Chapman and Hall, London.

Joint Nature Conservation Committee (1983 *et seq.*) *British Red Data Books.* JNCC, Peterborough.

Lorton, R. (2001) *A–Z of Countryside Law.* Stationery Office, London.

Moore, N.W. (1987) *The Bird of Time: The Science and Politics of Nature Conservation: A Personal Account.* Cambridge University Press, Cambridge.

Useful websites

www.jncc.gov.uk – Joint Nature Conservation Committee. Annual reviews and reports.

www.rswt.org/history – Royal Society for Wildlife Trusts, renamed from Royal Society for Nature Conservation in June 2004. Annual reviews and reports.

7

Landscape

S. Blackburn and P. Brassley

Introduction

Think of a lane in the English midlands on a bright spring morning, with the sun shining on masses of white hawthorn flowers, or the view across the water to the Isle of Skye with the Cuillin Hills disappearing into the clouds, the great wide skies of the East Anglian coast, or the sun setting on the Grand Canyon . . . we can all think of landscapes that we recall with enormous pleasure. Equally, other landscapes are less interesting or actively unpleasant. Tourists go in search of the more beautiful landscapes, and house prices are higher in areas noted for natural beauty, so these aesthetic factors can have economic implications. Some landscapes can become national icons: think of the white cliffs of Dover, or Death Valley, background to innumerable western films, or Ayers Rock (also known as Uluru). The sheer variety of the English landscape, in which wooded vales, rolling downs and towering sea cliffs can all be found within a few miles of each other, can be contrasted with the magnificence of, say, the Rocky Mountains, stretching endlessly onwards.

The purpose of this chapter is to examine the origins of landscape, taking some of those found in England as examples, then to consider why people respond differently to different landscapes and prefer some to others, and finally to look at the way in which we attempt, as a society, to control what happens to landscapes for our mutual benefit.

The origins of English landscapes

Landscapes change over time. The England of today looks different from the England of 50 years ago: the fields, villages and some of the roads are bigger, for example. On the other hand, within this changed land-scape there remain houses, churches, railways, canals, roads, woods, field boundaries and prehistoric monuments that have been there for hundreds, and in some cases thousands, of years. Since, as we shall see, people often prefer the patina of age to the rawness of youth in their landscapes, and longevity and tradition are often put forward as reasons for conservation and preservation, and indeed since we can only understand why we have our present landscapes if we know something about how they developed, we shall begin by looking at landscape history, taking the English land-scape as an example [although it is by no means the only landscape to have attracted the attention of historians, as Muir (1999, Chap. 1) points out].

The first problem that landscape historians have to deal with is how to organize their work. One of the earliest, most popular and influential histories of the English landscape (Hoskins, 1955) simply begins with prehistoric and Roman England and works its way, period by period, to the 1950s. One of the more recent (Thirsk, 2000) combines studies of the development over time of various landscape types, such as downs, wolds, lowland vales, moorlands and marshes, with case studies of individual villages and estates. Other authors (Rackham, 1986; Muir, 2000) have examined the development of individual landscape components, such as woods, settlements and fields, over time. It is this last approach that will be used in the following short account of landscape change in England.

In any period, woods, fields, settlements and other buildings, industrial sites and transport networks exist in varying proportions. At the end of the last ice age, from roughly 11000 BCE, as temperatures rose, the tundra vegetation would gradually have been invaded by birch woodland, succeeded in turn by pine (of which the Scottish pine woods are the last remnant), and then by a combination of oak and elm, with alder in wetter areas and an understory of hazel. This eventually covered most of the British Isles, but then, from about 5500 BCE, it was superseded in the lowlands by woodland in which lime was dominant, mixed with oak and ash (Rackham, 1976, pp. 40–42; Short, 2000, p. 131). Although it would be a mistake to imagine a completely tree-covered England at this point in time – there would always be clearings kept open by grazing animals, or marshland, or open mountain tops – it is clear that woodland was certainly the dominant vegetation and landscape. The first human occupants of the country would have had little impact (although their ability to light fires meant that they would at least have had some), but the emergence in the Neolithic period (from about 4000 BC) of people who grew crops and kept domestic animals gradually increased the rate of woodland clearance. Studies of pollen grains that have been preserved in bogs reveal a decline in elm pollen by 3000 BC, with a simultaneous increase in wheat, nettles, plantain and other arable weeds. Over the following millennia more land was cleared of trees by felling and burning, while the trampling and browsing of sheep and cattle prevented woodland regeneration. By the time the Romans arrived in the first century AD the English landscape was a mixture of woodland and farmland, with the woods reduced to between one-third and one-half of their original extent. Woodland cover at about AD 586 has been estimated to be about one-third of the area of the country, although it is important to realize that this was not evenly distributed. The same pattern can be seen 500 years later, at the time of Domesday Book (1086), which lists woodland in only half of the settlements mentioned. Some great woods remained, such as the Weald of Kent, Surrey and Sussex, which may have had as much as 70% tree cover, or the Forest of Dean, or the area of Somerset that a charter of 682 mentions as 'that famous wood which is called *Cantocwudu*' (Quantockwood). Conversely, some counties in eastern England, such as Lincolnshire and Cambridgeshire, had less than 5% of their countryside under trees. Overall, about 15% of the country was in woodland in 1086, and only 10% by 1350, after the expansion of cultivation in the twelfth and thirteenth centuries (Rackham, 1980; Short, 2000, p. 132). This is not much more than the present-day figure.

Thus by the beginning of the fourteenth century the balance of woodland and cultivation was not very different from that which we see today. On the other hand,

the woods themselves would have appeared very different. Obviously there would be no great blocks of non-native conifers, but also the management of the deciduous woods was different from that which is usually found today. Much of the fuel and building materials for medieval people had to come from their woodlands, and they managed them carefully to sustain their production. Timber for houses and ships came from standard trees, often growing further apart than we would expect to find in today's deciduous woods, while the coppiced understory of ash and hazel was cut over every few years (between 7 and 30 depending upon the desired product) to produce firewood, charcoal and fencing materials. This is clearly different from the present-day practice of forestry, in which an even-aged stand of timber trees is first thinned and then felled at maturity. This was little practised in England before the end of the seventeenth century, but then expanded until the middle of the nineteenth century (Rackham, 1976, pp. 96–7). Then railway transport brought down the cost of coal and shipbuilders increasingly turned to iron. By the end of the nineteenth century the demand for timber and coppice products had declined to the point where woods were more highly valued for their game than their timber content. World War I revealed a shortage of the kind of timber used in coal mining, and this was one of the principal reasons for the establishment, in 1919, of the Forestry Commission; thereafter the spread of softwood plantations into the uplands began.

If there was little regeneration of woodland after the departure of the Romans, it follows that the cultivated landscape remained in being. The first farmers may have been nomadic, cultivating a patch of land for a few years before moving on, but by the Bronze Age there is clear evidence of permanent field divisions, some of which still remain on Dartmoor, for example (Rackham, 1980, pp. 156–7; Fleming, 1988). The archaeological evidence suggests a landscape of hamlets and scattered farms with their arable fields and pasturelands, and often woods and rough grazing too. As the population expanded so would the arable expand into the pasture and rough grazing, and vice versa in periods of population decline. This appears to have been the pattern for hundreds of years, from the prehistoric, right through the period of Roman occupation, to the ninth century AD. In parts of England, especially in the highlands, it remains so today. However, much of the country, especially in the great belt of lowland

England stretching from Durham down through the midlands to the south coast, was affected by a period of major change between about 850 and 1100. It was then that individual farms and hamlets were abandoned as people gathered together into larger villages, each surrounded by their great open fields, which were divided into strips and farmed in common. Why it happened, and how, still remains a mystery. In some places it may have happened rapidly, within the course of a single year; elsewhere it may have been a matter of gradual evolution. It may be seen as a result of population growth, in which land, descending through families over many generations, gradually became split up to the point where farm fragmentation became just too complex to be worthwhile. At that point there may have been a communal decision to reorganize the lands of the parish. An explanation based on changing kinship networks has also been advanced (Muir, 1999, p. 59). Since this is a period for which little written evidence survives, we may never know precisely what happened and why. However, there is plenty of evidence, both on the ground and in maps and surveys, for the open fields that came to dominate much of lowland England. Even in parts of the north, west and East Anglia, where it used to be thought that open field arable was rare, there is now evidence that it existed for a time, even if it did not develop in quite the same way as in the midlands and disappeared earlier.

Almost half of England, according to one historian (Wordie, 1983), had either never been in common use or had been enclosed by 1500. A little more was enclosed in the sixteenth century, and almost a quarter of the country was enclosed in the seventeenth century. This radical change was brought about by agreement between farmers and landowners, and often no formal record was made of the decision. The enclosure of the arable fields of Middridge in Durham, which resulted in a case in the Bishop of Durham's court, sheds some light upon the tensions created by the process: the poor of the parish argued that 'the charge of the said division in hedgeing and other wayes would be soe great that it would beggar them', yet they admitted that it would produce great benefits in terms of increased crop yields, 'by reason of the plowing up of Fresh grounds and converting them into Tillage which heretofore were in Pasture' (Brassley, 1985, p. 104). Thus by 1700 only about a quarter of the country, much of it in the classic open field villages of the midlands, remained unenclosed. In these areas it was difficult to obtain agree-

ment to enclose, and from the middle of the eighteenth century landowners began to propose private Acts of Parliament to bring about enclosure. By the 1840s most of the farmland in the country was enclosed, and General Enclosure Acts were introduced to make subsequent enclosures easier. By 1914 only 4.6% of the country remained unenclosed. These post-1750 enclosures produced a different landscape from earlier enclosures by agreement. In the earlier period the decisions about field boundaries were made by the farmers and landowners involved. They often formalized existing divisions, so that hedges followed existing open field strip boundaries. The result was a landscape of small-scale, intimate enclosures. Even by the end of the nineteenth century progressive agriculturalists would be arguing that such landscapes contained far too many hedges to be efficient. The later enclosures, by Act of Parliament, were laid out by professional surveyors, sitting in offices and drawing lines on maps to produce an open landscape of big square fields that often paid little attention to existing landscape features.

The involvement of landowners in enclosure was just one of their contributions to the formation of the English landscape. It is important to remember that before the middle of the twentieth century much of the land in England was owned by people who did not farm it directly. They leased it out to tenants who did the farming, but the landowners themselves had enormous effects on the landscape through the way they managed their estates. In addition to their involvement in enclosure they promoted other forms of agricultural improvement such as the creation of water meadows or, especially between 1850 and 1870, the underdrainage of land with clay tiles. Not only did their great houses reflect their changing architectural tastes, but so too did their estate villages and farm buildings. They patronized the fashionable landscape designers such as William Kent (1684–1748), 'Capability' Brown (1716–83) and Humphrey Repton (1752–1818) (Rackham, 1986, p. 129), and in the nineteenth century they planted coverts for fox hunting and shelter belts for their game birds. The rural landscape of the eighteenth and nineteenth centuries was one of both production and pleasure. For much of the twentieth century production took precedence, and from 1940 onwards new buildings, new crops such as oilseed rape and maize (both of which had been grown much earlier, but not on such an extensive scale) and new machines appeared, and old hedges and woods disappeared. Even the cattle changed colour, as Friesians took over from Shorthorns as the dominant breed.

Although agriculture has usually been seen as the dominant rural industry, it is important to remember than it has never been the only one. Country people have also worked as miners and quarrymen, saltmakers, ironworkers, glassmakers, potters, charcoal burners and millers, in both wind and water mills, and all these activities have left their marks on the landscape. The remains of tin mining on Dartmoor and lead mining in the north Pennine dales can still be seen, as can the humps and bumps in the ground that are the remains of medieval quarries, at Barnack in Northamptonshire for example. Before the textile industry came to be concentrated in the steam mills of the north of England there were numerous medieval villages that became small towns in the fourteenth and fifteenth centuries as the rural cloth industry expanded in East Anglia, Wiltshire and Devon. Transport developments too brought change, as first the canals in the eighteenth century and then the railways in the nineteenth cut new lines across the landscape. So too did the motorways of the twentieth century. The new energy technologies of the twentieth century, of which the internal combustion engine was only one, had far-reaching landscape effects. The development of the national electricity grid led to a nationwide network of pylons whose size made them a prominent feature. Nuclear power stations were located in the countryside for safety reasons, but a series of conventional power stations were scattered along the Trent valley to be near to coal and cooling water. Electricity-generating windmills crowned Cornish hilltops. Military requirements too, as training grounds for the Army and military airfields, took increasing amounts of land. The development of earth-moving machinery extended the landscape influence of open-cast mining and quarrying, and some of the most dramatic new landscapes were produced by the earth-works of the new reservoirs, such as Kielder Water in Northumberland, which opened in 1982 and took two years to fill, Roadford in Devon, and Rutland Water, which covered a significant proportion of the old county of Rutland. Such mechanization meant that moving the volumes of earth that required the work of many people for many years in the prehistoric period could be accomplished relatively quickly by only a few. The pace had become faster, even if the effects were the same. The English landscape, therefore, is the result of thousands of years of changing human activity superimposed on the base provided by millions of years of geological and biological evolution. So too, of course, are the landscapes in all the other countries of the world.

Landscape perception

The long years of change have produced an infinite variety of landscapes. Many people have their favourite landscapes, views and places that they find comforting or fascinating or inspiring. What is perhaps more surprising is that many people agree that, for example, the landscapes of the Lake District are beautiful, and the Scottish Highlands magnificent, while the London suburbs are at best dull and at worst depressing. Since as a society we have decided, through the use of planning designations such as national parks, Areas of Outstanding Natural Beauty (AONBs) or Areas of Great Landscape Value, that some landscapes are more worthy of protection than others, we presumably need to discover why, as individuals, we prefer some landscapes to others, and why, surprisingly, many individuals agree on what is preferable and what is not.

Theories of landscape perception may be broadly divided into two types: the biological and the cultural. Biological theories suggest that people prefer landscapes that satisfy such basic biological needs as eating, sleeping and mating. They are based on the idea that humans spent much longer as scavengers or hunter-gatherers than in any other stage of development, so that over thousands of generations a process of Darwinian selection meant that people naturally prefer landscapes that they believe will promote survival, and feel uncomfortable in those that appear threatening in some way. The hunter-gatherers who correctly identified the benign landscapes survived to pass on their genes, and with them their preferences, and those that did not died out. Simplistic though this may appear, at first sight, it lies behind the work of theorists who have been influential for many years. The first of these is Jay Appleton, who argued that people would prefer landscapes that offered the possibility of seeing without being seen, because those would originally have allowed hunter-gatherers to identify food without being exposed to predators. This he called *prospect-refuge theory*; hence, perhaps, in modern times, all those serried ranks of people sitting in their cars at the viewpoints identified on road maps as offering a panoramic view of the surrounding countryside. Appleton later (1990) expanded these arguments to suggest that people would prefer landscapes that represented habitats with the capacity to satisfy their biological needs. These theories are consistent with those of the American environmental psychologists S. and R. Kaplan (1982), who argued that humans are by nature seekers of information, so that they prefer landscapes that offer a combination of intelligibility and stimulation, with appropriate levels of complexity, coherence and mystery.

Human geographers in particular have argued that these biological approaches to landscape are limited, in that most people's aesthetic judgements are much more likely to be affected by their cultural than by their genetic background. They suggest that landscapes are socially and historically constructed by people, and draw attention to the symbolic significance of some landscapes. Thus, for example, the patchwork of fields, woods, small villages and lanes found in the lowland counties around London are seen as an English heartland, associated with domesticity, security, agricultural productivity and social stability. It is an association more likely to be made by the socio-economic elite than by the agricultural labourer, who may simply see it as the background to underpaid work, but it is nonetheless powerful, having figured in recruiting posters in World War II and numerous advertising campaigns since then. The white cliffs of Dover perhaps form a similar national icon (Cosgrove, 1984; Cosgrove and Daniels, 1988; Muir, 1999, Chap. 7).

Landscape evaluation

Given that some landscapes are preferred to others, and that planning authorities are required to identify those that are most worthy of preservation, it is not surprising to find that several attempts have been made to evaluate landscapes. Moreover, given the potential for political controversy resulting from such judgements, it is equally unsurprising that attempts have been made to produce objective criteria for these evaluations. One type of approach, much investigated in the 1960s and 1970s, was to associate high landscape quality with individual landscape components, which could then be objectively measured, ideally from existing map resources. The problem with this was that it still required a subjective judgement of landscape quality.

Subsequently other methods, which still relied on quality judgements, but made by trained professionals, were developed. An alternative approach was to recognize the subjectivity of the process, but to attempt to reflect the preferences of society as a whole by asking a large sample of people to make the quality judgements. This approach in turn was criticized as likely to privilege past landscape values. In recent years the emphasis has been more descriptive, as in the Countryside Agency's Landscape Character mapping programme, and it seems fair to conclude that there is no widely accepted objective method of landscape evaluation (Muir, 1999, pp. 197–208).

Landscape planning

The overall perspective

Landscape planning is a broad topic. The start point for this part of the chapter is that landscape planning is concerned with mechanisms to protect or enhance scenic beauty. A simple way to approach this from the reader's perspective is by looking at the historical evolution of this in the national context over the last 50 years. Here it is considered from a policy perspective since other sources focus more closely on environmental design (see Marsh, 1997; Turner, 1998; Landscape Institute and Institute of Environmental Management and Assessment, 2002).

In order to appreciate the national context fully, one must first consider the leglislation, policies and agreements that dominate the international scene. There has been an increased level of supranational activity in the environmental area over the last five decades, and in particular in the last 30 years connected with the concept of environment and development and, latterly, sustainable development. The United Nations Educational, Scientific and Cultural Organization (UNESCO) convention concerning the Protection of the World Cultural and Natural Heritage was adopted in 1972, a World Heritage Committee was established in 1976 and the first sites were inscribed on the World Heritage List in 1978. At July 2003, the List comprised 754 properties (582 cultural, 149 natural and 23 mixed) in 129 countries (UNESCO, 2003).

At the European level, 18 countries became signatories to the European Landscape Convention on 20

October 2000. The general purpose of this convention is to encourage public authorities to adopt policies and measures at local, regional, national and international level for protecting, managing and planning landscapes throughout Europe. This is so as 'to maintain and improve landscape quality and bring the public, institutions and local and regional authorities to recognise the value and importance of landscape and to take part in related public decisions' (Council of Europe, 2003). The Council of Europe convention can be regarded as complementary to the UNESCO one, covering all landscapes rather than just those of universal value, and excludes historic monuments. The purpose of the convention is also to promote active management rather than just the passive listing of assets necessarily adopted by UNESCO.

Clearly, definitions are important in landscape planning and the IUCN (International Union for the Conservation of Nature and Natural Resources) now identifies six distinct categories of protected areas, of which one prioritizes landscape management, and the others may support it as a secondary objective either through maintenance of habitat or through traditional management regimes (IUCN, 2003):

I.   Strict Nature Reserve/Wilderness Area: protected area managed mainly for science of wilderness protection.
II.   National Park: protected area managed mainly for ecosystem protection and recreation.
III. Natural Monument: protected area managed mainly for conservation of specific natural features.
IV.   Habitat/Species Management Area: protected area managed mainly for conservation through management intervention.
V.   Protected Landscape/Seascape: protected area managed mainly for landscape/seascape protection and recreation.
VI.   Managed Resource Protected Area: protected area managed mainly for the sustainable use of natural ecosystems.

Landscape planning in the UK in outline

Turning particularly to the UK, the public protection of land partly for its visual resource value is now taken for granted, but this was not always the case. For example, it is only since 1987 that land managers in the first Environmentally Sensitive Areas (ESAs) were actively encouraged to think of themselves as stewards of these heritage landscapes. As with all agri-environment measures, payments are given to managers who enter into voluntary agreements to look after their land in traditional ways, keeping its characteristic landscape and wildlife value. (See Box 7.3 later in the text for more about ESAs.)

Public agencies like the Countryside Agency and local authorities are now legally responsible for identifying and encouraging the maintenance of important landscapes with other stakeholders. This forms part of their wider role in protecting the environment and ensuring sustainable development. In addition, there is considerable involvement from non-governmental organizations (NGOs) like the Council for the Protection of Rural England and the National Trust, who act as watchdogs for special areas on heritage and cultural grounds. This is important as most landscapes in the UK are controlled under the exclusive rights of individuals or corporations. These owners will have their own interests to consider and may be unable or unwilling to maintain and enhance those landscapes society values. This raises the important question of how society decides upon which parts of the country have important landscapes and how best to protect these.

As one example of this work, it is useful to look at landscape character assessment, an approach to categorizing the landscape devised for the then Countryside Commission by Carys Swanwick at the University of Sheffield and Land Use Consultants. (For the most recent guidance check the Countryside Agency's URL, and see also Swanwick and Land Use Consultants, 2002). This approach is essentially a two-stage methodology involving:

- *Landscape characterization*: this stage involves systematically describing landscape character in terms of generic types and then in terms of discrete spatial areas, progressively zooming in from the national, then regional to local scales. The end product is a map of landscape types or areas and a set of descriptions.
- *Making judgements*: this stage involves relating the results of the characterization to decisions about landscape planning, based on the purposes of the

Box 7.1   Mechanisms to control land use by increasing degrees of compulsion (based on Selman, 1988)

Mechanism One	Education of landowners: provision of information and advice.
Mechanism Two	Financial incentives and penalties, especially taxes and grants.
Mechanism Three	Restraints on undesirable uses: planning consents, site licences, etc.
Mechanism Four	Removal of property rights on the open market.
Mechanism Five	Removal of property rights by compulsion: compulsory purchase and nationalization.

initial characterization exercise. The most obvious applications would be to devise policy (expressed, for example, as a landscape strategy, or landscape guidelines), to assist in the distribution of resources (for example, grant aid) or to assess landscape capacity (for example, for new housing development).

The fundamental idea behind landscape planning is the modification of property rights to protect or enhance scenic quality (Selman, 1988). Box 7.1 gives a menu of potential mechanisms to do this. *Mechanism One* reflects the lightest brush of the state on private property rights, and *Mechanism Five* the heaviest, so mechanisms are used by the central or local state according to the nature of the issue at stake and the political acceptability. Legislation for nationalization of the right to develop land was consequent on the introduction of the Town and Country Planning Act (England and Wales) 1947 and the Town and Country Planning (Scotland) Act 1947. This gave central government and local planning authorities the power to control new development in terms of location, scale and design, and is an important tool in the protection or enhancement of special landscapes. Landscape planning in the UK has been equally concerned with *Mechanisms One, Two* and *Three* shown in Box 7.1. Boxes 7.2–4 interspersed later in the text are used to illustrate, by case studies, how these particular mechanisms have operated in the UK.

In order to understand how the state has intervened in landscape planning in the UK, it is helpful to consider the public bodies primarily involved in the regulatory mechanisms shown in Box 7.1. At the time of writing, the main public agencies that advise government (both central and local) on landscape are:

England:	Countryside Agency
Northern Ireland:	Environment and Heritage Service
Scotland:	Scottish National Heritage
Wales:	Countryside Council for Wales

These agencies are funded by central government and are primarily responsible for protecting and enhancing landscape quality at the national level. Their main role is to provide strategic advice and guidance to central and local government. They take a lead in championing a variety of landscape planning causes, alongside other aspects of their remits, as defined by 'amenity' legislation.

This public responsibility for advising on landscape issues can be traced through a number of forerunners of these agencies to the period immediately after World War II. The principles and practice of landscape planning at this level are therefore well established in the UK and based on a 50-year history. The earliest legislation to have a marked impact on the protection of fine scenery was the National Parks and Access to the Countryside Act 1947. This established the principle of designating areas of national significance in England and Wales as national parks or as AONBs. The Act also set up the forebear of the Countryside Agency, the National Parks Commission. The significance of this is that it marked the point in time at which the UK government intervened in private property rights on the grounds of scenic beauty.

The contemporary aims of the national agencies are quite similar[1] despite differing territorial interests, embracing outdoor recreation, cultural heritage and promoting sustainable rural areas. However, at the local level, it is generally the local authorities (especially the unitary, county and district councils) that are directly involved in landscape planning. The national agencies champion particular priorities for action and provide funding for research and development, often working closely with pro-active authorities. For example, in 2001–2002, Scottish National Heritage established two pilot projects in partnership with local authorities,

---

[1] The most significant difference relates to the treatment of scientific nature conservation, and in England, at the time of writing, a separate public body – English Nature – is responsible for this.

concerned with the development of management strategies for three National Scenic Areas. In each area an officer was employed to prepare the strategy in consultation with local stakeholders, including landowners and NGOs. This approach typifies the research and development function of the national agencies. An important role for local authorities in landscape planning is through their own countryside teams or ranger services. Some authorities employ, for example, tree wardens and landscape architects offering direct services. In reality, however, a larger part of the involvement of local authorities is indirect and strategic, through land use planning policy and decision making on individual planning applications. In Devon, for example, the development of policies in the South Hams Local Plan involved landscape character assessment to guide decision making on the strategic selection of land for development (for example, for selection of sites for new communities and major settlement expansion).

The requirement for environmental appraisal of all development plans also means that policies are considered against landscape impact criteria and can be modified to reduce or mitigate negative effects. This makes local authorities important actors in securing landscape protection and enhancement at the local level since their work with stakeholders is frequently wide ranging. It may involve, for example:

- consultation over a proposed landscape designation or the contents of a management plan;
- liaison over a planning application in a designated area, including negotiation of legal agreements;
- grant aid to implement a landscape management plan;
- provision of advice and guidance, both general (e.g. literature and publications) and site specific (e.g. as supplementary planning guidance or design briefs).[2]

Nevertheless, despite this potentially positive impact on scenic quality, landscape planning and appropriate design may be given a low priority compared to the

need for socio-economic development in many locations. This problem lies at the heart of the tension between national agencies that are involved in assessing and designating scenic resources and the local communities who are responsible for managing them.

The implementation of landscape planning in the UK

There have been significant changes to the implementation of landscape planning in the period since World War II. The early designations of the national parks (e.g. Dartmoor, 1951) and AONBs (e.g. Cannock Chase in 1958) in England and Wales illustrated the many difficulties of maintaining the quality of these fine landscapes in the face of socio-economic change. Development pressures in rural areas (for example, for road building, infrastructure and services), changes in the technology of agriculture and forestry, and changes in the societal expectations and use of fine landscape for recreation purposes combined to place particular pressure on these resources.

Tree planting became a popular solution to enhancing amenity from the early 1970s. Fairbrother (1974) was influential in promoting a new (functional) landscape for industrialized society by creating tree belts at the boundary of settlements to act as a visual container, a link to the general landscape and a buffering land use. The Countryside Commission for Scotland and the Countryside Commission for England and Wales supported tree planting in the 1970s, and in England local authorities were block-grant-aided to administer amenity tree planting grants to landowners. Experimental projects like the Central Scotland Woodlands Project (1978) (Box 7.2), the New Agricultural Landscapes Projects (1974) and the Lee Valley Project (1981) emphasized the need for locally based project teams to implement proposals, cascading financial resources and practical advice from national bodies to local places through grant aid. The publication of research findings from these types of project also helped the Commissions to spread ideas and promote innovative practice amongst local authorities.

By 1989 it was suggested that more emphasis should be given to the purpose of enhancing natural beauty and improving the landscape of the national parks, an active role of landscape creation rather than a passive one of protection (National Parks Review Committee, 1989).

---

[2] In practice, non-statutory policy documents can be important in guiding developers to be sensitive to local landscapes. This can range from building design guides, such as those produced by individual national parks, to more major analysis of vernacular styles and specialist topic-based publications. South Hams Conservation Officers produce specialist leaflets on fenestration (windows!), lime in buildings, etc.

Box 7.2   Countryside Around Towns: an illustration of *Mechanism One* – education of landowners: provision of information and advice

The Countryside Around Towns in Scotland Study by Dartington Amenity Research Trust looked at the effect of the proximity to urban areas on land uses and management. It defined two types of countryside: 'countryside within the city structure' and 'urban fringe'. The former accommodated development or activities emanating from the towns, the latter was land directly affected by activities in the town, not merely by accommodating activities exported from it. In the urban fringe, it was perceived that the normal countryside activities could not be sustained in some places, because of the strain of the other uses. In those areas that do suffer the effects, the land use and landscape quality tend to deteriorate rapidly; bitterness develops among those who manage or own the land and a cycle of abuse sets in. The report recommended the establishment of experimental projects to answer a series of research questions about this process.

The Clyde/Calders Urban Fringe Management Project was set up in the early 1980s in the Glasgow green belt, and looked in particular at the costs to agriculture of being in the urban fringe, and the opportunities for strategic work with local communities to tackle fly tipping and environmental improvements. A project officer catalysed the experimental work by offering advice and design services to landowners and local communities.

The Central Scotland Woodlands Project was set up in 1978 as an initiative to encourage tree planting and afforestation in the relatively unproductive soils of the central belt of Scotland. The aim was to demonstrate the feasibility of woodland as an alternative land use, with attendant benefits in landscape, recreational opportunities and land management. Project officers worked with landowners and local communities to improve the appearance of transport corridors and in the urban fringe by strategic planting schemes using existing grant aid. The project showed how tree planting could be a cost-effective alternative to traditional land restoration approaches and how farmers could be encouraged to consider woodland planting with appropriate support and guidance.

Significance

Selman (1992) suggests that there are three broad purposes for project management in practice:

(1)   'area management' or the resolution of small-scale conflicts within a clearly defined stretch of countryside;
(2)   the implementation of management agreements (either informal or legally binding) for specific sites;
(3)   the production of management plans or similar broad policy statements for tracts of countryside

The value of using project officers or animateurs to facilitate landscape planning is dependent on their training and ability to undertake extension work effectively. Clearly it can provide a user-friendly approach to the administration of grant aid, and offer tailored advice and guidance. However, it is an expensive approach compared to literature and may be difficult to support long term. The Central Scotland Woodlands Project became a separate not-for-profit company at the end of its sixth year, and is now known as the Central Scotland Forest Trust.

The Committee made a strong link between the environmental attributes and cultural traditions that contribute to their high quality, stressing the importance of the local community in conserving the quality of the overall resource (National Parks Review Committee, 1989, p. 25):

'Although national park authorities control major changes in the appearance of buildings (except agricultural buildings), their upkeep is the responsibility of the local people who live and work in them; their actions will affect how this important element of the park will look in years to come . . . National Park authorities have a measure of control through their planning powers, but many unfortunate alterations have been made without any planning consent, and others have been allowed that should have been stopped.'

The Committee also recommended that the means of protecting these features should include:

- a search for new uses;
- grant aid for maintenance;
- use of rural conservation areas;
- incorporation of building conservation into farm plans;
- promotion of good design, design guides, development briefs and design advice;
- monitoring of the status of listed buildings.

The same problems were facing managers of landscape resources in AONBs. These were as prone to change as both the national parks and the wider countryside, but worse off from the point of view of controlling change. Confirmation of this concern began to emerge in 1978 when the Countryside Commission began a systematic survey of the AONBs (Countryside Commission, 1980), and the Countryside Commission for Scotland looked at its Scottish counterparts, the National Scenic Areas (NSAs), in Loch Rannoch and Glen Lyon (Centre for Environmental Management and Planning, 1986). The study for the Countryside Commission found that there was insufficient co-ordination between various authorities and interest groups in the AONBs. This resulted in inadequate management and planning, and it was recommended that Joint Advisory Committees were established with an identified local authority officer to take an overview and co-ordinate forward planning and management. It was conceded that in most AONBs there was a lack of any clear statement as to the purpose of designation and the subsequent role of the AONBs. The recommendation was that statements of intent and management plans should be prepared for each AONB by the local authorities, relating to the statutory structure and local plans. The Countryside Commission believed that development control should emphasize the conservation of natural beauty, and that any development permitted should be sympathetic to the objectives of conservation. It was considered that, in the majority of cases, large-scale developments should be excluded (including mineral extraction and road construction). An increased programme of expenditure was planned for the AONBs by the Countryside Commission to support the preparation of plans and appropriate capital works.

Clearly, the problem facing the AONB designation at this time was the same one as faced the national parks: the reconciliation of conserving natural beauty and safeguarding agriculture, forestry, other rural industries and the socio-economic needs of the local communities.

The Countryside Commission for Scotland's study showed that lack of management of landscape was a problem, associated with the lack of prosperity (Centre for Environmental Management and Planning, 1986). In Scotland at this time, 0.7 million ha of NSAs lay in the Highlands and Islands where new investment was desperately sought. The designations were frequently perceived as negative, bringing disincentives to development, and providing no mechanisms for introducing additional finance to assist beautiful but disadvantaged areas. Thus the early, static conceptualization of designation (drawing a line round an area and treating it differently to the areas beyond) proved faulty. The use of the planning system in isolation to protect landscape

then gave way to a more dynamic and strategic approach to landscape planning both in protected areas and also in the wider countryside. This embraced broader concerns about encouraging active landscape management and the issues of working alongside other stakeholders.

The change in viewpoint had been partly facilitated by the move to forward planning by local authorities themselves, and their more active role in environmental management. For example, for the larger councils, development plans from the early 1970s, as Structure Plans and Local Plans, were based on strategic assessment of landscape resources in a local authority area (see, for example, Gilg, 1979). These plans often identified the need to protect more locally significant landscapes such as Areas of Great Landscape Value. They also set out policies for the wider countryside including landscape improvement of problem areas like the urban fringe and the green belts, recognizing the need to enhance some environments by more active management and by introducing more appropriate landscape elements.

By the end of the 1980s the Countryside Commission was promoting the idea of rural strategies to focus more broadly on the wider countryside. It was also promoting the idea of landscape assessment (now called landscape character assessment), rather than the more basic landscape evaluation. The latter was concerned with establishing which areas were 'better' landscapes for designation and protection. However, the former was concerned with the characterization of landscape (identifying, mapping, classifying and describing) and the process of making judgements (based on landscape character) to inform decision making in a wide range of situations (Swanwick and Land Use Consultants, 2002). Landscape character assessment is therefore more likely to provide a robust and effective framework for the protection, maintenance and enhancement of areas, as well as allowing decisions to be made about the capacity of landscapes to absorb particular types of development (for example, settlement expansion, road building and reservoir creation).

Another more recent change in implementation of landscape planning results from the embedding of environmental objectives in government policy – led by planning for sustainable development and the wider restructuring of agriculture. The Wildlife and Countryside Act 1981 set an agenda for conservation based on voluntary agreement and compulsory compensation,

Box 7.3   Agri-environment schemes: an illustration of *Mechanism Two* – grant aid

Environmentally Sensitive Areas (ESAs) were introduced as a result of amendments by the UK to Article 19 of the EC Structures Regulation (797/85), enabling financial incentives to be given from the EC via member states to encourage appropriate farming practices, *'where the maintenance or adoption of particular agricultural methods is likely to facilitate the conservation, enhancement or protection of the nature conservation, amenity or archaeological or historic interest of an area'*. By the end of 1994, Winter (1996) commented that 15% of the agricultural area of the UK was covered by designated ESAs and the budget was some £43 million, making it the single most important agri-environment measure. By 2003 there were 22 ESAs in England, covering some 10% of agricultural land (DEFRA, 2003a).

The following criteria were used to define ESAs (Winter, 1996):

• They must be of national environmental significance.
• Their conservation must depend on the adoption, maintenance or extension of a particular form of farming practice.
• They should be areas where the encouragement of traditional farming practices would help to prevent damage to the environment.
• They should comprise a discrete and coherent unit of environmental interest.

Within an ESA, landowners and tenants can put their holdings forward on a voluntary basis, receiving tiered payments. The amount of money they receive is based primarily on the submitted hectarage and the level of conservation work they are willing to undertake. Each ESA is unique in terms of its landscape character, so the schemes are tailored to the specific area and a wide range of landscape elements may be covered including management of hay meadow, flood meadow, hedgerow, woodland, etc. The take-up of the scheme has in most cases been high. The attraction appears to be its flexibility, and whilst payments are not high, farmers can continue (in many cases) to farm as they have done before, since the voluntary principle gives them the primary control.

The Blackdown Hills ESA was launched in 1994 and aims to protect and enhance the landscape, wildlife and historic interest of the area through the maintenance and adoption of traditional management practices associated with livestock farming systems, particularly those based on permanent grassland (DEFRA, 2003b). Applicants volunteer to enter the programme and are required to enter their whole farm (including woods and heathland) into a 10-year management plan. Grants and maintenance are payable under a three-tier scheme. Current details of the scheme and targets can be accessed from the DEFRA website (www.defra.gov.uk).

Countryside Stewardship was designed by the Countryside Commission and launched in 1991, targeting landscape grants to smaller and more coherent sites than under the ESA approach. A mixture of annual payments and capital payments was provided, designed with precise and demanding management requirements compared to the ESA approach. Targets were set for eight specific landscape types (chalk and limestone grassland, coastal land, historic landscapes, lowland heath, old meadows and pastures, uplands, waterside landscapes and hedgerow landscapes) as well as for hedgerow restoration, traditional walls restoration, creation of field margins and public access. In the early stages, Countryside Stewardship was criticized for its small budget, the expectation that public access should be provided, and the fact that any impact was likely to be very localized compared to the wider landscape planning possible under the ESAs. However, since 1994 the two schemes have been administered jointly and are now under the England Rural Development Programme, resulting in a more coherent market image from the user's perspective.

Significance

In the case of both schemes there has been a clear intention of planning for retention of particular landscape characteristics and monitoring of the effects of grant aid. The ESAs have been monitored using aerial photography and interview surveys to determine the land cover changes and public perceptions of the effects of policy. Garrod *et al*. (1994) used a contingent valuation approach based on visitor spend in the Somerset Moors and Levels to measure resident and visitor valuations of landscape conservation. Overall it was concluded that the benefits of the ESA in financial terms were four to five times its real cost, although visitors valued the conserved landscape more than local people. A similar ratio pertained in the South Downs ESA (Willis and Garrod, 1992). The cost of landscape conservation measures is now linked with economic development for the green tourist market. This may be seen as a way of reconciling the environmental needs of an area with the socio-economic needs of the local people – an important principle in sustainable development.

and this was paralleled by greater commitment and resources for environmental protection in the countryside as a whole. This was fostered by the major changes in grant aid to agriculture under the Agriculture Act 1986. It removed many of the conservation-damaging improvement grants (like the removal of hedges and stone walls) and substituted the requirement for whole farm plans where business development grants were given, thus allowing ADAS and MAFF the opportunity to modify conservation-unfriendly proposals. The significance of the agri-environment schemes as a tool for landscape planning is outlined in Box 7.3.

Latterly, greater emphasis has been placed upon landscape management following the restructuring of the agricultural industry. This is reflected in the modification of the sectoral remit of the former Ministry of Agriculture, Fisheries and Food to the Department of Environment, Food and Rural Affairs (DEFRA) following the foot-and-mouth disease outbreak in 2001. Landscape management is an important responsibility of landowners and managers today compared to 50 years ago when agriculture and forestry production were central, and is now supported by grant aid and active education and training programmes.

Environmental impact assessment (EIA) of projects is a further mechanism to ensure that potential development is appropriate to local landscape and to identify mitigating measures. Only major projects like power stations and new road building are subject to mandatory EIA, but designated landscapes – especially

---

Box 7.4   Community forest: an illustration of *Mechanism Three* – restraints on undesirable uses, planning consents, etc.

Community forest schemes were introduced in 1989, and are now run jointly by the Countryside Agency and Forestry Commission. Community forests are supported by local project teams who work up a plan for the designated area. Tree planting is supported by existing grant aid packages and planning consents are used to obtain wider benefits associated with development under Section 106 agreements, made under the Town and Country Planning Act 1990. The UK's National Forest is an example of this approach and covers 200 square miles in the English Midlands [see their website for more information (www.nationalforest.org)]. Under the forest plan, planting has been focused on strategic locations adjacent to the M42 where there would be maximum visual impact. In all cases the process of changing land use is controlled by a landowner's willingness to plant, or willingness to release land for planting, and the development pressure to gain other land uses within the forest matrix which will require planning permission. This is likely to be easier near existing settlements if the open countryside is degraded, and is also likely to be easier in situations where derelict land is available or mineral permissions are sought. Bishop (1991) looked extensively at the implementation mechanisms for the community woodland concept. Planning conditions attached to consents, particularly those involved in voluntary Section 106 agreements, can be used to get public goods from private development gains. These agreements are legally binding and are carried forward with changes of ownership, so are potentially powerful mechanisms. Mineral operators in the Central Scotland Woodlands Project (see Box 7.2) were particularly keen to consider landscape enhancement involving wider geographical areas than actual extraction sites. This was because they owned mineral rights to substantial areas and the profitability of extracting the mineral resources made it possible to offer substantial community benefits. Mineral plans and mineral planning guidance notes also emphasize the benefits of getting environmental gains in impoverished landscapes by new land uses and long-term management commitments under mineral consents.

Thames Chase Community Forest is intended to provide extensive opportunities for:

- community involvement;
- landscape restoration;
- a thriving forestry and farming industry with increased scope for diversification;
- recreation: including walking, riding, sports, and artistic and cultural events;
- education: as an outdoor classroom;
- nature conservation: including the creation of new habitats;
- protection of areas of high landscape quality or of historical or archaeological interest.

The Forest Plan identified eight landscape character zones in which the degree of emphasis on forestry varies. The overall aim is to plant 2000 ha of new woodland over 50 years, representing approximately 26% of the forest area, including the restoration of some 120 ha of mineral workings to woodland. It is intended to tackle the urban fringe problems by reducing the 'hope value' of land, by achieving integration of landscape design, planning policy and land management, by encouraging responsible recreational use of the countryside and by assisting the viability of agriculture.

Significance
The use of the planning system in this context is dependent on development pressure, since changes of use of land and development of land need planning permission. Only certain locations, e.g. the peri-urban fringe, have sufficient development value to make the approach workable, but clearly this offers a mechanism to introduce new landscape elements to enhance the scene. An additional attraction for central or local government is that private owners invest in and manage these new landscapes rather than the public sector. However, traditionally the urban fringe has been a problem zone for management and only time will establish whether the new landscapes thus created are actively protected, maintained and managed.

those of national or regional significance – are potentially protected from development by EIA regulations first introduced by EU directives. Guidance on where environmental assessments are required is now well tested by best practice and case law, and potentially sensitive landscapes subject to projects involving large land take (e.g. 20+ ha) are subject to discretionary assessments required by the local planning authority (LPA). The LPAs take guidance from statutory agencies over the desirability of requiring an EIA and the quality of the studies when published.

The changes to recognize the particular context of landscape and target resources more fully to the wider countryside as well as nationally significant fine landscape are paralleled by the move to active celebration of custom and tradition relating to local place, landscape character and regional distinctiveness; for example, through the making of parish maps, village design statements, etc. This is clearly an alternative to centrally imposed landscape designation, since part of the emphasis is on the empowerment of local communities to identify and 'claim' their valued resources. The voice of the voluntary sector is therefore central to this approach and ties in closely with New Labour's vision of a 'Third Way' of delivering public services (Blackburn *et al.*, 2003). The idea of creating new landscapes closely related to the aspirations of local communities can be seen in the community forest concept outlined in Box 7.4. This also shows how the planning system can be used creatively to deliver landscape planning.

References

Appleton, J. (1990) *The Symbolism of Habitat*. University of Washington Press, Seattle.

Bishop, K. (1991) Community forests: implementing the concept. *Planner*, **77** (18), 6–10.

Blackburn, S., Skerratt, S., Errington, A. and Warren, M. (2003) *Rural Communities and the Voluntary Sector*. Final Report for DEFRA, University of Plymouth.

Brassley, P. (1985) *The Agricultural Economy of Northumberland and Durham, 1640–1750*. Garland Press, New York.

Centre for Environmental Management and Planning (1986) *A Landscape Review of the Loch Rannoch and Glen Lyon National Scenic Area*. Report to the Countryside Commission for Scotland. CEMP, Aberdeen.

Cosgrove, D. (1984) *Social Formation and Symbolic Landscape*. Croom Helm, London.

Cosgrove, D. and Daniels, S. (eds) (1988) *The Iconography of Landscape*. Cambridge University Press, Cambridge.

Council of Europe (2003) *European Landscape Convention*. www.nature.coe.int (accessed 17 Oct 2003).

Countryside Commission (1980) *Areas of Outstanding Natural Beauty: A Policy Statement*. Countryside Commission, Cheltenham.

DEFRA (2003a) *Environmentally Sensitive Areas*. www.defra.gov.uk/erdp/schemes/esas (accessed 22 Nov 2003).

DEFRA (2003b) *England Rural Development Programme: Somerset Levels and Moors Environmentally Sensitive Area (ESA)*. The provisions and details of this ESA are available at www.defra.gov.uk/erdp/docs/national/annexes/annexx/slmrex2.htm (accessed 22 Nov 2003).

Fairbrother, N. (1974) *New Lives, New Landscapes*. Penguin, Harmondsworth.

Fleming, A. (1988) *The Dartmoor Reaves: Investigating Prehistoric Land Divisions*. Batsford, London.

Garrod, G.D., Willis, K.G. and Saunders, O.M. *et al.* (1994) The benefits and costs of the Somerset Levels and Moors ESA. *Journal of Rural Studies*, **10** (2), 131–45.

Gilg, A.W. (1979) *Countryside Planning: The First Three Decades, 1945–76*. Methuen, London.

Gilg, A.W. (1996) *Countryside Planning*, 2nd edition, Chaps 3–5. Routledge, London.

Hoskins, W.G. (1955) *The Making of the English Landscape*. Hodder and Stoughton, London. (See also revised edition, 1988, ed. C. Taylor.)

IUCN (2003) *World Conservation Protected Areas*. www.iucn.org (accessed 17 Oct 2003).

Kaplan, S. and Kaplan, R. (1982) *Cognition and Environment: Coping in an Uncertain World*. Praeger, New York.

Landscape Institute and Institute of Environmental Management and Assessment (2002) *Guidelines for Landscape and Visual Impact Assessment*, 2nd edition. SPON Press, London.

Marsh, W.M. (1997) *Landscape Planning: Environmental Applications*, 3rd edition. Wiley, Chichester.

Muir, R. (1999) *Approaches to Landscape*. Macmillan, Basingstoke.

Muir, R. (2000) *The New Reading the Landscape: Fieldwork in Landscape History*. University of Exeter Press, Exeter.

National Parks Review Committee (1989) *Fit for the Future*. CCP334. Countryside Commission, Cheltenham.

Rackham, O. (1976) *Trees and Woodland in the English Landscape*. Dent, London.

Rackham, O. (1980) *Ancient Woodland: Its History, Vegetation and Uses in England*. Edward Arnold, London.

Rackham, O. (1986) *The History of the Countryside*. Dent, London.

Selman, P. (ed) (1988) *Countryside Planning in Practice: the Scottish Experience.* Stirling University Press, Stirling.

Selman, P. (1992) *Environmental Planning.* Chapman, London.

Short, B. (2000) Forests and wood pasture in lowland England. In: *The English Rural Landscape* (ed. J. Thirsk). Oxford University Press, Oxford.

Swanwick, C. and Land Use Consultants (2002) *Landscape Character Assessment: Guidance for England and Scotland.* Ref AAX 84F. Scottish Natural Heritage and the Countryside Agency, Cheltenham.

Thirsk, J. (ed.) (2000) *The English Rural Landscape.* Oxford University Press, Oxford.

Turner, T. (1998) *Landscape Planning and Environmental Impact Design.* UCL Press, London.

UNESCO (2003) *World Heritage Site Index.* whc.unesco.org (accessed 17 Oct 2003).

Willis, K.G. and Garrod, G.O. (1992) Assessing the Value of Future Landscapes. *Landscape and Urban Planning*, **23**, 17–32.

Winter, M.F. (1996) *Rural Politics.* Routledge, London.

Wordie, J.R. (1983) The chronology of English enclosure, 1500–1914. *Economic History Review*, **36**; 483–505.

Further reading

Appleton, J. (1986) *The Experience of Landscape.* Wiley, Chichester.

Bell, S. and Lucas, O. (1989) Farm woodland design. *Landscape Design*, Sept 17–20.

Blacksell, M. and Gilg, A.W. (1981) *The Countryside: Planning and Change.* Allen and Unwin, London.

Brooke, D. (1994) A countryside character programme. *Landscape Research*, **19** (1), 128–32.

Brotherton, I. and Devall, N. (1988) On the acceptability of afforestation schemes. *Land Use Policy*, **5**, 245–52.

Countryside Commission (1981) *Countryside Management in the Urban Fringe.* CCP136. Countryside Commission, Cheltenham.

Countryside Commission (1983) *The Changing Uplands.* CCP153. Countryside Commission, Cheltenham.

Countryside Commission (1984) *A Better Future for the Uplands.* CCP162. Countryside Commission, Cheltenham.

Countryside Commission (1989) *Forests for the Community.* Countryside Commission, Cheltenham.

Countryside Commission (1990) *Advice Manual for the Preparation of a Community Forest Plan.* Countryside Commission, Cheltenham.

Countryside Commission (1991) *Assessment and Conservation of Landscape Character: The Warwickshire Landscapes Project Approach.* Countryside Commission, Cheltenham.

Countryside Commission (1994) *Countryside Character Programme.* Countryside Commission, Cheltenham.

Countryside Commission (1994) *The New Map of England: A Celebration of South Western Landscapes.* Countryside Commission, Cheltenham.

Countryside Commission (1994) *The New Map of England: A Directory of Regional Landscape.* Countryside Commission, Cheltenham.

Countryside Commission for Scotland (1976) *The Countryside Around Towns in Scotland.* Darlington Amenity Research Trust/CCS Battleby, Perth.

Countryside Commission for Scotland (1987) *The Countryside Around Towns in Scotland: A Review 1976–85.* Darlington Amenity Research Trust/CCS Battleby, Perth.

Crowe, S. (1978) *The Landscape of Forests and Woods.* FC Booklet no. 44. HMSO, London.

DETR (1992) *Planning Policy Guidance Note No. 2 Greenbelts.* HMSO, London.

Feist, M.J. (1978) *A Study of Management Agreements.* CCP114. Countryside Commission, Cheltenham.

Lucas, O.W.R. (1990) *The Design of Forest Landscapes.* Oxford University Press, Oxford.

Mather, A.S. (1988) Agriculture and forestry. In: *Countryside Planning in Practice: The Scottish Experience* (ed. P. Selman), pp. 69–87. Stirling University Press, Stirling.

Mather, A.S. and Thompson, K.J. (1995) The effects of afforestation on agriculture in Scotland. *Journal of Rural Studies*, **11**, 187–202.

O'Riordan, T., Wood C. and Sheldrake, A. (1991*) Interpreting Landscape Futures in the Yorkshire Dales National Park.* University of East Anglia, Norwich.

Parnell, B. (1988) Urban fringe and green belt. In: *Countryside Planning in Practice: The Scottish Experience* (ed. P. Selman), pp. 225–46. Stirling University Press, Stirling.

Parry, M.L. (1982) *Surveys of Moorland and Roughland Change.* Birmingham University.

Parry (1985) *Mid Wales Uplands Study.* CCP177. Countryside Commission, Cheltenham.

Sinclair, G. (1992) *The Lost Land: Land Use Change in England 1945–90.* Council for the Protection of Rural England, London.

Tompkins, S. (1993) Conifer Conspiracy. *Ecos*, **14** (2), 23–7.

Willis, K.G. and Garrod, G.D. (1991) *Landscape Values: A Contingent Valuation Approach and Case Study of the Yorkshire Dales National Park.* ESRC Countryside Change Initiative. Working Paper 21. Department of Agricultural Economics and Food Marketing, University of Newcastle-upon-Tyne.

Useful websites

www.communityforest.org.uk – National Community Forest Partnerships.

www.countryside.gov.uk – UK Countryside Agency.

www.cpre.org.uk – Council for the Protection of Rural England.

www.csct.co.uk – Central Scotland Countryside Trust, daughter of the Central Scotland Woodlands Project, and recently renamed as the Central Scotland Forest Trust. An example of a community forest.

www.defra.gov.uk – Government's Department of Environment, Food and Rural Affairs. View the Wildlife and Countryside Directorate for landscape conservation in the countryside.

www.iucn.org – International Union for the Conservation of Nature, an international partnership of governments, NGOs and scientists assisting in the development and implementation of policy. View in particular the parts of the site concerned with the World Conservation Protected Areas theme.

www.nationaltrust.org.uk – National Trust for England and Wales, an important UK NGO. View in particular the environment and conservation web pages.

www.nationalforest.org – the UK's National Forest company.

www.nature.coe.int – Council of Europe portal giving access to detailed information on landscape conservation.

www.nts.org.uk – Scottish National Trust.

www.snh.gov.uk – Scottish Natural Heritage.

whc.unesco.org – UNESCO. View the World Heritage Sites Index.

8

Building conservation

P.C. Child

Introduction

The following chapter discusses the topic of rural building conservation essentially from an English standpoint. Although approaches and principles in this field are similar in Scotland, Wales and Northern Ireland to those that apply in England, there are significant differences, particularly legislatively, between each country, and readers outside England will need to check what is set out here against their own national practices. To complicate things further, the legislation in England relating to building conservation and agri-environment grant schemes is currently being reviewed. These changes are discussed below, but the reader should be aware that it is a scene in flux. This chapter does not attempt to cover the related and very large topic of rural archaeology – the evidence of mankind's past which survives as earthworks or buried remains within the man-made landscape.

Most people would agree that it is essential to conserve the most important historic structures – the cathedrals and great houses – which are some of the most tangible major artistic and cultural achievements of our ancestors. For many, however, this feeling extends to cover most of the buildings that have survived from the past, not simply the exceptional ones, so that the proposed demolition of even commonplace structures can sometimes provoke considerable local opposition. This emotion is not simply based on the fact that many old buildings are in themselves objects of beauty and historic interest, but because they represent a very tangible link with the past – something that in a fast-changing world becomes more and more important. In recent years too, the growing recognition that sustainability should be an underlying principle where all development is concerned adds more support to the argument for the retention and re-use of old buildings rather than their replacement with wholly new structures.

Historic buildings

The UK is fortunate in possessing a vast resource of historic buildings. This is the consequence of a peaceful and relatively affluent past; in England, for instance, there survive many thousands of farmhouses built in the Middle Ages. Rural historic buildings cover a complete spectrum from castle to cottage. The grandest buildings tend to be built in national styles, but even those historic buildings of a relatively high status (such as parish churches) possess regional characteristics. At a domestic level these differences are more marked, so that a farmhouse in Kent bears little resemblance to one in Devon or Lancashire, particularly in their building materials and constructional methods, but also in their arrangements of rooms. On the other hand, within particular regions firm adherence to strong traditions produced buildings that conform consistently to local patterns. Similarly, differing agricultural practices in the past have produced different types of farm buildings in different areas of the country. Today, a uniformity of design, combined with the mass production and distribution of materials, has eroded the local distinctiveness which is very evident in older buildings. The way in which historic buildings are distributed in the rural landscape also varies considerably and in a manner that reflects past patterns of land ownership and tenure. For instance, isolated farmsteads are more characteristic of upland areas of Britain such as the South West and Wales than of lowland England where historically buildings tended to be concentrated in villages.

The legal background

Scheduled ancient monuments

Popular pressure in the UK to ensure the preservation of its built heritage has resulted in a well-established and detailed legislation intended to provide its protection. The earliest legislation dates back to the early twentieth century and relates to the 'scheduling' of ancient monuments; these are generally restricted to archaeological sites and to ruins and structures not capable of viable or economic use. These monuments are designated selectively and any work to them, however minor, needs consent in England from the Department of Culture Media and Sport (DCMS), who use English Heritage, a government agency, as its adviser for this purpose. English Heritage also looks after those monuments that have been taken into 'guardianship' by the state (such as Stonehenge and Dover Castle) under the same legislation as a means of ensuring their protection.

Listed buildings

Although some buildings (as opposed to monuments or sites) were listed in the past as the only available means formally to protect them, it was recognized after World War II that it was not a system that was appropriate for the majority of buildings, particularly those that were occupied and in use. From 1946 onwards therefore the 'listing' system (i.e. inclusion in a List of Buildings of Special Architectural or Historic Interest) was developed. 'Buildings' in this context covers almost any man-made structure from milestones to castles, although the majority of items on the list are private houses. Although buildings can be listed for their historic interest (such as having had a famous owner) this is very rare and most are included solely for their architectural interest. If they pre-date 1840 and are largely complete they will generally qualify for listing; after 1840 they have to be of more particular interest. There are three grades of listed building: 1, 2* and 2; there used also to be a grade 3, but this grade has now been abolished. Grade 1 buildings are of 'exceptional interest' and make up 2.5% of the total. Grade 2* buildings are 'particularly important' and make up 6.0%. The overwhelming majority (91.5%) are Grade 2 and simply of 'special' interest. These differences in grading, apart from highlighting which are the most important buildings, can affect eligibility for grant aid as well as necessitating different procedures for approving alterations (see below). In England in 2004 there were 371 591 listed items, but a proportion of these items include more than one building, so that the overall total of individual buildings and structures is probably well over 400 000.

Despite the already large numbers of listed buildings in the UK, it can be argued that the lists are not as complete as they should be; farm buildings, for instance, are not as well covered as other categories of building. Some local authorities construct their own 'local lists' to supplement the official, statutory ones, as they feel that the latter are too exclusive. Such local lists are not statutorily binding, but they can be used as considerations in the planning process. The same informal status is accorded also to the Registers of Historic Parks and Gardens and of Historic Battlefields which English Heritage has drawn up for every county in England.

Consent for changes

Decisions upon proposals to alter or demolish listed buildings are made by councils in their role as local planning authorities, although they cannot approve changes to the higher two grades without clearance by the Office of the Deputy Prime Minister which has ultimate responsibility for planning matters in England. Consent is needed to do any work to a listed building that would affect its character. In contrast to scheduled monuments, permission is not needed to carry out repairs, although the dividing line between a 'repair' and an 'alteration' is not always easy to define. There is no set of black and white rules which set out what one might or might not be permitted to do in terms of altering or extending a listed building, although if it is proposed to demolish one there is a series of strict tests to be met if consent is to be granted – something that rarely happens today. The acceptability of a proposed alteration or extension to a listed building has to some extent to be a matter of judgement but should be based on the guidelines set out by the Government in PPG 15 *Planning and the Historic Environment*. Councils usually employ specialist conservation officers to advise them on these matters; English Heritage has inspectors who perform a similar role. If consent is refused for any proposal, then the applicant can use an appeal procedure similar to that which applies in the planning process.

Conservation areas

In recent decades the protection of individual buildings through the listing system did not provide adequate protection for the character of historic areas in which many of them were located. Consequently in 1968 legislation was introduced to establish 'conservation areas', intended to be areas of towns or villages of particular historic interest, although in reality any area could be so designated. Some rural features such as canals or old industrial sites have therefore been made conservation areas, although the majority are the historic centres of settlements. Being made a conservation area does impose some limitations (particularly as regards demolition) on what development can be carried out within it and councils can take up special powers to control unsuitable alterations to buildings in such areas if they so choose. They are also able to ask for a high standard of design for any new building in the area and to ensure that such new building is in keeping with its surrounds.

Proposed changes to the law

As can be seen from the very brief summary above, the legislative system concerned with building conservation, because it has evolved over many years, is extremely complex, confusing even to the professionals who use it on a day-to-day basis. Consequently, the DCMS has recently carried out a major consultation to see if a new unified system of protection might provide a better and more easily understood method to enable the protection of the historic environment in England. The consultation suggests a new system within which all statutory (and some non-statutory) existing categories of protection would be brought under one umbrella, although retaining some form of the existing separation within this. The actual mechanics of achieving this are not yet designed, and new legislation may be needed to make some of the proposed changes.

Building conservation in the countryside today

High property prices and a continuing desire by the British to live in the country mean that problems experienced in the relatively recent past of rural historic houses being neglected for lack of funds have largely vanished. Increasing affluence is almost a problem in itself as wealthy owners seek to give these houses all the trappings of twenty-first century life, undermining their traditional character with large extensions, swimming pools and the like.

Farm buildings and conversion

If there is not a problem in conserving rural houses, the same cannot be said for the traditional farm buildings which accompanied every farmhouse. Effectively these buildings are all redundant for all but low-key modern farming uses and are wholly redundant in many instances where holdings have been amalgamated and farmhouses sold off. Planning policy has since its beginnings in the 1940s been opposed to the construction of new houses in the countryside except where there was a specific proven need, so that the conversion of farm buildings to houses which has been allowed since around 1980 is curiously at variance with mainstream planning philosophy. The argument has been that the need to find a new use for these buildings outweighs the long-established policy against new houses in rural locations; converting them is the only way to conserve them. Government advice has been that it would be preferable to see such buildings converted to industrial or holiday use than to housing, as this would better benefit the rural economy, but in reality the majority end up as new homes for good economic reasons. The Government in its latest (2004) draft advice on development in the countryside (PPS 7 *Sustainable Development in Rural Areas*) has reiterated its previous support for the conversion of farm buildings to houses.

Planning authorities have generally pursued policies of minimal alteration in the conversion work to reduce the visual impact of the change, and the 'barn conversion' with its roof lights and glazed threshing doors is now a feature of this century's countryside (Figure 8.1). There are no available statistics on the number of farm

(a)

(b)

Figure 8.1   (a) Before and (b) after photographs of a barn converted to a residential dwelling, Devon, UK.

buildings that have been converted to houses in this way nor on the number of traditional farm buildings left that might be suitable for such conversion. The Countryside Agency and English Heritage are attempting to carry out an audit of the remaining resource of farm buildings from which more specific policies can be derived. Such policies might, for instance, be able to distinguish which farm buildings should *not* be converted because of their importance in historic terms.

Financial help

Historic building grants

Grants for the repair of buildings are generally not easy to obtain and there are no tax concessions available to

the owners of listed buildings except (bizarrely) VAT relief on alterations or extensions – *not* repair – to residences. Although local authorities are empowered to give grants for historic building repair, many no longer operate grant schemes, which are often victims of financial cutbacks. Nevertheless, it is still worth enquiring of your local council whether any grants are available if repair work is needed, especially to a listed building. English Heritage makes grants for the repair of only the more important grades of listed buildings and for scheduled monuments, but its budget is limited and diminishing, so that it rigorously prioritizes its grants to the most needy cases and generally offers them only where there is a 'conservation deficit', i.e. where the cost of repairing a building exceeds its market value. The Heritage Lottery Fund is another potential source of grant aid and operates various schemes at different levels, but all need to show some community benefit. Privately owned buildings would not be eligible for grant under its rules.

Building preservation trusts exist in many parts of the country whose objectives are to rescue, repair and restore to a viable use buildings that proved difficult to conserve for financial or other reasons. Some country towns benefit from area grant schemes run in conjunction with local councils either by English Heritage (Heritage Economic Regeneration Schemes, HERS) or by the Heritage Lottery Fund (Townscape Heritage Initiatives, THIs). These schemes are similar and choose areas (usually the historic cores of settlements) within which a programme of targeted and grant-aided building repair is run, generally for three years. A programme of enhancement of the surrounding buildings and streets accompanies the repair work.

England Rural Development Programme grants

One of the most important sources of grant aid for rural buildings can be found in the agri-environment schemes run by DEFRA's Rural Development Service (RDS) under the England Rural Development Programme. Currently these schemes are divided into those within designated Environmentally Sensitive Areas (ESAs) and those outside these areas which fall into the category of Countryside Stewardship Schemes (CSSs). ESA schemes have operated since 1987 in 22 areas of particular environmental interest, particularly national parks and Areas of Outstanding Natural Beauty (see Chapter 7). Besides grants for improving the natural environment of farm holdings, there is provision in both schemes for restoring traditional (non-domestic) buildings on farms, essentially on a like-for-like basis, and using appropriate traditional materials. The rate of grant for building repair in ESAs varies from 40–80% depending on the ESA (although a few offer no buildings option); within CSS the principal grant rate is 50%, although in exceptional cases it can be higher. Only in special cases would a repair grant be offered for a building in isolation; it would normally only be eligible as part of a scheme of wider environmental management for the holding. In 2002–2003 more than £8 million was spent on building repair grants under these schemes, but the budget is limited so that CSS grants at least are offered on a competitive basis. From 2005 all holdings in England will be eligible to join an 'entry level' environmental stewardship scheme receiving a flat rate payment per hectare in return for appropriate environmental management. In parallel with this new scheme will run an enhanced 'higher level' scheme based on Farm Environmental Plans for each holding involved. This will include grants for the repair of traditional farm buildings, and it is likely that the rate of grant for this purpose will rise to as much as 80%. The budget for higher level schemes will be limited so that agreements under them will still be made on a competitive basis.

The England Rural Development Programme also includes a Rural Enterprise Scheme (RES) which can offer funding (at a rate of 30–50%) for the renovation and adaptation of farm buildings to new commercial uses, and higher rates of grant are available for non-profit-making projects that contribute towards the objectives of the RES of supporting the development of more sustainable, diversified and enterprising rural communities. Such grants can potentially be combined with agri-environment grants for the restoration of the historic character of a building, so as to both achieve the conservation of traditional farm buildings and give them a viable use.

Conservation techniques

Philosophy

In the UK the approach to building conservation has been massively influenced by William Morris who founded the Society for the Protection of Historic Buildings in 1896. His advocacy of preventative maintenance and his philosophy of repairing 'as found' in preference to speculative restoration have been the cornerstone of official conservation policy over the last century.

The conservation of historic buildings is a highly developed skill and extremely sophisticated approaches are available for the resolution of particular issues and problems. However, most conservation matters are not complex and given a reasonable level of knowledge and interest can be addressed without too much difficulty. Buildings generally present problems either as a consequence of neglect or because they have been unsympathetically treated in the past. With all buildings, both old and new, failure to carry out routine maintenance is by far the most common cause of trouble, but with historic buildings it can be compounded by the use of unsuitable modern materials in attempts to remedy these problems. There are considerable differences between traditional construction methods and modern ones and the latter are not always suitable for the former.

Materials and techniques

Historic buildings were generally built before the advent of damp-proof membranes and cavity walls; they were built using solid walling bedded in lime mortar and coated, sometimes externally and usually internally, in lime plaster. This sort of traditional construction is highly vapour permeable, allowing the absorption of moisture, in the form of rain or vapour from normal habitation, and its easy evaporation from these absorbent surfaces. This contrasts with modern building construction methods, which concentrate on keeping moisture out of the building, using damp-proof membranes, cavity walls and impermeable materials such as concrete and cement-based mortars. Problems arise in historic buildings when these impermeable materials are used to repair traditional construction. A common example is the use of cement-based mortar or render to point or plaster a stone/cob/timber-framed external wall. The rigidity of the cement means it often cracks when the more flexible structure below moves with thermal expansion or contraction. Water then gets into the structure through the resultant hairline cracks, but cannot evaporate out again through the impermeable mortar joint or render. Moisture is therefore trapped in the structure, and can cause decay of structural timbers such as beam ends or lintels, and of internal finishes. The use of lime-based mortars greatly helps to avoid such problems and is to be encouraged for all work in the repair and maintenance of historic buildings

Advice about repair

There are various sources of advice for owners of historic buildings as well as for those professionally involved. The most accessible sources are probably local authority conservation officers who can provide help on technical issues (as well as on listed building and conservation area matters) and can usually point owners in the right direction. They may also be able to recommend suitable professional advisers and contractors. There are a variety of courses on traditional building and repair techniques, particularly on the use of lime in repairing historic buildings, which are now run all over the country, by specialist builders, local historic building and lime centres and organizations such as the Society for the Protection of Ancient Buildings (SPAB). The SPAB, as well as some of the other societies concerned with historic buildings, also runs courses for specialists, and produces a wide range of excellent technical leaflets on different aspects of building repair. English Heritage produces highly researched technical documents on specialist subjects, such as wall paintings, as well as guidance on subjects such as thatch, and runs campaigns on issues such as pointing of brickwork, repairing timber windows, etc.

Professionals and craftsmen

There has been a heightened awareness of the issues of unsuitable materials and techniques in recent years and the use of lime-based materials has become much more common. However, the lack of understanding of tradi-

tional building construction remains a real problem and there are still many builders wedded to the use of inappropriate but more convenient modern materials. There is also a shortage of those who are proficient in traditional practices. There are currently no recognized qualifications for craftsmen working on historic buildings, although NVQs are being developed in some areas such as thatching. Anything other than simple repairs may require the employment of a professional surveyor or architect. The Royal Institute of Chartered Surveyors has a specialist building conservation qualification for qualified members, as does the Royal Institute of British Architects (AABC, 'architects accredited in building conservation'). There are no legal requirements to use a qualified architect or surveyor to specify or supervise work on historic buildings even when they are listed, except where the work to that building is being grant-aided by English Heritage.

Further reading

Ashurst, J. and Ashurst, N. (1988) *Practical Building Conservation, Vols 1–5.* Gower, Aldershot.

Brereton, C. (1991) *The Repair of Historic Buildings.* English Heritage, London.

Brunskill, R.W. (2000) *Illustrated Handbook of Vernacular Architecture,* 4th edition Faber and Faber, London.

Pearce, D. (1998) *Conservation Today.* Routledge, London.

Peters, J.E.C. (1981) *Discovering Traditional Farm Buildings.* Shire, Princes Risborough.

Useful websites

www.ahfund.org.uk – Architectural Heritage Fund.

www.cadw.wales.gov.uk – CADW (Welsh Historic Environment Agency).

www.ehsni.gov.uk – Northern Ireland Environment and Heritage Service.

www.english-heritage.org.uk – English Heritage.

www.hlf.org.uk – Heritage Lottery Fund.

www.historic-scotland.gov.uk – Historic Scotland.

www.ihbc.org.uk – Institute of Historic Building Conservation.

www.spab.org.uk – Society for the Protection of Ancient Buildings.

Part 4

THE RURAL ECONOMY

9

The Common Agricultural Policy of the European Union

P. Brassley and M. Lobley

Background, institutions and the legislative process

The need for a Common Agricultural Policy

When the European Community was created, it needed a Common Agricultural Policy (CAP) for two reasons: first, the agricultural industries in each of the member states were subject to government intervention, and, second, if this intervention were to continue, it had to be compatible with the other provisions of the Community.

Left to themselves, in the absence of government intervention, the markets for agricultural products in developed economies will produce:

- fluctuating prices, and, as a result, fluctuating incomes for producers; and
- gradually declining prices, in terms of purchasing power, and, as a result, declining real incomes for producers in the long run.

The price of agricultural products, like that of any other product in a free market, depends upon the balance of demand and supply. If demand increases less than supply, the price will fall.

The demand for agricultural products in a developed country does not usually increase rapidly, unless those products can be sold cheaply on the world market. Technical change produces increased output per hectare and per unit of labour, so supplies tend to increase more rapidly than demand. Thus, in the long run, prices tend to fall. Demand is also relatively unresponsive to price changes, so short-run supply variations, caused by changes in weather and disease, produce short-run price fluctuations.

Short-run price fluctuations in a free market may result in the demise of businesses that would be viable at average levels of input and output prices, and the possibility of this event creates a disincentive to investment on such farms. In the long run, falling prices produce lower incomes for those farms that are unable either to reduce costs or to increase output. If such low incomes are unacceptable, they may cease trading altogether. It is usually found that farm incomes have to fall to very low levels before farmers leave the agricultural industry, and in any case the land that they no longer farm is often taken over by another farmer. The total output of the industry is therefore maintained, so there is no tendency for prices to rise. While farmers with low incomes may be found in all areas, some regions may be particularly disadvantaged by physiographical or structural factors (i.e. infertile land and small farms), and if agriculture is the major industry the whole region may be affected by income and outmigration problems.

In a free-trading economy, farm income problems may be further intensified by the availability of imported agricultural produce at prices below the domestic supply price. Imports will not only restrict domestic price rises but also increase the total import bill, which may be considered a problem in countries with balance of payments deficit problems. However, the government of a country that has no difficulty in exporting enough to pay for food imports may still consider it unwise to be totally reliant on imported food supplies: unforeseen circumstances may give rise to difficulties in obtaining supplies, or something may reduce the price advantage of imports in the long run. A capacity for rapid expansion of domestic agricultural output may therefore be desirable even though considerable reliance is placed on imports. In short, free markets for agricultural products in developed free market economies may produce low farm income, and balance of payments and potential supply security problems.

Whether or not anything should be done about these problems is a political issue which different member states have approached in various ways. From the middle of the nineteenth century until the mid-1930s the UK essentially operated a free trade policy which resulted in a relatively small agricultural population and a high level of food imports. From the 1870s onwards, most continental European countries, including all of the larger ones, took the view that their interests were best served by maintaining a large rural population and a high level of food self sufficiency. It was not simply an economic issue: the small family farm was perceived to have a valuable social role, and latterly it was also seen as a guardian of the rural environment. Thus for many years the European model has stressed the desirability of addressing several rural policy issues through the medium of agricultural policy, a principle recently recognized under the term 'multifunctionality'. In the continental European countries, agricultural policies usually involved a system of import controls and price support, which kept price levels higher than they would have been in a free market and so maintained agricultural incomes and self-sufficiency levels. When the discussions that led to the formation of the Community were held in 1955–56, it was agreed that the exclusion of agriculture from the general common market was impossible: without free trade in farm products, national price levels could differ and those countries with the lowest food prices would have the lowest industrial costs, thus undermining the common policies that would be introduced for other industries. Although there was a general similarity between the existing agricultural policies, the differences in detail were so numerous that the simple continuation of existing policies would have been impossible. Not only did the Community countries require an agricultural policy, they required a *common* agricultural policy.

Origins and development of the European Union

The treaty establishing the European Community (EC), as it was then called, was signed in Rome on 25 March 1957, and the Community came into being at the begin-

ning of 1958. The main objective of the original six member states was to promote economic development by removing the barriers to trade between them. As they gradually succeeded in bringing this about they were joined by other countries:

Date of accession	Member states
1958	Belgium, France, West Germany, Italy, Luxembourg, The Netherlands
1973	Denmark, Ireland, the UK
1981	Greece
1986	Portugal, Spain
1995	Austria, Finland, Sweden
2004	Cyprus, the Czech Republic, Estonia, Hungary, Latvia, Lithuania, Malta, Poland, Slovakia, Slovenia

Bulgaria and Romania are also expected to join the European Union (EU), possibly in 2007. Their accession, together with that of the previous ten new entrants, will increase the agricultural area of the EU by about 50% and the number of agricultural workers by about 100%, but the size of the whole economy by less than 5%. This clearly has major implications for the future of the CAP. The Special Accession Programme for Agricultural and Rural Development (SAPARD) has been established to assist the accession process.

The process of decision making in the EU

The EU must make decisions about many areas of policy without, on the one hand, being bogged down in the process of consultation, or, on the other, neglecting the needs and views of 25 member states, over 300 million individual citizens, and numerous lobbies and interest groups. Decisions must be made about what policy shall be (primary legislation) and how it should work in detail (secondary legislation). Several institutions are involved in this process, of which the two most important are the Commission and the Council.

The *European Commission* is presided over by 20 Commissioners. They are appointed by agreement between the member governments, but are required to act in the interests of the EU and not individual member states. The Commission employs roughly 16 000 officials, of whom about one-quarter are translators, and they work in special services such as the legal department and the statistical office, or in Directorates General which are concerned with policy areas. Thus there is a Directorate General for Economic and Monetary Affairs, another for the EU Budget, another for the Environment, and so on, and each is the responsibility of one of the Commissioners. The Agriculture Directorate General, with a staff of about 1000, is among the larger ones. In some ways, therefore, the Commission is like a national civil service, but it also has extra functions: it is responsible for proposing primary legislation; it handles the day-to-day administration of EU laws and policies resulting from this primary legislation, and may enact secondary legislation in order to do so; and it represents the EU in its relations with non-member states and other international bodies such as the World Trade Organization (WTO).

The *Council of the European Union* is the major legislative body of the Community, and consists of a minister from each member state, depending on the subject under discussion: agriculture ministers for agriculture matters, finance ministers for financial matters, and so on. The chair is held by an individual country for a 6-month period and rotates from country to country in alphabetical order, according to the name of each country in its national language: Greece (E for Ellas) therefore follows Denmark. It is normally the final decision-making body for all primary legislation, although major constitutional issues or especially insoluble problems may be passed on to the *European Council*, a meeting of the heads of government of the member states which has no legal basis in any treaty and which developed from earlier summit meetings. Its decisions have to be passed back to the Council of the European Union to be given legal validity.

Since the Community is a partnership of the member states, it makes its decisions by negotiations in a series of committees. The annual review of agricultural prices is a good illustration of this process. The initiation of primary legislation is normally the responsibility of the Commission. The first moves are made by the appropriate department in the Agriculture directorate, usually in consultation with national civil servants, members of trade associations and independent experts, all of whom may serve on expert working groups chaired by a member of the Commission staff. Other Commission departments with a legitimate interest in the proposals are consulted at an early stage, since agricultural policy

measures might affect negotiations on, for example, external trade or monetary affairs. By the time the draft proposals are nearing completion the pressure groups make their views known to the Commission. There is a wide variety of these groups, from BEUC, the European Consumer organization, to the European Environmental Bureau (EEB), and various European trade associations such as the EDA (representing the dairy industry), CIAA (the food industry) and, perhaps most influential, the Committee of Professional Agricultural Organizations (COPA), which acts for farmers' unions in the Community. When the Agriculture directorate considers that the draft is complete, they submit it to a full meeting of the 20 Commissioners for their approval.

Once this approval is given, the draft becomes a Commission proposal and is submitted to the Council of the European Union and consultative bodes such as the European Parliament, the Economic and Social Committee and the relevant Management Committees (which are mostly concerned with secondary legislation – see below).

For most proposals the Council, before it makes a decision, is required to receive the opinion of the *European Parliament*. The Parliament carries out this part of its work by nominating a committee which produces a report on the proposal which may then be debated by a plenary session of Parliament before becoming the Opinion which is passed on to the Council. In practice this opinion carries little weight and may often be ignored. The Parliament also has considerable indirect impact on the formulation of policy through its questioning of members of the Commission, both formally and informally.

Meetings of the Council are often complicated and lengthy, and so require detailed preparation. This preparation is the task of the *Committee of Permanent Representatives* (COREPER) for non-agricultural matters, and of the *Special Committee for Agriculture* (SCA) for agricultural legislation. Both of these committees consist of senior civil servants from the member states, meeting with members of the Commission. Detailed consideration of the proposals is carried out in working parties of national civil servants. The purpose of this procedure is to identify and, if possible, to resolve points of conflict. If it is possible to produce a draft proposal that is acceptable to all member states, it is returned to the Council of Ministers on the 'A' list, which can be passed by the Council with no further discussion. Otherwise it returns to the Council as a 'B' point, for further negotiation.

While this process of sorting out the agenda for the Council meeting is going on, the business of lobbying continues. Pressure groups, national government ministers and other politicians express their views on the Commission proposals, both at a European and at a national level. The President of the Council of Ministers is required to hold meetings with both COPA and, since 1980, BEUC, and to report on them to the rest of the Council. Nevertheless, most of the political pressure on individual ministers in the Council comes through national lobbying channels, so that national political problems, such as a forthcoming national or even local election, can have significant effects on the decisions made.

Thus any proposal, before a decision is taken on it in the Council, will have been the subject of comment by a wide variety of formal and informal, corporate and individual, Community and national, expert and lay sources. The issues that remain to be resolved by the Council should be the basic political ones. After debating a proposal, a decision must be made. Some legislation requires the unanimous approval of the Council, and a quasi-formal agreement, known as the 'Luxembourg Compromise', provides for a member state to exercise a veto if it believes that its 'vital national interests' are at stake. However, increasingly, and especially since the signing of the Treaty of Nice in 2001, most decisions are taken on the basis of a *qualified majority*, the size of which varies with the number of member states. Each country has the number of votes shown in Table 9.1.

By this process the Council is said to 'adopt a common position'. Since the adoption of the Single European Act this is not the end of the decision-making process. The common position is then sent to the Parliament, which has 3 months to carry out a second reading and make its opinion known. It may:

(1) approve or take no decision, in which case the Council adopts the measure in question and it is effectively passed;
(2) reject the common position by an absolute majority, in which case the Council may maintain it and adopt it, but only if it can do so unanimously; or

Table 9.1   Votes in the Council of Ministers.

Country	Number of votes
France, Germany, Italy, the UK	29 each
Spain, Poland	27 each
Romania	14
The Netherlands	13
Belgium, Greece, Portugal, the Czech Republic, Hungary	12 each
Austria, Sweden, Bulgaria	10 each
Denmark, Ireland, Finland, Lithuania, Slovakia	7 each
Luxembourg, Cyprus, Estonia, Latvia, Slovenia	4 each
Malta	3

(3) amend the common position by an absolute majority, in which case the Commission revises its proposals within 1 month and re-submits them to the Council, which may then:
  • adopt the Commission proposal without change by a qualified majority;
  • adopt the Commission proposal after amendment, but only if it can do so unanimously; or
  • reject the proposal, in which case it lapses.

The proposals thus adopted fall into three categories. A *Regulation* is directly applicable to all people and governments in all member states. A *Directive* is binding as to its intention on governments, which must then pass legislation to give it effect, so that it is more flexible than a Regulation. A *Decision* is as binding and immediately applicable as a Regulation, but only on the people or governments to whom it is addressed.

Secondary legislation, which is concerned with the day-to-day running of the CAP (e.g. the administration of tenders to intervention stores, or setting the level of export refunds), is largely the responsibility of the Commission. In formulating its proposals, it may or may not take the advice of an advisory committee of representatives from all sides of the appropriate industry. The proposal is then considered by a *Management Committee* made up of national civil servants from the relevant division of the national ministries and Commission officials from the relevant Directorate General. There is a Management Committee for each of the main commodities, and others for agri-monetary affairs, plant health, structural policy and so on. The frequency of the meetings reflects the amount of work to be done: the beef committee may meet every fortnight, whereas the research committee may only meet a few times a year. The Committee gives its opinion on the Commission proposal using the same qualified majority system as the Council. After a proposal has passed through the Management Committee and been adopted by the Commission it becomes, in effect, law.

Community law is thus the primary and secondary legislation produced by the Council and the Commission, together with the law embodied in the Treaty of Rome and the various treaties of accession to the Community. It takes precedence over national law. It is the responsibility of the *Court of Justice* to decide whether or not Community legislation has been correctly applied, is being applied or is flawed. Some of the decisions of the Court have had a substantial effect on the way in which the Community is run, from deciding upon the powers of Parliament to defining what means may legitimately be used to prevent trade in food products.

The *Court of Auditors* exists to audit the expenditure of the Community and to ensure that its finances are properly managed. In this context it has published a number of reports critical of the operation of the CAP and has highlighted the problems of CAP fraud.

Formation and development of the CAP

The problems of the CAP

The CAP has proved a remarkably resilient policy. Its original objectives set out in the 1957 Treaty of Rome still remain (although see discussion of Agenda 2000 below) and the main mechanism of internal market support persisted largely unchanged until the 1990s. The main problems of the CAP also have a long history, notably the financially and politically costly surpluses and, later, widespread environmental problems. More recently, moves toward the liberalization of agricultural trade and the need to realign the CAP to more easily facilitate expansion of the Union can be added to the list of 'challenges' for the CAP in the twenty-first century. In order to understand these problems and the imperative for reform created by trade liberalization and EU expansion, it is necessary to understand the

original purpose of the CAP and the main policy instruments employed.

The original objectives of the CAP as laid down in Article 39 of the Treaty of Rome are:

(1) to increase agricultural productivity by promoting technical progress and by ensuring the rational development of agricultural production and the optimum utilization of the factors of production, in particular labour;
(2) thus to ensure a fair standard of living for the agricultural community, in particular by increasing the individual earnings of persons engaged in agriculture;
(3) to stabilize markets;
(4) to ensure the availability of supplies;
(5) to ensure that supplies reach customers at reasonable prices.

Article 40 of the Treaty lays down the broad guidelines for the various policy instruments by which these objectives are to be met. Detailed policy instruments developed later and it was not until 1968 that common prices were applied.

Outline of original price mechanisms

The underlying principle of the CAP price mechanism was that a producer should receive a price determined by market forces, but that these market forces were subject to control so that the market price was only allowed to fluctuate between predetermined upper and lower limits. Thus, the farmer was protected against excessively low prices, and the consumer against excessively high prices.

This price support system recognized that farm products might be produced both within and outside the EU. As the demand for farm products is relatively constant, if markets were supplied only from within the EU the major reason for low prices would be oversupply. The CAP's *intervention* mechanism was designed to overcome this problem by purchasing farm products for storage when prices were low. If prices remained low, this produce could be sold on the world market with the aid of *export refunds*. The export refund acts to lower the cost of a product in order to make it competitive with other products outside the single market. The existence and use of this mechanism has caused most criticism of the CAP beyond the borders of the EU and it came as no surprise that export refunds were a key target in the GATT negotiations of the early 1990s (Fennell, 1997).

Many farm products could also be supplied from countries outside the EU (known as third countries). If these imports were available at less than the EU market price, that market price would be reduced. This effect could be mitigated by artificially increasing the price of products from third countries by applying variable levies. Conversely, when EU prices were high, perhaps as a result of shortages, the entry of produce from third countries would serve to reduce prices by increasing supply.

For most of its history market control and internal price support have been relied upon to achieve the diverse objectives of the CAP although it is generally agreed that they have proved incapable of simultaneously satisfying all objectives (Fennell, 1997). In addition, the intervention mechanism has effectively insulated producers from the market, encouraging the production of products for which there is no genuine demand. Open-ended price support stimulated the long-term expansion of supply. Self-sufficiency in wheat, for example, rose from 89% in the late 1950s to 101% in the mid-1960s (Fearne, 1991), while by the end of the 1970s milk production had increased by 1.7% annually and beef by 2% (Fennell, 1997). The result was persistent surpluses across a growing range of commodities.

Although the precise meaning of 'surplus' can be debated, the European Commission is largely concerned with 'structural surpluses', i.e. long-term surpluses which arise from oversupply compared to demand rather than from accidental annual fluctuations due to weather conditions, etc. Even though surpluses had been in existence since the very inception of the CAP, a general unwillingness to take radical measures to deal with them (other than the ill-fated 'Mansholt Plan' – see below) resulted in major problems by the 1980s.

The problems and costs of surpluses stimulated through the operation of the CAP have received considerable public attention. In the UK surplus storage and handling costs increased nearly four-fold between 1980 and 1994 and, when surpluses peaked in 1986, 7 billion ECUs of the CAP budget was spent on handling and storage costs (Winter *et al.*, 1998). Some of the surpluses that could not be stored or converted into other products (such as industrial alcohol) were simply destroyed. Much of the surplus, however, was disposed of on world markets through the use of the EU export subsidy programme to deflate prices. Rather than dealing with the surplus problem internally, the heavy subsidization of exports essentially shifts the problems to 'third countries' (i.e. outside the EU), with impacts on food security within importing countries and on the markets of traditional food exporters. At the same time, the impact is also felt in the domestic budget as spiraling protectionism results in large parts of agricultural support throughout the world being mutually neutralizing. For example, it was estimated that in the early 1990s, 25% of world spending on agricultural support merely offset the effects of other countries' policies, while in the US 40% of agricultural support in 1986/87 offset the loss in profits deriving from low prices as a result of surplus dumping (Potter, 1998).

In addition to budgetary and surplus problems, during the 1980s, an environmental critique of the CAP developed, pointing to a direct link between the operation of the CAP, which encouraged intensification and specialization, and widespread environmental change in the countryside. It was argued that the operation of the CAP created conditions of confidence (through guaranteed prices and investment aid) in which farmers were encouraged to specialize, increase output and intensify production through new technology embodied in capital which was increasingly substituted for labour. Farmers were operating within 'a system which systematically establishes financial inducements to erode the countryside, offers no rewards to prevent market failure and increases the penalties imposed . . . on farmers who may want to farm in a way which enhances and enriches the rural environment' (Cheshire, 1985). High guaranteed prices were necessary if small and marginal family farms were to be maintained, but, since the level of support received by an individual farmer was largely determined by output, there was an incentive to bring uncultivated land into production (while at the same time larger farmers benefited more). High prices combined with reduced capital costs 'led to the logical choice of a specialised enterprise. . . . This specialisation, in turn, meant that hedges served little purpose, and the larger machinery that capital substitution and larger throughputs had induced, demanded larger fields' (Bowers and Cheshire, 1983).

The catalogue of post-war environmental change is now well rehearsed. Between 1949 and 1984 80% of chalk and limestone grassland, 40% of heaths, 50% of lowland wetlands and 30% of upland moors were either lost or significantly damaged (Nature Conservancy Council, 1984), while by the mid-1980s, the area of species-rich, unimproved lowland grass was estimated to have declined by 97% over a 50-year period. Similar changes have been recorded throughout Europe.

Reforming the CAP

Early attempts to reform the CAP were typically piece-meal, resulting from budgetary crises. During the 1980s a number of attempts were made to deal with the escalating surplus problem and associated budgetary issues. A ceiling on agricultural spending was introduced in 1984, whereby the increase in the farm budget was limited to 74% of the rate of the economic growth of the then EC. At the same time, a stabilizer system was introduced establishing a production limit beyond which support prices would fall. The crucial threshold, however, was set too high to have a significant impact on surpluses. Another important development in 1984 was the introduction of milk quotas. Milk quotas effectively introduced a super levy at the farm level on all milk delivered to dairies over the quota level. Although originally met with shock and dismay from the farming community, quotas soon became popular as production within quota continued to benefit from high prices and because quota was tradable and soon became a valuable capital asset. Towards the end of the decade, in 1988, a new and experimental approach was adopted to control production in the arable sector. A voluntary 5-year set-aside programme was established which provided compensatory payments to farmers for not cultivating arable land. In practice, uptake rates were low and the impact on total production consequently limited. Importantly though, the voluntary set-aside experiment demonstrated that such an approach could have an impact on production levels and, significantly, if managed appropriately, set-aside land offered important environmental benefits (Potter, 1998; Winter *et al.*, 1998).

Despite these efforts, surpluses were at a peak in 1986–87, the budget was rising and EU agricultural support polices were coming under the spotlight of the GATT negotiations. At the same time it was clear that not only did much of the farm budget fail to reach farmers, but also that which did failed to satisfy concerns regarding the support of smaller and more marginal farms. In 1991 the European Commission published proposals for the reform of the CAP including the now famous 80:20 formulation: that is, that 80% of farm spending was going to only 20% of farmers and that these tended to be the larger, more efficient producers. Despite budgetary pressures and moves towards greater liberalization, the Commission also clearly stated a major unwritten objective of the CAP that:

> 'sufficient numbers of farmers must be kept on the land. There is no other way to preserve the natural environment, traditional landscapes and a model of agriculture based on the family farm as favoured by society generally.' (CEC, 1991)

The original reform proposals were put forward in a document entitled *The Development and Future of the CAP* (CEC, 1991) – also known as the MacSharry proposals after Ray MacSharry, the agriculture commissioner at the time. The reforms proposed by MacSharry explicitly stemmed from the failure of earlier measures to tackle the problem of surplus production, the encouragement of intensification and environmental damage through the coupling of farm support to output and, importantly, the perceived need to bring about a more equitable distribution of funds.

The distributional issue was particularly important and led to the promotion of 'modulation' (positive discrimination in favour of smaller farmers who would receive a greater degree of compensation). The reforms also had wider implications: in addition to redressing the distribution of agricultural support, the proposals reaffirmed the role of farmers as joint producers of food and managers of the environment. This role was to be enhanced explicitly through a series of agri-environmental schemes and was implicitly strengthened by a continued commitment to keeping farmers on the land – particularly the smaller family farmers who had been disadvantaged by the operation of earlier price support policies.

Almost inevitably the original proposals were watered down in the final agreement, notably in terms of the removal of the redistributive modulation proposal. Nevertheless, for the first time sharp price reductions were achieved combined with the partial decoupling of support from output. Briefly, the final 1992 MacSharry reform package introduced substantial cuts in support prices for certain commodities (for example, a 35% cut in the intervention price for grain and 15% for beef) combined with direct producer payments to partially compensate for the cut in institutional prices (through arable aid payments and headage payments in the livestock sector). In addition, a new annual set-aside scheme was initiated, participation in which was a condition of receipt of arable aid payments. Finally, a number of Accompanying Measures were introduced designed to facilitate early retirement from agriculture, the afforestation of agricultural land and, importantly, the improved environmental management of farmed land.

Despite the significant change in policy encompassed in the 1992 reforms it soon became apparent that this was simply a stage (albeit significant) in a much larger and longer reform process. The 1994 Uruguay Round Agreement (URA) of the GATT negotiations committed signatories to further reform in the future, particularly related to subsidies linked to production. Although the 1992 reforms had gone some way to making the CAP GATT-compatible, in 1995 the European Commission itself acknowledged that the CAP was still overly protectionist, bureaucratic and that support was insufficiently decoupled from production to meet the likely future demands of the WTO (the successor to GATT). In addition, if countries from central and eastern Europe were to be accommodated within the EU, further reform of the CAP was necessary. Following this analysis contained in a 1995 Agricultural Strategy paper (CEC, 1995) and the 1996 European Conference on Rural Development (popularly referred to as the Cork Conference), the Commission began to outline a 'third way' which emphasized the 'European model of agriculture' and its attendant support needs.

In his opening address to the Cork conference, agriculture commissioner Franz Fischler argued that 'rural society is a socio-economic model in its own right which must be preserved in the interests of European society as a whole' (Fischler, 1996). The Declaration that emerged from the conference signalled broad policy principles, recognizing that the justification of the 1992 reform payments was increasingly being thrown into question. In their place:

'the concept of public financial support for rural development, harmonised with the appropriate management of natural resources and the maintenance and enhancement of biodiversity and cultural land-scapes, is increasingly gaining acceptance.' (CEC, 1996)

The principles set out for the EU's future rural development policies in the conference declaration reflected a continuing concern with the management and development of rural economy and society, albeit encompassing a shift away from an agriculturally centred notion of rural land management based on a sectoral policy towards a more territorially based rural policy. Thus, it was argued that a growing share of resources should be devoted to rural development and environmental objectives, and that an integrated policy framework be established encompassing 'agricultural adjustment and development, economic diversification, the enhancement of environmental functions and the promotion of culture, tourism and recreation' (CEC, 1996). Despite the clear intention of retaining rural-based EU spending, the declaration provoked criticism from those who saw a Cork-style policy as a threat to traditional agricultural policy entitlements (Lowe *et al.*, 1996).

Nevertheless, when, in 1997, the European Commission published its Agenda 2000 reform proposals, rural development was introduced as a 'second pillar of the CAP' while further price cuts were designed to pre-empt the next round of trade negotiations. In Agenda 2000 the Commission set out new objectives for the reformed CAP and in doing so began to outline the essential characteristics of 'the European model of agriculture' and its associated support policy:

- a *competitive* agricultural sector capable of exploiting the opportunities existing on world markets without excessive subsidy;
- *safe* production methods capable of supplying *quality products* that meet consumer demand;
- a *diverse agriculture*, reflecting the rich tradition of European food production;
- maintenance of *vibrant rural communities*;
- an agricultural sector *sustainable* in environmental terms which contributes to the preservation of natural resources and the natural heritage and maintains the visual amenity of the countryside;
- a *simpler, more comprehensible* policy which establishes clear dividing lines between the decisions that have to be taken jointly at Community level and those that should remain in the hands of the member states;

- an agricultural policy that establishes a *clear connection between public support and the range of services* which society as a whole receives from the farming community.

Agenda 2000 was based on a different model to the EU's major competitors and according to the European Commission a key difference could be found in the concept of

'the multifunctional nature of Europe's agriculture and the part it plays in the economy and the environment, in society and in preserving the landscape, whence the need to maintain farming throughout Europe and to safeguard farmers' incomes.' (CEC, 1998)

Agenda 2000 displayed a considerable degree of continuity with the approach adopted under the Mac-Sharry reforms. Price cuts were designed to improve competitiveness alongside increased direct aid payments designed to contribute towards a fair standard of living for farmers. The market reforms were also intended to contribute towards environmental enhancement whereby direct compensation payments could be made conditional on the observance of environmental requirements (cross-compliance). Non-compliance would result in a proportionate reduction or cancellation of payments which could then be redirected to finance agri-environmental or rural development measures. However, few member states took up this option.

A range of additional measures were also introduced to exert a positive influence over environmental management. National discretion over a proportion of direct payments to the beef sector (the so-called national envelope) could be used to support extensive grazing, while the ability to modulate direct payments has been employed to boost agri-environmental spending. In the UK, modulation means that cuts in direct payments (of initially 2.5%, rising to 4.5%) yield funds that are then diverted into agri-environmental and rural development spending. In addition, under the Rural Development Regulation (RDR) all member states are required to implement agri-environment measures (see below). Despite these developments, spending on agri-environment and rural development remains a minor element of the CAP budget.

However, following agreement on 26 June 2003, the architecture of the CAP has been fundamentally reformed. The key concept at the centre of the agreement is that the link between farm support payment and production should be broken. This is known as decoupling. The 'default' policy model to be implemented from 2005 is that most direct payments will be delivered in the form of a new Single Farm Payment (SFP) which is decoupled from production (although the agreement provides a number of alternative options to only partially decouple – see below). In addition, the budget for the RDR will be enhanced through compulsory modulation, and member states have the option of introducing a new 'national envelope' by diverting 10% of the value of SFPs to support environmentally friendly farming systems and to address some of the potentially negative social and environmental impacts of decoupling.

The CAP in practice

Price mechanisms

The underlying principle established when the CAP came into being in the 1960s was that producers should receive a price determined by market forces, but that those market forces should be controlled so that prices fluctuated only between predetermined upper and lower limits. Thus the farmer was protected against excessively low and the consumer against excessively high prices. This principle remains in force, but now operates in association with the concept that producers should receive part of their income as a Single Farm Payment, independent of output.

The most basic of all CAP support schemes (or regimes as they are often called) is that for cereals. The regimes for other commodities may be seen as variations of the cereals regime, the legal basis of which is set out in Regulation 1251/99.

An intervention price for cereals, proposed by the Commission and decided upon by the Council of Ministers, is established for a period of years. For example, the Berlin agreement of 1999 decided on the cereal intervention price until 2006. In principle, therefore, once the market price falls below the intervention price it will pay producers or traders to sell to intervention stores. In practice the system is more complex, because minimum quantities and quality standards must be observed, intervention stores are only open from

November to May, and there are small monthly increments in the prices paid. The prices are fixed in euros, so they can also change from day to day in countries with non-Euro currencies.

The second major component of the cereals regime is import control. Since world prices are often lower then EU prices it would be worthwhile, in the absence of any control, to import simply for sale into intervention. In practice, imports are controlled by the issue of import licences, and to obtain such a licence import duty must be paid. The level of duty is determined by the difference between the duty-paid import price and the world price. Again, complexities arise at this point. The duty-paid import price is fixed by international negotiation: at the time of writing it was the intervention price multiplied by 1.55, but future WTO negotiations could result in its variation. 'World price' is an imprecise term, and in practice the EU Commission has identified several reference cereals which are used to specify prices for import duty calculations. Variations in insurance and freight costs and currency exchange rates must also be taken into account (for details see the *CAP Monitor*, vol. 10A, pp. 13–14).

Licences are also required for cereal exports, and the Commission uses export licences, and the refunds to which they entitle traders, to stabilize EU markets. Since EU prices are normally higher than world prices, EU traders will only find it worthwhile to export to non-EU countries with the aid of these export refunds. The Commission also has to take into account the commitment, under the Uruguay Round Agreement, to reduce both the volume of subsidized exports and the budgetary expenditure on export subsidies. In the unusual event of world prices being higher than EU prices the Commission may impose export levies.

There are also support regimes involving variations of these principles for other arable crops, such as oilseeds (oilseed rape, sunflowers and soya beans), protein crops (peas, broad beans, horse beans and lupins), linseed and sugar beet. Fruit and vegetables prices are largely supported by aid to producers' organizations. The basic support principles applying to arable crops are used for livestock and livestock products too: the farmer receives a return from the market, in which prices are controlled to some extent, and also receives compensatory payments, for some products at least. The most complex of these regimes applies to dairy products, prices of which are supported by inter-

vention arrangements for butter and skimmed milk powder. Individual producers are affected by milk quotas until (at least?) 2014/15, and from the 2004 marketing year there will also be direct payments to producers in the form of dairy aid payments, and, in some cases, additional payments. Beef and sheep meat producers are mainly supported by headage payments, although there are provisions for intervention buying (in the case of beef) or payments for the storage of sheep meat in private cold stores. Pig meat prices are occasionally supported by aids to private storage, but in practice pig meat, like poultry meat and eggs, is an unsupported product.

From 2005 these compensatory payments, such as arable area payments or livestock headage payments, will be subsumed within a *Single Farm Payment (SFP)*. This will be based upon the payments received by a farm in the years 2000–2002 (the 'reference years'). The principle of the system is that it will remove the direct relationship between the output of a farm and the level of support it receives, a process known as 'decoupling'. There are several additional features of the system, not all of which have been worked out or agreed in detail at the time of writing:

- The SFP scheme was brought into effect on 1 January 2005, although member states may opt to delay implementation of the scheme for 2 years, or to adopt various partial decoupling schemes which would maintain some area or headage payments.
- There is some discussion over whether SFPs should be calculated on some basis other than receipts in the reference years, the most simple of these being a flat rate per hectare.
- Assuming that SFPs are historically based, they will only be paid on eligible hectares, which do not include land used for growing potatoes, fruit and vegetables, and may be transferred, with or without land, between farmers within the same member state.
- SFPs will only be paid if the recipient observes various 'cross-compliance' regulations concerned with environmental, food safety, animal and plant health and animal welfare standards.
- Farmers receiving more than €5000 in SFPs will be subject to 'modulation', which means a reduction (at a rate of 5% from 2007) in SFP payments, the

money so released being used for rural development (pillar 2) schemes.

- A new 'national envelope' created through a 10% reduction in the total value of the SFP. The national envelope, which is optional for member states, can be used to assist environmentally important farming systems and to address some of the negative social, economic and environmental implications of decoupling.
- A 'financial discipline' mechanism, also applicable only to those receiving more than €5000 in SFPs, that will reduce SFPs if overspending on them is forecast.

The eventual effects of the whole system will probably depend upon the decisions made on these detailed points. However, DEFRA estimates that the impact of the reforms will include a 12% increase in total income from farming (TIFF), partly as a result of 'dynamic adjustment' as labour is reduced and farm businesses restructure.

Monetary arrangements

Institutional prices in the CAP, such as intervention prices, together with headage payments, arable area payments and so on, are denominated in euros. With the introduction of the Economic and Monetary Union (EMU), 11 member states adopted the Euro as their currency on 1 January 1999, and Greece joined the Euro zone at the beginning of 2001. For these countries, therefore, there is no need for any special monetary measures for CAP purposes. For the non-participating countries – the UK, Denmark and Sweden – a system is needed to convert monetary amounts denominated in Euros into national currencies.

For non-Euro currencies, the value of prices, direct aids, export refunds and so on changes according to (1) the exchange rate published each day by the European Central Bank (ECB), and (2) the date of the relevant 'operative event'. The nature of the operative event depends upon the payment concerned. For intervention payments, it is defined as the beginning of the delivery or the acceptance of the tender; for export refunds or import tariffs it is the acceptance of the customs declaration; for SFPs it is the beginning of the year. The exchange rate used for any of these payments is the most recently published ECB rate prior to the operative event, i.e. for an event taking place on Friday, Thursday's rate will be used, but events falling on Saturday, Sunday or Monday will use the previous Friday's exchange rate.

Exchange rate fluctuations will therefore affect intervention prices and, perhaps most importantly for most farmers, arable area payments and livestock headage payments. If a national currency strengthens against the Euro, such payments would decrease: suppose that €1 is worth 65 p; if sterling then increases in value so that €1 = 60 p, clearly the sterling value of a headage payment would decrease. Equally, weakening sterling would increase the value of such payments. There are provisions for compensating producers for currency revaluations in some circumstances, but the EU only contributes to half of the amount, the rest being paid by the national treasury.

Environmental and structural policy under the CAP

When the architects of the CAP were drawing up its objectives and policy instruments there was no explicit mention of the environment, although, if pressed, most would have argued that maintaining family farming would itself maintain the 'physical fabric' of the countryside. The earliest explicit mention can be found in the 1975 Less Favoured Areas (LFAs) Directive, designed to ensure the continuation of farming in order to maintain a minimum population level or conserve the countryside. Despite these objectives, the main emphasis on LFA policy has been to support farm incomes rather than the farmed environment (indeed the operation of LFA policy became a major target of the environmental critique of the CAP).

It was not until 1985 that a UK proposal designed to allow support to be given to environmentally sensitive farming practices in certain areas was incorporated into European legislation (Article 19 of Regulation 797/85 on Agricultural Structures), although cofinancing was not forthcoming until 1987. Specifically, Article 19 provided support 'in order to contribute towards the introduction or continued use of agricultural production practices compatible with the requirements of conserving the natural habitat and ensuring an adequate income for farmers' (CEC, 1985). In the UK following the 1986 Agriculture Act a series of Environmentally Sensitive Areas (ESAs) were designated in areas of national environmental importance where the continuation or introduction of certain farming practices was necessary for landscape and environmental reasons.

With the exception of the Netherlands and Denmark (and later France) other member states were slow to take up the provisions of Article 19 which was seen to reflect a narrow British preoccupation with landscape (Baldock and Lowe, 1996), and by 1990 few member states had made provisions under Article 19 although some 60 000 farmers were receiving EU premiums. This limited uptake contributed towards pressure for further agri-environmental reforms and in 1988 the European Commission produced two important discussion papers, *The Future of Rural Society* (CEC, 1988a) and *Environment and Agriculture* (CEC, 1988b). The latter argued that 'society has to accept the fact that the farmer, as manager of the environment, is rendering a public service which merits an adequate remuneration' (CEC, 1988b). The mould was set for more wide-ranging, multi-faceted agri-environmental polices to assist in the maintenance of agricultural incomes, improve market balance (i.e. contribute to the reduction of surpluses) and contribute to the maintenance and conservation of the farmed countryside. Environmental concern had become an important justification for agricultural support.

The 1992 agri-environment regulation

In addition to introducing significant changes to the system of market support, the 1992 CAP reforms introduced an ambitious programme of agri-environmental reforms. The main environmental reforms were contained in one of three Accompanying Measures. Regulation 2078 on '*agricultural production methods compatible with the requirements of the protection of the environment and maintenance of the countryside*' was designed to complement the extensification expected to follow from the market reforms and 'encourage farmers to serve society as a whole' by providing incentives to introduce or maintain farming practices that contributed to the maintenance of natural resources and the landscape. Significantly, aid was also to be provided to encourage farmers to reduce agricultural pollution, although this was only supposed to be available for schemes that went beyond good agricultural practice (Scheele, 1996). Regulation 2078 was compulsory on all member states and was designed to

overcome some of the shortcomings of previous legalization which was seen as limited in its scope and had received only limited application, being confined mostly to the northern member states where the idea had developed. Overall, the emphasis of the Regulation was on stimulating and maintaining extensification with additional support for organic farming, long-term (20-year) set-aside, public access and the upkeep of abandoned farmland and woodland.

The Regulation can be seen as an exercise in subsidiarity and has consequently been implemented in strikingly different ways in different member states (see Potter, 1998; Buller *et al.*, 2000). The acceptability of this approach to farmers can be seen in the rapid expansion in the area enrolled. By 1999 an estimated 20% of the agricultural area of the EU was enrolled in a scheme operated under Regulation 2078.

In England the then MAFF originally operated a range of schemes under the Regulation involving an expansion of existing initiatives (such as ESAs and the later Countryside Stewardship Scheme, CSS) alongside the introduction of new schemes such as the Habitat scheme, the Moorland scheme and the Organic Aid scheme. Different combinations of schemes operated across the UK, with some restricted to particular counties such as Tir Cymen (and later Tir Gofal) in Wales and the Countryside Premium scheme in Scotland. Despite the expansion in the number of schemes, overall spending on the agri-environment measures remained low in comparison to mainstream CAP spending and in the UK the bulk of spending was devoted to just two schemes – ESAs and CSS. Nevertheless, ESAs have contributed to the maintenance of landscapes in designated areas, while the CSS has been more successful in bringing about positive environmental change and stimulating employment.

The rural development regulation

Under Agenda 2000 the European Commission declared its aim to establish rural development as the 'second pillar of the CAP'. This 'second pillar' is mainly linked to the Rural Development Regulation (RDR) agreed as part of the reform package and also the reorganization of structural policy to concentrate on three new objectives (see below).

The Regulation aimed to establish a coherent and sustainable framework for the future of Europe's rural areas and to complement the reforms introduced into the market sectors, by promoting a competitive, multifunctional agricultural sector in the context of an integrated strategy for rural development. The RDR is highly discretionary, strengthening subsidiarity and promoting flexibility through a 'menu' of actions to be targeted and implemented according to member states' specific needs. The RDR has three main objectives:

(1) to create a stronger agricultural and forestry sector, the latter recognized for the first time as an integral part of the rural development policy;
(2) to improve the competitiveness of rural areas;
(3) to maintain the environment and preserve Europe's rural heritage.

Agri-environmental measures are the only compulsory element of the RDR. In principle this represents a decisive step towards the recognition of the role of agriculture in preserving and improving Europe's natural heritage. That said, the RDR is largely based on a repackaging of previous measures and only represents a small proportion of overall CAP spending.

In Britain, the RDR is implemented through a series of Rural Development Plans (one each for England, Scotland and Wales). Each plan contains an overview of the measures to be adopted plus an analysis of needs and policy responses at a more local level. Under the England Rural Development Plan (ERDP) the regional element is based upon the English Government Office regions. The ERDP is based on a significant expansion (and consolidation) of existing agri-environmental schemes, a reorientation of LFA support and the introduction of new schemes supporting training, diversification and rural development. The agri-environmental schemes expanded or continued under the ERDP are the:

• Countryside Stewardship scheme
• Organic Farming scheme
• Woodland Grant scheme
• Farm Woodland Premium scheme
• Environmentally Sensitive Areas scheme

Together these schemes will absorb over 70% of spending under the ERDP between 2000 and 2007 (see Table 9.2).

The system of support for farmers in LFAs, long a target of environmentalists, has been significantly realigned under the ERDP. A new Hill Farm Allowance (HFA) scheme has been introduced designed to:

Table 9.2 England Rural Development Plan indicative expenditure. Source: MAFF data.

Scheme	Seven-year total (£ million)	Percentage of total budget	Percentage increase in annual spending by 2006/2007 compared with 1999/2000
Agri-environment[1]	1052	64.2	126
Of which:			
Environmentally Sensitive Areas scheme	335	20.5	19
Countryside Stewardship scheme	576	35.2	260
Organic Farming scheme	141	8.6	99
Woodland Grant scheme[2]	22	1.3	–
Farm Woodland Premium scheme	77	4.7	87
Hill Farm Allowances[3]	239	14.6	–
Processing and Marketing grant	44	2.7	New
Energy Crops scheme	31	1.9	New
Rural Enterprise scheme	152	9.3	New
Vocational Training scheme	22	1.3	New
Total	**1638**	**100.0**	**105**

[1] Expenditure shown is on a financial year basis (2000/2001–2006/2007) and therefore differs from that in the ERDP which is on a FEOGA year basis (16 October 1999–15 October 2006).
[2] Excludes baseline expenditure of £117 million.
[3] Excludes expenditure in 2000 on HLCAs.

- contribute to the maintenance of the social fabric in upland communities through support for the continued agricultural use of land;
- help to preserve the farmed upland environment by ensuring that land in LFAs is managed sustainably.

Under the HFA, support for hill farmers is explicitly justified as a social and environmental measure. Unlike the old HLCA scheme which offered livestock headage payments, aid under the HFA is paid on an area basis, differentiated according to the type of land and 'modulated' according to the area claimed for. Claims for the first 350 ha attract the full payment rate, 50% of payments are made on land between 350 and 700 ha and no payments are made above 700 ha. Top-up payments of 10 or 20% are available to farmers meeting additional environmental criteria.

In addition to incentives for environmentally friendly land management, new schemes have been initiated under the ERDP aimed at enabling the farming and forestry sector (and to some extent other rural businesses) to adapt to the changing rural economy. The *Rural Enterprise Scheme* (RES) is the most substantial in terms of its budget share and is designed to support the development of sustainable and diversified rural economies and communities. The scheme is regionally managed, and although a wide range of different categories of project can be funded, each application is assessed on a competitive basis against regional priorities. A sliding scale of grant aid is offered depending on the commercial return expected. Projects with little or no commercial return but that offer community or environmental benefits attract the highest rate of grant aid. The development of processing facilities is assisted through the *Processing and Marketing Grant* (PMG) aimed at projects with eligible costs in excess of £70000. As part of the rural development programme, MAFF and the Forestry Commission jointly run an *Energy Crops Scheme* (ECS) providing establishment grants for short-rotation coppice and miscanthus (elephant grass). Investment in human capital is provided through a *Vocational Training Scheme* (VTS) which is again managed on a regional basis and offers grants of up to 75% for training in areas such as information technology, processing and marketing and conservation land management. (Further details on all these schemes are available from the DEFRA website.)

Following the 2003 CAP reform agreement there will be some changes to the RDR and potentially changes to the ERDP. As well as increasing the EU budget for rural development and agri-environmental schemes, a new regulation (amending the earlier Rural Development Regulation) broadens the scope of rural development to encompass improvements in animal welfare. Under Article 24 of the new Rural Development Regulation support can be used to promote:

- ways of using agricultural land that are compatible with the protection and improvement of the environment, landscape and its features, natural resources, the soil and genetic diversity;
- environmentally favourable extensification and management of low-intensity pasture systems;
- conservation of threatened high-natural-value farmed environments;
- upkeep of the landscape and historical features on farmland;
- the use of environmental planning in farming practice;
- the improvement of farm animal welfare.

Additional new measures also include the option of providing support to producers engaged in quality food production, time-limited aid for farmers adapting to new legislative requirements and aid for the management of integrated rural development strategies by local partnerships. Although the latter, in particular, has been welcomed by many commentators as a means of addressing some of the shortcomings of the RDR, DEFRA has indicated its reluctance to implement new measures given that additional funding will not be forthcoming in the existing programming period.

Structural policy

Although market support has historically dominated CAP spending (and direct support will continue to do so), a parallel range of instruments known collectively as structural policy have been designed to speed up agricultural restructuring, not just in terms of the number of farms but also in terms of investment, marketing, etc. An early, and infamous, attempt to develop a radical approach to structural policy was presented in a *Memorandum on the Reform of Agriculture in the EEC* (CEC, 1969), more popularly referred to as the Mansholt Plan. In its essentials, the Plan envisaged a leaner European agriculture with a smaller number of larger production units created through the removal of 5 million of the Community's small and marginal farmers. The process of restructuring would be based on a series of 'carrots' and 'sticks' – the former in the shape of grants, pensions for farmers aged over 55 and assistance for younger farmers in finding alternative employment, while the latter took the form of a proposal for price reductions sufficient to force small marginal producers from the land (CEC, 1969). It was envisaged that around half of those leaving would retire and the other half take up employment in industry, the land released being used to create larger 'modern agricultural enterprises' and to reduce the agricultural area of the Community by some 15 million ha through afforestation and the creation of reserves and recreational parks.

There was sympathy with the desire to improve the structure of agriculture, although the proposal to 'force' small and marginal farmers from the land was greeted with hostility. In the event, Mansholt's specific proposals were not acted on and a series of structural measures implemented in 1972 represented a watered-down version of the 'Mansholtian' vision. Rather than a 'new agriculture', a policy was now prosecuted that put the emphasis on the modernization of the existing community of farmers, reinforcing the political and symbolic importance of maintaining a large number of farmers on the land. Under the emergent farm structural policy, Directive 72/159 provided assistance for investment on farms that could be shown to be capable of reaching viability in a short period of time, while Directive 72/160 introduced a system of payments to those wishing to leave the industry, the idea being that land released could be used to achieve restructuring and viability in conjunction with Directive 72/159. There were few applicants for the outgoers scheme and little land released went to farmers receiving aid under the modernization plan (due to differences in the spatial pattern of uptake of the two schemes).

Early structural policies were criticized on the grounds that they concentrated too much on agricultural investment in northern member states. Subsequently, therefore, attempts were made to divert more of the funds available to Mediterranean countries, and to recognize that development in rural areas might involve other industries as well as agriculture. In the 1980s, for example, there was an 'integrated programme' for the

Western Isles of Scotland; EU structures funds supported an integrated rural development programme in England's Peak District; and there were also integrated Mediterranean programmes. Following the 1988 reform of structural funds, the programming approach was developed further by combining spending from the European Regional Development Fund and the European Social Fund with FEOGA (European Agricultural Guidance and Guarantee Fund) guidance section funding targeted at five key objectives, only some of which were relevant to rural areas.

New arrangements for structural funds

Although the 'second pillar' of the CAP is mostly pursued through the RDR, rural development in a broader sense is facilitated through structural policy which has been reorganized following Agenda 2000 to concentrate on just three 'Objectives':

- *Objective I* is essentially the same as previously (areas with less than 75% of the average EU GDP) but now also includes special measures targeted at the EU's sparsely populated arctic regions.
- *Objective II* combines former Objective 5b areas with former Objective 2 areas (which were declining industrial areas) which meet new Objective II criteria. Support is provided for economic diversification, training, environmental protection and improvement, etc.
- *Objective III* is a 'horizontal' measure, that is, it applies to all non-Objective I and II areas and supports training and education.

Further reforms to the structural funds are expected to be introduced after the current programming period ends in 2006.

Problems and possibilities of the CAP

It would be a mistake to see the agreement of 26 June 2003 as the end of the process of CAP reform; rather the constant flux of external influences and internal circumstances exerts continuing pressures to change.

The internal circumstances may be divided into three main groups: those affecting the economic environment; those affecting the political environment; and the evolution of existing policy. Changes in the economic environment most obviously include short-term market developments producing shortages or, more usually, surpluses. As there appears to be little evidence for the cessation of technical change, the underlying trend to produce more persists, even in the face of the tendency for farm size to increase as the older generation of farmers is only partially replaced by the younger. The people, and sometimes their houses, leave the industry, but the land remains, incorporated into neighbouring farms. At the same time, the rest of the food chain continues to evolve. A food processing and retailing industry increasingly dominated by large firms may be expected to exercise its market power whenever possible; equally society as a whole may require the whole food chain to respond to food safety concerns. All these factors therefore affect the economic environment within which the agricultural industry works. Similarly, changing public attitudes to the environment and rural society impact upon the political environment, which is also likely to be affected by differing views of monetary union and EU enlargement in the first decade of the century. These factors would be likely to modify even a CAP that was perfectly in tune with policy requirements, and few would claim that it was so, since there are still unreformed policy sectors (most notably sugar), and the policy as a whole remains largely agricultural when there is an increasingly shared perception that the problems it addresses should be seen as more broadly rural. It has also been argued that the costs of the new SFPs will be easier to identify, so that the financial burden of support for farmers will, in political terms, become increasingly sensitive.

The external influences on the CAP are partly economic and partly a result of international politics. The economic issues arise from world market fluctuations: high world prices decrease the costs of export refunds and reduce pressures for third-country access to EU markets. Low prices, which have historically been more common, have the opposite effect. More predictable is the impact of the Doha Round negotiations of the WTO. Whether or not the CAP changes agreed in June 2003 prove sufficient to produce agreement in these talks remains to be seen. In the long run they may well be seen as simply one more stage in a more or less continuous process of reacting to long- and short-term, internal and external, pressures for change in the CAP.

References

Baldock, D. and Lowe, P. (1996) The development of European agri-environmental policy. In: *The European Environment and CAP Reform: Policies and Prospects for Conservation* (ed. M. Whitby), pp. 8–25. CAB International, Wallingford.

Bowers, J. and Cheshire, P. (1983) *Agriculture, the Countryside and Land Use*. Methuen, London.

Buller, H. Wilson, G. and Höll, A. (2000) *Agri-Environmental Policy in the European Union*. Ashgate, Aldershot.

CEC (1969) *Memorandum on the Reform of Agriculture in the EEC*. Supplement to Bulletin no. 1. Office for Official Publications of the European Communities, Luxembourg.

CEC (1985) *Council Regulation (No. 797/85) on Improving the Efficiency of Agricultural Structures*. Official Journal of the European Communities, Office for Official Publications of the European Communities, Luxembourg.

CEC (1988a) *The Future of Rural Society*. COM (88) 601. Office for Official Publications of the European Communities, Luxembourg.

CEC (1988b) *Environment and Agriculture*. Office for Official Publications of the European Communities, Luxembourg.

CEC (1991) *The Development and Future of the CAP*. COM (91) 100. Office for Official Publications of the European Communities, Luxembourg.

CEC (1995) *Study on Alternative Strategies for the Development of Relations in the Field of Agriculture Between the EU and the Associated Countries with a View to the Future Accession of These Countries*. COM (95) 607. Office for Official Publications of the European Communities, Luxembourg.

CEC (1996) *The Cork Declaration: A Living Countryside*. Office for Official Publications of the European Communities, Luxembourg.

CEC (1998) *Explanatory Memorandum on the Future for European Agriculture*. Office for Official Publications of the European Communities, Luxembourg.

Cheshire, P. (1985) The environmental implications of European agricultural support policies. In: *Can the CAP Fit the Environment?* (eds D. Baldock and D. Conder), pp. 9–18. Institute for European Environmental Policy/Council for the Protection of Rural England/World Wide Fund for Nature, London.

Fearne, A. (1991) The history and development of the CAP, 1945–85. In: *The CAP and the World Economy: Essays in Honour of John Ashton* (eds C. Ritson and R.D. Harvey). CAB, Wallingford.

Fennell, R. (1997) *The Common Agricultural Policy: Continuity and Change*. Clarendon Press, Oxford.

Fischler, F. (1996) Europe and its rural areas in the year 2000: integrated rural development as a challenge for policy making. Opening speech to European Conference on Rural Development, 7 Nov, Cork.

Lowe, P., Rutherford, A. and Baldock, D. (1996) Implications of the Cork Declaration. *Ecos*, **17** (3/4), 42–5.

Nature Conservancy Council (1984) *A Nature Conservation Strategy for Great Britain.* NCC, Peterborough.

Potter, C. (1998) *Against the Grain: The Environmental Reform of Agricultural Policy in the United States and European Union.* CAB International, Wallingford.

Scheele, M. (1996) The agri-environmental measure in the context of CAP reform. In: (ed. M. Whitby) *The European Environment and CAP Reform: Policies and Prospects for Conservation*, pp. 3–7. CAB International, Wallingford.

Winter, M., Gaskell, P., Gasson, R. and Short, C. (1998) *The Effects of the 1992 Reform of the Common Agricultural Policy on the Countryside of Great Britain.* Rural Research Monograph Series no. 4. Cheltenham and Gloucester College of Higher Education, Cheltenham.

Further reading

Agra Europe Weekly and *CAP Monitor* (continuously updated reference guide) are both published by Agra Europe (London) Ltd.

Brassley, P. (1997) *Agricultural Economics and the CAP: An Introduction.* Blackwell Science, Oxford.

CEC (2000) *The CAP: 1999 Review.* Office for Official Publications of the European Communities, Luxembourg.

Ingersent, K.A. and Rayner, A.J. (1999) *Agricultural Policy in Western Europe.* Edward Elgar, Northampton, Massachusetts.

Lowe, P., *et al.* (1998) *CAP and the Environment in the UK.* Centre for Rural Economy, Newcastle.

Ministry of Agriculture, Fisheries and Food (1999) *Restructuring the Agricultural Industry.* Working Paper 1. Economics and Statistics Group, MAFF, London.

Ritson, C. and Harvey, D.R. (1998) *The Common Agricultural Policy*, 2nd edition. CABI, Wallingford.

Useful websites

http://europa.eu.int/comm/agriculture/index_en.htm – DG Agriculture and Rural Development at the European Commission.

www.defra.gov.uk – Department of Environment, Food and Rural Affairs (DEFRA).

10

Livestock production

R.A. Cooper

Livestock in Europe

Livestock production is an important component of EU agriculture. The relative importance of individual species, the size of enterprises and the systems of production adopted vary between both countries and regions, being influenced by climate, topography and custom. Some idea of the size of the main livestock industries is given by the data in Tables 10.1–10.5, which detail production, consumption and trade. It is obvious from these data that the existence of an export market is crucial to the very survival of some sectors of the industry in some countries. Ireland and The Netherlands are obvious examples.

Overall, the EU is more than self-sufficient in beef and veal (103%) and in pig meat (105%). Although systems of beef husbandry in EU countries are broadly similar to those in the UK there are differences in the type of beef and veal produced. In particular, there are differences in the contribution of the sexes. Bull beef is predominant in Italy (60%) and Germany (50%), and is important in France (20%). In France over 50% of beef comes from cows and heifers whilst in the UK almost 50% comes from steers. France (15%), Italy (13%) and The Netherlands are the main producers of veal. Pig meat production is similar across the EU, being characterized by large unit size and increasing intensification. Increasing environmental demands, and concerns over animal welfare, have imposed ever more stringent controls, especially in terms of planning regulations and environmental protection legislation. As a result, much development is now taking place in Eastern Europe (Poland and Hungary in particular) where welfare requirements are less stringent.

On the other hand, the EU produces only 83% of its requirement for sheep and goat meat. Most of this is sheep meat, with recorded goat meat production only 81 500 t (out of a total of 11.4 million tonnes). Of this, Greece produces 47 500 t, Spain 19 000 t, France 7500 t and Italy 3700 t. In some ways these figures underestimate the size of this sector of the industry since, in most cases, kids are slaughtered at young ages and low liveweights in order to increase the availability of goat milk. Goat milk production in the EU totals 1.444 million tonnes, of which France produces 0.46, Greece 0.45, Spain 0.32 and Italy 0.14 million. These same countries also account for most of the EU output of sheep milk with figures of 250 000, 670 000, 306 000 and 850 000 t respectively. These figures represent only 2.8% of France's total milk output, the figures for Italy, Spain and Greece being 5.2, 7.7 and 59% respectively.

Livestock production in the UK

Despite the impact of bovine spongiform encephalopathy (BSE) and of the 2001 foot-and-mouth disease outbreak, livestock production is still the major contributor to UK agricultural output (see Table 10.6). Milk and milk products account for over 20% of output and beef accounts for 15%. However, it must be recognized that almost half of the current beef output (46%) is made up of support payments (subsidies). Whilst the figures for sheep are lower, subsidies still represent 30% of sheep output.

Milk production

There are about 17 000 dairy farms in the UK. Average herd size is 86 cows in England, 75 in Wales and 102 in Scotland, with a total dairy cow population of 2.27 million. Some 95% of dairy cows are Holstein/Friesian. An average lactation yield is 6300 litres. As yields

Table 10.1 Beef in the EU (including veal) (tonnes). (Data taken from FAOSTAT (2002) with permission of the Food and Agriculture Organization of the United Nations.)

Country	Percentage self-sufficiency	Produced	Imported	Exported	Consumed
Austria	127	203 489	20 721	64 091	160 119
Belgium/Luxembourg	134	274 300	51 907	122 950	203 257
Denmark	116	153 900	95 328	116 972	132 257
Finland	97	90 204	6 513	4 529	92 118
France	99	151 400	328 420	314 936	1 530 958
Germany	119	1 303 000	226 661	433 564	1 096 096
Greece	27	65 665	203 181	6 407	241 010
Ireland	1022	576 900	13 312	548 610	56 409
Italy	81	1 155 588	446 197	171 403	1 430 382
The Netherlands	152	470 600	156 564	355 663	310 348
Portugal	59	99 980	71 010	234	170 755
Spain	109	631 784	86 514	137 167	581 132
Sweden	86	149 800	31 335	6 218	174 917
UK	69	708 000	328 664	12 798	1 023 866
EU total	103	7 397 210	2 066 358	2 295 541	7 203 721

Table 10.2 Sheep and goat meat in the EU. Year 2000–2001 (tonnes). (Data taken from FAOSTAT (2002) with permission of the Food and Agriculture Organization of the United Nations.)

Country	Percentage self-sufficiency	Production	Imported	Exported	Consumed
Austria	81	8 077	1 973	45	9 969
Belgium/Luxembourg	20	4 490	29 079	11 051	22 518
Denmark	23	1 453	5 304	451	6 306
Finland	40	750	1 126	15	1 861
France	47	140 000	170 432	10 522	299 910
Germany	56	45 306	40 545	4 282	81 569
Greece	87	125 000	18 625	463	143 162
Ireland	260	82 900	1 621	52 630	31 891
Italy	75	69 051	24 976	2 546	91 481
The Netherlands	134	18 600	7 544	12 285	13 859
Portugal	72	26 852	10 421	90	37 185
Spain	102	251 132	11 079	15 217	246 994
Sweden	53	3 910	3 562	110	7 362
UK	95	359 000	108 880	88 822	379 058
EU total	83[1]	1 136 522	435 131	198 529	1 373 125

[1] To meet the shortfall, the EU imports, mostly from New Zealand. In 2001 import licences covered 275 000 t, of which 216 000 were for NZ.

Table 10.3 Pig meat in the EU. Year 2000–2001 (tonnes). (Data taken from FAOSTAT (2002) with permission of the Food and Agriculture Organization of the United Nations.)

Country	Percentage self-sufficiency	Production	Imported	Exported	Consumed
Austria	102	562 604	109 863	122 610	549 857
Belgium/Luxembourg	254	103 100	119 943	758 470	414 573
Denmark	431	1 624 500	59 699	1 307 103	376 608
Finland	99	172 790	20 936	19 301	174 425
France	103	2 312 000	501 667	562 352	2 251 315
Germany	90	3 981 000	938 500	501 660	4 417 849
Greece	38	143 100	253 979	5 653	373 425
Ireland	154	226 400	41 859	120 796	147 463
Italy	64	1 478 450	975 368	153 388	2 300 430
The Netherlands	198	1 622 800	115 982	911 311	827 471
Portugal	76	329 095	122 359	18 611	432 843
Spain	112	2 912 390	116 629	135 022	2 606 219
Sweden	87	277 000	60 728	19 816	317 913
UK	63	923 000	784 988	240 127	1 467 861
EU total	105	17 618 229	4 204 509	5 226 220	16 758 253

Table 10.4 Poultry meat in the EU. Year 2000–2001 (tonnes). (Data taken from FAOSTAT (2002) with permission of the Food and Agriculture Organization of the United Nations.)

Country	Broiler	Turkey
Austria	87 052	23 794
Belgium/Luxembourg	380 000	6 866
Denmark	185 000	11 000
France	1 100 000	735 000
Germany	480 000	340 000
Greece	151 800	2 100
Ireland	78 500	34 000
Italy	816 000	340 000
The Netherlands	700 000	42 000
Portugal	220 000	46 000
Spain	1 012 000	22 000
Sweden	95 000	2 200
UK	1 257 500	256 000
EU total	6 562 852	1 860 960

Table 10.5 Egg and milk production in the EU. Year 2001 (tonnes × 10^3). (Data taken from FAOSTAT (2002) with permission of the Food and Agriculture Organization of the United Nations.)

Country	Eggs	Milk
Austria	97	3 340
Belgium/Luxembourg	197	3 700
Denmark	77	4 660
Finland	60	2 500
France	1047	24 890
Germany	890	28 300
Greece	120	770
Ireland	37	5 416
Italy	707	11 900
The Netherlands	658	10 500
Portugal	1008	1 860
Spain	560	6 294
Sweden	98	3 300
UK	629	14 717
EU total	5284	122 147

increase, the size of the national herd decreases (by 1% in 2001). At the same time the number of producers is going down while herd size is increasing. (In 1975, 3.25 million cows produced 13.5 million litres, averaging 4100 litres per cow). With an output of over 14 000 million litres (see Table 10.5), the UK is the third largest milk producer in the EU (and seventh in the world). Milk production is concentrated in the wetter,

grass-producing regions in the west of the UK. Main areas include Dumfries and Galloway in Scotland, Carmarthenshire in Wales, and Cumbria, Cheshire and Devon in England.

Small amounts of milk (less than 3%) are processed on farm. The majority is sold to processors, either directly or through a purchaser group. There are more than 100 processors in the UK, but the 20 biggest

Table 10.6 UK agricultural output (2001). (Reproduced from DEFRA website www.defra.gov.uk/statistics, with permission.)

Value of agricultural output (main commodities)	£ million	£ million
Cereals	2019	
Potatoes	600	
Other crops	1017	
Vegetables	970	
Flowers	958	
Total value of all crops		5564
Milk	2818	
Beef	1000	
Beef 'subsidies'	870	
Pig meat	740	
Lamb/mutton	442	
Sheep 'subsidies'	187	
Eggs	406	
Poultry meat	1251	
Total value of all livestock		7714
Total value of major commodities		13278
Total value of all agricultural output		15126

Table 10.7 Target reproductive parameters for a dairy herd.

Calving index (days)	365
Interval of calving to first service (days)	60
Interval of calving to conception (days)	80
Services per conception	1.6
Calves per 100 cows per year	96

account for over 85% of the output. As in the rest of the EU, milk production is limited by quota. Current UK production quota is 14.2 billion litres, but at the level of the individual farm figures may be moderated by changes in butterfat content of the milk, being decreased as fat levels rise. Approximately 50% of milk produced is used for liquid consumption [with semi-skimmed (55%) and skimmed milk (15.5%) now the great majority]. The remainder is processed, mainly into cheese (51%), milk powder, (24%) and butter and cream (8%).

In order for a mammal to lactate she must first give birth. In terms of the cow (dairy or beef) the target should be to have every animal produce a calf every 365 days. In practice this target (called the calving index) is seldom achieved, with the average in dairy herds being closer to 390 days and many herds doing significantly less well even than this. The main targets are shown in Table 10.7. Infertility, the main factor influencing calving performance, is one of the main causes of culling (the disposal of underperforming animals). The other reasons are lameness and mastitis (udder infection). Incidence of all these has increased as milk yields have gone up, reflecting increased stress on the cow, so that the average herd life of a dairy cow is now less than four lactations. Beef cows, on the other

hand, being under less stress, may well go on for 10–12 years.

Traditionally, most dairy cows in the UK were calved down in the autumn, optimizing milk yields and simplifying management. This was relatively easy to achieve with small herds and lower yields. Nowadays, despite best efforts, many herds have calving periods that stretch over most months of the year. Automated feeding systems and larger herds, which allow for sub-grouping of cows, reduce the adverse effects of these extended calving periods. Some milk-buying groups may require specific production levels at different times of the year and will adjust their payment structures to reflect this, in an attempt to encourage farmers to better manage their calving programmes, thus producing the milk when it is required.

Dairy cow feeding

Cows calving in 'autumn' (September to December) will usually go straight on to winter feeding. This will normally be based on silage (grass, maize or whole-crop cereal) supplemented with a compound feed (a mixture of cereals, cereal by-products and sources of vegetable protein). Silages are often evaluated in terms of their ability to support some milk production in addition to meeting the maintenance (M) needs of the animal. For example, a sample rated as M + 10 should, if fed alone, provide for production of up to 10 litres of milk. Cows yielding above 10 litres would then receive additional nutrients from other feeds. These might include root crops such as kale, by-products such as brewers' grains or sugar beet pulp, or compound feeds. High-yielding cows will struggle to eat enough for their needs and may 'live off their backs' (i.e. utilize their body fat reserves) for the first 60–90 days of lactation. This is often a cause of infertility, cows only returning to oestrus once weight loss has ceased. These effects may be mitigated by ensuring that cows have adequate body fat reserves when they calve down (steaming up) and by stimulating appetite by providing a variety of palatable, high-quality feeds. Regular feeds of small amounts of compound, rather than two large amounts, fed at milking time, will also help. Response to compound feeding varies, but on average 0.4 kg will be fed for every litre of milk produced. Total compound feed usage will normally be between 1000 and 1500 kg per cow per year, with average silage requirements between 6 and 10 t per cow. Spring-calving cows may well be able to go straight out to grass after calving, but those calving in late winter will need a period of indoor feeding first. Lower-yielding cows may need little in the way of additional feeding, once grass growth is adequate, but the higher-yielding animal will require some extra high-quality, non-rumen-degradable protein. On average, spring-calving cows will yield 750–1000 litres less than their autumn-calving equivalents. Additionally, the calves of spring calvers are often more difficult to fit into a beef production system.

Lactation

After calving, the milk production of a cow follows a regular pattern (Figure 10.1). Yield will rise to a peak somewhere between 6 and 10 weeks post-calving, a figure that will vary with breed, age and nutrition. Beyond peak, yields will decline at between 2 and 2.5% per week and, if mating targets have been achieved, the animal will be dried off after approximately 300 days, allowing a 60-day dry period before next calving. An estimate of potential total lactation yield may be obtained by multiplying daily yield at peak by 200. Thus, an 8000-litre cow would be expected to have a peak yield of 40 litres/day and an 'average' cow one of about 30 litres. As individual yields increase, so drying off becomes more difficult since cows may well still be yielding 20 litres or more per day after 300 days. In such circumstances a conscious decision may be taken to go for a longer calving interval in these animals. Sometimes, however, this decision is taken by the cow herself, who may delay returning to oestrus, post-calving, if suffering the nutritional stress associated with high yields.

Most cows are milked twice daily, with an ideal (though seldom achieved) interval of 12 hours between each milking. The adverse effect of intervals of over 12 hours increases with increasing yield. Milking cows three times each day can increase yields by up to 15%. This is partly due to reduced pressure within the udder but also to the improved feeding that three-times milking normally implies.

Milk compositional quality varies between breeds (Table 10.8) and within lactation (Figure 10.1). In crude terms, yield and quality are inversely related. Thus both fat and protein levels drop as peak yield approaches, and they recover towards the end of lactation. Composition may also be affected by nutrition. In particular, fat levels depend upon the provision of adequate fibre in the diet. As a result they may be depressed when high levels of compound feed are being used to support high yields, or when cows are turned out to lush, low-fibre grass in the spring. On the other hand, milk protein levels are affected mainly by the levels of fermentable

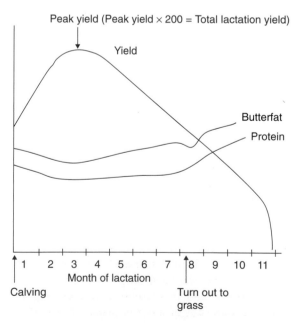

Figure 10.1   Changes in milk yield and composition over a lactation, for a September calving cow (not to scale).

Table 10.8   Mean milk composition by breed.

Breed	Mean yield (litres)	Fat (g/kg)	Protein (g/kg)
Ayrshire	5940	40.8	33.4
Guernsey	4898	48.1	35.9
Jersey	4611	54.8	38.7
Friesian/Holstein	6846	40.4	32.8

carbohydrate (sugars and starch) in a diet, especially at medium to low yields. Ruminants have the ability to convert dietary nitrogen into useful protein, provided that they have a supply of energy that rumen micro-organisms can use. A shortage of energy-rich material limits this ability and may cause dietary nitrogen to be lost in urine. That said, high yielders will be unable to meet all their protein needs from this source and will require additional, non-rumen-degradable protein (DUP) in their diets. In such an animal, insufficiency of DUP is likely to lead to reductions in yield, rather than reductions in milk protein *per se*. Low milk proteins are often encountered in late winter (February to March) and tend to reflect long-term deficiencies in winter feeding (especially the under-feeding of energy feeds). Payments for milk reflect changing consumer demand. Since milk fat is no longer favoured, payments are designed to encourage production of milk with higher protein content. In addition, in response to consumer demands, much research work is now directed towards the production of 'designer' products such as butter that will spread straight from the refrigerator.

Grassland utilization

Like all ruminants, dairy cows are major utilizers of grassland. On well-managed, intensive grassland it may be possible to run 2.5 cows/ha (0.4 ha/cow). This allows for 0.2 ha to be utilized as grazing and 0.2 ha to be conserved as hay or silage for winter feeding. At these levels, stocking densities in early grazing may be as high as 6 cows/ha, allowing for maximum areas to be cut for conservation. Average stocking rates are lower than this, especially in areas where summer rainfall is limited, with many herds achieving no better than 2.0 cows/ha. As the grazing season progresses the proportion of land allocated to grazing will increase at the expense of conservation, reflecting the reduced rate of grass growth as the year progresses. Silage cuts may be taken every 6–8 weeks, but after a second or sometimes third cut, all the grass will be dedicated to grazing (see Table 10.9).

Breeding policy

Given the high cull rate in most dairy herds, self-contained units will need to mate at least 50% of their cows to a bull of the same breed if they are to have

Table 10.9 Allocation of grassland to cutting or grazing. (Assumes an overall stocking rate of 0.5 ha/cow (2 cows/ha) and expects three cuts of silage to be taken.)

Period	Stocking density (cows/ha)	Proportion of grassland for	
		Grazing	Cutting
April–May	5	40%	60%
June–mid July	4	55%	45%
Late July–Aug	3	70%	30%
Sept	2	100%	

enough heifer calves from which to select replacements. Problematically, this also produces 25% pure-bred bull calves that are often of dubious value as beef animals, especially if the breeding is Holstein type rather than the beefier Friesian. The remaining 50% of cows may be bred to a beef bull. Traditionally this would have been a Hereford, but more recently Charolais, Limousin and Belgian Blue (breeds with better beef characteristics) have been prevalent. This has been made easier by the fact that a majority of dairy cows are now bred by artificial insemination (AI). Perversely, this might be one reason why fertility levels are falling, since accurate identification of oestrous animals by the stock person, a prerequisite for successful AI, is not always easy.

Replacement heifers will be reared in ways similar to those described later for beef, though perhaps at lower rates of growth. Conventionally it is expected that heifers will reach breeding weight (330 kg for a Friesian/Holstein) by 15 months of age, so that they may calve for the first time at 24 months. In practice these targets are often modified so that the heifer calves at an appropriate time (usually August or September, ahead of the main herd) rather than at a specific age. As a result, mean age at first calving is probably closer to 30 than 24 months. Bulls to be used on heifers should be chosen for their ability to produce small, easily born calves. The Aberdeen Angus is often the breed of choice. The animals being reared to come into the herd are known as 'followers'. They will often be used to utilize grassland that is too far away from the milking parlour for the dairy cows to use. Target stocking rate for one follower 'unit' (i.e. one yearling plus one in-calf heifer) should be as high as that attained by the dairy herd itself (though it seldom is!).

Beef production

Prior to the BSE crisis, which began in March 1996, the UK was more or less self-sufficient in beef, producing about 1 million tonnes/year. At that time almost all cull cows from beef or dairy herds entered the food chain, accounting for 23% of all beef produced. At the same time a majority of dairy calves were reared for beef, despite the poor conformation of many carcases.

In 1996, as a result of BSE, the Over Thirty Month Scheme (OTMS) was introduced. This banned the entry into the food chain of any animal more than 30 months old and thus effectively removed all cull cows from the food chain. At the same time it also removed many older, larger beef steers that were being reared on extensive, grass-based finishing systems. These systems have since largely disappeared. Animals over 30 months of age were slaughtered, and their carcases rendered down. By June 2002 some 5.8 million animals had been disposed of in this way. In addition, in March 1996 the Calf Purchase Aid Scheme (CPAS) was developed. This was designed to remove excess calves from the market, so putting a floor under prices. Large numbers of calves (1.975 million) were slaughtered under this scheme before it was discontinued in July 1999. A majority of these calves (91%) were dairy-bred animals, with poor beef potential, so it is possible to argue that this improved the average conformation score of animals slaughtered in the years thereafter.

Beef production in 2002 was 692 000 t. This represents about 72% of UK consumption (20% of all meat eaten). In the absence of any cow beef, 53% of UK production derived from the beef herd and 47% from the dairy herd. In addition to 2.27 million dairy cows, the UK cattle herd contains 1.67 million beef cows, kept on some 26 400 farms. These are either pure-bred beef animals or crosses whose sole output is a calf that will normally go on to slaughter for meat, having been reared on a traditional beef system. These cows are often referred to as suckler cows, since they remain with and suckle their calves for up to 10 months (depending upon calving date). This contrasts with the dairy cow that remains with her calf for 2–3 days at most.

Outlines of the main calf-rearing systems

Bucket-fed calves

These are usually dairy-bred animals. Having remained with the cow for up to 3 days, to maximize intake of colostrum, the calf is removed and taught to drink from a bucket. Feed will consist of a milk 'substitute', based on skimmed milk powder and reconstituted with warm water at about 100 g/litre. These 'substitutes' are high-energy, high-fat (16–18%) and high-protein (24–26%) feeds. Target intake will be 400 g powder in 4 litres of water, split into two feeds each day. The feed should be offered at blood heat (about 40°C). Water, hay or straw and appropriate concentrate feed (calf pellets of 18% protein) should be on offer from 7 days, the target being to wean the calf off milk once it is consuming at least 0.75 kg of solid feed each day (usually at about 6 weeks of age). From 6 weeks onwards a cheaper, 15% protein concentrate may be fed, to a maximum of 2 kg/day, along with hay, silage or straw. Target liveweight at 6 months is 150–170 kg.

Alternatives to this system may involve offering milk substitute *ad libitum*. This may be achieved by using a machine that mixes on demand, or by mixing large volumes of an acidified substitute (often fed cold). The low pH of such material (5.5–5.8) inhibits bacterial growth and lengthens storage life. These systems reduce labour costs and may lead to fewer health problems but require higher feed inputs. They also demand a high level of management. Autumn-born calves approaching 6 months of age in spring may be turned out to grass for the summer, continuing to receive some concentrates for a few weeks after turnout. Spring-born animals will not be big enough to be able to make good use of grass and will usually remain indoors, often going on to one of the intensive systems of finishing.

Suckled calves

Suckling is the traditional method of feeding calves born to the beef cow. The calf will remain with the cow for 5–10 months (the shorter period for spring-born calves). Nutrition of spring-born calves will be based upon milk plus, as calves grow, some grazing. Autumn-born calves, usually housed, may receive additional compound feed fed through a creep, a device that restricts access by the cows. Spring-calving cows may receive no compound feed at any stage, but if winter forage is not high quality then autumn-calving cows may need to be supplemented with small amounts (1–2 kg/day) of an appropriate compound feed. Suckled calves are usually weaned in autumn and 'stored' or put into a finishing system (depending on their weight). Many suckler herds are found on hill and upland farms that do not have facilities to keep the weaned calves on over the winter. Large numbers of animals therefore change hands at 'store' markets in October and November. Autumn-calving cows will generally need to be housed and will require higher-quality feed than 'dry' spring calvers, which may remain outdoors over winter and be fed on forage-only diets. As a result, many hill or upland farms, lacking both adequate housing and the ability to produce good-quality conserved forage, calve their cows down in spring or early summer, with the majority of autumn calvers being found on lowland units. Stocking rates for suckler cows vary considerably, but are generally lower than for dairy herds. The best spring-calving lowland units may stock at 1.8 cows/ha, compared to 1.5 cows/ha for autumn calvers, but on hill units the stocking rate may be no better than 1.0 cows/ha. Since the only output from a suckler cow is her calf, calving percentage is important. A target of 95 calves reared per 100 cows should be set, but the average on lowland farms is only 91% and on hill units this falls to 89%. Additionally, a tight calving period (<10 weeks) makes subsequent management easier. Few suckler herds achieve this target, many effectively having a calving spread that runs from autumn to spring (or *vice versa*). Under the Common Agricultural Policy (CAP), support is offered to suckler-cow units. The Suckler Cow Premium is payable at £128.42 (€200) per head, according to a quota allocation, with a further £25.68 (€40) per head 'Extensification Payment', where animals are stocked at less than 1.8 livestock units (LU)/ha and £51.37 (€80) if stocking is less than 1.4 LU/ha.

Veal production

Veal production is common in France, Italy and The Netherlands, but very little is produced in the UK. Traditional systems use male calves (either castrate or entire) fed a high-fat milk substitute *ad libitum*. Consumption may be up to 15 litres/day in three feeds. Animals are individually penned, in slatted-floor pens, with no bedding and with no solid feed on offer. These animals effectively remain monogastric, the rumen failing to develop. The target is a 140-kg animal at 16 weeks of age, with a daily gain of up to 1.5% of lightweight at a feed-converting efficiency of 1.3 : 1 and a killing-out percentage of 62% or better.

As a response to welfare concerns, UK legislation now permits only the production of 'humane' veal. Animals must be group housed, in daylight, and have bedding and access to hay, silage or straw. These animals grow more slowly and produce a 'pink' veal that is less attractive to the continental consumer.

Beef finishing systems

Beyond the calf stage all animals, whether of dairy or beef origin, will go on to one of three finishing systems, designed to produce a slaughter-ready animal (see Figures 10.2 and 10.3): intensive or cereal-based finishing; silage/concentrate feeding (18-month beef); or grass finishing. Choice of appropriate system will depend upon the breed and age of the animal and on the resources available. Some animals, especially spring-born suckler calves, may first be put through a 'store' period. Storing involves keeping animals at below their full capacity for growth (usually 0.4–0.7 kg/day). This allows them to develop in terms of skeletal and muscle growth while adding little fat tissue. When such animals are put back onto full feed they respond with a period of 'compensatory growth', during which they perform better than would be predicted. Animals are often stored over winter, using conserved forage and little or no compound feed, so that they may make optimum use of grass (a cheaper feed) the following spring.

Intensive finishing (cereal beef)

Intensively finished animals will generally go from birth (or soon after) to slaughter on a consistent diet and at high growth rates. Most will be housed throughout. Feed (cereal-based compound or high-quality grass or maize silage) is fed *ad libitum*. Compound feeds will also be fed to silage-fed beef. Feed consumption in a cereal-beef system will be 1500–1750 kg/head. Silage-fed animals will consume 5 t silage and up to 1000 kg of compound. Such high-performance diets suit only animals capable of gaining weight quickly without

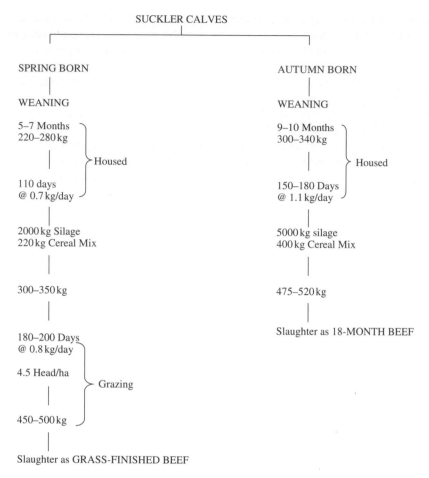

Figure 10.2 Outline of systems for finishing suckled calves.

going to fat. Late-maturing breeds are ideal and entire bulls are better than steers. Slaughter age will be between 11 months (cereal systems) and 15 months (silage beef), with target gains of 1–1.5 kg/day, leading to final liveweights of 420–480 kg. Such systems demand good housing but have the advantage that they do not require animals to go out to graze. Thus, they may fit well onto arable farms, where the grassland rotates around the farm and where fencing may be absent. They also allow for higher stocking rates, 1 ha of grass producing enough silage to finish seven to eight animals. For obvious reasons, suckler calves do not easily fit into these systems. However, spring-born animals, especially of later-maturing types, may be housed at about 5 months of age and finished on an intensive, silage-based diet. (Silage-fed beef is some-

times known as Rosemaund beef, in recognition of its development at the Rosemaund Experimental Husbandry farm in Herefordshire.)

Silage/concentrate feeding (also known as 18-month beef)

Animals going onto this system will be coming into housing, off grass, at 10–15 months of age and weighing 300–350 kg. Average feeding period will be 150–180 days, during which time they will be fed varying amounts of silage and compound feeds (depending on sex, breed type and silage quality). Target daily gains will be between 0.9 and 1.2 kg to give a final liveweight of 450–500 kg. The degree of control over diet that is possible means that this system can be

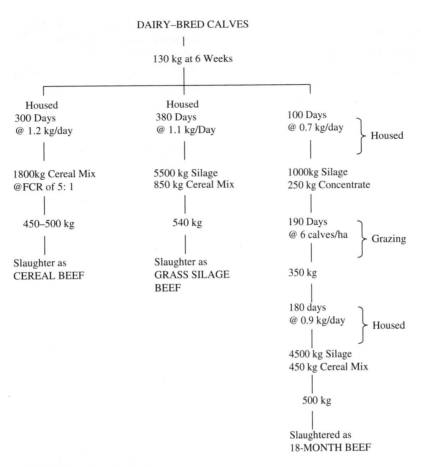

Figure 10.3 Systems for finishing dairy-bred calves.

tailored to the needs of any breed type or sex of animal, provided that they are housed in groups of similar animals. It is thus used both for dairy-bred animals and for suckler calves. A typical animal will consume 4.5 t of silage and 450 kg of concentrate feed. Overall stocking rate (summer grazing + winter finishing) will be 3.5–4.0 cattle/ha.

Grass finishing

Grass finishing tends to produce a more variable product, since there is less control over food quality and, particularly, over intake. Additionally, there is a conflict between the need to achieve good individual animal performance and to maintain high stocking rates. Daily gains will be slower at between 0.7 and 0.9 kg/day, but the animals are older at slaughter so that they may achieve final liveweights of 500–550 kg or

even higher. Most animals entering this system will have been 'stored' over the previous winter and will have consumed 2.0 tonnes of silage and 350 kg of concentrates. During the grass-finishing phase they will be stocked at up to four animals per hectare. The system particularly suits animals of early-maturing types and is most commonly used for finishing suckler cattle. Some autumn-born animals going through the system may well be 24 months old at slaughter.

For male beef animals, the EU offers production support in the form of the Beef Special Premium (BSP). For steers the BSP may be claimed at any age between 7 and 20 months, and for a second time for any animal that is older than 20 months at slaughter. The current rate of payment is £90.13 (€150) per head (up to a maximum of 90 head per holding per year). For bulls, only a single payment is possible, at 7 months or above, but the value is increased to £126.18 (€210).

Carcase composition

The proportion of liveweight that is represented by the carcase is known as the killing-out or dressing percentage (KO%). This can vary from 48–64%, with breed, sex and weight all having an influence. Fatter animals tend to have a higher KO%. However, one of the major components 'lost' during butchering is the rumen, gut and gut contents. The weight of this can vary substantially according to whether the animal is full or empty. Holding an animal indoors overnight before slaughter reduces liveweight but increases KO%. It also makes for easier (and potentially cleaner) butchery.

Carcases may be described using a system that is universally applied across the EU. Carcases are evaluated visually, according to the level of fat cover and amount and distribution of lean tissue (see Figure 10.4). Fatness is assessed over five classes from 1 (very lean) to 5 (very fat), with classes 4 and 5 being subdivided into L (leaner) and H (fatter) bands. The overall shape of a carcase (its conformation) is described by a five-class scoring (E, U, R, O, P), with U and O subdivided into higher (+) and lower (−) bands. An E classification describes a carcase of outstanding shape whilst class P is reserved for thinly muscled, inferior carcases, usually from extreme dairy-type animals. Classification is always expressed by giving the conformation score first. Thus a U + 4L carcase is significantly better than average (R) for conformation and average for fat. The percentage of classified carcases falling into each category in 2001 is shown in Figure 10.4. Target output would normally be −U or above for conformation and 4H or below for fat. Conformation is an inherited characteristic, largely unaffected by husbandry and management, but fat class is almost entirely a reflection of level (or perhaps appropriateness) of feeding. A carcase description may also include its weight, sex and, optionally, age.

Sheep production

In 1980 the EU introduced a new support system, the Sheepmeat Regime, which offered relatively generous support to UK sheep farmers. As a result, the UK breeding flock rose from 14.9 million ewes to a peak of 20.3 million in 1998. It now stands at 16.4 million (2002), with 11.5 million lambs/hoggets slaughtered annually. Before the 2001 outbreak of foot-and-mouth disease (FMD) the UK was the biggest producer of lamb in the EU (approximately 360 000 t in 2000). In 2001 production fell to 265 000 t (8% of total meat production). Export of live lambs and carcases has always been an important feature of the UK sheep picture, serving to hold up market prices. Exports are more than balanced by imports of about 110 000 t, mostly from New Zealand, though in 2001 these were insufficient to compensate for losses due to FMD and UK consumption fell by 60 000 t to 235 000 t.

One of the main problems currently faced by the industry is the fact that the type of carcase favoured for export is the same as that required by the UK supermarkets – the major retailers. A majority of carcases have no better than average conformation and many tend to be too fat. The structure of the industry

	1	2	3	4L	4H	5L	5H	Overall
E			0.1					0.1
U+		0.2	0.6	0.5	0.1			1.4
−U	0.1	0.7	2.8	4.9	1.5	0.1		10.1
R	0.2	1.9	8.9	23.0	8.8	0.7	0.1	43.6
O+	0.2	2.1	8.6	15.9	4.7	0.5	0.1	32.1
−O	0.2	1.6	3.9	3.9	0.7	0.1		10.4
P+	0.2	0.6	0.6	0.3				1.7
−P	0.1							0.1
Overall	1.0	7.1	25.5	48.5	15.8	1.4	0.2	

Leanest ⟶ Fattest

Figure 10.4 Beef carcase classification and proportion of carcases in each grade in 2001 (reproduced with permission from *Beef Yearbook* (MLC, 2002a)).

makes it unlikely that this situation will improve in the short term.

The structure of the UK sheep industry

Sheep are found on the majority (over 41 000) of farms in the UK. On some they are important contributors to income, but on many they are seen as a useful grass-land management tool, often used to 'tidy up' the grass-land after the dairy herd. In broad terms, sheep production may be classified as hill (40% of ewes), upland or marginal (20%) or lowland (40%).

The interrelationship and interdependence of hill, upland and lowland systems in the UK is generally referred to as stratification. Traditionally, each sector has depended for its viability on the presence of the other two. Changes in market requirements and in the support schemes mean that this is not necessarily the case any more. An outline of the relationships is shown in Figure 10.5.

Hill farms

Hill farms provide a harsh environment. Climate and topography often limit herbage production, with indige-nous species dominant and growing seasons short. Con-servation of forage for winter feeding is minimal. Most hill flocks consists of small, pure-bred hill breeds, with the ewes only remaining in the flock for four to five lambings. Older animals are routinely drafted out and replaced by home-bred ewe lambs.

Target lambing percentage (lambs reared per 100 ewes put to ram) is 100%. Lambing is usually late (late April/May), giving a short period for growth. Many ewe lambs are retained for breeding and the majority of male lambs are sold as stores. In order to survive, hill farms depend upon there being a market for these store lambs, and for their draft ewes. Even so, they would not be economically viable were it not for the support available through subsidies. Until 2001 these were offered through the Hill Livestock Compensatory Allowance (HLCA) scheme, which paid on a headage basis and which without doubt led to overstocking in many areas. Since 2001 hill support payments have been based on area farmed, rather than on flock size. Stocking rates on hill farms are very variable. A farm's carrying capacity will be determined by the relative proportions of grass and heather grazing available, and on any ability to conserve feed for winter. Carrying capacity will range from 0.75–4 ha/ewe (not ewes/ha).

Upland farms

Upland farms may be no lower than hill farms but have a better climate and flatter fields, allowing for greater use of cultivations, lime, fertilizer and improved species of grass. Additionally, they are more likely to allow for conservation of forage as silage or hay, making winter feeding easier and increasing both lambing percentage (130–140%) and the proportion of lambs sold finished. A management strategy known as the 'two-pasture system' is often used. This attempts to match availability of 'improved grazing' to the specific needs of the flock, especially just before mating and in early lactation. For this strategy to work, at least 20% of available grassland needs to be 'improved'. Farms with this facility can stock at rates approaching those achieved by lowland units.

Upland farms may still carry hill breeds, but, since they can buy draft ewes from the true hill farms, they do not need to breed pure. This allows greater choice of ram breed. Traditionally longwool breeds such as the Teeswater or Blue-faced Leicester have been used. The cross-breeding that this involves brings benefits in terms of improved reproductive performance in the progeny, known as heterosis or hybrid vigour. It is this trait, reflected in higher lambing percentage (up to 200%), better milk production and improved lamb vigour, that makes the females from these crosses sought after by lowland shepherds. This provides upland farmers with a good and regular source of income. There are many such crosses, but the most common are the Mule (Blue-faced Leicester × Swaledale), the Masham (Teeswater × Swaledale), the Scotch Halfbred (Border Leicester × Cheviot) and the Welsh Halfbred (Border Leicester × Welsh Mountain). Unfortunately, the males from these crosses are less attractive. They tend to be 'leggy', with poor to average conformation, and their presence helps explain the poor appearance of many UK lamb carcases.

Lowland farms

The majority of lowland flocks are used to produce finished lambs for slaughter. Lowland farmers buy their replacements from upland producers, and thus do not

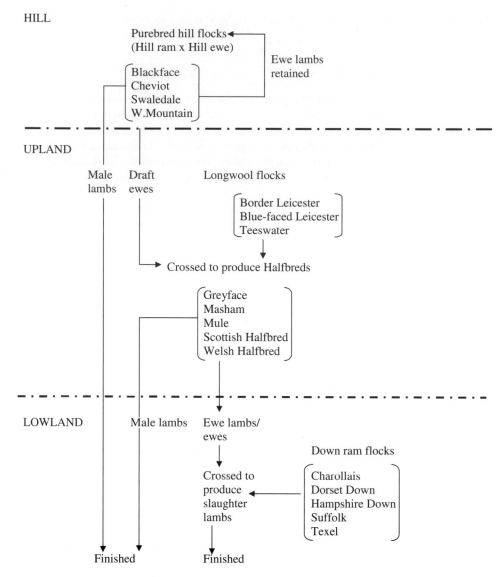

Figure 10.5 Stratification of the UK sheep industry.

need to breed replacements. This allows the use of terminal sires, breeds and individuals chosen for their ability to produce a carcase of better conformation. Terminal breeds include the Suffolk, the Dorset Down and, more recently, the Charollais and the Texel. Lowland shepherds will have a target lambing percentage of 200%, though few achieve it. Lowland flocks have a wider potential lambing window, thanks to better climatic conditions and a greater use of winter housing.

Early lambing flocks may begin lambing in December or early January, aiming to be marketing lambs before Easter. The majority will lamb March/April, whilst a few may deliberately delay lambing until May. Choice will depend upon target market, grassland management, other enterprises and housing availability. Average stocking rate on lowland farms is about 13 ewes/ha, whilst the best units may carry as many as 18 ewes/ha. (These are figures averaged over a year. On any par-

Leanest ——————————————————▶ Fattest

	1	2	3L	3H	4L	4H	5	Overall
E		0.2	0.8	0.3	0.2	0.1		1.6
U		1.6	9.0	5.2	1.8	0.7	0.4	18.7
R	0.3	10.0	32.7	12.1	3.1	1.1	0.5	59.8
O	0.5	6.6	8.4	2.0	0.4	0.1	0.1	18.1
P				2.1				2.1
Overall[1]	0.8	25.4	50.9	19.6	5.5	2.0	1.0	

[1] Excluding carcases graded P.

Figure 10.6 Sheep carcase classification and proportion of carcases in each grade in 2001 (reproduced with permission from *Sheep Yearbook* (MLC, 2002c)).

ticular day the stocking density of a flock may be as high as 40 ewes, and their lambs, per hectare.)

Lamb finishing systems.

Some lowland lambs will be finished 'off grass', going direct from grazing to the abattoir or market. Growth rates will depend upon the milk yield of the ewe, the sex of the lamb and the number of lambs suckling each ewe. A single, male lamb may put on up to 500 g/day at peak and average nearly 3 kg gain/week for the first 10 weeks of life. Average performance is closer to 250 g/day. Many, however, will not be fit to market so soon and about 50% will still be on farm at the end of September. These will need to undergo a period of 'finishing' (fattening), using silage, root crops or cereals. Home-bred lambs may be joined by others purchased from hill or upland farms. Additionally, some farmers may buy and finish such hill/upland stores even if they do not run a ewe flock. Roughly 50% of these lambs will be marketed before Christmas. The remainder will remain on farm for longer, being sold as 'hoggets' in the period January–March. Given the wide range of genotypes, weights and ages represented by these animals, matching the animal to the finishing system is important (but not easy) if over-fat carcases are to be avoided. Simplistically, early maturing types, such as the wether (castrate) lambs of the hill breeds, will need low energy-density diets, such as grass or silage alone, and may need only to put on 5–7 kg to produce an acceptable carcase. This they will do over an 8 to 10-week period. On the other hand, the potentially larger, half-bred wethers (e.g. Mule or Scotch Half-bred) will require a concentrate feed, such as sugar-beet pulp, to supplement their silage or may be fed on the more traditional 'root' crops such as swedes or rape. Target growth for these animals will be 80–100 g/day over a 16-week period. Target stocking rates for lambs on such feeds are shown in Table 10.10. With swedes, in particular, there is a danger that lambs get fat too quickly. Animals on these diets need to be weighed, or at least handled weekly, so that those ready for market may be drawn out in good time.

Table 10.10 Stocking rate targets for lambs on forage crops.

Crop	*Yield (t/ha)*	*Stocking (lamb days/ha[1])*
Rape	30–35	2500
Stubble turnips	40–50	2500
Swedes	65–100	8000
Grass	—	600

[1] 100 lambs for 1 week = 700 lamb days.

Carcase composition

As in beef cattle, lamb carcases are evaluated for conformation on the scale E, U, R, O, P and for fat on the scale 1–5. However, given that over-fat lambs are not wanted by the trade (or the consumer) it is fat classes 3 and 4 that are subdivided into low (L) and high (H) bands. Figure 10.6 shows a classification grid and the proportion of the UK kill falling into each category in 2001–2002. The relatively low proportion in the desired range is obvious. Lowland lambs will generally be slaughtered at liveweights of 36–46 kg, leading to carcase weights of 16–23 kg. Purebred lambs of some of the smaller hill breeds may well be significantly lighter. Slaughter weights tend to go up as the season progresses, being lowest for some of the very early, pre-Easter lambs and heaviest for the 'hoggets' sold after Christmas. Killing-out percentages vary, but will usually be in the range 43–47%. The proportion of carcases in the higher fat-score ranges tends to go up with increases in slaughter weight.

Winter housing

A majority of lowland and many upland ewes now lamb indoors. However, a smaller percentage (perhaps 40%) is housed for longer periods. Housing improves the shepherding environment and can lead to better lambing percentage, as a result of lower lamb mortality. Additionally, it may reduce replacement costs, allowing older ewes to be kept for an extra year or two, and can facilitate earlier lambing, but it is not a cheap option. Houses can cost anything between £15 and £50/m^2 to build, giving an annual cost per ewe housed of at least £3.50. On top of this, additional variable costs (feed, bedding, etc.) will add an extra £5/ewe. The major advantage of flock housing is that it rests and protects grassland, leading to better early growth (up to 1 month earlier and 25% more grass) and higher stocking rates. Housed ewes may be sheared at housing, rather than in May/June, which produces a cleaner fleece. Shorn sheep have increased appetites (+10%) and produce heavier lambs (+0.5 kg), but can complicate turnout if the early spring weather is inclement. A minimum of 10 weeks is necessary between shearing and turnout. Small ewes will need a floor area of at least 1.0 m^2 (1.2 m^2 is better). Larger ewes will require between 1.4 and 1.5 m^2. Ventilation is critical if respiratory disease is to be avoided. The temperature inside a sheep house should be no higher than that outside, but draughts must be avoided.

Sheep and the environment

Over 60% of UK breeding ewes are found on hill and upland farms, many of which are suitable for very little other than sheep farming. The majority of vegetation on these farms is indigenous grasses, heather and, increasingly, bracken. With appropriate stocking rates, sheep are an important tool in the management and maintenance of the landscape. For example, the trampling effect of sheep's feet can be effective in reducing the spread of bracken. Heather moor is semi-natural, but in the absence of controlled grazing/burning it will quickly revert to scrub woodland. Undergrazing, or burning too infrequently, initially favours heather development, but over time the heather will deteriorate. Overgrazing, or frequent burning, on the other hand, can lead to reduction in areas of heather and an increase in native grassland. Grassland has a higher carrying capacity than the heather hill in summer but lower in winter. In many areas the survival of the ewes over winter (outdoors on the hill) depends upon the availability of heather grazing, which is often accessible even under lying snow and which also provides ewes with some shelter. One of the major problems for hill farmers is the disparity between summer and winter carrying capacity. Most hill farms sell all lambs, and their draft ewes, in late autumn so as to minimize stock numbers over winter.

There is no doubt that incorrect stocking rates have an adverse effect on the ecology of an area. The problem lies in determining what the appropriate stocking rate should be. It will vary according to the type of indigenous vegetation and the presence or absence of rare species such as orchids. It is clear, however, that sheep may be stocked at levels that adversely affect the herbage without there being any detrimental effect on sheep productivity. It is also true that managing an area of hill at an optimum for sheep may have adverse consequences for other species, such as grouse or wading birds, many of which nest on upland moors. Government policy of relating support payments such as the Hill Livestock Compensatory Allowance to ewe numbers exacerbated problems by effectively encouraging overstocking and some areas became severely overgrazed as a result. This situation changed in 2001–2002 since which time support for hill farmers has been paid per unit area of land.

In addition, traditionally hill sheep were shepherded on a daily basis, being driven up/down the hill, or kept away from newly burned areas, as necessary. Such close shepherding is no longer feasible. However, shepherds may exert some control over grazing patterns, especially where sheep are overwintered on the hill, by varying the location and distribution of feed blocks or big bale silage.

Pig production

The UK pig herd is getting smaller. In 2002 breeding sow numbers were 457 000 head and the UK accounted for 6% of the EU pig herd. (It is worth noting that some US units manage more sows than are in the whole UK herd!) At any one time there are some 6 000 000 pigs, of various ages, on UK farms. Pig meat (pork, ham and bacon) accounts for 25% of the UK meat market, with production in 2001/2002 at 827 000 t. Approximately 200 000 t of this was bacon/ham, meeting 40% of UK demand for these products. The 627 000 t of pork produced represented 90% of demand. There are about 4600 pig producers in the UK, with an average herd size of over 350 sows, but there are still over 40% of producers with fewer than 10 sows.

In the relatively recent past the vast majority of sows were kept intensively and housed throughout their lives. More recently, however, welfare concerns and the cost of establishing new intensive units have seen more and more outdoor units being created, so that in 2002 up to 30% of sows were outdoors. With good management, outdoor units can compete, in technical performance terms, with intensive units and, given their lower fixed-cost base, can be profitable. However, not all farms are suitable for outdoor units and it has to be said that some units have been established on less-than-ideal sites. A free-draining site can be stocked at up to 25 sows/ha. There is a perception that outdoor pig keeping is more welfare friendly, but some sows, outdoors over winter on inappropriate sites, might dispute this!

Weaner production

The target for a weaner producer must be to optimize the number of piglets produced for sale per sow per year. Each sow will be expected to produce 2.3 litters per year and to rear 10 piglets per litter (24 piglets per sow per year). To achieve this, piglets will be weaned at about 24 days of age and the sows re-mated within 7 days (see Figure 10.7). A productive sow will be pregnant or lactating for 340 days in every 365. As a result, productive life is short (average 2.5 years), with sow replacement rates approaching 40 per annum.

In most parts of the world, pregnant sows are held in individual sow stalls, often tethered, but this system was outlawed in the UK in 1999. Pregnant sows are

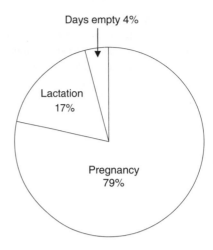

Figure 10.7   The reproductive cycle in the pig (reproduced with permission from *Pig Yearbook* (MLC, 2002b)).

now housed in a variety of systems, ranging from low-tech outdoor arks to indoor, group-housed units where individual feeding is controlled by computer. These units require at least $1.5\,m^2$ bedded area/sow and demand generous straw provision. Outdoor sows will be left to farrow (give birth) in their arks, but indoor animals will generally be moved to farrowing accommodation (normally farrowing crates) 3–5 days before parturition. Farrowing crates are designed to minimize piglet mortality, especially that caused by crushing (over-lying), and often contain a creep area where piglets may be introduced to solid feed before weaning. In the absence of the straw present in outdoor arks, farrowing crates require additional heat. This is usually provided by an infra-red lamp or by a heated pad against which the piglets may snuggle. Once the danger of piglet mortality is passed, sow and piglets may be moved to 'follow-on' pens that still provide heat and creep feed for the piglets but afford the sow greater mobility. At weaning the sow will be moved to a service area whilst the piglets will be grouped and moved to a weaner unit. It is at this time that they are most susceptible to gastrointestinal problems. Antimicrobials and antibiotics are commonly used in pig diets and give some protection at this time, but in response to consumer pressures these materials are likely to be phased out. This will necessitate the later weaning of piglets, if post-weaning scour is to be avoided, and 4 to 5-week weaning may become the norm. From the weaner unit, batches of piglets will be moved to grower/finisher accommodation, where they will remain until slaughter. Nowadays the distinction between pork and bacon pigs has largely disappeared. Better feeding and more consistent genotypes mean that most markets may be satisfied by pigs slaughtered to produce carcases of between 60 and 80 kg.

Sow nutrition

Sows are fed to try to maintain body weight throughout the production cycle. Pregnant sows will be fed about 2.5 kg/day of a high-quality concentrate (depending on body weight). Most lactating sows will receive 6–7 kg/day (depending upon litter size) throughout, but systems that increase feed on offer from 2.5 kg on day 1 to 10 kg on day 18 (i.e. +0.5 kg/day) appear to obtain better results. Total feed usage will be about 1500 kg per sow per year.

Grower/finisher nutrition

The nutrient requirements and appetite of the pig change as it gets heavier. Feed may be offered dry, in meal or pellet form, or wet. Pelleted feed reduces dust, minimizes wastage and is easier to handle. Wet feeds permit the inclusion in the diet of materials such as skimmed milk, whey or potato starch, allowing the higher capital cost of liquid-feed systems to be balanced by lower ingredient costs. Consumption of dry feed will go up from 1 kg/day at 20 kg liveweight to 2.9 kg/day at 90 kg. Most units will use three or four different diets, of decreasing nutrient density, during the finishing cycle. Feeders will go from 20–90 kg lightweight in an average of 110 days, gaining 622–650 g/day at a feed conversion ratio of 2.6 : 1 and consuming less than 200 kg of feed in the process.

Carcase quality

There is much less variability between pig carcases compared with beef or lamb, mainly because slaughter weights, genotypes and feeds are more consistent. The EU classifies carcases on a six-point scale related to lean tissue percentage (see Table 10.11). In the UK, the Meat and Livestock Commission calculates lean meat percentage from an equation relating lean meat percentage to back fat thickness and carcase weight. Using optical or ultrasound probes, back fat is measured over the last rib at 4.5 cm (P_1), 6.0 cm (P_2) or 8.0 cm (P_3) from the mid line.

Table 10.11 Pig carcase gradings and percentage lean in the carcase. (Data from *Pig Yearbook* (MLC, 2002b), with permission.)

Grade	Percentage lean	Percentage UK carcases in 2002
S	>59	32.7
E	55–59	57.3
U	50–54	9.2
R	45–49	0.8
O	40–44	0.1
P	<40	0.0

Pigs and the environment

Intensively housed pigs make minimal direct demands on land. However, they do produce large amounts of slurry, which must be disposed of. Each sow will produce up to 2000 kg of slurry per year and each fattening pig up to 4 kg/day. Thus, one sow and her progeny will produce up to 10 t of slurry per year. To meet the recommended maximum loading on land of 250 kg N per hectare per year from organic manures the output of one sow/finisher unit requires at least 0.8 ha of land! For finishing units the story is complicated by the inclusion of copper, as a growth stimulant, in the diet. Copper is toxic, especially to sheep, which should not be grazed on pastures previously treated with pig slurry. Much research is currently being directed towards optimizing the nutrient content of diets (especially amino acids and minerals) in an attempt to reduce water consumption and hence slurry output.

Goat production

There are about 12 million goats in Europe, the majority in Greece (5.3 million), Spain (2.8 million), Italy (1.4 million) and France (1.2 million). In France, as in the UK, the main product is milk, but across southern Europe the goat is used as a dual-purpose animal and is an important source of meat, especially in rural areas.

In the UK there are currently 73 000 goats. Of these, perhaps 20 000 are milking does, kept in herds that range in size from 50 to over 1000. There are also many smaller herds, including those of the 'minority' breeds, which may milk some animals and sell their produce at the farm gate or locally. There are about 7000 fibre-producing Angora goats, again with relatively few 'large' and many smaller producers.

Milk from goats

The majority of milking does are Saanen or Saanen ×, with much smaller numbers of Toggenburg, British Alpine and Anglo-Nubian. Lactation characteristics are very similar to those of the dairy cow. A good milking animal will produce in excess of 1000 litres in a lactation, but herd averages are closer to 800 litres. Goat milk is similar in composition to cows milk but tends to be lower in protein (29–32 g/kg). It has small fat globules (28% < 1.5 μ), which makes it easier to digest. It also contains more short-chain fatty acids (capric, caproic and caprillic). It is these fatty acids that give goat milk products their characteristic flavour. The make-up of the proteins in goat milk is also different. In particular, the casein particles are smaller and the proportion of β casein higher (67% versus 43%). These differences mean that the milk is often well tolerated by those who are allergic to cows milk.

The goat is a browser, rather than a grazer, and has 'catholic' tastes in foodstuffs, often being willing to eat materials (such as wild garlic) that might taint its milk. This, together with the problem of parasitic gastroenteritis associated with the grazing animal, leads to most milking does being housed throughout lactation, especially in the bigger units. The traditional feed for goats is hay, supplemented by compound feeds, which are often fed at a flat rate of 1–2 kg/day, depending upon average yield. A doe will require 750 kg of hay per year together with 400–600 kg of concentrate. Silage feeding is becoming more common. Silage, especially good grass silage, is higher in nutrients than hay. In particular, protein content may be double. This reduces the requirement for concentrate feed, in terms both of quantity and composition. An allowance of 3500 kg of silage per year might be supplemented with 200–300 kg of concentrates. As with all lactating animals, an adequate supply of clean water is important. Care must be taken when feeding goats on 'big-bale' silage. Levels of the pathogenic organism *Listeria monocytogenes* are often quite high in such material and the goat is particularly susceptible to the organism, which can cause a range of conditions, including meningitis.

Because goats, like sheep, are seasonal breeders that normally kid in spring, there is naturally a shortage of milk over winter. Most commercial units therefore use controlled breeding techniques (day-length manipulation or hormone therapy) to kid a proportion of their flock down in autumn, thus giving continuity of production.

Fibre from goats

The goat produces a 'double-coat' fleece with longer guard hairs over a fine undercoat. It is these undercoat fibres that are of use. The cashmere goat retains a permanent top coat but produces an annual undercoat that is shed and lost if not harvested. This undercoat may be combed out, in early spring, with up to two combings at 4-week intervals. Yields are small (52–250 g) and fibres short (<6 cm) and fine (12–16 μ) but of high value. The majority (60%) of world production of cashmere comes from China (7500 t).

The angora goat has few, if any guard hairs and a longer undercoat. This may be sheared off on a bi-annual basis (usually April and September) as mohair. Yield is much higher (up to 5 kg) and fibres longer (up to 15 cm) and coarser (18–40 μ). Mohair is graded according to fibre diameter. The finest fibres (Superfine Kid) must be less than 23 μ in diameter. The other grades are Kid (23–26 μ), Young goat (26–30 μ), Adult (30–34 μ) and Strong Adult (>34 μ) As animals get older their yields tend to increase, as does fibre diameter, leading to lower quality. As raw materials, both cashmere and mohair are subject to the vagaries of the fashion industry. When they are 'in' then even coarser fibres may sell well. In poorer years even the finest fibres may not find a ready market. For example, in 1999 Kid fibre was worth £8.42/kg whilst in 2000 its value increased to £16.75/kg. In that same year Young Adult made only £3.60/kg.

Goats and the environment

As already noted, goats are browsers rather than grazers. As a result they can, if permitted, do considerable damage to the foliage and even the bark of trees and shrubs. In some parts of the world there is no doubt that poor management of goats has contributed to erosion and desertification problems. Under more controlled conditions goats may be used to reduce or remove plants such as rushes (*Juncus* spp.) and bent grass (*Molinia*) and can be valuable in hill pasture improvement programmes. They are also increasingly being used as 'environmental managers' in land reclamation projects and in nature reserves. For such uses the feral cashmere is better suited than other types. On the other hand, cashmere goats are less likely to respect fences!

Red deer

Red deer run wild in several areas of the UK and many farmers began their herds by capturing wild animals or by buying from deer parks. Over 75% of all deer units in England use red deer, the remainder using fallow deer. In Scotland almost all units use red. Fallow deer are about half the weight of a red, and consequently require half the feed and space. A red deer hind stands about 1 m tall and weighs 100–120 kg, whilst stags will grow to 160 kg. Stags shed their antlers in April and grow a new, larger set each year before the rut.

In the UK the industry is still in its infancy, with fewer than 300 producers holding some 35 000 animals. It is estimated that in New Zealand, where the industry began at much the same time, there are now 700 000 head!) The main output is the meat (venison), but sales of breeding stock are also important. As with beef and sheep, many hill or upland units find it necessary to sell their calves as stores, to be finished elsewhere.

Minimum economic size is about 100 hinds, there being substantial capital costs involved in providing fencing and handling systems. Fencing alone may cost as much as £500/ha. Stocking rate, depending on grass-growing ability, may be up to 8 breeding hinds/ha (or up to 15 fallow). Supplementary feeds (roots, hay, cereals) will be needed over winter, especially for young stock that are normally housed. Red deer appetite responds to day length, reducing with shorter days and increasing again in spring.

One stag may cover 30–40 hinds. Mating begins in late September and a 233-day gestation means that calving begins in mid April and may go on until August.

Yearlings, provided that they weigh at least 70 kg, may be bred but not all will calve. Older animals should produce 95 calves per 100 hinds and will go on breeding for at least 12 years, with many still breeding regularly at 15 or 16. As in sheep or cattle the quality of hind milk increases as lactation progresses, fat percentage increasing from 85 g/kg on day 3 to 130 g/kg on day 100 and protein going up from 71 to 86 g/kg. This high-quality feed leads to growth rates in the calves of up to 300 g/day for the first 150 days of life.

Most meat animals are slaughtered as yearlings (15–17 months). A well-fed stag will weigh up to 110 kg, with a killing-out percentage of 55%. There is a 'closed season' for taking deer from the wild (1 May to 31 July for stags), but this does not apply to farmed deer. However, welfare regulations forbid the transporting of stags when their new antlers are still soft (when they are said to be 'in velvet'). Since this occurs during the period May to July this effectively rules out the slaughter of stags at an abattoir during this period. On the other hand, it is still permissible to shoot yearling stags in the field (provided that they are then transported to an abattoir and dressed within an hour).

Red deer and the environment

Red deer are a native species and have been present in the UK at least since the 13th century. They are, by nature, a forest species. For farmed deer, availability of an area of trees in which they can shelter can significantly reduce calf mortality while not damaging the trees. Weaned hinds may be outwintered satisfactorily on heather hill at much lower cost than using in-bye (fenced and improved pasture). However, stocking density is critical. At less than 1 hind/ha the percentage of heather on the hill should be maintained, but at 1.5 hinds and above, heather cover may be dramatically reduced over a relatively small number of years. The Highlands of Scotland are estimated to carry as many as 400 000 wild deer, numbers that are beginning to cause concern as evidence grows of damage to heather moors and to trees. It may be that the red deer may soon need to be viewed as a pest!

Poultry production

Egg production

Until relatively recently most UK farms kept poultry as part of a mixed farming system. The development of intensive, specialist poultry farms was facilitated by the introduction of battery cages, and of vaccines, which allowed increased intensification. In 1975 there were over 130 000 'specialist' units, each keeping about 2000 birds. Today over 50% of all UK eggs are produced from fewer than 400 flocks and over half of all layers are held in flocks of more than 50 000 birds. The current national flock stands at 33.5 million hens.

Improved genetics, nutrition and environmental control have led to increases in egg production. In 2001 production was 913 million dozen eggs for consumption and a further 803 million dozen for hatching. Mean egg output per hen was 262 for free-range birds and 282 for battery cages. In 1975 the average production of a free-range hen was 198 eggs per year while that of a cage-housed bird was 245. Initially almost all intensive layers were white, and produced white eggs, but consumer demand in the UK was such that by 1975 over 60% of layers were brown, and by 2000 the figure was nearer 100%. There are currently about 70 breeds of fowl and bantam in the UK, but most commercial units use hybrids, based upon the Rhode Island Red. Over most of Europe and North America the preference is still for white eggs and the white bird derived from the White Leghorn. Some of the improvements in output have been due to the hybrid vigour demonstrated by these crossbred birds.

A developing egg spends about 25 hours in the reproductive tract, a new ovum (yolk) being shed 15 minutes after the last egg was laid. Most of this time (21 hours) is spent in the shell gland (uterus). Egg laying is controlled by the hormone melatonin, which is secreted by the pineal gland during the hours of darkness, lay being initiated as day length increases. In controlled conditions, day length may be manipulated so as to maximize this effect. Pullets will normally begin to lay at about 22 weeks of age and will reach peak production (98 eggs per 100 birds per day) in the next 4–6 weeks. By week 70 or thereabouts production will have dropped to 70 eggs per 100 birds and at this time the whole flock will normally be replaced, culled birds going for processing into soups or meat pastes. It is possible to put hens through a non-productive 'moult' and then to allow them to go through another laying season. The eggs so produced will be bigger but overall production lower and this method is rarely used.

Feeding

The hen has no effective teeth. Swallowed food first enters the crop where it is soaked in saliva and where starch digestion begins. It then passes into the proventriculus (stomach) when digestion begins, before moving into the gizzard. The gizzard effectively acts as the teeth of the hen. It is a thick, muscular structure that is lined with a hard keratin membrane. This, together with any grit that the hen has eaten, serves to grind up the feed, exposing grains, etc. to the digestive juices. For free-range birds, having access to seeds, grains and other materials, this mechanism is important, but intensively housed birds are fed a pre-ground material, so that the gizzard is effectively redundant. As a relic of its jungle fowl ancestry, a hen's appetite is affected by temperature. The hen is most comfortable at 24–26°C. Below these temperatures it will use some feed energy to keep warm and will increase feed intake accordingly (by about +1.5%/1°C). Equally, at temperatures over 27°C feed intake will fall to the point that egg output may suffer. Thus, hens housed at 10°C will need some 15 g of feed per day more than those held at 24°C (that is, an extra 16 kg of feed per 1000 birds). On a 100 000 bird unit that is over 1.5 t of feed per day. Average intake for caged birds will be between 110 and 120 g/day. Deep-litter animals will take about 130 g while those on free-range will need at least 140 g. The feeding in larger units is fully automated, with feed and water available continually. One of the advantages of smaller, free-range or deep-litter systems is that they allow birds to make use of a range of feeds, such as table scraps or vegetable waste, on an *ad hoc* basis.

Housing

Poultry housing has welfare implications and the keeping of hens in battery cages is an emotive issue. Indeed, in response to consumer demand, the last few years have seen an increase in the number of smaller units producing free-range or conservation-grade eggs. Free-range units bear some similarity to the farmyard-fowl systems of old. Birds have access to pasture, even if they receive the same processed feed as their housed cousins, and may be stocked at up to 1000 birds/ha. Unless management is of a high order there can be problems with bullying, with cannibalism and with dirty eggs. Feed usage and labour inputs will be higher and output lower. On a small scale, fold units may be used. These consist of a combined pen and run, which may be moved regularly to give birds access to fresh grass. It is reckoned that outdoor systems may account for up to 20% of current egg production.

In a deep-litter unit birds are housed continually, at up to 6 birds/m^2 (EU legislation sets a maximum of 7 birds/m^2). Droppings mix with the litter and bacterial action breaks them down, the heat produced keeping the litter dry and friable. However, poor ventilation, especially over winter, or leaking drinkers can turn the floor into a sticky mess. By adding perches to the house one can create a sort of 3-D deep-litter unit. Such units are commonly called percheries and can increase stocking rates to as many as 15 birds/m^2. (Almost all meat-bird systems are deep litter, but in these stocking rates may be higher.) In 2002 deep-litter or perchery systems accounted for about 10% of UK egg output.

Battery cages

Cages account for over 70% of UK production. Most cages hold between two and five birds. They are usually 45 cm deep and 45 cm high, their width depending upon the number of birds housed. A three-bird cage will normally be 35 cm wide and a five-bird cage 55 cm. In some countries the minimum space requirement is 300 cm^2 per bird, but in the EU the minimum is 450 cm^2 for five-bird cages. Under new legislation it is proposed that by 2012 all cages should be of an improved, 'enriched' design. These will offer more space (470 cm^2 per bird), perches and a laying box. However, animal welfare groups argue that even this does not go far enough and it is possible that eventually all cages may be banned.

The full production cycle

While small numbers of eggs may be hatched using a 'broody' hen, the majority are incubator-hatched. Tem-peratures are maintained at 38.7°C and relative humidity set at about 61%. In a fertilized egg, development begins before it is laid, but stops at below 20°C. Eggs may be stored for up to 7 days before being incubated. (Eggs for consumption are produced from unmated hens.) In the incubator development is rapid, with the head and eyes visible at the end of day 1 and limb buds, brain and nerves present by day 4. By day 10 bones have calcified. From egg set to hatch takes 21 days in the hen, 28 in the turkey and 30 in most ducks.

Once hatched, chicks will be sexed and females moved to a brooder area. (Because only the female hen lays eggs, and since 50% of all eggs hatched produce male birds, it is necessary that all male chicks are humanely destroyed at day old.) A brooder area may be heated overall or by using localized heat sources. Overall heating will run at 32°C, reducing by 3°C per week to 21°C (broilers) or 18°C (pullets). Lights will initially be left on for 23 hours/day to encourage feeding, but after 7 days will be reduced to 8 hours/day. Pullets may weigh 40 g at day old, 140 g at 14 days and 600 g at 56 days, and feed consumption may also be expected to increase, from 10 to 50 g/day over the same period. Floor space requirement at 8 weeks will be 10 m^2 per 100 birds. At 8 weeks, birds move into the rearing phase, which will last until point of lay (POL), which will begin at about 22 weeks. Some will remain on deep litter, but many, especially in bigger units, will be put into cages at this point. Feeding will continue *ad libitum*, but a cheaper diet specification will be used. Consumption at 20 weeks will be 110 g/day, at which time the birds will be switched to the final, layers' ration. Up until this point birds will have been exposed to 8 hours of daylight daily. From now on, light exposure will be increased steadily, by about 15 minutes/day each week. This will trigger lay at about 23 weeks, but light will continue to increase until it reaches 17–17.5 hours at about week 40. On small-scale, farmyard systems these levels of sophistication are neither possible nor necessary, but even here a simple light source, such as an electric bulb or a kerosene lamp may be used to extend day length (and thus lay) into the winter.

Broiler production

Poultry meat production (broiler chicken or turkey) is now a separate sector of the poultry industry. Many units are fully vertically integrated, with the same company responsible for breeding management, hatching, rearing, feed formulation and milling, and processing and marketing. In such units production levels of 500 000 birds per week are not unusual. The bulk of UK broiler chicken production is in the hands of 12 production companies. In 2002 the UK slaughtered an estimated 800 million broilers and 26 million turkeys, producing over 1.2 million t of broiler meat and 160 000 t of turkey. In addition, the UK imported 350 000 t mostly from within the European Union. Comparative production figures for EU countries are shown in Table 10.4.

At day old, broiler chicks may or may not be sexed. Sexing allows female birds, which grow more slowly and mature at lighter weights, to be housed and fed separately. Broilers are taken from day old to slaughter in the same pen and housed on deep litter. Stocking density is now measured in terms of bird weight rather than bird number. Current maximum is 34 kg/m^2, equivalent to 17 birds/m^2 at slaughter weight. Broilers grow quickly; a bird that weighs 40 g at day old will be 150 g at 7 days, 1250 g at 28 days and 2400 g at 42 days (if allowed to live that long!) Given this rate of growth, feed requirements change over time. Many birds are fed *ad libitum*, though some may be rationed during the starter phase (1–21 days). Diets will be high energy and will contain 22% protein (starter), 20% protein (grower) and 18% protein (finisher). Antimicrobials, present in most feeds, will be withdrawn for the last 7 days.

An average broiler, slaughtered at 42 days of age, will weigh 2200 g liveweight and yield a 1600 g carcase. During its life it will have eaten an average of 3800 g of feed (600 g starter, 1800 g grower and 1400 g finisher) at a feed converting efficiency of 1.9 : 1 or below.

As with egg production, consumer concerns have led to an increase in free-range meat birds, reared in relatively small numbers on free range. Such birds receive diets not dissimilar to those of conventional broilers but without the antimicrobials and using organically certified ingredients. Growth rates are inevitably lower, with birds reaching 1400–2200 g at 56–70 days, and prices must be that much higher to compensate.

Glossary

Broody hen: a hen in an hormonal state that causes her to 'sit' on a clutch of eggs.

Browsing: eating of shrubs and branches, rather than grass (grazing).

BSE: bovine spongiform encephalopathy. A degenerative brain disease, similar to Creutzfeldt-Jakob disease in humans. Affected animals are slaughtered.

Bull beef: an entire (uncastrated) animal being reared for meat.

Calving index: herd average for calving intervals (measured backwards from the most recent calving).

Calving interval: number of days between consecutive calvings in the same cow.

Concentrate feed: high dry-matter feed based upon cereals and vegetable protein sources. (Also called a compound feed.)

Conservation: storage of grass or other forage in the form of hay or silage.

Draft ewe: productive ewe at the end of her useful life in a particular environment.

DUP: digestible undegradable protein. Undegraded in the rumen but digested in the abomasum.

FCE: feed converting efficiency. Amount of feed (kg) required to produce 1 kg liveweight gain.

Follower: young animal destined to enter the herd/flock as a replacement breeding animal.

Genotype: genetic make-up of an organism.

Gilt: young, female pig (usually up to the end of her first lactation).

Heifer: young, female cow (usually up to the end of her first lactation).

Hogget: lamb sold for meat after 1 January of the year following its birth. (Normally only applied to spring-born lambs.)

Killing-out percentage (KO%): proportion of an animal's liveweight that is represented by its carcase.

Lactate: to produce milk (hence lactation).

Nutrient density: concentration of nutrients in 1 kg of feed (normally expressed in dry matter terms).

Oestrus: 'heat'. Period of sexual receptivity in the female.

RDP: rumen degradable protein. Protein (or other nitrogenous material) capable of being degraded by rumen micro-organisms.

Rut: breeding season in the red deer.

Steaming up: feeding of non-lactating animal, in late pregnancy, so as to allow her to lay down body fat for utilization in early lactation.

Steer: castrated bull.

Stocking density: number of animals per unit area of land at any particular moment.

Stocking rate: average number of animals carried by one unit area of land over a 12-month period.

Store: animal being kept at less than its full potential for growth.

Terminal sire: male of a breed whose main attribute is one of high carcase quality (usually in terms of conformation).

Wether: castrated male sheep.

References

MLC (2002a) *Beef Yearbook*. Meat and Livestock Commission, Milton Keynes.

MLC (2002b) *Pig Yearbook*. Meat and Livestock Commission, Milton Keynes.

MLC (2002c) *Sheep Yearbook*. Meat and Livestock Commission, Milton Keynes.

Further reading

Allen, D. (1990) *Planned Beef Production and Marketing*. BSP Professional Books, Oxford.

Boden, E. (1991) *Sheep and Goat Practice*. Baillière, London.

Cardell, K. (1998) *Practical Sheep Keeping*. Crowood Press, Marlborough.

Croston, D. and Pollott, G. (1994) *Planned Sheep Production*. BSP, Oxford.

De Nahlik, A.J. (1992) *Management of Deer and Their Habitat*. Wilson Hunt, Gillingham.

Dennett, M. (1995) *Profitable Free Range Egg Production*. Crowood Press, Marlborough.

HMSO (1998) *Codes of Recommendations for the Welfare of Livestock*. CABI, Wallingford.

Leeson, S. and Summers, J.D. (1997) *Commercial Poultry Production*, 2nd edition. University Books, Ontario.

Mowlem, A. (1992) *Goat Farming*. Farming Press, Ipswich.

Sainsbury, D. (1998) *Animal Health*. BSP, Oxford.

Webster, J. (1987) *Understanding the Dairy Cow*. BSP, Oxford.

Whittemore, C. (1998) *Science and Practice of Pig Production*. Blackwell, Oxford.

Useful websites

www.deer.org.uk – British Deer Farming Association.
www.allgoats.com – British Goat Society.
www.defra.gov.uk – Department of Environment, Food and Rural Affairs (DEFRA).
http//:apps.fao.org/faostat – FAOSTAT.
www.mdc.org.uk – Milk Development Council.
www.nationalsheep.org – National Sheep Association.
www.pigsuk.com – Pigs UK.
www.pigworld.org – Pig World.

11

Cropping in the UK

A. Samuel

Introduction

Crops other than grass are grown on approximately 4.5 million ha or a quarter of the area of agricultural holdings in the UK. Cereals (wheat and barley) are the most commonly grown crops followed by oilseed rape (Figure 11.1). Over the last few years there have been changes in the relative area grown of some crops such as barley and linseed. Some of the reasons will be discussed in this chapter.

Factors affecting UK cropping

Many factors affect the types of crops and areas where they are grown in the UK; because of this, many parts of the country have very different farming systems and are more or less suited to growing arable crops.

Soil type

The UK has a very varied geology and hence soil type. Soil type or texture describes the amount of sand, silt and clay in a soil. A light-textured soil contains a large percentage of sand and is easy to cultivate but can be very susceptible to drought, nutrient losses and soil erosion. With irrigation a range of crops can be grown on these soils, particularly root crops such as potatoes and sugar beet.

The soils that contain a high percentage of clay are often called heavy soils. They usually need drainage before arable crops can be grown successfully. Even after drainage only a limited range of crops can be grown. Suitable crops include winter cereals, oilseed rape and field beans. Yield potential can be high.

Some of the most suitable soils for crop production are the light silts found in Cambridgeshire and south Lincolnshire. A wide range of crops can grow on these soils, including root crops and field vegetables.

There is a small amount of lowland organic/peat soils which when drained are also suitable for growing field vegetables.

Climate

The climate or average annual weather conditions vary across the UK and can affect the types of crops grown and their yield potential. In the far south-west of England there are areas that are rarely affected by ground frosts; these areas are important for some crops such as early potatoes and winter cauliflowers.

In eastern England rainfall can be 30% lower than in the west and this can increase the available working days (important when growing crops such as potatoes and sugar beet) and reduce grass growing days. Disease incidence can also be affected by climate. Crops grown in western counties are much more affected by wet-weather diseases than in the east, e.g. *Septoria* leaf blotch in wheat and potato blight.

Some crops such as forage maize require a soil temperature of 10°C before growth begins and high accumulated temperature to reach maturity by autumn. Forage maize yield potential is greatest in crops grown in the more southerly counties of the UK.

Markets

Proximity to the market can affect crop or type of crop grown. For example, many areas of the UK are suitable for growing sugar beet, but due to the small number of processing factories (only one in western England) the main areas of production are the eastern counties of England and the West Midlands. The proximity to flour mills, feed mills, maltings or export ports often can influence the type of cereal or variety grown.

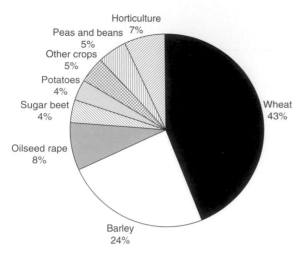

Figure 11.1 Cropping in the UK, 2003 (reproduced with permission of HMSO).

Political influences

In the past the different crop payments, e.g. through the arable area aid scheme, made the growing of some crops such as linseed economically viable and the hectareage grown increased dramatically. In the future the areas grown of some arable crops may be affected by the EU mid-term review (see Chapter 9). The area grown of potatoes used to be (up to 1997) governed in the UK by quotas. Quotas restrict sugar beet production across the EU.

Availability of labour, machinery and storage

Crops such as potatoes require specialist machinery for planting and harvesting. A store usually needs to be available to provide a quality crop to the market for a large part of the year. To start growing potatoes a large capital outlay is usually required unless there are good contractors available. Growing some of the horticultural crops requires a lot of seasonal labour which can be problematic some years.

Economics

One of the main factors governing areas of crops grown is of course crop returns. Some crops such as field beans

have varied in popularity depending on not just yields but commodity and support prices. Many soil types are more suited to growing barley rather than wheat, but because the wheat price tends to be higher than for barley and there have been greater improvements in wheat yields than barley, wheat is now the most commonly grown cereal in the UK. Some of the current environmental schemes (e.g. the Countryside Steward-ship Scheme) may pay farmers either not to grow crops or to grow alternative crops.

Rotations

The need for a crop rotation or sequence of crops means that only a limited number of crops can be grown continuously. Rotations can help to reduce the incidence of some weeds, diseases and pests. Growing a break crop such as beans in a cereal rotation can boost yields of the following wheat crop by about 1 t/ha. Some crops such as maincrop potatoes should not be grown more frequently than one year in four to reduce the risk or incidence of the pest potato cyst nematode. Farmers with a sugar beet contract (British Sugar plc) are not allowed to grow sugar beet in a field where any beet has been grown in either of the two previous years. The aim is to reduce the risk of the viral disease rhizomania, the pest beet cyst nematode and even problems with weed beet. Oilseed rape is recommended only to be grown one year in five to avoid problems with diseases such as clubroot. Integrated crop management systems rely on good crop rotations to reduce reliance on pesticides. Organic systems also rely heavily on good rotations to maintain soil fertility as well as reducing some pests and diseases.

Cereals

Cereals are the main arable crop grown in the UK. In 2002, 3.2 million ha were grown and the UK was the third largest grower of cereals in Europe (out of the original 15 member states) after France and Germany. Of the cereals, wheat (2 million ha), barley (1.1 million ha) and oats (126000 ha) are the most commonly grown cereals. Rye, durum wheat, triticale (a cross between wheat and rye) and grain maize are very minor (occupying 23000 ha), though more commonly grown in other parts of Europe. There has been quite a change over the last one hundred years in the relative popularity of the main cereals (Figure 11.2). The relative importance of wheat, barley and oats has been affected by mainly market requirements and now by yield potential (Figure 11.3). All the cereals can be easily distinguished by their grain, ears and their vegetative characteristics. Wheat and rye have naked grain, i.e. they

do not have any husk (lemma and palea) covering the grain once they have been harvested (Figure 11.4). Barley and most oat varieties have covered grains (Figure 11.4). The ears of wheat, barley and rye are in spikes unlike the loose open panicle or ear of oats (Figure 11.5). Barley and rye have awned ears unlike wheat (usually not awned). When no ears are present

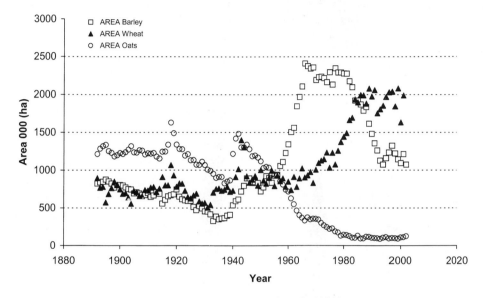

Figure 11.2   Cereal areas grown in the UK (reproduced with permission of HMSO).

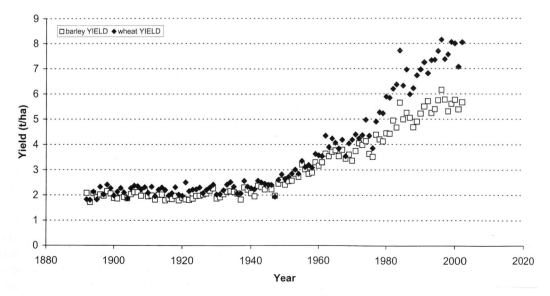

Figure 11.3   Wheat and barley yields (reproduced with permission of HMSO).

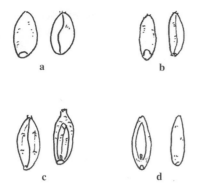

Figure 11.4 Cereal grains: **a** wheat; **b** rye; **c** barley; **d** oats (adapted from Finch *et al.*, 2002).

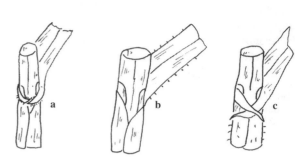

Figure 11.6 Identification of cereal leaves: **a** wheat; **b** oats; **c** barley (adapted from Finch *et al.*, 2002).

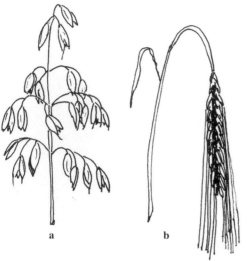

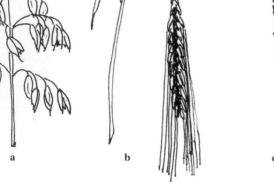

Figure 11.5 Cereal ears: **a** oats; **b** barley; **c** wheat.

the different cereals can easily be distinguished by their leaf characteristics (Figure 11.6). Oats have no auricles, wheat has small whiskery auricles and barley has large bare auricles at the leaf base.

Cereal markets

The UK currently produces around 23 million tonnes of cereals, 16 million of which is wheat. Some grain is still imported for the specialist markets such as milling for bread making. Figure 11.7 highlights the main markets for usage of wheat and barley. Just less than half the cereals are used for animal feed. Quality requirements for the other markets such as malting, flour milling and seed are very specific and the grain

will command a premium. Many growers for these specialist markets grow on contract. Achieving the correct quality can be very difficult some years, usually because of inclement growing conditions. Since the 1980s the UK has been self-sufficient in cereals and now is an important exporter both in and outside Europe. Quality requirements of grain for export are often different from those for the home market.

Most cereals are now certified by one of the various assurance schemes such as the Assured Combinable Crop Scheme (ACCS). These schemes help to ensure that the crop is grown and stored using good agricultural practices and is traceable. Most buyers of UK grain will only buy assured grain.

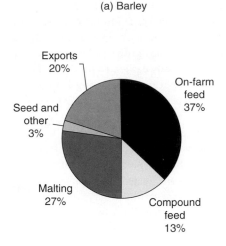

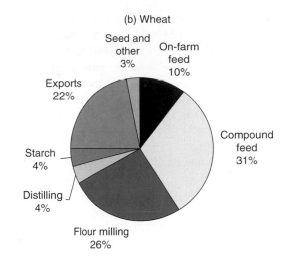

Figure 11.7   Cereal usage in the UK.

Cereal husbandry

- Grow Home Grown Cereals Authority (HGCA)-recommended varieties for the chosen market.
- Sow winter cereals from mid September and spring cereals as soon as soil conditions are suitable in the spring.
- Nitrogen is the main nutrient applied followed by phosphorus, potassium and sulphur.
- Pesticide usage varies with crop and variety; winter wheat has the greatest requirement and spring barley the least.
- Harvest is from mid July with winter barley, ending at the beginning of September with spring wheat.
- Yields are usually highest for winter wheat, averaging 8 t/ha compared with 5 t/ha for spring barley.

Variety choice

One of the major reasons for yield increases over the last 30 years, especially in wheat, has been improvements in varieties. Currently the HGCA levy on all sold cereals helps support variety testing. There are recommended lists provided by the HGCA and NIAB. These lists are produced from data from a large number of cereal trials carried out across the UK on a range of soil types. Recommendations are given both for the UK and for different parts of the country. Information is pro-

vided on yield potential with and without full disease control, as well as details of market potential, straw production and disease resistance. Popular feed varieties change fairly regularly, and not many stay on the recommended list for more than 5 years. Quality wheat varieties for breadmaking or barley varieties for malting tend to change more slowly.

Time of sowing and seed rates

Winter cereals are mainly planted in the autumn from the middle of September and spring cereals in the spring as soon as soil conditions are suitable. The earliest drilled crops in the autumn are usually first wheat crops (sown after a break crop) and winter barley. The last winter cereals to be drilled are usually second and third wheat crops (at risk from take-all disease), in fields where there is a grass weed problem such as brome or blackgrass, and following late-harvested root crops. Winter cereals unlike spring cereals need a period of cold in order for the ear to be initiated (vernalization). The majority of wheat crops are winter not spring varieties and just over half the barley crops are winter varieties.

When cereal prices were high, seed rates used tended also to be high (188 kg/ha was commonly drilled for wheat). Now that prices for cereals are low, seed rate is being more finely tuned. Seed rates of 100 kg/ha or lower are used for many early drilled winter wheat crops where the seed bed is very good and there are no serious weed problems. The highest seed rates of over 200 kg/ha may be used for the latest drilled crops in November or December where the seed bed is very poor, or the crop has to be broadcast rather than drilled, or for organically grown crops. Many people drill by seed number rather than by weight as seed size can vary widely according to year. The average seed rates used are usually between 140 and 180 kg/ha.

Fertilizer use

Recent results from a British survey of fertilizer practice show that nitrogen is the main fertilizer applied to cereals (Table 11.1). Amounts of the main nutrients nitrogen, phosphorus and potassium applied have changed little in recent years; the only change has been the requirement for sulphur in areas with low depositions. Crops for the bread-making market require high grain nitrogen (protein) content and an extra 40 kg/ha of nitrogen is commonly applied. Malting barley requires low grain nitrogen so that less is usually applied than for a feed crop. A large number of nitrogen response trials have been undertaken over many years. From these trials the Ministry of Agriculture, Fisheries and Food (now DEFRA) produces its standard fertilizer recommendations (RB209) published by the Stationery Office. Applications above the RB209 recommendations are considered not to be good agricultural practice. Fifty-five percent of the UK is now designated a nitrate-vulnerable zone (NVZ). In these NVZs there are restrictions on time and application of organic manure as well as timing of application of inorganic fertilizers. No nitrogen is ever required in winter cereal seed beds. The main time for nitrogen application is when the crop is growing rapidly in the spring. Winter wheat will commonly have a small application

in February/early March followed by the main application (usually split twice) in April. Too early applications of nitrogen can lead to too lush a crop, prone to lodging and disease. As winter barley grows earlier than wheat, applications may be slightly earlier. Depending on time of drilling nitrogen applications in spring barley may be applied to the seed bed or split between seed bed and early emergence.

On farms where straw is baled there is a larger requirement for potassium.

The other major nutrient required by cereals in some areas (18% of cereals are treated) is sulphur. Average rates of application from fertilizer surveys for winter wheat are 50 kg/ha. Trace elements (nutrients required in grams) such as manganese and copper are only required where there is a known deficiency.

Cereals especially barley are prone to crop damage if the soil is acidic. Most arable soils should have the pH kept around or slightly above 6.5. On light sandy soils lime may be required every 4 years, less frequently on heavier soils.

Crop protection

The application of pesticides has changed over the last 30 years with the introduction of better products and the removal of outdated ones. Wheat tends to have the highest pesticide requirement and spring cereals the lowest. Table 11.2 highlights the main timings for some standard applications in winter cereals. The latest pesticide survey shows that on average a crop of winter wheat receives three applications of fungicides, two herbicides, one insecticide and one growth regulator spray. Most farmers do not rely on pesticides alone for crop protection especially as there are an increasing number of cases of pesticide resistance. For example, the use of a more integrated approach such as ploughing, later drilling of cereals or changing the rotation can help reduce blackgrass problems. Several foliar diseases are now showing resistance to some of the commonly applied fungicides. Growing resistant varieties as well as using mixtures of fungicides with different modes of action can help reduce the likelihood of the problem developing further. Most farmers rely on qualified (BASIS) advisers for their agrochemical advice unless qualified themselves. In 2003 various bodies produced a voluntary initiative for pesticides. It is hoped that the majority of farmers will join. The key

Table 11.1 Fertilizer use in cereals (reproduced with permission of HMSO).

Nutrient	Winter wheat	Winter barley
Nitrogen (N)	189	149
Phosphorus (P_2O_5)	66	65
Potassium (K_2O)	72	82

Table 11.2 Winter cereals – chemical calendar (adapted from Finch *et al.*, 2002).

Month	Crop growth stage[1]	Herbicide growth regulators	Insecticides/molluscicides[2]	Fungicides
September	Crop sowing (from mid-September)	Weed control including couch and volunteers post-harvest/pre-drilling	Seed dressing. Control of slugs	Seed dressing
October	Crop emergence	Grass weed control (blackgrass, wild oats, etc.) and broadleaved weeds		
	Leaf emergence		Control of aphids (BYDV)	
November	Tillering			
December				
January			Control of wheat bulb fly	
February				
March				Pre-T1 if foliar diseases such as yellow rust or mildew are present
	Stem extension	Broadleaved weed control		
April		Growth regulators and control of wild oats and cleavers		Eyespot and foliar diseases – T1
May	Flag leaf emerging			Foliar diseases – T2
June	Ear emergence			Foliar and ear diseases – T3
July	Flowering		Orange blossom midge (wheat).	
	Ripening		Aphids (wheat)	
August	Harvest	Couch control pre-harvest		

[1] Will be affected by sowing date, crop (wheat or barley) and climate.
[2] Tend to treat when necessary or in high-risk situations, not on a routine basis.
BYDV: Barley yellow dwarf virus.

aim is for farmers to consider the environmental impact of using pesticides and take steps to reduce the risk of any damage. If successful the initiative should avoid the need for the government to introduce a pesticide tax.

Yields and harvest

Winter barley is the first cereal to be harvested from mid July. Winter wheat and spring barley are usually harvested from the middle of August, followed by spring wheat. Winter oats are normally combined just before winter wheat.

Yields are usually highest from winter wheat (see Table 11.3). Grain can be stored for up to a month at 16–17% moisture, but for long-term storage grain should be cooled to below 5°C and dried to 14% moisture content. Insecticides may be used against storage pests such as mites and grain weevils. Grain is sold with a passport giving details of application of any insecticides in store. If farmers do not have their own drying facilities they may join a co-operative with dryers and

Table 11.3 Average cereal yields (reproduced with permission of HMSO).

Crop	Yield (t/ha)
Total cereals	7.1
Wheat	8.0
Barley	
Total	5.6
Winter	6.3
Spring	5.0
Oats	6.0

grain stores or hire a mobile dryer. Grain prices are usually lowest for grain sold straight off the field and then normally rise throughout the season.

Other cereal crops

A very small hectareage of other cereal crops are grown in the UK. Such crops are very specialized and currently the hectareage is unlikely to increase.

Durum wheat

Durum wheat is mainly grown in southern Europe. The grain is used for milling to produce semolina flour for making pasta. In the UK it is often very difficult to achieve satisfactory quality. Most success has been achieved in south-eastern counties of the UK, though yields are much lower than for conventional wheat. Most durum wheat is spring sown.

Rye

Rye can be successfully grown on poor, light, drought-prone soils. In the UK it is mainly grown on sandy soils. Europe is self-sufficient in rye. In the UK rye is autumn sown usually with hybrid varieties. As the crop is very disease resistant, fungicide inputs are low. The crop grows very tall and is susceptible to lodging, so growth regulators are required.

Triticale

Triticale, though a high-yielding cereal on the more marginal soils in the UK, has not really taken off. One of the problems has been marketing the grain. Production of the crop is as winter wheat though fungicide requirements are often lower.

Grain maize

Grain maize is the third most important cereal crop grown in Europe. Unfortunately the climate is not warm enough in the UK for the crop to reach maturity. A small number of farmers in south-east England have been successful, though yields are usually less than the average yield of between 8 and 9 t/ha achieved in France and Italy.

Oilseed crops

Soya, palm, oilseed rape (canola) and sunflowers are the most important oil crops grown in the world. In the UK oilseed rape is the main oilseed crop. Before the UK joined the EU it was a very minor crop. On joining the EU, with the good support prices and with new variety development, the crop area expanded rapidly (Figure 11.8).

Oilseed rape

- The most popular arable break crop in cereal rotations (see Glossary).
- Most oilseed rape varieties are grown for crushing for oil for human consumption.
- Majority is sown end of August to early September (winter oilseed rape).
- Similar input costs as winter cereals.
- Herbicides different from those used in cereals.
- High residual N left after harvest.
- Crop usually harvested in July.
- Yields can be variable. Average winter oilseed rape yields are 3 t/ha.

Initially varieties were high in erucic acid (HEAR) and only suitable for industrial use. Most varieties are now low in erucic acid so that the seed can be crushed for oil for human consumption. The oil content of rape is usually between 38 and 40%. Price received for oilseed rape is based on oil content. (A small area of HEAR is grown on set-aside land.) Once the oil is extracted the protein-rich meal can be used in animal

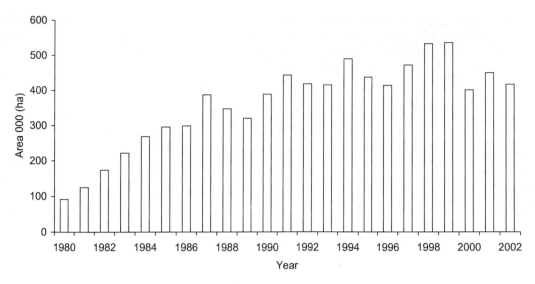

Figure 11.8   UK Oilseed rape area.

Table 11.4   Winter oilseed rape crop husbandry calendar.

Month	Crop growth stage	Herbicides	Insecticides/ molluscicides	Fungicides
August	Crop	Volunteer cereals	Slugs	Seed treatment
	Sowing	plus other weeds		
September	Emergence		Flea beetle	
October	Leaf formation	Broadleaved weed control plus	Cabbage stem flea beetle	Light leaf spot and *Phoma*
November	Five true leaves	weed grasses		
December				
January	Rosette stage			
February		Specific weed problems		
	Stem elongation			
March				Light leaf spot and *Phoma*
April	Green/yellow bud stage		Pollen beetle	
May	Full flowering Pods forming		Seed weevil/ bladder pod midge	*Sclerotinia Alternaria*
June	End of flowering Seed development			
July	Ripening	Couch control Desiccation		
	Harvest			

feed. Initially amounts were restricted due to high glucosinolate levels. Now the varieties grown are low in glucosinolates (called double low varieties – low in erucic acid and glucosinolates) so that greater amounts of meal can be included in livestock feed rations.

The majority of oilseed rape grown is winter rather than spring sown. Winter oilseed rape can return the same margins as some cereal crops particularly at current prices. The winter crop usually costs between winter barley and winter wheat input costs. There are many advantages of growing rape, not just the yield advantage in the following crop.

Winter oilseed rape is sown at the end of August/beginning of September, just before cereals are drilled. Because of the small seed only between 5 and 7 kg/ha is required to give a population of 50–70 plants/m^2 in the spring. Hybrid varieties require an even lower seed rate. Chemical inputs such as herbicides are very different from those used in cereals so there is a chance to control some difficult or herbicide-resistant weeds such as the bromes and blackgrass. Pest and disease incidence increased during the 1980s. Now commonly the crop receives two herbicide treatments, two fungicide sprays and one insecticide. See Table 11.4 for approximate timings. The only major pest not mentioned in the table are pigeons, which can cause serious damage in some years. Oilseed rape is very responsive to sulphur, and currently about a quarter of crops are treated with sulphur (61 kg SO$_3$/ha). Average fertilizer rates are 196 kg N/ha, 64 kg P$_2$O$_5$/ha and 68 kg K$_2$O/ha. Fertilizer rates have remained fairly static over recent years. A small amount of the nitrogen may be applied in the seed bed, but the rest is applied in early spring, usually one dressing in February followed by the main dressing in March. Nitrogen applications are earlier than in cereals as the crop is usually more advanced at that time of year. The characteristic yellow fields of flowering oilseed rape are usually seen from the end of April.

Harvest comes just before or after winter barley depending on the region where grown. Harvest is a critical time when growing oilseed rape, as the winter crop does not ripen all at the same time. The crop is indeterminate unlike cereals which are determinate, and pods are developing when the last flowers at the top of the raceme are still flowering (Figure 11.9). Only in some years does the crop ripen evenly enabling it to be combined directly off the field. Spring rape can often

Figure 11.9 Oilseed rape.

be combined directly. The farmer usually evens up ripening either by desiccating with a herbicide such as diquat and combining 7–10 days later, or the crop is cut (swathed) and then left to dry in swaths on top of the stubble for 14–20 days before combining. Desiccation is not recommended on exposed sites as there can be very high seed losses before combining. Swathing requires a specialist swathing machine to be available.

Harvest of the winter crop is usually from the middle of July to early August, with average yields of 3 t/ha. Spring crops are not harvested until the end of August to September, with average yields of 2 t/ha. At harvest the seed is usually between 8 and 15% moisture content and will need to be dried. Oilseed is usually sold at a moisture content of 9% or below.

Linseed

- Crop mainly grown for industrial use.
- Area grown significantly affected by amount of financial support.
- Suitable for a range of soil types.
- Mostly spring sown.
- Harvested September.
- Average yield 1.5 t/ha.

Figure 11.10   Linseed.

Linseed has been grown in the UK for many centuries. Linseed (Figure 11.10) has characteristic pale blue flowers and it is the seeds in the seed cases or bolls that are harvested for industrial use. Flax which is the same species has been bred for production of stem fibres. The area grown has fluctuated very dramatically over the last few years mainly due to changes in support prices. With the reduction in aid the area grown fell to only 12 000 ha in 2002. Without support this crop would be grown on an even smaller area as it would not be viable in the UK at current world prices.

A limited number of varieties are available, and most are spring sown; winter varieties have given variable results. Plant breeders are trying to breed varieties with higher oil contents, currently about 38%. Varieties are also being developed that are suitable for human consumption (Linola). Linseed is a very adaptable crop and can be grown on a wide range of soil types across the UK. The crop should not be sown more than one year in five due to some soil-borne diseases such as *Sclerotinia*. Results tend to be better from April-sown crop rather than drilling in March.

The crop requires fairly low inputs; seed is often the major cost. A plant population of 400–500 plants/m^2 needs to be established; depending on seed size this may mean a seed rate of between 40 and 60 kg/ha. Weed control is very important as the crop is very uncompetitive. As the crop is minor only a very limited range of herbicides are approved for use.

When the hectareage increased in the 1990s there was an increase in pest and disease problems, though farmers still do not tend to grow this crop with pesticide prescriptions (a programmed approach). If weather conditions are dry at establishment then the flax flea beetle can cause serious damage. Occasionally diseases such as *Pasmo*, *Alternaria* and *Botrytis* have caused problems. Fertilizer inputs are generally low; on average only 69 kg N/ha, 43 kg P$_2$O$_5$/ha and 56 kg K$_2$O/ha are applied.

Harvest of spring linseed is normally in September. Desiccation may be required especially if weed populations are high. Combining can be difficult as the straw is very wiry. The use of stripper headers has proven quite successful. The seed is normally sold at 8% moisture content so in most years it will require some drying. Care must be taken at harvest as the seed is very smooth and slippery and tends to find any holes in trailers, etc. Average yields for spring linseed are 1.5 t/ha.

Other oilseed crops

Plant breeders are developing varieties of soya bean (a legume) and sunflowers that can be grown and reach maturity in the UK. The area grown is still small and yields can be variable depending on the season.

A limited area of borage and evening primrose for oil extraction for the pharmaceutical industry are also grown. These crops are normally grown on contract with the end user.

Grain legumes

- Crops require no nitrogen as nitrogen-fixing bacteria are associated with root nodules.
- Harvested crops contain high protein content.
- The EU is a major importer of vegetable proteins.
- Are useful break crops in a cereal rotation.

A large number of grain legumes or pulse crops are grown across the world. The EU is a major importer of vegetable protein and through the arable area payment scheme (AAPS) supported their production. Field beans (Figure 11.11) and combinable peas (Figure 11.12) are the most commonly grown in the UK. In 2002, 164 000 ha of field beans and 85 000 ha of peas were grown. The average grain protein content of field

Figure 11.12   Peas.

Figure 11.11   Field beans.

beans is around 28%. Peas tend to have slightly lower protein content of 24%. There is much interest in growing lupins as the protein content is much higher at 40%. The problem with lupins is that results have been very variable, though currently there are some new better varieties and experimental work is being undertaken. Development work with new earlier-maturing varieties is also being carried out on navy beans (for baked beans) and lentils. No legume crops require nitrogen fertilizer because the *Rhizobium* bacteria associated with the root nodules can fix atmospheric nitrogen. Some crops such as lupins and navy beans need to be sown with a suitable *Rhizobium* inoculum unlike peas and beans where there is enough soil inoculum.

Peas and beans are good break crops in a cereal rotation and will leave some residual nitrogen for the following crop. They should not be grown more than one year in five in order to reduce the risk of some diseases and pests. Peas tend to yield better in dry years compared with field beans which tend to yield better in wet seasons. Commonly field beans are grown on the heavier soils and/or in higher rainfall areas, whereas peas tend to be grown mainly in drier areas on lighter soil types.

Field beans

- The most commonly grown combinable legume in the UK.
- Winter crops are planted from October on heavier soil types.
- Spring crops are sown from February on lighter soils.
- Low amount of inputs required, making them a cheap crop to grow.
- Harvest from end of August.
- Winter beans yield on average 3.5 t/ha compared with 3.25 t/ha for spring beans.

The majority of field beans are grown for animal feed. There is a small export market and market for small (tick) beans for pigeon feed. Most field beans have coloured flowers. There are a limited number of white-flowered varieties that have low tannin content in the seed coat so that greater quantities can be included in animal feed. Field beans are one of the cheapest combinable crops to grow, and inputs are usually low. The hectareage is split between winter beans planted in October/November and spring beans sown from February.

Winter beans are established very simply by broadcasting onto bare ground and then ploughing in to a depth of 10–15 cm. The earliest sown spring crops may also be ploughed in but more usually are drilled to a depth of 7.5 cm. Seed rates are between 180 and

250 kg/ha depending on the crop and seed size. Winter crops need an established population of 18 plants/m^2 whereas the spring crop requires about 40 plants/m^2.

Average fertilizer use is no nitrogen, 35 kg P$_2$O$_5$/ha and 40 kg K$_2$O/ha. As there is only a small area grown of field beans few pesticides have approval for use on the crop. The majority of herbicides are applied pre-crop emergence and are residual. There are very few post-emergence products for use in the winter crop. The winter crop is most susceptible to the disease chocolate spot and two fungicide treatments are commonly applied at early flowering in May. Average pesticide use in winter beans is normally one herbicide treatment and two fungicide sprays. The spring crop is usually not affected by chocolate spot though black bean aphid, bruchid beetle and rust can be serious problems and are commonly treated against.

Winter beans are usually harvested in August/ September. Spring beans are harvested slightly later, from the end of August. Occasionally, if ripening is uneven or there is a high weed population a desiccant may be required. Care must be taken drying the bean crop. Yields for winter crops are usually slightly higher than for spring crops. Average yields for winter beans are 3.5 t/ha compared with 3.25 t/ha for spring beans.

Combinable peas

- Several different types are grown for either human consumption or animal feed.
- The majority are spring sown (from February onwards).
- Recommended for sowing only if soil is relatively stone free.
- Higher input requirements than with field beans.
- Harvest in August.
- Average yield 3.75 t/ha.

Combinable peas are either grown for animal feed or for human consumption. All varieties can be grown for feed but only specific types for human consumption. Marrowfat varieties are the main type for dried and canned peas for human consumption (mushy peas). Marrowfat varieties are green/blue, large and dimpled. Large and small blues are blue/green, but the seed coat is smooth. White peas are white/yellow with a smooth seed coat. Maple peas have a coloured flower unlike the other varieties which are white flowered. Maple pea

seed is brown and is usually grown for harvesting green for wholecrop forage.

Fresh harvested green peas or vining peas for the frozen market are different varieties again. Vining peas are a very specialized crop and are harvested with a pea viner not a combine. Farmers normally grow these crops under contract, often in a co-operative. Vining peas are harvested much earlier than combinable peas, but if the crop becomes too ripe before harvesting green it may be left and combined for animal feed.

Most pea varieties are spring rather than autumn sown. Drilling usually starts as soon as ground conditions are fit from the middle of February. Peas are very sensitive to soil compaction so drilling date is very dependent on soil conditions.

Seed costs are one of the main inputs when growing peas. Drilling rate is usually between 200 and 250 kg/ha though it is dependent on seed weight, time of drilling and soil conditions. Optimum plant populations are dependent on pea type and price of seed. Required plant populations vary between 65 and 80 plants/m^2.

Fertilizer requirements are low: on average no nitrogen, 35 kg P$_2$O$_5$/ha and 40 kg K$_2$O/ha. Sulphur and manganese may need to be applied in some areas. Peas are very uncompetitive against weeds. A number of pre- and post-emergence products are available for most weed problems. Care must be taken with some varieties as they can be damaged by a number of products. Various pests and diseases may require treating in some years, but peas tend not to be treated as routine. Average inputs are two herbicide and two fungicide treatments plus one insecticide spray. Pea moth can be an important pest of peas for human consumption and additional insecticide treatments may be required.

Harvest in August is a critical time for dried peas as they invariably go flat. It is important that peas are grown in a stone-free soil or that the stones are rolled into the seedbed at drilling to avoid them being picked up with the combine. Plant breeders have developed semi-leafless varieties that stand better. (In a semi-leafless variety the leaflets are tendrils rather than leaves.) Yields of peas are usually higher than for field beans, but the variable or input costs are also higher. Average dried pea yield is 3.75 t/ha.

Root crops

The most important cash root crops in the UK are sugar beet and potatoes. Both crops are very mechanized and require specialized machinery such as planters and harvesters; potatoes also require a store. Depending on area grown farmers will either have their own machinery or rely on contractors. The tonnage grown of sugar beet is governed by EU quotas. There used to be potato quotas organized by the Potato Marketing Board, but these were withdrawn in 1997, and subsequently the area grown has fallen possibly due to some years when prices have been very poor. Neither crop was included in the AAPS. The contract sugar beet price is decided annually in the EU and does not fluctuate dramatically unlike potato prices! Though growing costs for these two crops are much higher than for the combinable crops, there is the potential for much higher returns.

Sugar beet

- Area of sugar beet grown is restricted by EU quotas.
- Sown from mid-March.
- Precision drilled.
- Herbicides are one of the major input costs.
- Fertilizer requirements are average except for high potassium and sodium.
- Harvest date affected by delivery date to factory.
- Average yields of 50 t/ha.

Around 170 000 ha of sugar beet (Figure 11.13) are grown in the UK each year on 7000 farms. Most growers farm near the sugar beet factories. The UK has agreements with a number of sugar cane countries (developing countries), and half the sugar consumed in the UK is from sugar cane. The UK is unusual in this respect and subsequently is the major sugar importing country in the EU. Because of price support for sugar in the EU, sugar beet has not suffered some of the large fluctuations in price seen in potatoes. The EU sugar regime is in place until 2006.

Husbandry

Sugar beet yields have steadily been improving with development of new varieties and technology. Sugar beet husbandry is now very mechanized; 40 years ago it was very labour intensive. The use of monogerm

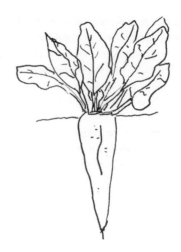

Figure 11.13 Sugar beet.

seed, precision drilling, development of pesticide technology and mechanization of harvesting have reduced labour requirements from 500 to only 50 man hours/ha today.

Sugar beet should only be grown one year in three, and most growers have longer rotations. Problems such as soil acidity should be corrected the year before establishing the crop. Soil cultivations should be kept to a minimum though the seedbed needs to be fairly fine with no compaction. Sugar beet is normally drilled from mid March. Earlier drilling can cause the beet to go to seed (bolting), so reducing yield and building up potential weed beet problems. Monogerm seed is drilled in 50-cm rows with a precision drill to establish 80 000 plants/ha. Seed spacing is about 18 cm within the row. Most seed is coated with an insecticide/fungicide dressing to reduce losses caused by pests and diseases.

Crop protection

As the crop is drilled to a low population it is critical to ensure that an adequate weed control (herbicide) programme is used. Herbicides are one of the most expensive inputs when growing beet. Commonly there are four applications of herbicide usually applied at very low rates when the weeds are just emerging. Separate measures such as steerage hoeing may need to be undertaken to try to reduce the increasing problem of weed beet.

The main pest problems are those that cause damage to the establishing beet crop and the aphids that spread virus yellows. The use of insecticidal seed dressings has reduced these problems. Foliar diseases such as powdery mildew may require treatment in some years. The fungal-transmitted virus *Rhizomania* has only recently appeared in the UK. Up to 2002 this was a notifiable disease in the UK (though widespread in Europe) and growers were unable to grow beet on affected fields. Depending on the strain UK growers are now able to grow *Rhizomania*-tolerant varieties. There is a *Rhizomania* outgoers scheme which enables affected growers to permanently transfer their contract requirement.

Fertilizers

Sugar beet is different from many crops, with a requirement for sodium (salt) and larger amounts of potassium than most other arable crops. Nitrogen rates are based on achieving maximum sugar yields. High nitrogen tends to encourage top growth and high amino nitrogen in the root. Recently average nitrogen application rates have fallen. On some soils there may be a requirement for magnesium as well as the trace elements boron and manganese. Phosphorus, potassium and magnesium are commonly applied in the autumn, whereas nitrogen is applied to the seed bed and at early post-emergence of the beet. Average rates of fertilizer used are now 106 kg N/ha, 76 kg P_2O_5/ha and 124 kg K_2O/ha.

Irrigation

Sugar beet is a deep-rooting plant, and it is mainly grown on the lighter soils in the drier parts of the UK. Irrigation can give an economic return in some years if there is a soil moisture deficit in June, July or August.

Harvest

Due to a number of sugar beet factories closing in England the sugar beet harvest or 'campaign' is starting earlier and ending later in the season. Delivery to the factories begins in September and has in some years finished as late as the end of February. Farmers are issued with dated loading permits; they then adjust their harvesting and storage (in clamps) so that the correct deliveries can be made. Prices paid will reflect time of harvest and effect on yield as well as dirt tare, purity and sugar content. Sugar beet yields on average 50 t/ha with a sugar percentage of 17%.

After harvest the sugar beet tops can be grazed by stock when the tops have wilted.

Once the sugar has been extracted from the beet the remaining pulp is sold for animal feed, e.g. as sugar beet nuts.

Potatoes

- Potatoes have no price support system and returns can be very variable.
- High input costs and labour requirements.
- Specific varieties and quality requirements for various markets.
- Preferably planted into a deep stone- and clod-free soil.
- Earlies and second earlies planted first, followed by maincrop varieties from April.
- Very high fertilizer requirements.
- Fungicides to control potato blight are main pesticides applied.
- Earlies are harvested from May and maincrop from September.
- Average maincrop yield 43 t/ha.

Between 150 000 and 160 000 ha of potatoes (Figure 11.14) are grown in the UK, though this dropped dramatically in 2003 partly due to the low prices the

Figure 11.14   Potatoes.

previous year. There is currently no support for potatoes in the UK. There has been quite a change in potato growing in recent years. Few growers now grow potatoes without first confirming the market. There has been a trend for some potato growers to expand and for many of the smaller growers to stop growing the crop. Just ten companies now buy 40% of the UK crop.

Overall potato consumption per person in the UK has stayed fairly constant at 108 kg/head/year, the third highest in Europe. What has changed is what is bought. Twenty years ago 70% of the potatoes were bought fresh and 30% were processed. Now equal quantities of potatoes are purchased fresh as processed. The fresh market itself has also changed with many more potatoes being washed and pre-packed rather than sold unwashed. With the availability of more controlled environment stores, potatoes of the same quality are available all the year round. Depending on the year, a certain amount of potatoes are usually imported into the UK. Other European countries also use potatoes for starch production.

All growers of potatoes pay a levy to the British Potato Council (formed from the Potato Marketing Board). As well as carrying out research work, its aim is to promote the British potato industry.

Potato husbandry

There are three main types of potatoes, first earlies, second earlies and maincrop.

First earlies are grown for the market from May until July. The crop is grown in parts of the country that are virtually frost free such as Cornwall and south Pembrokeshire. Many crops are grown under polythene so that the crop matures earlier when prices are high. Yields of earlies are much lower than for maincrop, but price is usually much higher. Production of the early crop is being affected by earlier imports and the introduction of salad potatoes. Earlies are traditionally sold with loose skins (slip skins) that need to be scrubbed. The small salad potatoes are sold with a hard or set skin all the year round and are taking a lot of the traditional earlies market.

Second earlies are planted in March for harvest in July and August and maincrop are usually planted from April and harvested from September.

A deep, medium- or light-textured soil is most appropriate for growing potatoes. If stones are present they can damage the tubers at harvest so most growers now use stone/clod separators pre-planting. Seed bed preparation often starts in the autumn to reduce the number of cultivations required in the spring.

Variety choice

Variety choice is very important as varieties have different cooking characteristics, yield potential and susceptibilities to several diseases and pests. Only a limited number of varieties are suitable for some markets, for example crisp production. Most manufacturers and retailers specify variety and quality requirements. Because of difficulties in plant breeding, few new varieties are currently being introduced.

Seed and seed rates

There are various categories of 'seed', most of which is grown in Scotland or The Netherlands. Note 'seed' is not true seed but small potato tubers grown for the seed market and complying with DEFRA seed regulations. The areas where seed is grown traditionally do not have as many aphids to spread viruses. Some farmers will save a certain amount of their own seed (once grown) as seed is a major cost when growing potatoes. For early crops the seed is often sprouted or chitted before planting.

Virtually all potatoes are planted mechanically with automatic or semi-automatic planters. Seed rates depend on the type of crop, variety, seed size, price and proposed market. Earlies, salad and seed potatoes require the highest seed rate and maincrop the lowest.

Fertilizer

Fertilizer requirements for phosphate and potash are normally higher than for nitrogen. Fertilizer rates are based on the soil nitrogen supply and soil analysis results as well as crop variety. Commonly, organic manures such as farmyard manure are applied. Average fertilizer rates for maincrop and second early potatoes are 175 kg N/ha, 163 kg P_2O_5/ha and 231 kg K_2O/ha. These rates do not include applications of farmyard manure or slurry. Organic manures are usually applied during the winter; the other nutrients are applied during seed bed preparation or placed in the ridges.

Potatoes are not affected by soil acidity.

Crop protection

Potatoes are a very competitive crop and after early weed control with one or two herbicide applications there are usually no further herbicide requirements. Few growers now cultivate the ridges to control weeds pre-emergence or at early crop emergence.

Many diseases attack potatoes, though the major requirement for pesticides is for the control of potato blight. Most growers use a 7- to 10-day spray programme as soon as blight warnings are issued. About nine fungicide treatments are commonly applied. There is a problem with resistant strains of blight to some of the currently available fungicides. Mixtures of fungicides with different modes of action are applied to reduce the risk. Currently no varieties are fully resistant to blight. Other diseases such as black scurf are controlled by fungicides applied to the seed before planting. Storage diseases are controlled either by store management or use of fungicides in store.

A few pests such as cutworm, wireworm and slugs can cause serious damage to potato tubers. The other main pest problem in the main potato growing areas is potato cyst nematode (PCN) which is controlled by use of resistant varieties, widening the rotation and application of nematicides.

Most potatoes going into packhouses supplying the supermarkets have to be 'assured' with one of the certifying bodies such as the Assured Produce Scheme. The assurance schemes have protocols for all aspects of crop husbandry including crop protection and fertilizer applications and ensure that the crop is traceable. Some of the protocols restrict the use of some approved pesticides.

Irrigation

Potatoes are fairly shallow rooting and can give large economic yield increases from application of water. Amount of irrigation is usually based on soil moisture deficits and many commercial irrigation model/programmes are available. Irrigation at the early tuberization stage can help reduce the incidence of common scab disease.

Harvest and storage

Before the crop can be harvested the tops or haulm have to be flailed or more commonly burnt off, for example with a desiccant or propane gas burner. Burning the tops will enable mechanized harvesting. Earlies are usually harvested soon after burning off the tops when the skins are still not set. Maincrop are usually left for 3 weeks before harvesting to ensure the potato skins are set or hard. Great care must be taken at all stages of harvesting to avoid damaging the potatoes. Potato damage can just be superficial cutting or scuffing of the skin or more seriously internal bruising which can mean a crop is rejected. Now that a large proportion of potatoes are sold washed, visual appearance is all important.

Some potatoes are sold straight off the farm, though many are stored either short term in an ambient store or in a cold store. Temperature control is important to ensure that sprouting does not take place even if the potatoes are not sold until May. Potatoes for crisping or chips should not be stored at too cold a temperature as it can affect the cooking quality of the potatoes. If necessary a sprout suppressant can be applied as the potatoes are loaded into store.

Yields

Average potato yields are 40 t/ha, though earlies may only yield 7 t/ha at the start of the season and a very good maincrop may yield up to 65 t/ha.

Minor crops

There has been a small increase in area grown of alternative crops for fibre or as a biomass energy source. With the aid of various support schemes the production of these crops has become more viable.

Fibre crops

Flax

During the 1990s there was increased interest in growing flax for short fibres for industrial use. A number of processing plants were set up. Unfortunately due to support changes the economics of the crop make it much less viable. Husbandry for flax is similar to that for linseed except that it is combined earlier and then the straw is left in the field to rett for 10–21 days before baling. All crops are grown on contract.

Hemp

There has been increased interest in growing hemp for fibre with the inclusion of the crop in the IACS scheme. A licence must be obtained from the Home Office to grow the crop and all crops are grown on contract. Hemp is an annual spring-sown crop with low input requirements. The crop is cut in August and left to rett in the field before baling. About half the fibre is used for industrial textiles; the remainder is used for animal bedding.

Energy crops

Miscanthus

Miscanthus is included in the Energy Crops Scheme. The crop originates from Africa and Asia and is really only suited to growing in southern England. Miscanthus is established by planting rhizomes. Once planted a crop should be harvested annually for the following 15–20 years. Crop inputs are low after establishment.

Conclusions

Crops that are grown in the UK have very different yield potentials (Table 11.5), input and labour requirements (Table 11.6). Final commodity prices and subsequently crop profitability have for many crops in the past been significantly affected by political influences. This has led to the relative importance of some crops changing. In the future with the EU mid-term review and the decoupling of support prices these effects should become less important.

Table 11.5 UK crop yields (reproduced with permission of HMSO).

Crop	Yield (t/ha)
Winter wheat	8.0
Spring barley	5.0
Winter oilseed rape	3.0
Linseed	1.5
Winter field beans	3.5
Combinable peas	3.7
Sugar beet	50
Potatoes	
Maincrop	42
Earlies	18

Table 11.6 Relative inputs (seed, fertilizers and crop protection chemicals) and crop labour requirements.

Crop	Relative inputs compared with potatoes (9)	Relative labour requirements compared with potatoes (9)
Winter wheat	1.0	1.0
Spring barley	0.7	0.9
Winter oilseed rape	1.0	1.0
Linseed	0.6	0.9
Winter field beans	0.6	0.9
Combinable peas	0.9	1.0
Sugar beet	3.0	3.5
Potatoes		
Maincrop	9.0	9.0

Glossary

Arable: field or farm growing crops.

Break crop: a crop grown in between cereal crops, e.g. breaks a rotation of continuous cereals. Example crops include oilseed rape and combinable peas.

Cash root crop: potatoes and sugar beet are examples of cash root crops. In some years, particularly for potatoes, returns can be much higher than for combinable crops such as cereals. Capital investment in machinery can be very high.

Catch crop: a crop grown between two main crops, for example stubble turnips grown after winter barley and utilized before drilling a spring-sown crop.

Crop rotation: describes the sequence of cropping for a field.

Cultivations: field operations using tractor-mounted machinery (e.g. a plough or harrows) that alter the soil structure. One of the main objectives of cultivations is to produce a suitable seed bed in which to sow the crop.

Drilling: the sowing of a range of seeds using a mechanized drill which ensures uniform depth of planting.

Fungicide timings, T1, T2 and T3: in winter cereals there are three main timings when fungicides may be applied. T1 is at the second node detectable stage, T2 is at the flag leaf stage and T3 at the ear fully emerged stage.

Holding: parcel of farmed land (rented or owned).

Integrated crop management: a system of crop production that produces safe, wholesome food in an environmentally sensitive way but which is still economically viable.

LERAP: Local Environment Risk Assessment for Pesticides. This is an assessment that needs to be done when applying certain pesticides that have a buffer zone (or no spray area) requirement when sprayed next to watercourses.

Lodging: when a crop has fallen over and is either lying or nearly lying flat on the ground.

Malting barley: all cereals can be malted to produce the substrate for fermentation to produce beer; but in the UK it is mainly barley that is malted. The maltster requires consistent grain characteristics and will dictate variety and grain quality including germination percentage.

Milling wheat: there are different variety and quality requirements for wheat according to whether it is to be milled to produce flour for biscuit making or for production of bread.

Packhouse: a place where produce such as potatoes are washed, graded and packed (pre-packed) ready for distribution to the major retailers (supermarkets).

Pesticides: since the introduction of the Control of Pesticide Regulations 1986, the term pesticide embraces all agrochemicals including fungicides, insecticides, herbicides, molluscicides and nematicides.

Pulse crops: combinable grain legume crops including peas, field beans and lupins.

Set aside: introduced under the CAP Reform Scheme. Anyone growing more than 15.5 ha of eligible crops had to set aside a minimum area of land. A number of different options, including non-cropped or cropped with industrial crops, are available for management of set-aside land.

Soil nitrogen supply: crop nitrogen recommendations are based on the soil nitrogen supply (SNS). The available soil nitrogen can be analysed to give the SNS, or it can be calculated from previous crop, soil type and winter rainfall using the Stationery Office publication *RB209 Fertiliser Recommendations*.

Soil pH: describes the acidity or alkalinity of the soil. Agricultural soils in the UK range from 4–8, with 7 being neutral. Some crops are more susceptible to acidity than others. The target is to keep soil near pH 6.5.

Soil structure: describes the arrangement of individual soil particles in aggregates.

Soil texture: describes the relative amount of mineral material (sand, silt and clay) in a soil.

Stripper header: a rotary header attached to the combine so that ears rather than ears and a lot of straw are harvested.

Stubble: the remains of the crop stalks that are left after the crop is harvested and any straw is removed.

Sugar beet quotas: the UK has a production quota for a certain tonnage of sugar set by the EU; each grower is then given a tonnage quota. In the past there was little or no movement of quotas between farms until 2002 when there was an outgoers scheme.

Vernalization: a cold period is required during the winter by some plants in order to trigger flower development.

Viner: a specialist harvester for fresh peas for freezing that separates the peas from the pods and stems (vines).

Voluntary initiative: the government wishes to introduce a pesticide tax in order to reduce the impact of pesticides on the environment. The voluntary initiative has been introduced by the industry as an alternative approach. One of the key elements is a commitment by users of pesticides to consider potential environmental impact and take steps to reduce the risk.

Further reading

Alford, D.V. (2000) *Pest and Disease Management Handbook*. Blackwell Science, Oxford.

Ashman, M. and Puri, G. (2002) *Essential Soil Science*. Blackwell Science, Oxford.

Biddle, A.J., Knott, C.M. and Gent, G.P. (1988) *The PGRO Pea Growing Handbook*. PGRO, Peterborough.

Cooke, D.A. and Scott, R.K. (1993) *The Sugar Beet Crop – Science Into Practice*. Chapman and Hall, London.

Davies, B.D., Eagle, D.J. and Finney, J.B. (2001) *Resource Management: Soil*. Farming Press, Ipswich.

DEFRA (2001) *Planting and Growing Miscanthus – Best Practice Guidelines*. DEFRA, London.

DEFRA (2002) *The British Survey of Fertilizer Practice – Fertilizer Use on Farm Crops* (published annually). DEFRA, London.

Finch, H.J.S., Samuel, A.M. and Lane, G.P.F. (2002) *Lockhart and Wiseman's Crop Husbandry*. Woodhead, Cambridge.

Gooding, M.J. and Davies, W.P. (2000) *Wheat Production and Utilisation*. CAB International, Wallingford.

Harris, P.M. (1995) *The Potato Crop: The Scientific Basis for Improvement*. Chapman and Hall, London.

HGCA (2003) *UK Recommended Lists for Cereals and Oilseeds* (updated annually). HGCA, London.

Henry, R.J. and Kettlewell, P.S. (2000) *Cereal Grain Quality*. Chapman and Hall, London.

Knott, C.M., Biddle, A.J. and Mckeown, B.M. (1994) *The PGRO Field Bean Handbook*. PGRO, Peterborough.

MAFF (2000) *Fertiliser Recommendations for Agricultural and Horticultural Crops (RB209)*. The Stationery Office, London.

NIAB (2003) *Pocket Guide to Varieties of Cereals, Oilseeds and Pulses, Spring and Autumn* (updated annually). NIAB, Cambridge.

NIAB (2003) *Pocket Guide to Varieties of Potatoes* (updated annually). NIAB, Cambridge.

Pettit, I. (2002) *Sugar Beet: A Grower's Guide*. British Beet Research Organisation, Norfolk.

Soffe, R. (ed.) (2002) *Primrose McConnell's The Agricultural Notebook*. Blackwell Science, Oxford.

Ward, J.T., Basford, W.D., Hawkins, J.H. and Holliday, J.M. (1985) *Oilseed Rape*. Farming Press, Ipswich.

Weiss, E.A. (2000) *Oilseed Crops*. Blackwell Science, Oxford.

Welch, R.W. (1995) *The Oat Crop – Production and Utilisation*. Chapman and Hall, London.

Useful websites

www.agricentre.co.uk – BASF.
www.britishsugar.co.uk – British Sugar plc.
www.defra.gov.uk – Department of Environment, Food and Rural Affairs (DEFRA).
www.hgca.com – Home Grown Cereals Authority (HGCA).
www.iacr.bbsrc.ac.uk – Biotechnology and Biological Sciences Research Council, Rothamsted.
www.niab.com – NIAB.
www.pesticides.gov.uk – Pesticides Safety Directorate (PSD).
www.pgro.co.uk – Processors and Growers Organization (PGRO).
www.potato.org.uk – British Potato Council (BPC).
www.ukagriculture.com – UK Agriculture.

12

Grassland

R.J. Wilkins

Introduction

Grassland is the most widespread land use in the UK, covering 13 million ha, or 54% of total land area. Of that, 11 million ha are used for agriculture, 64% of the total agricultural area. Table 12.1 provides details of grassland area in the UK. The proportion of the agricultural area that is grassland is particularly high in the north and west of the country, exceeding 75%, but it is less than 25% in East Anglia (Hopkins, 2000). The nature of grassland and its use will have large impacts not only on agricultural productivity, but also on biodiversity, landscape and the quality of water and the atmosphere.

This chapter will consider the use of grassland for:

- animal production;
- sustaining the productivity of overall mixed farming systems;
- the protection of water and air quality;
- recreation, biodiversity and landscape;
- multifuctional use to satisfy both agricultural and environmental objectives.

There is large potential for grassland to be used for industrial products including biomass energy, plant fibre and fine chemicals, but these uses are currently of little importance in the UK and will not be discussed in this chapter.

Grassland for animal production

This is the traditional use of grassland and still represents the major use. The costs of energy and protein in grassland feeds are normally lower than in grain and concentrate feeds, producing a strong economic incentive to maximize the use of grassland in feeding rumi-

nant animals. This is particularly the case with grazed grass, because grazing avoids the extra costs associated with conservation as hay or silage. Limitations to the feeding value of grassland feeds, however, mean that grain and concentrate feeds will be used to supplement grass and increase levels of animal performance. Nevertheless, grassland supplies some 60, 80 and 90% of the nutrients consumed by dairy cows, beef cattle and sheep respectively.

Ruminant animals evolved together with grassland. The rumen is of crucial importance, providing a capacious fermentation vat in which much of the fibre in the grass can be broken down by bacteria, fungi and protozoans with the production of volatile fatty acids and microbial protein. The volatile fatty acids are absorbed from the rumen to provide energy to the animal, whilst the microbial protein is broken down by enzymes in the small intestine and the amino acids produced are absorbed to provide a protein source to the animal.

The structure of the grass plant when it is not flowering is illustrated in Figures 12.1 and 12.2. The region of active plant growth (the stem apex) remains close to the ground and is protected from grazing or cutting. Grasses are thus able to survive and regrow after defoliation, and grassland can persist for many years without reseeding.

Extensive grassland systems using unsown semi-natural grasslands and no additional feeds occupy many millions of hectares throughout the world, with these extensive grasslands generally known as 'rangelands'. Grasslands of this type constitute much of the 5.6 million ha of 'rough grazing' in, mainly, upland and hill areas of the UK. Such grassland will, however, be of low productivity, giving some 1–4 t dry matter (DM)/ha annually, and low quality. Consequently the stocking rates (number of animals maintained per hectare) will be low and levels of animal production will be low.

Table 12.1 Grassland area in the UK. Data is estimated from official statistics except for 'Other grassland' which is taken from NERC (1977).

Category	Area (thousand ha)
Agricultural grassland	
Grass under 5 years	1 226
Grass 5 years and above	5 364
Sole right rough grazing	4 375
Total	10 965
Other grassland	
Common rights rough grazing	1 229
Sports grounds	110
Domestic lawns	90
Urban parks	132
Road verges, railway embankments, airfields	208
Nature reserves, common land, country parks	305
Total	2 074
Overall total	13 039

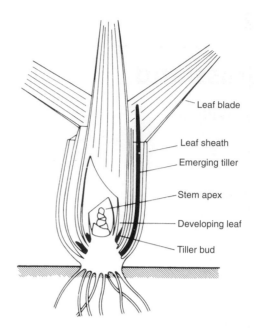

Figure 12.2 Position of stem apex and development of leaves and tillers in grassland (from Brockman and Wilkins, 2003).

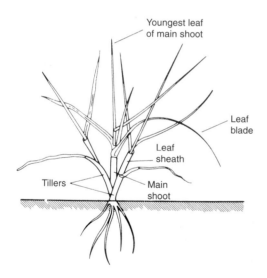

Figure 12.1 Vegetative development in grasses (from Brockman and Wilkins, 2003).

Grassland is very responsive to increases in the supply of plant nutrients and yields of over 15 t DM/ha can be obtained in the lowlands with intensive management. The input of fertilizers, the sowing of grasses such as ryegrass with high productivity and feeding value, and improved methods of grass conservation were all important in increasing animal production

from grassland. There was a ten-fold increase in the use of mineral fertilizers on grassland from 1950 to 1985 (Wilkins, 2000), but agricultural grassland in the UK is still characterized by a large range of fertilizer inputs. Even when 'rough grazings' are excluded, 37% of grassland receives less than 50 kg N/ha, whilst 16% receives more than 200 kg N/ha, with the highest applications being over 400 kg N/ha. Some grassland is regularly reseeded, whilst much grassland is never reseeded. Green (1982) found that 49% of grassland in England and Wales was more than 20 years old. Some 29% of this was on land that had serious impediments to cultivation (e.g. slope, drainage status), but 37% was on land free from physical impediments to cultivation or to sward management.

Factors determining grassland production

Climate

Figure 12.3 illustrates the typical seasonal pattern of grass production in south-east England. Little grass growth occurs when temperatures at the growing point near soil level are below 6°C, so that there is little if

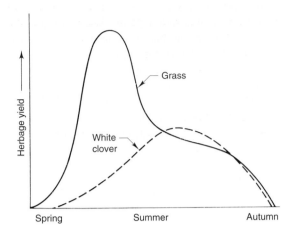

Figure 12.3 Seasonal pattern of grass production (modified from Brockman and Wilkins, 2003).

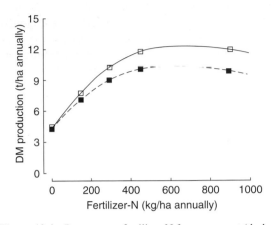

Figure 12.4 Response to fertilizer N for permanent (dashed line with solid squares) and reseeded (line with open squares) grassland (modified from Hopkins *et al.*, 1990).

any growth from November to early March. The start of the growth season results from increase in temperature, but during much of the growing season growth is limited by solar radiation. The particularly rapid growth in late spring is associated with many of the tillers in the sward being stimulated to produce flowers. The sward then has a higher efficiency of conversion of solar radiation to harvestable DM, because of both the change in sward structure and the small investment of DM in stubble and roots that occurs at that time. In the south and east of the country water shortage will often limit growth from June to August resulting in the loss, on average, of some 25% of potential annual production, with greater reductions on soils with low water-holding capacity such as sands. The loss of production in wetter areas of the country is much less, but in dry periods the supply of grass for grazing may become limiting.

Nutrients

Grass requires substantial quantities of nitrogen (N), phosphorus (P), potassium (K) and sulphur (S) in order to maximize growth. These nutrients may be supplied from:

- mineralization of organic and inorganic components in the soil;
- manures such as slurry and farmyard manure;
- returned excreta from grazing animals;

- rainfall (particularly important for S and to a more limited extent N);
- biological fixation of N, particularly by legumes;
- fertilizers.

Whilst the supply of all these nutrients is important, grass is particularly responsive to N, with many experiments demonstrating yield responses of more than 20 kg DM per kg of N fertilizer applied, as shown in Figure 12.4. However, most farmers still apply less fertilizer N than the rate of 300 kg/ha that has often been considered optimal (MAFF, 2000). A number of factors may reduce the response to fertilizer N:

- In hill or mountain conditions and in areas with very low rainfall and soils of low water-holding capacity, potential production may be severely limited, thus restricting the quantity of N required for maximum (or economic) grass growth.
- The supply of N from soil organic matter and applied manures may approach the quantity of N required for maximum growth.
- In grazed swards, some 80% of the N in the herbage grazed will be returned in dung and urine.
- In swards with substantial quantities of legumes, such as white clover, biological fixation can provide N equivalent to at least 200 kg of fertilizer N/ha.

There are, however, many situations where a farmer will not be seeking to maximize the rate of herbage production. Almost invariably grassland is used to provide nutrients for animals on the farm, rather than the grass being sold. Thus, if more grass is grown, the farmers must have sufficient livestock to consume the extra grass produced. This may bring with it a requirement for additional labour and for substantial capital investment, not only in additional livestock, but also in buildings, silos and machinery. If these factors are constraints and the farmer does not have alternative opportunities for profitably using his or her land, then he/she may sensibly opt for a lower-input more extensive system.

Type of grassland

There is much variation in the age of grassland and in its botanical composition. These factors though have less impact on level of production than climate and the supply of plant nutrients.

There are differences in yield between the grass species that are commonly sown, with Italian ryegrass normally having the highest yield, followed by perennial ryegrass, with somewhat lower yields for cocksfoot, timothy and meadow fescue (Spedding and Diekmahns, 1972). The yields of pure swards of grasses such as Yorkshire fog, rough-stalked meadow grass and creeping bent that are common components of old grassland, but not normally sown, are also generally less than those of the ryegrasses, particularly at high rates of N fertilizer (Frame, 1989; Sheldrick et al., 1990).

Progress has been made by grass breeders in increasing yields. The DM yields of the best varieties of perennial ryegrass entering Official Testing in the UK have increased by some 0.5% per year (Camlin, 1997). It appears though that there are important interactions between grass species and varieties and that swards containing a number of species and varieties often give yields higher than would be expected from the productivity of the components when grown separately. Smith and Allcock (1985) found that mixtures of perennial ryegrass, white clover and a number of normally unsown grasses gave yields similar to those for a simple perennial ryegrass–white clover mixture, despite the normally unsown grasses giving lower yields when grown pure and making a major contribution to the yield of the complex mixture. Hopkins et al. (1990) compared the productivity of old permanent swards containing a number of grass species with that of a reseed of perennial ryegrass at 16 sites in England and Wales. Reseeding gave a large increase in yield in the first year after sowing, but thereafter increase in yield occurred only with the highest inputs of N fertilizer (450 and 900 kg N/ha) and averaged only 9%. It appears that the productive potential of much of the grassland in the UK is high and that, once nutrient deficiencies are corrected, the production of old grassland may approach the highest levels achieved with sown perennial ryegrass.

Effect of legumes

Legumes, particularly white clover, are commonly present in old grassland and also included in many seed mixtures. Before the advent of N fertilizers, legumes were particularly important in sustaining high production, because of their ability to fix atmospheric N. This characteristic is now particularly important in organic systems, but also has wider relevance. When red and white clover and lucerne are grown pure, the annual rate of N fixation may be above 300 kg/ha. All three legumes are, however, often grown in mixture with grass and this is almost invariably the case with white clover. The use of the mixture will normally reduce total N fixation, but may increase DM yields, even without the use of N fertilizers. Legumes make an important contribution in low-input systems, particularly as feeding value tends to be higher with legumes than with grasses. The contribution of legumes in mixed swards with grasses tends to fall with increase in N fertilizer inputs and may be reduced with severe grazing. This has led to white clover being less used in intensive systems, but varieties are now available that maintain higher clover contents at high levels of N input from fertilizers and from grazing returns. The use of red clover and lucerne is discussed in more detail in Wilkins and Paul (2002).

Feeding value

Grassland feeds alone will rarely support milk yields of above 30 kg/day or growth rates of beef cattle above 1 kg/day. These figures are considerably below the genetic potentials for milk production and growth. In some circumstances, grassland feeds may be so low in quality that they barely support maintenance. Table 12.2 indicates the range of metabolizable energy (ME) – the best measure of energy value to the animal –

Table 12.2 Range in nutrient contents of different grassland feeds and grain (adapted from Wilkins and Kirilov, 2003).

	Metabolizable energy (MJ/kg DM)	Crude protein (g/kg DM)
Fresh grass	8.5–13.0	80–220
Grass silage	8.5–12.0	80–180
Lucerne silage	8.0–10.0	140–220
Maize silage	10.0–12.0	60–120
Whole-crop cereal silage	8.0–10.0	60–120
Grain	12.0–14.0	80–120

the CP in grassland feeds is often broken down very rapidly in the rumen, leading to high levels of ammonia absorption and high losses of N in urine. Grassland feeds make important contributions to the supply of minerals and vitamins, but these factors are much less likely to limit animal performance than energy or protein supply.

The feeding value of grasses is much affected by the growth stage at which they are cut or grazed. Figure 12.5 illustrates the changes in yield and quality that occur during uninterrupted spring growth in perennial ryegrass. The fall in ME arises from an increase in stem content and the very rapid fall in the ME of the stem as it matures and becomes indigestible. Much of the skill of grassland management is in achieving the best compromise between high quality (requiring cutting at a young stage of growth) and high yield (achieved with cutting at a more mature stage of growth).

There are quality differences between grass species. The ryegrass species are normally the highest in ME content, commonly having an advantage of 0.5 MJ/kg DM over other grass species harvested at the same

and crude protein (CP) contents in grassland feeds in comparison with grain. Values for ME are generally lower than those in grain and are highly variable. Furthermore, with a fall in ME, there will normally be a reduction in the quantity of DM that the animal will consume, with both these factors leading to reduction in animal production. Contents of CP may be above the nominal requirements of highly productive animals, but

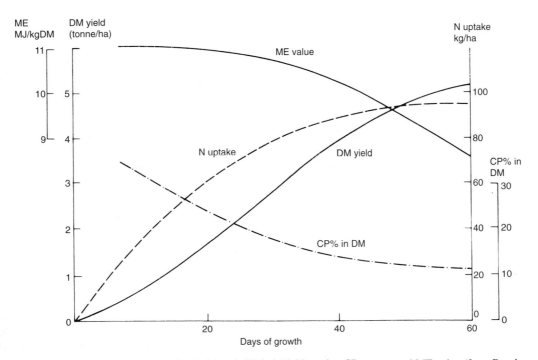

Figure 12.5 Relationship between days of grass growth, DM yield, N uptake, CP content and ME value (from Brockman and Wilkins, 2003).

growth stage. This gives increased animal performance with ryegrass. The quality of some of the grasses common in upland grassland (e.g. *Nardus stricta* and *Molinia caerulea*) is particularly low and ME contents may be 2 MJ/kg below those for perennial ryegrass. The quality of white clover is extremely high and ME concentration falls only slowly as the growth period is prolonged, because there is little stem formation with white clover. Both fresh and ensiled legumes are characterized by higher levels of feed intake at a particular ME concentration than grasses.

Fertilizer treatments have relatively small effects on ME content of grass, but high rates of N fertilizer will increase CP content.

Grass utilization

Animals normally graze during the growing season as this avoids the extra costs associated with harvesting for zero grazing (feeding cut fresh grass) or with conservation as hay or silage (Box 12.1). The quantity of DM conserved in the UK in 2000 was 2.4 million t of grass hay, 10.2 million t of grass silage and 1 million t of silage from other forage crops. The quantity of DM grazed was around 18 million t.

Grazing

As noted earlier, grassland is generally well suited to surviving grazing by ruminants and other herbivores. There are, however, difficulties in managing grazed swards to achieve the objectives of (1) efficient harvesting of the grass that has grown and (2) high rates of animal performance. In order to maximize utilization of the grass present, it is necessary to graze severely (to leave little uneaten herbage), but in order to maximize the intake (and performance) by the individual grazing animals it is necessary to graze laxly with animals being

Box 12.1   Grass utilization

- Ruminants are normally fed grazed grass during the growing season.
- Silage and hay for winter feeding are made from grass in excess of requirements for grazing.
- Special crops such as Italian ryegrass, maize, wholecrop cereals, red clover and lucerne are also grown for silage.

presented with a tall sward and not being forced to eat all the herbage available. The intake by grazing animals is a function of the time that they spend grazing and intake rate (grass DM consumed per minute of grazing). Intake rate is, in turn, determined by the number of bites taken per minute and bite mass (DM consumed per bite). The effects of sward height on these attributes are indicated in Figure 12.6. The major effect of short swards is to reduce bite mass because of difficulties in prehending sufficient herbage. Grazing time may increase, but not sufficiently to compensate for lower bite masses. However, if swards are grazed only laxly, a large proportion of the grass grown will not be consumed and will be lost through senescence and decay. Table 12.3 illustrates the effects of differences in grazing severity. The sward was grazed by beef cattle, with the numbers of grazing animals adjusted to maintain the swards at three different heights throughout the grazing season. The most severe grazing treatments had the highest stocking rates and resulted in the largest quantity of herbage being consumed per hectare, but the grazing animals grew at only 410 g/day. With lax grazing, individual animal performance was doubled, but the number of stock carried was much lower. In this example animal production per hectare was maximized at the intermediate sward height.

The maintenance of a target sward height is a key to efficient grazing systems. The optimum height will vary between different classes of animals according to their size and physiological condition (affecting the requirements for nutrients) as indicated in Table 12.4. The normal approach to maintaining sward height at the target level is to integrate grazing with cutting for silage. If herbage growth is rapid and sward heights are increasing above the target, the grazing area can be reduced with more of the land used for cutting for silage, with the reverse occurring during periods of slow growth. A range of different grazing schemes are followed in practice, as discussed by Brockman and Wilkins (2003), but the principles of efficient grazing are similar to those discussed above.

Silage

Silage is the predominant method of grass conservation in Britain and throughout western and northern Europe. Preservation depends on the maintenance of anaerobic conditions during storage and the achievement of a suf-

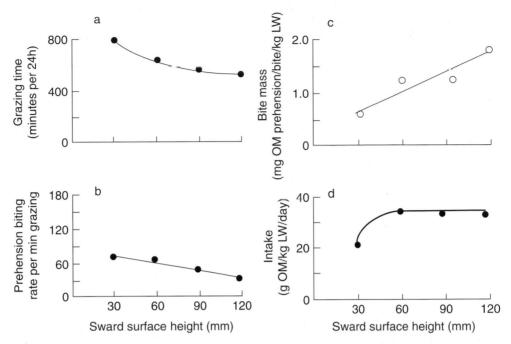

Figure 12.6   Effects of sward height on **a** grazing time, **b** prehension biting rate, **c** bite mass and **d** intake (from Penning *et al.*, 1991).

Table 12.3   Animal production and herbage utilization for steers continuously grazing a permanent sward maintained at different heights (reproduced from Wilkins *et al.*, 1983 with permission from the Institute of Grassland and Environmental Research).

	Grazing height (mm)		
	41	*58*	*78*
Grazing days (per ha)	1220	740	530
Liveweight gain (g/head/day)	410	820	900
Liveweight gain (kg/ha)	500	600	480
Utilized metabolizable energy (GJ/ha)	57	46	36

Table 12.4   Suggested target ranges of sward surface height for continuous stocking (from Hodgson *et al.*, 1986 with permission from the British Grassland Society).

	Sward height (mm)
Sheep	
Ewes and lambs	40–60
Dry ewes	30–40
Cattle	
Dairy cows	70–100
Dry cows	60–80
Finishing cattle	70–90

ficiently low pH to prevent the degradation of acids and protein by, mainly, clostridial bacteria (McDonald *et al.*, 1991). The pH is generally reduced by the fermentation of water-soluble carbohydrates (WSC) in the grass (or other forage crop) by indigenous bacteria to produce lactic acid and acetic acid. Figure 12.7 illustrates that if the reduction in pH is not sufficient, there will be a subsequent increase in pH with breakdown of the lactic acid formed initially and breakdown of

protein in the crop to ammonia, amines and other simple nitrogenous compounds. These changes will result in substantial lowering in the intake and protein value of the resulting silages, as well as increased losses of DM and nutrients. Table 12.5 indicates the sources of loss and their magnitude during silage making.

A key to loss reduction is preventing access of air during the storage period. The ready availability of cheap polythene sheeting had a revolutionary impact

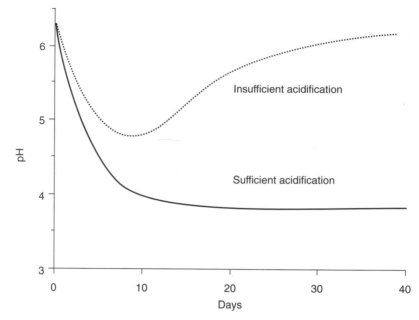

Figure 12.7   Course of pH changes in silages differing in extent of acidification (reproduced from Weissbach, 1996 with permission of the Institute of Grassland and Environmental Research).

Table 12.5   Sources of dry matter loss (%) during ensiling (from Brockman and Wilkins, 2003).

Source	Range of loss	Comment
Field		
Respiration	0–8	About 2% per day in field
Physical	0–8	Increases at high DM content
In silo		
Plant respiration	0–8	Increases with slow filling, poor consolidation and heating
Fermentation	0–10	No loss if completely homolactic; increases with clostridial fermentation
Surface waste	0–30	Increases with poor sealing
Aerobic deterioration after opening	0–20	Generally higher in high DM silages; heating indicates high losses
Effluent	0–10	Closely related to DM content and silage density

on silage making by providing a cheap and convenient means of preventing the access of air. It is, however, vital that the sheeting is protected from damage. This can be achieved by covering the sheeting with a layer of soil, straw bales or old tyres placed so they are in continuous contact. If a big-bale technique, in which the bale of grass is wrapped in several layers of polythene, is followed, there is a need to protect the polythene from damage by wind, vermin and birds.

The risks of clostridial fermentation are very low if the grass entering the silo has more than 3% of the fresh material as WSC, but most grasses have a less WSC content than this at cutting. In order to ensure good preservation it is usually necessary to either (1) increase DM content, by leaving cut grass in the field in order for some drying to occur before it is picked up for ensiling – this wilting will increase WSC concentration in the DM – or (2) add a chemical or biological additive at ensiling. The approach with additives is generally to add:

- additional fermentable material, such as sugar in molasses;
- acid, normally formic acid, at ensiling to reduce the pH of the material entering the silo;
- an inoculum of lactic acid bacteria in sufficient quantity to dominate the indigenous microflora and achieve more efficient conversion of WSC in the crop to lactic acid.

Bacterial inocula are widely used with partially wilted grass, ensiled at about 25% DM. Inocula are less likely to give reliable results with either very wet grasses or with crops such as legumes with very low WSC, unless they are wilted to higher DM contents.

Hay

The quantity of hay made has been much reduced in the last 30 years. Silage-making techniques have improved, whilst hay-making remains an extremely weather-sensitive method of grass conservation. Efficient silage systems involving wilting to 25% DM need only one day of dry weather, but with hay-making a DM content of 80% must be reached requiring, in the UK, at least four days of good dry weather. The longer the period in the field, the larger will be the losses of DM and of nutrients through continued plant respiration and through the losses of, particularly, leaf during mechanical treatment and pick up. Consequently DM losses will often exceed 30%, with reductions in ME content by 2 MJ/kg DM and substantial reduction in CP content.

Very few intensive production systems are now based entirely on hay, but many farms make small quantities of hay. If hay-making is used as an adjunct to silage-making, it becomes feasible to concentrate hay-making only within the period with the best weather forecast and have much lower weather risk than if the entire quantity of conserved forage is made as hay. Making some hay in this way may be attractive in order to provide (1) feed for young animals not accustomed to silage, (2) a dry feed to small groups of animals or animals remote from the farm buildings, to take advantage of the greater stability of hay compared to silage during feedout or (3) material more suited for sale and off-farm transport.

In making hay efficiently, there is need to:

- plan the work according to weather forecast;
- increase drying rates during the early stages of drying by mechanical conditioning and spreading

the cut grass across the total field area in order to intercept radiation efficiently;
- avoid vigorous mechanical treatments during the final stages of drying, when the grass is fragile and leaf loss most likely to occur.

The field drying period may be substantially reduced in barn drying systems in which the hay is moved to the barn at 60% DM and drying is completed by forced ventilation with cool or slightly heated air. Although widely practised in some central European countries, there is little use of this approach in the UK.

Grassland in crop rotations

Grass and forage crops with their associated animals played a key role prior to the industrial production of mineral fertilizers in maintaining soil fertility and the productivity of mixed farming systems involving both crop and livestock production. Red clover and fodder roots were part of the Norfolk four-course rotation which was followed from the eighteenth century in much of the arable areas in the east of England. The N fixed by red clover and nutrients in the manure from animals fed red clover, roots and straw contributed nutrients to subsequent crops in the rotation.

In the twentieth century ley farming systems were developed, with the alternation between leys (short-term grass or grass – clover mixtures grown for about three years) and arable crops grown for a roughly similar period. This approach was strongly advocated by Stapledon and Davies (1942) and was widely practised until the increase in fertilizer N use in the 1960s. In addition to providing feed for ruminants the grass ley was important because:

- soil organic matter accumulates under grass and mineralization of this organic matter that occurs following ploughing and cultivation supplies nutrients to the subsequent arable crops;
- alternation between different crop types provides control for weeds, pests and diseases;
- manure from animals on the grass ley can be returned to the arable crops;
- soil structure improves with grassland, thus improving cultivability for subsequent crops.

The production of cereals following a grass ley has been shown to be closely related to the quantity of mineralizable N in the soil which was in turn greatly influenced by the management of the ley (Clement and Williams, 1974). It increased with (1) the duration of the ley, (2) its legume content and (3) when swards were grazed rather than cut. The industrialization of farming in the second half of the twentieth century reduced the importance of ley farming. Plant nutrients could be provided in fertilizers and technical chemicals could be used to control weeds, pests and diseases. This set the scene for progressive specialization in farming and economic systems for continuous production of arable crops, without grassland or animals.

The recent increases in organic farming and interest in sustainable systems have created a new interest in grassland in mixed farming systems. The prohibition of use of mineral fertilizers in organic systems puts a strong focus on the use of legumes and organic manures to increase nutrient supply to crops, and grassland can play an important part in weed and disease control. The exploitation of N fixation in forage legumes and the use of N accumulated under grassland can be important components in other systems relying on low levels of external inputs.

For such systems to be efficient there is need for care in the transition from grassland to the subsequent crop, because of the risk that the N released by mineralization will be lost by leaching. Shepherd *et al.* (2001) found that 60–350 kg N/ha was lost by leaching following the ploughing of grassland and autumn reseeding. In order to minimize this loss (and to maximize the benefit from the released N) actions should be taken to ensure that there is good crop cover and a strong demand by the crop for N directly after cultivation. Losses are likely to be minimized if the grassland is cultivated in spring or early summer, giving greater opportunities than in the autumn for the rapid establishment of a subsequent crop and build-up of its demand for N. Whilst it is likely that there will be some increase in mixed ley farming, this will probably be rather slow as not only livestock but also livestock management skills and the whole infrastructure (e.g. buildings, fences, water supply) for livestock farming have been lost from much of the country.

Grassland management for air and water quality

There was little attention given to the effects of agriculture on air and water quality until the 1970s. It was then realized that agriculture was making major contributions to both point source pollution through organic matter escaping to watercourses and the diffuse pollution of watercourses and aquifers with nitrate. More recently, widespread concern has developed for:

- the pollution of watercourses with phosphorus and nitrate leading to eutrophication;
- emissions of ammonia to the atmosphere causing, on deposition, undesirable nutrient enrichment of fragile plant communities;
- the emission of the greenhouse gases nitrous oxide and methane.

Farming systems are having to change to reduce these adverse effects on the environment. Changes are being driven by legislation and a whole raft of regulations have now been implemented and are described on the DEFRA website (www.defra.gov.uk) and in the Codes of Good Agricultural Practice for air, soil and water (MAFF, 1998a, b, c), together with guidelines on ways to reduce environmental pollution. Legislation to protect water is particularly important, with over half of England being designated Nitrate Vulnerable Zones. Although extra costs may be involved, there are potential benefits for production efficiency, because actions taken to reduce the loss of nutrients to the environment will generally increase their recovery in the crop or the animal, with the result that production can be sustained with reduced inputs.

Although agriculture is being constrained by the requirements to improve air and water quality, new opportunities may emerge. In particular, the advent of carbon credits could lead to payments being made to farmers for systems that will increase C storage. When arable land is converted to grassland the quantity of C in the soil may double over a 20-year period (Tyson *et al.*, 1990). The equilibrium quantity of C per hectare in grassland soils is as large as that in temperate forests and some four times that in tropical forests (Goudriaan, 1990).

Table 12.6 Nutrient budgets (kg/ha) for dairy and hill sheep farms in the UK (from Jarvis *et al.*, 1996 and Haygarth *et al.*, 1998).

	Nitrogen		Phosphorus	
	Typical dairy farm	*Mixed grass–clover swards*	*Typical dairy farm*	*Hill sheep farm*
Inputs				
Fertilizers	250	0	16.0	0.4
Fixation	10	144		
Rain	25	25	0.2	0.1
Feed and straw	52	41	27.2	0.2
Total	337	210	43.4	0.7
Outputs				
Milk, meat and wool	67	54	16.2	0.2
Surplus (kg/ha)	270	156	27.2	0.5

Opportunities for mitigating environmental losses

Nitrate and ammonia

Nitrogen is lost to water as nitrate and to the atmosphere as ammonia and as the greenhouse gas nitrous oxide. Grassland and its associated ruminants represent a major contribution to all three of these sources of loss. This arises largely through the inefficiency of the ruminant animal; less than 20% of the N consumed will normally be retained by the animal or secreted in milk, with the remainder excreted in urine and faeces. In order to mitigate losses, the whole production system should be addressed in order to identify ways in which N efficiency in the soil, the plant, the animal and the manure can be increased (see Jarvis and Aarts, 2000). Any increase in efficiency means that less N needs to be introduced into the system in the form of fertilizers and feeds to sustain the required level of production. Whole-farm nutrient budgets, in which the inputs of N to the farm and the outputs from the farm are calculated, provide a strong focus for considering overall efficiency (see Table 12.6). In The Netherlands such budgets provide the basis for legislatively fixed limits for acceptable levels of N surplus.

Large losses can occur from grassland receiving high rates of fertilizer N, with these losses being greater with grazing than cutting and greater in well-drained conditions (Scholefield *et al.*, 1993). In grazed swards much of the N is lost from 'hot spots' resulting from the return of very high concentrations of N in urine. Potential approaches to reduce N leaching include:

- reduction in N fertilizer input, but this would in many cases result in the need to reduce stock numbers;
- altering the pattern of N application and, in particular, avoiding applications late in the growing season;
- utilizing grassland as silage rather than by grazing; this would avoid the creation of hot spots following urination, as the animal manures can be returned to the sward much more evenly and at the time when risks of leaching are low;
- reducing the length of the grazing season.

Brown and Scholefield (1998) developed models for the prediction of N transfers in grassland systems and impacts on production. These models can be used to identify N fertilizer inputs and patterns of input that will optimize production, whilst satisfying constraints in relation to nitrate leaching and ammonia emissions.

In Nitrate Vulnerable Zones, there is (1) a limit of 250 kg/ha to the quantity of N that can be applied to grassland in fertilizer plus manure, (2) a specified requirement for slurry storage capacity, (3) prohibition of slurry spreading to grassland from 1 September to 1 November on sandy or shallow soils and (4) requirement for farm records of cropping, livestock and the use of manures and fertilizers.

Agriculture is responsible for over 80% of the national emissions of ammonia, with much of this arising from grassland systems (Misselbrook *et al.*, 2000; DEFRA, 2002). The loss of ammonia directly from fertilizers applied to grassland is low, but 60% of the agricultural losses result from feeding ruminant animals. The urea in urine is rapidly converted to ammonia and liable to be lost from grazing returns, from animal buildings, from manure stores and on application of manure to land. Actions that can be taken to mitigate emissions, discussed in DEFRA (2002), include:

- adjusting fertilizer and feeding strategy to reduce N inputs to the animal and optimize efficiency of feed utilization;
- increasing grass utilization by grazing rather than through housed animals, because losses are less with grazing (note though that this change will increase losses of N through leaching);
- adapting building design to recover ammonia (an expensive option, but incorporated in some new buildings in The Netherlands);
- producing solid farmyard manure rather than liquid slurry;
- covering slurry stores to reduce ammonia volatilization;
- ploughing in manure to arable land directly after application;
- application of slurry to grassland by shallow injection or by trailing shoe applicator rather than by aerial application.

Whilst the adoption of these actions should allow reductions in N fertilizer use, a requirement to satisfy stringent limits for ammonia emission will increase production costs.

Phosphorus

Phosphorus is lost to water through leaching and through the transfer of P in particulate form from soil or manures. Losses of P that are insignificant agriculturally can cause eutrophication of watercourses and lakes. P balances for contrasting grassland farms are given in Table 12.6. There may be substantial inputs in fertilizers and imported feeds. It appears that in many circumstances these inputs can be reduced without adverse effects on plant or animal production. The possibilities of P loss are reduced if:

- inputs in fertilizers and feeds are reduced;
- fertilizers and manures are applied at times when heavy rainfall is unlikely over the days following application;
- if the risks of particulate flow to water are reduced by maintaining good crop cover over the winter and maintaining vegetated buffer strips adjacent to streams and rivers.

Greenhouse gases

Agriculture does not make a major contribution to the emission of carbon dioxide in the UK, but is responsible for 64% and 42% of the emissions of nitrous oxide and methane respectively.

Grassland is directly responsible for much of the emission of nitrous oxide (Brown *et al.*, 2001). The losses tend to increase with the total quantity of N cycling in the system and the losses, particularly through denitrification, are particularly high when soils are anaerobic through waterlogging. The main approaches to reduce nitrous oxide production are to (1) improve the overall efficiency of N utilization, so that animal production can be maintained with lower inputs of N, particularly as fertilizers, (2) reduce N inputs and stocking rates, (3) avoid application of fertilizers and manures at times when conditions are likely to become waterlogged and (4) improve the drainage status of the grassland (although this may increase nitrate leaching).

Fermentation in the rumen is responsible for 90% of the agricultural emissions of methane, with the remainder being produced from manures. The proportion of the feed DM that is emitted from the rumen as methane varies from 40–100 g/kg DM and values tend to be higher with feeds of relatively low ME, such as many grassland feeds. When methane emissions are expressed per unit of animal output, the losses generally fall with increase in animal performance. This arises because methane output relates closely to DM intake and the feed required for maintenance (and the methane produced from it) is a lesser proportion of the total when intake and animal production is high. Levels of intake and animal performance are often lower with rations containing high proportions of grassland feeds than those based on grain or concentrate feeds, leading to high methane production per unit of animal output with the grassland feeds. Current research seeks to reduce methane outputs by identifying feeds and animals that produce less methane or by directly manipulating rumen fermentation.

Grassland for recreation, biodiversity and landscape

The large proportion of the country covered by grassland has ensured that grassland has over the centuries played a key role in the provision of recreational facilities, the maintenance of wildlife and the appearance and aesthetics of the countryside. There is now a strong focus on these aspects because (1) intensification of grassland farming has led to reductions in biodiversity, particularly in the numbers of plant species, insects and birds in grassland, and (2) society has increased concern for recreation and the environment associated both with the decline in environmental quality and with the increased affluence and leisure time of the population.

Recreation

Substantial areas of grassland are managed intensively for recreation on sports pitches, golf courses and lawns (Table 12.1). The characteristics of grasses that give good persistence under grazing management also give grass the ability to survive well with the intensive cutting that occurs in the management of these amenity areas. The grass needs also to wear well when subjected to scuffing and trampling, to grow over a long season and to maintain a green colour. Many grass varieties have been bred specifically for amenity use and this whole area is thoroughly reviewed by Thorogood (2000).

Grassland also provides grazing and conserved forage for the increasing population of horses kept for recreational purposes. Bax and Lane (2005) estimate that 0.8 million ha of grassland is used by horses. Many of the principles of grassland management discussed earlier are also applicable to management for horses, but there are additional factors to consider relating both to grazing and to conservation. Fields tend to be grazed very irregularly by horses, with some areas being grazed extremely severely, producing 'lawns', whilst other parts become latrine areas with the accumulation of large quantities of ungrazed mature herbage. Approaches to mitigate this problem, discussed by Bax and Lane (2005), include:

- the movement of feeding and watering points;
- the adoption of rotational grazing;
- the alternation between grazing and cutting;
- and, in extreme circumstances, sward renewal.

Hay has traditionally been the main conserved forage given to horses, but it has long been realized that there is a premium on the hay being free of mould and dust in order to prevent respiratory diseases. It is crucial to use good haymaking techniques, in which the cut grass is exposed to bad weather for a minimum period and it is sufficiently dry for mould-free storage. Barn-dried hay is particularly well suited to horses. There is increased interest in the use of silage for horses. Most attention has been given to the ensiling of grass wilted to a high DM content in wrapped large or small bales. With many horses being maintained in small groups, the use of small bales (consumed within a day or two of opening) avoids the problem of mould growth and aerobic deterioration, but the cost of plastic for wrapping is high.

Biodiversity and landscape

The general effects of grassland management on agricultural production and on different classes of biota are summarized in Table 12.7. Most of the actions that increase agricultural output have negative effects on plants, insects and birds. It is not surprising that intensification of grassland use in the second half of the twentieth century had a negative effect on wildlife and increased concerns for actions to maintain or enhance biodiversity.

Actions to protect species and communities and the countryside more generally are discussed in Chapter 6. Grassland is a major component of many national nature reserves and Sites of Special Scientific Interest (SSSI), which were set up from the 1950s to protect particularly prized species or communities.

Management to protect biodiversity on farms received a major stimulus in 1985 with the provision in the EU for the introduction of 'agri-environmental schemes' in which farmers would be paid for following practices that were intended to either maintain or enhance the environment. These schemes are now incorporated in the Rural Development Programme and over 2.5 million ha of land in the UK is committed to the schemes. The major ones are the Environmentally

Table 12.7 Generalized effects of agricultural management options on utilized output and on wildlife (numbers of) (adapted from Wilkins and Harvey, 1994).

	Utilized output	Higher plants	Insects	Birds	Small mammals
Increase fertility	++	−−	−−	−	0
Increase level of utilization[1]	++	−−	−−	−	−−
Increase reseeding	+	−−	−−	−−	−
Improve drainage	+	−−	−−	−−	?
Use herbicides	+	−−	−−	−	−
Use pesticides	+	0	−−	V	−
Earlier cutting	+	−−	−−	−−	−
Later cutting	−	++	++	++	++
Graze cattle rather than sheep	−	+	++	+	+
Graze intermittently rather than continuously	0	V	++	++	V

The direction of change is indicated by + and −; V indicates variable effects.
[1] Particularly through increase in grazing severity.

Sensitive Areas (ESA) scheme and Countryside Stewardship (CS) (and their equivalents in Wales, Tir Gofal, Scotland, Rural Stewardship, and Northern Ireland, Countryside Management scheme). These are voluntary schemes, with ESA available only in designated areas of the country, but CS is available throughout. The Hill Farm Allowance scheme and the Organic Farming scheme are also relevant to grassland.

There are different tiers in the ESA schemes, with the higher tiers involving more severe restrictions and higher levels of payments. In relation to grassland, the provisions relate principally to restrictions on:

- the use of fertilizers, manures and technical chemicals;
- permissible stocking rates;
- the times at which swards can be grazed or cut;
- ploughing and reseeding.

These restrictions are imposed with the intention of increasing biodiversity, particularly through the protection and improvement of species-rich grassland, and reducing the risk of diffuse pollution. Many farmers have joined the ESA schemes with 1.6 million ha included in the UK, at least 80% of which would be grassland.

Countryside Stewardship encompasses a range of possible actions, with objectives including (1) sustaining the beauty and diversity of the landscape, (2) improving and extending wildlife habitats, (3) improving opportunities for enjoying the countryside and (4) creating new habitats and landscapes. It is more targeted than ESA and applications for CS agreements are subject to individual scrutiny for 'value for money'. Some 0.7 million ha is involved in the UK in CS and the equivalent schemes in the other countries, with, again, much of this being grassland.

Support from the EU and the UK governments for broadly agri-environmental measures will increase with reform of the Common Agricultural Policy (CAP). In 2005 the ESA and CS schemes in England will be replaced by Environmental Stewardship with two main strands. Entry Level Stewardship will be open to all farmers and land managers who want to deliver a basic level of environmental management above that of good farming practice and satisfying cross compliance requirements. Those who want to deliver higher levels of environmental management will be able to apply for Higher Level Stewardship.

There has been much recent research to establish approaches that will increase biodiversity on farms. In grassland, the focus has been on identifying ways of increasing the number of plant species in grassland. This is important in its own right, but also because increased diversity of plant species will increase the range of food sources for butterflies and other insects and birds and thus increase biodiversity more generally.

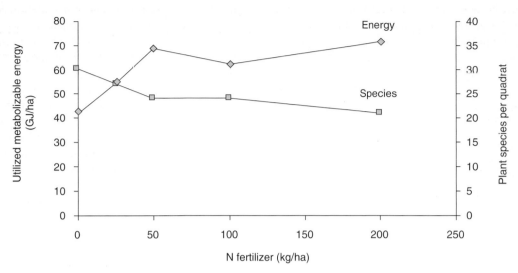

Figure 12.8   Output and number of plant species with different fertilizer N inputs (from Tallowin, 1996 and Brockman and Wilkins, 2003).

Where existing grasslands already contain a large number of plant species, the input of fertilizers will reduce species numbers. This is shown for a species-rich grassland on the Somerset Levels that had previously had no input of mineral fertilizers (Tallowin, 1996). After application of fertilizers for 4 years, there was a significant reduction in species number even with the lowest input of fertilizers used – 25 kg N/ha plus P and K to replace that removed in hay (Figure 12.8). Fertilizer inputs were then discontinued, but there was only a slow recovery in species numbers. It was calculated that it would take 10 years for species numbers to recover to the original numbers following the input of 200 kg N/ha for only 4 years. These results confirm that it is particularly important to prevent intensification of inputs to grassland that is already species rich, because damage to the flora occurs rapidly, but it is only slowly restored.

Techniques are being developed for increasing the rate at which species numbers may be restored in grassland that has undergone intensive management. Janssens *et al.* (1998) examined the relationship between soil phosphorus content and number of plant species in old grassland in a number of European countries. Soil P concentrations are normally increased in intensive systems and none of the grassland with high

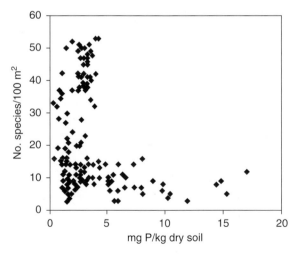

Figure 12.9   Relationship between extractable phosphorus in top 15 cm of soil and number of plant species (reproduced from Janssens *et al.*, 1998, with permission of Kluwer Academic Publishers).

soil P levels had a large number of species (Figure 12.9). When soil P contents were low, a wide range of species numbers were present. It is probable that species numbers were low in many of these swards because of a deficiency in the supply of propagules of new species from either the seed bank or from surrounding vegetation. This was well illustrated in an experiment in Wales in which the cessation of fertilizer inputs gave a rapid reduction in agricultural output, but even after 8 years there was little increase in the number of broadleafed species present (see Wilkins, 2002).

In order to increase species numbers rapidly it is necessary to:

- achieve low levels of soil P;
- ensure the presence of propagules of the desired species;
- provide suitable conditions for their establishment.

Hopkins *et al.* (1999) investigated a range of restoration treatments. All treatments involved the cessation of fertilizer inputs. In soils with low P content, high species numbers were achieved without cultivation and without seed addition. With higher soil P content it was necessary to introduce seed by slot seeding, and more rapid increases in species numbers occurred following deturfing and sowing.

Multifunctional grassland

The increasing importance of environmental factors provides new constraints and opportunities for grassland farmers. It is no longer appropriate to solely target agricultural output efficiency with no regard for the environmental effects of the managements imposed. Also, if only agricultural output is targeted, a farmer will be ignoring the possibility of increasing his or her income by providing environmental services through participation in agri-environmental schemes. There is now much discussion on how multiple objectives may best be met in grassland farming.

It is clear that there will be increasing legislation to restrict losses of N compounds, P and, probably, greenhouse gases to water and air. Ways in which losses to the environment can be limited have already been discussed. It will be important to recoup as much as possible of the extra cost through savings in inputs of fertilizers and feeds. Systems are being developed that are much more 'nutrient tight', but it is likely that imposition of strict emission standards will lead to some increases in the cost of production. The onset of payments for carbon credits and the demand by water companies in some areas for supply of water satisfying

particularly stringent water purity standards may open up new sources of income for grassland farmers.

It is clear that there is some conflict between efficient agricultural production and the enhancement of biodiversity. There is not a well-developed market for biodiversity. However, at least in the short to medium term, the market for biodiversity is represented by the monetary payments made in the various agri-environmental schemes. It is unfortunate that these schemes generally provide payments to farmers for following rules that it is believed will increase biodiversity, rather than making direct payments for achieved increases in biodiversity. There is much debate on whether support in agri-environmental schemes should be spread broadly across a wide area (with a low payment per hectare) or be more precisely targeted (with high payment per hectare on fewer hectares). In relation to grassland, there is a strong case for targeting support. Farming at an intermediate level of intensity throughout gives the risk of producing systems that are agriculturally inefficient, but still do not give real advantages in biodiversity (Wilkins and Harvey, 1994). It is, however, often possible to identify areas on farms in which management specifically for biodiversity will give rapid results (areas where the number of plant species are already reasonably high, where soil P is low and where there is a good seed bank or other adjacent sources of new species). These areas could be managed specifically for increased biodiversity, whereas other areas on the farm could be managed for agricultural objectives, constrained to restrict pollution of water or the atmosphere. There would be some agricultural output from the areas managed for biodiversity, from grazing or conserved forage, and the integration of the feed resources produced from areas managed principally for biodiversity with those areas managed for agricultural production efficiency will provide new challenges for farmers (see Wilkins, 2002).

References

Bax, J. and Lane, G.P.F. (2005) *Emerging Equine Science*. Proceedings of British Society of Animal Science Occasional Symposium, No. 32, Nottingham University Press.

Brockman, J.S. and Wilkins, R.J. (2003) Grassland. In: *The Agricultural Notebook* (ed. R.J. Soffe), pp. 131–76. Blackwell Science, Oxford.

Brown, L. and Scholefield, D. (1998) A model to optimize nitrogen use according to grassland management, soil conditions and weather patterns for economic and environmental targets. *Grassland Science in Europe*, **3**, 651–4.

Brown, L., Armstrong Brown, S., Jarvis, S.C., *et al.* (2001) An inventory of nitrous oxide emissions from agriculture in the UK using the IPPC methodology: emission estimate, uncertainty and sensitivity analysis. *Atmospheric Environment*, **35**, 1439–49.

Camlin, M.S. (1997) Grasses. In: *Proceedings of British Grassland Society Occasional Symposium, Seeds of Progress* (ed. J.R. Weddell), no. 31, pp. 2–14.

Clement, C.R. and Williams, T.E. (1974) Leys in arable rotations. *Silver Jubilee Report 1949–1974*. Grassland Research Institute, Hurley, pp. 18–26.

DEFRA (2002) *Ammonia in the UK*. AEQ Division, DEFRA, London.

Frame, J. (1989) Herbage productivity of a range of grass species in association with white clover. *Grass and Forage Science*, **45**, 57–64.

Goudriaan, J. (1990) Atmospheric CO_2, global carbon fluxes and the biosphere. In: *Theoretical Production Ecology: Reflections and Prospects* (ed. R. Rabbinge), pp. 17–40, Simulation Mongraphs, 34, Pudoc, Wageningen.

Green, J.O. (1982) *A Sample Survey of Grassland in England and Wales 1970–1972*. Grassland Research Institute, Hurley.

Haygarth, P., Chapman, P.J., Jarvis, S.C. and Smith, R.V. (1998) Phosphorus budgets for two contrasting grassland farming systems in the UK. *Soil Use and Management*, **14**, 160–7.

Hodgson, J., Mackie, C.K. and Parker, J.W.G. (1986) Sward heights for efficient grazing. *Grass Farmer*, **24**, 5–10.

Hopkins, A. (ed.) (2000) *Grass: Its Production and Utilization*. Blackwell Science, Oxford.

Hopkins, A., Gilbey, J., Dibb, C., Bowling, P.J. and Murray, P.J. (1990) Response of permanent and reseeded grassland to fertilizer nitrogen. 1. Herbage production and herbage quality. *Grass and Forage Science*, **45**, 43–55.

Hopkins, A., Pywell, R.F., Peel, S., Johnson, R.H. and Bowling, P.J. (1999) Enhancement of botanical diversity of permanent grassland and impact on hay production in Environmentally Sensitive Areas in the UK. *Grass and Forage Science*, **54**, 163–73.

Janssens, F.A., Peeters, A., Tallowin, J.R.B., *et al.* (1998) Relationship between soil chemical factors and grassland diversity. *Plant and Soil*, **202**, 69–78.

Jarvis, S.C. and Aarts, H.F.M. (2000) Nutrient management from a farming systems perspective. *Grassland Science in Europe*, **5**, 363–73.

Jarvis, S.C., Wilkins, R.J. and Pain, B.F. (1996) Opportunities for reducing the environmental impact of dairy farming managements: a systems approach. *Grass and Forage Science*, **51**, 21–31.

MAFF (1998a) *Code of Good Agricultural Practice for the Protection of the Air*. MAFF, London.

MAFF (1998b) *Code of Good Agricultural Practice for the Protection of the Soil*. MAFF, London.

MAFF (1998c) *Code of Good Agricultural Practice for the Protection of Water*. MAFF, London.

MAFF (2000) *Fertilizer Recommendations for Agricultural and Horticultural Crops (RB209)*. The Stationery Office, London.

McDonald, P., Henderson, A.R. and Heron, S.J.E. (1991) *The Biochemistry of Silage*. Chalcombe Publications, Marlow.

Misselbrook, T.H., Van der Weerden, T.J., Pain, B.F., *et al.* (2000) Ammonia emission factors for UK agriculture. *Atmospheric Environment*, **34**, 871–80.

National Environment Research Council (1977) *Amenity Grassland – The Needs for Research*. NERC Publication Series 'C', no. 19. NERC, Swindon.

Penning, P.D., Parsons, A.J., Orr, R.J. and Treacher, T.T. (1991) Intake and behaviour responses by sheep to changes in sward characteristics under continuous stocking. *Grass and Forage Science*, **46**, 15–28.

Scholefield, D., Tyson, K.C., Garwood, E.A., Armstrong, A.C., Hawkins, J. and Stone, A.C. (1993) Nitrate leaching from grazed grassland lysimeters: effects of fertilizer input, field drainage, age of sward and patterns of weather. *Journal of Soil Science*, **44**, 601–13.

Sheldrick, R.D., Lavender, R.H. and Martyn, T.M. (1990) Dry matter yield and response to nitrogen of an *Agrostis stolonifera*-dominant sward. *Grass and Forage Science*, **45**, 203–13.

Shepherd, M.A., Hatch, D.J., Jarvis, S.C. and Bhogal, A. (2001) Nitrate leaching from reseeded pasture. *Soil Use and Management*, **17**, 97–105.

Smith, A. and Allcock, P.J. (1985) The influence of species diversity on sward yield and quality. *Journal of Applied Ecology*, **22**, 185–98.

Spedding, C.R.W. and Diekmahns, E.C. (eds) (1972) *Grasses and Legumes in British Agriculture*. Commonwealth Agricultural Bureaux, Farnham Royal.

Stapledon, R.G. and Davies, W. (1942) *Ley Farming*. Penguin, Harmondsworth.

Tallowin, J.R.B. (1996) Effects of inorganic fertilizers on flower-rich hay meadows: a review using a case study on the Somerset Levels, UK. *Grasslands and Forage Abstracts*, **66**, 147–51.

Thorogood, D. (2000) Amenity grassland. In: *Grass: Its Production and Utilization* (ed. A. Hopkins), pp. 317–42. Blackwell Science, Oxford.

Tyson, K.C., Roberts, D., Clement, C.R. and Garwood, E.A. (1990) Comparison of crop yields and soil conditions during 30 years under annual tillage or grazed pasture. *Journal of Agricultural Science*, **115**, 29–40.

Weissbach, F. (1996) New developments in crop conservation. *Proceedings of the 11th International Silage Conference*, Institute of Grassland and Environment Research, Aberystwyth, pp. 11–25.

Wilkins, R.J. (2000) Grassland in the 20th century. In: *IGER Innovations 2000*. Institute of Grassland and Environmental Research Aberystwyth, pp. 26–33.

Wilkins, R.J. (2002) Environmental sustainability of wool production. *Wool Technology and Sheep Breeding*, **50**, 705–23.

Wilkins, R.J. and Harvey, H.J. (1994) Management options to achieve agricultural and nature conservation objectives. In: *Grassland Management and Nature Conservation* (eds R.J. Haggar and S. Peel). British Grassland Society Occasional Symposium, no. 28, pp. 86–94.

Wilkins, R.J. and Kirilov, A.P. (2003) Role of forage crops in animal production systems. *Grassland Science in Europe*, **8**, 283–91.

Wilkins, R.J. and Paul, C. (eds) (2002) Legume silages for animal production – LEGSIL. *Landbauforschung Voelkenrode*, **234**.

Wilkins, R.J., Newberry, R.D. and Titchen, N.M. (1983) *The Effects of Sward Height on the Performance of Beef Cattle Grazing Permanent Pasture*. Annual Report 1982, pp. 82–83. Grassland Research Institute, Hurley.

Useful websites

www.britishgrassland.com – British Grassland Society, includes details of meetings and publications.

www.ceh-nerc.ac.uk – Centre for Ecology and Hydrology, includes Annual Report, details of research projects and publications

www.defra.gov.uk – Department of Environment, Food and Rural Affairs, includes statistics, details of Rural Development Programme schemes and environmental legislation and the Codes of Good Agricultural Practice.

www.iger.bbsrc.ac.uk/igerweb/ – Institute of Grassland and Environmental Research, includes Institute's Annual Report, details of research projects and publications.

www.mluri.sari.ac.uk/ – Macaulay Institute, includes details of research projects and publications on land use.

13

Organic farming

N.H. Lampkin

Definitions and historical perspectives

Since the early 1990s, organic farming has expanded rapidly in the UK, other parts of Europe and around the world. The expansion has been fuelled by strong interest from consumers and policy-makers, reflecting the perceived potential of organic farming to contribute to environmental, animal welfare, social and nutritional goals. More recently, increasing attention has also been paid to the rural development potential of organic farming in the face of declining incomes from many conventional[1] farming systems. However, organic farming as a concept is not new, dating back to the beginning of the twentieth century, and the concept is much more than the common perception of no use of artificial fertilizers and pesticides.

What is organic farming?

Organic farming can be defined as an approach to agriculture where the aim is to create integrated, humane, environmentally and economically sustainable production systems. This encompasses key objectives relating to achieving high levels of environmental protection, resource use sustainability, animal welfare, food security, safety and quality, social justice and financial viability.

Maximum reliance is placed on locally or farm-derived, renewable resources (working within closed cycles) and the management of self-regulating ecological and biological processes and interactions (agro-ecosystem management – see Altieri, 1995), in order to provide acceptable levels of crop, livestock and human nutrition, protection from pests and diseases, and an appropriate return to the human and other resources employed. Reliance on external inputs, whether chemical or organic, is reduced as far as possible. In many European countries, organic agriculture is known as ecological agriculture, reflecting this reliance on ecosystem management rather than external inputs.

The term 'organic' refers not to the type of inputs used, but to the concept of the farm as an organism, in which all the component parts – the soil minerals, organic matter, micro-organisms, insects, plants, animals and humans – interact to create a coherent and stable whole.

In recent years, the market for organic food has developed strongly and is now often seen as a main feature of organic farming. However, the market was initially developed as a means to support the broader goals of organic farming, rather than an end in itself, at a time when official support was non-existent. This allowed organic producers to be compensated financially for restricting their production practices, effectively internalizing costs that could be considered as externalities of conventional agriculture (Pretty *et al.*, 2000).

Detailed descriptions of the principles and practices of organic farming can be found in various publications (see 'Organic farming in practice' below), as well as the detailed codes of practice contained in the standards documents of the various certification bodies operating in each country (see 'Role of the market and regulations' below).

Why organic farming?

The development of organic farming can be traced back to the 1920s, although many of the underlying ideas of

---

[1] The term conventional is used to refer to a wide range of non-organic systems that reflect the majority or normal practice in a particular region. Both high and low intensity holdings can be included in this usage of the term.

self-reliance and sustainability feature also in earlier writings (for reviews see Boeringa, 1980; Merrill, 1983; Conford, 2001; Reed, 2001). Steiner (1924) laid the foundation for biodynamic agriculture (Sattler and Wistinghausen, 1992), grounded in his spiritual philosophy of anthroposophy, which later was to have a significant influence on the development of organic farming. At about the same time, Dr Hans Müller founded a movement for agricultural reform in Switzerland and Germany, centred on Christian concepts of land stewardship and preservation of family farms. Later, Dr Hans-Peter Rusch contributed important ideas relating to soil fertility and soil microbiology, which led to the further development of organic-biological agriculture in central Europe (Rusch, 1968).

In the English-speaking parts of the world, King (1911) in *Farmers of Forty Centuries* used the long history of Chinese agriculture with its emphasis on recycling of organic manures as a model of sustainability, while McCarrison (1936) focused on nutritional issues and the influence that methods of food production might have on food quality and human health. Stapledon's work on alternative husbandry systems in the 1930s and 1940s (see Conford, 2001) and Sir Albert Howard's work on the role of organic matter in soils and composting (Howard, 1940) were also of key importance. These writers provided a powerful stimulus to Lady Eve Balfour (Balfour, 1943/1976), who founded the Soil Association in 1946. The key emphasis at that time, as the name suggests, was on soil fertility and soil conservation, with the dust bowls of the 1930s a recent event. The links between a healthy, fertile soil and crop and livestock health, food quality and human health were central to the mission of the organization.

Since the 1960s, wider concerns have influenced the development of organic agriculture, including food safety concerns due to pesticide residues, BSE and other issues, social concerns over working conditions in agriculture, the loss of jobs and rural population decline, as well as animal welfare and environmental concerns over the loss of wildlife species and habitats, pollution and the use of non-renewable resources.

These issues have come to be reflected in the broad concept of sustainable agriculture (Pretty and Hine, 2001), which emphasizes the use of systems and practices that maintain and enhance food supplies, safety and quality; financial viability of farm businesses; resource use sustainability; ecological impacts (including impacts on other ecosystems, pollution, biodiversity); and social and cultural well-being of rural communities.

The objective of sustainability lies at the heart of organic farming and is one of the major factors determining the acceptability or otherwise of specific production practices and technologies. However, organic farming does not represent the only approach to achieving agricultural sustainability, with integrated crop production in Europe (LEAF, 2001) and low-input sustainable agriculture (LISA) in the USA (Francis *et al.*, 1990) providing less-restrictive options, while permaculture (Mollison, 1990) and agro-forestry (Young, 1997) place more emphasis on the potential of integrating perennial crops (trees and shrubs) to achieve sustainability.

Although the concept of sustainable agriculture is now widely used, a perfectly sustainable agriculture is not achievable since all human activities have some negative impacts. Also, sustainability is a multi-objective concept and there are inevitably conflicts and trade-offs between the different objectives. A key issue, therefore, is the relative contribution that the different approaches can make to sustainability goals.

This is subject to considerable debate (see Stockdale *et al.*, 2000, and Stolze *et al.*, 2000, for contrasting views) – not just because of the lack of evidence in some areas, but also owing to the need to take account of the weightings placed on individual sustainability objectives by different parts of society, with clear differences observable, for example, between the food industry and environmentalists.

Growth of organic farming

Recent years have seen very rapid growth in organic farming (for detailed statistics see Foster and Lampkin, 2000; Hamm *et al.*, 2002). In 1985, certified and policy-supported organic production accounted for just 100 000 ha in the EU, or less than 0.1% of the total agricultural area. By the end of 2002, over 5 million ha, 4% of the total EU agricultural area, was managed organically, representing a 50-fold increase in 17 years (Lampkin, 2003). Over the same period, the number of certified holdings increased from 6000 to over 150 000 (Figure 13.1).

These figures hide great variability within and between countries. Several countries have now

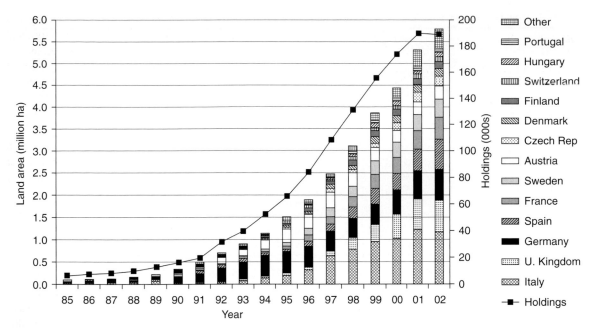

Figure 13.1 Certified and policy-supported organically managed and in-conversion land area and holdings in Europe, 1985–2002 (year end).

achieved 7–15% of their agricultural area managed organically, and in some cases more than 30% on a regional basis. Countries like Austria, Italy, Spain, Sweden and Switzerland, and more recently the UK, have seen the fastest rates of growth.

In the UK, the organic land area grew from 6000 to 50000 ha between 1985 and 1996, but increased dramatically to 750000 ha or 4% of UK agricultural land area by March 2003 (Table 13.1). The number of farms has increased from 865 in 1996 to 3500 in early 2003.

Alongside the increase in the supply base, the market for organic produce has also grown, but statistics on the overall size of the market for organic produce in Europe are still very limited. Recent estimates (Hamm *et al.*, 2002; ITC, 2003; Soil Association, 2003) suggest that the retail sales value of the European market for organic food was over 10 billion Euro in 2002, representing about 40–50% of the total global organic market of 23–25 billion Euro.

The UK share of the European market is approximately 10%, with the UK organic market valued at £260 million retail sales value in 1997/98, rising to £390 million in 1998/99, £600 million in 1999/2000 and over £1 billion in 2002/2003 (Soil Association, 2003). However, the rate of land conversion in the UK did not keep pace with demand until recently, leading to increased reliance on imports, estimated at 70% of retail sales in 2003 (Soil Association, 2003). This trend is beginning to be reversed as a result of greater availability and policies to encourage greater uptake of UK organic produce.

Organic farming in practice

Detailed descriptions of the principles and practices of organic farming can be found in the organic farming standards documents (e.g. IFOAM, 2002; ACOS, 2003) as well as in a wide range of books (e.g. Lampkin, 1990; Blake, 1994; Newton, 1995), magazines (e.g. *Organic Farming*, published by the Soil Association), conference proceedings, fact sheets and technical guides available from ADAS, EFRC, FIBL, HDRA, OCW, SA and SAC (see respective websites for further details); what follows is a necessarily brief summary of the key characteristics of organic farming practice.

Table 13.1 Number of organic holdings and land area (ha) in the UK, January 2004 (data from DEFRA).

Region	Holdings	Land area		
		March 2003	*January 2004*	*Percentage of total*
North East	73	27 748	28 883	5.0
North West	165	22 804	22 507	2.5
Yorkshire and Humberside	132	9 225	9 754	0.9
East Midlands	217	14 858	17 737	1.4
West Midlands	320	29 401	29 180	3.1
East	250	11 893	12 672	0.9
South West	1007	96 059	97 159	5.4
South East	406	39 848	41 038	3.5
England	2570	251 836	258 930	2.8
Wales	610	55 101	58 246	4.0
Scotland	687	428 608	372 562	6.7
Northern Ireland	150	4 115	5 882	0.6
United Kingdom	4017	741 174	695 619	4.0

Soil fertility and soil management

The soil is one of the most important resources in agriculture and the maintenance of the long-term fertility of soils is a key objective of organic farming. Here a distinction needs to be made between fertility, in the sense of the inherent capacity of the soil to sustain production, and productivity, which could entirely be due to the input of external resources. Soil biological activity (microbes, fungi, nematodes, earthworms, arthropods, etc.) plays an essential role in the maintenance of soil fertility. Organic matter plays a critical role in maintaining a stable soil structure and soil biological activity, by providing soil micro-organisms, fungi and earthworms with the energy to make nutrients available to crop plants. Organic farming favours supportive practices such as green manuring, straw incorporation and the utilization of other crop and livestock residues. There is now a substantial body of research evidence to indicate that these management practices, as well as the restrictions on fertilizer and pesticide inputs, have a significant beneficial effect on soil biological activity (e.g. Reganold *et al.*, 1993; Mäder *et al.*, 1996).

Soil cultivations in organic farming should aim to provide a soil structure that will allow for deep rooting by crops, providing adequate aeration and drainage so that plant roots can exploit available nutrients in the full soil profile. This is particularly important as highly soluble nutrient sources applied to the soil surface are avoided. At the same time, cultivations should aim to maintain the biologically active surface layers in the top 15–20 cm; therefore shallow ploughing or minimal cultivations in combination with subsoiling-type operations may be preferred, but the chosen method of cultivation will also need to fulfil other objectives, in particular weed control requirements.

Manure management

Farmyard manures and slurries, often referred to inappropriately as livestock 'wastes', represent a valuable resource in organic farming. If mismanaged, they represent not only a significant environmental problem, but also a productive loss, with financial consequences for the farm business.

Great care has to be taken to avoid pollution by manures or slurries (MAFF, 1998; ADAS, 2001). Storing manure outside in a field is unsatisfactory due to the risk of nutrient leaching and pollution, but this may be minimized by considering siting and covering with a plastic sheet during winter. Careful composting and slurry aeration can enhance the quality of farmyard manures and slurries by improving the smell, nutrient stability and weed control.

Manure and slurry applications should be planned into the rotation with applications directed at points of greatest nutrient off-take, for example where green matter is harvested and sold off the farm or conserved as forage.

Crop nutrition

The principle of reliance on farm- or locally derived renewable resources in organic farming means that, where possible, fertility should be obtained from within the farm system. In the case of nitrogen, which is freely available from the atmosphere, biological nitrogen fixation provides the basis for nitrogen self-sufficiency. This requires an appropriate proportion of legumes in the rotation and good management of crop residues and livestock manures.

In principle, energy self-sufficiency through solar energy capture by photosynthesis and other forms of renewable energy should also be possible, but in practice some reliance continues to be placed on fossil energy for mechanization and other purposes. The development of bio-fuel crops, in particular for bio-diesel, might in future provide a means of achieving the energy self-sufficiency objective in practice.

Despite the aim to improve the recycling of other crop nutrients, there will always be some net export from the holding as crops and livestock are sold. The first step is therefore to minimize unnecessary losses from the system, by avoiding leaching and erosion. Attention should also be paid to the sales of nutrients off the farm, for example to avoid high potassium losses from the system through selling of straw and conserved forage crops. Second, some reliance can be placed on the release of nutrients from soil minerals through normal soil formation processes. Purchases of livestock feeds and straw can also add a substantial amount of nutrients to the cycle.

If sales exceed the purchases and rate of natural regeneration, then it will be necessary to supplement nutrients such as potassium, phosphate, calcium, magnesium and trace elements from external sources. These should preferably be in an organic form, or in a low-solubility mineral form, so that the nutrients are released and made available to plants indirectly through the action of soil micro-organisms. In this way, luxury uptake by crop plants can be avoided. In some cases, such as potassium, this may not always be possible. Production standards therefore allow restricted use of potassium sulphate in cases of demonstrated need and provide a list of inputs for soil improvement and crop nutrition that organic farmers can use.

Nutrient budgeting can be used to check whether the returns of manures and/or slurry are adequate to compensate for nutrient removals in a balanced rotation (Lampkin, 1990).

Crop production

Rotations

The rotation is the core of most organic farms, based on the principle that diversity and complexity provide stability in agricultural as in other ecosystems. Beneficial interactions between enterprises, and between the farm and the external environment, should be exploited fully. The role of the rotation is therefore to:

- ensure sufficient crop nutrients and minimize their loss;
- provide a self-sustaining supply of nitrogen through legumes;
- minimize and help control weed, pest and disease problems;
- maintain soil organic matter and soil structure;
- provide sufficient livestock feed where necessary;
- maintain a profitable output of cash crops and/or livestock.

There is no 'blueprint' organic rotation. There are as many different types of organic system as there are conventional systems, and each uses rotation in different ways to satisfy the requirements listed above. In some cases, such as all-grass farms or perennial cropping, there may be no rotation at all. Instead, diversity is achieved through species mixtures in space rather than over time.

Polycultures

Polycultures provide an important alternative to rotations in many situations, but may also be included within a rotational context. The term includes a variety of approaches including strip cropping, intercropping, under-sowing and crop and variety mixtures. In many contexts, these are more productive than growing single crops, due to improved exploitation of available space, light and growth factors, complementarity between nitrogen-fixing and nitrogen-demanding species, and structural/shade support offered by specific species. Polycultures form the core of many tropical organic farming systems, and polycultures including perennial tree and shrub species are a key element of permaculture (Mollison, 1990) and agro-forestry (Young, 1997) systems.

The main perceived disadvantage of polycultures in the context of commercial UK agriculture is that they tend to be less easy to mechanize and more labour intensive. However, many crop mixtures can be mechanized, for example cereals and grain legumes or variety mixtures of the same species.

Weed control

Herbicides are not permitted under organic production standards. Weed control is achieved primarily through preventive measures, including rotation design, soil cultivations, crop variety selection (e.g. tall cereal varieties to shade out weeds), under-sowing, use of transplants, timing of operations and mulching with crop residues or other materials. Direct intervention to correct weed problems when they occur should be seen as secondary. Options include mechanical, thermal and biological controls of different types. In practice, the need for significant additional weed control (in addition to techniques of cultivation around crop establishment) in arable crops and grassland is relatively low, although in some cases perennial weeds such as some grass weeds, docks and thistles can become problematic. For horticultural crops, mechanical control and hand-weeding may be needed, the extent of which depends on the extent of mechanization of the holding and on weather conditions and timeliness.

Pest and disease control

Pest and disease control should be achieved primarily through preventive measures also, including balanced crop nutrition, rotation design, variety selection for resistance, habitat management to encourage pest predators, and organic manuring to stimulate antagonists to soil-borne pathogens. Where direct intervention is required, organic production standards permit a number of biological controls such as *Bacillus thuringiensis* for cabbage white caterpillar control, as well as certain non-synthetic pesticides and fungicides, several of which require prior approval by the certifi-

cation body. These products should be used as a last resort as they still have the capacity to disrupt beneficial insects and the ecosystem interactions on which the stability of the organic farming system depends. The general requirement for undressed seed may increase pest and disease risks at germination, particularly for forage crops, so seed quality and hygiene become very important.

Livestock production

Role of livestock on organic farms

Except in the case of stockless horticultural and arable farms, livestock form an integral part of organic farming systems. Ruminant livestock are able to utilize and provide a financial return to the legume and other herbage species that contribute to nitrogen fixation and the fertility-building phase of the rotation. They can digest cellulose and hence obtain energy from herbage that would not otherwise be available for human consumption, and the return of livestock manures provides a nutrient source for subsequent crops. Livestock thus play an important role in nutrient and energy cycling. They can also contribute to weed, pest and disease control through grazing and forage conservation. As with crop production, an emphasis on single species can lead to problems with internal parasites or weeds (e.g. bracken in grassland). The mixing of species, such as sheep and cattle or sheep and poultry, can contribute to the control of parasites and improve grassland management.

As far as possible, livestock enterprises should be land-based and supported from the farm's own resources. This effectively excludes intensive, permanently housed pig, poultry and feedlot cattle production, which depend heavily on bought-in feeds and lead to livestock 'waste' disposal problems. The manures produced should be capable of being absorbed by the agricultural ecosystem without leading to disposal or pollution problems. Stocking rates should reflect the inherent carrying capacity of the farm and not be inflated by reliance on 'purchased' hectares. The EU regulation defining organic livestock production (1804/1999) stipulates a maximum stocking rate of ca. 2.0 LU (livestock units)/ha, although farms may stock more intensively if they are able to form an agreement with another organic farm to take the surplus manure and the combined stocking rate on the two holdings does not exceed the specified limit.

Livestock nutrition

The aim in organic livestock feeding is to rely primarily on home-produced feeds which are suited to the animal's evolutionary adaptation and which, as far as possible, avoid the use of feedstuffs that are suitable for direct human consumption. In the case of ruminants, this means the ration must be predominantly forage (>60% DM). This normally comprises grass/clover grazing and conserved forage, although there is clearly potential for other forage crops such as arable silage, fodder beet and maize. The use of cereals and pulses in ruminant diets is aimed at balancing the diet rather than stimulating additional production. Pig and poultry rations will inevitably rely on cereals and pulses, although there is also a need to provide some green material and indeed considerable reliance can be placed on forage crops in pig diets.

European organic livestock regulations aim to achieve fully organic diets for all animals by August 2005. In the interim, annual limits on the quantities of dry matter fed from conventional sources have been specified as 10% in the case of herbivores (i.e. ruminants and horses) and 20% for all other species (i.e. pigs and poultry). All other feedstuffs must be organically produced, although up to 30% may come from land still in conversion. This proportion can increase to 60% if the feed is produced on the farmer's own holding. The standards contain a positive list of permitted feed materials and only components listed can be used. Pure amino acids and coccidiostats are not permitted.

Trace elements and mineral supplements are not fed as a routine but are supplemented where specific deficiencies occur, usually in the form of rock salt or seaweed meal. Problems such as hypomagnesaemia are less common on organic farms because of reduced fertilizer use, but pastures may be dusted with calcined magnesite or magnesium supplements fed if necessary.

Livestock health

Animal health in organic farming is based on preventive management and good husbandry. The aim, through good stockmanship and attention to detail, is to optimize breeding, rearing, feeding, housing and general management in order to achieve stability and balance in the farming system, maximize the natural health of the animal and minimize disease pressure and stress. All organic livestock farms are required to have an animal health plan as part of the livestock management, which should be drawn up with the help of a specialist organic advisor and a veterinary surgeon.

Preventive treatment is restricted to limited use of vaccination and homoeopathic nosodes for known farm problems or strategic use in the context of a health plan. Growth promoters, hormones and the routine (prophylactic) use of antibiotics (e.g. dry cow therapy) are not allowed. Where possible, treatment of ailments is approached by aiding the animal's own resistance and the use of complementary therapies such as homoeopathy. Conventional treatment should be used in all cases where it is necessary to prevent prolonged illness or suffering, but a maximum limit of three courses of treatment within a year and longer withdrawal periods are imposed under organic production standards. A medicine book recording all cases and treatments (including complementary medicines) must be kept.

Hovi *et al.* (2000) produced a comprehensive guide to preventive management strategies and treatments for livestock diseases in organic farming. Further information on animal health and welfare issues in organic farming can also be found on the Network for Animal Health and Welfare in Organic Agriculture (NAHWOA) website.

Livestock behaviour and welfare

Animal welfare is a key objective of organic farming and organic production standards contain a number of provisions designed to achieve this, including, as a minimum, that the DEFRA welfare codes for different livestock species should be applied, and that animals should be free to exhibit their normal behaviour patterns. However, the animal welfare case is often disputed (e.g. House of Commons, 2001), in part because welfare is often seen in the narrow context of animal health and the restrictions placed on the use of prophylactic treatments. An understanding of basic animal ethology and welfare considerations (e.g. Foelsch, 1978) is therefore fundamental to the interpretation and implementation of production standard requirements.

The Farm Animal Welfare Council's 'Five Freedoms' (covering hunger, thirst, pain, disease, stress and freedom to exhibit natural behaviour patterns) provide a basis for assessing the welfare provisions and impacts of organic farming – on many, the organic standards provide for more than the legal minimum, in particular with respect to housing, outside access and freedom to exhibit normal behaviour patterns, but some areas are open to debate, for example:

- Do the restrictions on feedstuffs permitted in organic farming cause a welfare problem in high genetic merit animals due to insufficient nutrient concentrations leading to hunger? Many organic producers are reluctant to use high productivity/high genetic merit breeds for this reason.
- Do the restrictions on the use of prophylactic medications, and on the number of treatments before an animal loses its organic status, result in a failure to protect animals from pain and disease? Organic standards require farmers to implement health management plans and to treat to avoid prolonged suffering, but is this sufficient?
- Do restrictions on mutilations such as beak clipping represent a greater welfare threat because of the risk of feather-pecking and cannibalism? Are the alternative management procedures proposed for organic free-range poultry production sufficient to minimize the risks of a problem developing?

These questions are not simple to answer and there are many different views on the subject. Further research is underway to assess the animal health and welfare implications of organic farming (see the NAHWOA website), but there is also a need to keep organic standards under review in order to ensure that the animal welfare goals are achieved in practice.

Livestock housing

Outside access and housing appropriate to animal welfare and behavioural needs are required under organic production standards. Battery cages, tethering, fully slatted floors, etc. are prohibited, with the emphasis instead on free-range systems, particularly for pig and poultry production. Where livestock have to be housed, group housing is required (for calves after one week), natural bedding materials should be used and the standards specify minimum space allowances for different livestock species. Housing enrichment, for

example scratching posts, and overhead shelter on poultry ranges are recommended.

Livestock breeding and rearing

Emphasis is placed on maintaining closed herds and flocks, i.e. breeding replacements on the farm, so as to minimize the risk of importing diseases from elsewhere. Other breeding objectives may include high lifetime yields from forage for dairy cows, suitability for outdoor systems for pigs, and internal parasite resistance in sheep. Breed choice will also be related to the quality requirements of target markets. Breeding males may be brought from non-organic sources as part of the 10% non-organic livestock replacement allowance and the use of artificial insemination is permitted. Calves have to be fed on organic whole milk or organic milk replacer up to the age of 12 weeks, and, therefore, multiple suckling by a nurse cow is often preferred in dairy herds, but rearing at foot may also be an option in some cases. In others, natural weaning may be practised. Although these practices add to production costs, it is often argued that the benefits in terms of subsequent health, production and longevity more than compensate.

Organic farming as a business

Financial viability is as important for an organic farmer as for any other farmer. A business that is not viable is not sustainable. In order to maintain financial viability while restricting the use of practices and technologies that could enhance yields, organic producers have had to focus more on quality, adding value and marketing of their products. In recent years, the strong market for organic products has resulted in substantial differences in the prices for organic and conventional products, although in fact organic prices have remained relatively stable, while conventional prices have declined significantly. Conventional and organic markets have effectively become decoupled, so that the notion of a fixed percentage premium for organic products is outdated.

The financial and physical performance of organic farming in Europe has been reviewed by Lampkin and Padel (1994) and Offermann and Nieberg (2000). Lampkin *et al.* (2004) provide detailed information on typical gross margins for organic enterprises in the UK, while Fowler *et al.* (2000, 2001) provide survey results for different organic farm types in England and Wales.

Enterprise gross margins

In a northern European context, organic crop yields can be substantially lower (40–50%) than conventional, particularly where intensive production methods are used. However, this does not necessarily apply elsewhere. In the United States and Australia, where conventional methods are typically less intensive than in northern Europe, the relative differences are less significant (10–20%), and Pretty and Hine (2001) have shown clearly that in many resource-poor, developing countries, sustainable agricultural approaches (including organic farming) have the potential to actually increase yields through more effective utilization of the farm's own resources.

However, the yield reductions experienced in the UK are such that higher prices are essential to compensate, despite the savings on input costs. Current organic prices for crops normally more than compensate for these losses, resulting sometimes in substantially higher gross margins on organic farms (Table 13.2). For milk, higher gross margins per cow (and per litre) can be achieved, but the price premium would need to be at least 6p per litre (ppl) to achieve similar margins per

hectare, which has been difficult to achieve for many organic dairy farmers in recent years.

Whole farm performance

Despite the relatively good performance at gross margin level, the rotational and other constraints on enterprise mix on organic farms means that high gross margins for individual enterprises do not necessarily translate into high whole farm gross margins and incomes. In particular, the financial returns to the fertility building phase of the rotation may be significantly lower than to the main cash crops, but account needs to be taken of the hidden benefits of interactions between the enterprises (non-financial internal transfers), for example with respect to weed control and parasite control. It would therefore be a mistake to use gross margin performance as a main criterion for selecting production enterprises in organic farming.

Recent surveys of organic farms in England and Wales (Fowler *et al.*, 2000, 2001) and other European studies (Offermann and Nieberg, 2000) have confirmed that organic farms can make incomes that are comparable to similar conventional farms, although in the UK much of this has been due to the declining profitability of conventional farms in recent years rather than an increase in the profitability of organic farms.

Table 13.2 Comparative gross margins for selected organic and conventional (*Conv.*) enterprises, 2004 prices (Nix, 2003; Lampkin *et al.*, 2004).

	Unit	Milling wheat		Potatoes			Unit	Milk	
		Organic	Conv.	Organic	Conv.			Organic	Conv.
Yield	t/ha	4	8	25	42	Yield	l/cow	6000	6650
Price	£/t	165	80	220	80	Price	ppl	22.5	18
Output	£/ha	900[1]	880[1]	5500	3350	Output	£/cow	1325	1150
Seed	£/ha	85	40	940	500	Feed	£/ha	300	225
Fertilizer and sprays	£/ha	40	185	190	600	Forage	£/ha	60	80
Input	£/ha	140	225	2600	2125	Input	£/cow	525	450
Margin	£/ha	760	655	2900	2225	Margin	£/cow	800[2]	700[3]

[1] Including £240/ha arable area payment.
[2] £1200/ha @ 1.5 LU/ha (£1425/ha @ 25 ppl, £1875/ha @ 30 ppl).
[3] £1400/ha @ 2.0 LU/ha.

Various studies indicate that labour use is higher on organic farms, typically in the range 10–25% (Lampkin and Padel, 1994; Jansen, 2000; Offermann and Nieberg, 2000). These increases are associated with the introduction of more labour-intensive (but high-value) crops and/or production techniques, on-farm cleaning, grading, processing and marketing of produce, and the diseconomies of scale associated with a greater diversity of enterprises. There is little evidence that labour requirements for *existing* enterprises increase substantially, although labour use per animal may increase as a result of reduced stocking, preventive health management, increased observation and where intensive livestock enterprises are converted to free-range systems.

Conversion to organic farming

The conversion (or transition) from conventional to organic farming systems is a complex process subject to several physical, financial and social influences which differ from those associated with established organic farming systems, involving a significant number of innovations and restructuring of the farm system as well as changes in production methods. The time required and the difficulties associated with the necessary changes depend on the intensity of conventional management and the condition of the farm before conversion, the extent to which new enterprises and marketing activities are introduced, and any yield and financial penalties related specifically to the conversion process. As a consequence, the conversion period may well be longer than the statutory 2-year conversion period prescribed by regulation before full organic certification and premium prices are obtained.

Prior to and during conversion, farmers face a range of challenges, including personal (family), social (peer pressure) and institutional resistance to the decision to convert (Padel, 2001), in addition to the direct financial and technical impacts on the farm business. Intensive preparation, including visits to and contacts with other organic farmers, as well as detailed planning, can help address many of these potential problems (Padel and Lampkin, 1994; Lampkin *et al.*, 2004). Once the decision to convert has been taken, two main options exist within the context of organic production standards:

(1) *Staged* conversion involves the conversion of parts of the farm, typically 10–20%, in successive years, using a fertility-building legume crop as an optimal entry point into organic management. The learning costs, capital investment and risks can be spread over a longer period, sometimes up to 10 years or the full length of a rotational cycle, and are more easily carried by the remainder of the farm business.

(2) *Single-step* or crash conversion involves converting all the land on the farm at one time. This enables the farm to gain access to premium prices sooner, but means that all the risks, learning costs and financial impacts of conversion are concentrated into a short period of time. Rotational disadvantages can arise because not all of the farm can be put down to fertility-building crops at the same time. If mistakes are made, the impacts are likely to be more severe and the approach may turn out to be more costly, despite the earlier access to premiums.

There is some evidence of a decline in yield during the conversion period greater than that which would be expected in an established organic system (Dabbert, 1994). This is because biological processes such as nitrogen fixation and rotational effects on weeds, pests and diseases take time to become established. In many cases, avoidable conversion-specific yield reductions may be due to mistakes or inappropriate practices, such as the removal of nitrogen fertilizer without taking action at the same time to stimulate biological nitrogen fixation using legumes. Forage crops are the one area where conversion-specific yield decline may be inevitable, with production lost as a result of reseeding grassland with new mixtures containing legumes, or waiting for clover to establish naturally following the withdrawal of nitrogen fertilizer. The loss of output can place significant pressure on stocking rates for livestock on predominantly grassland farms, particularly where permanent grassland is involved.

The costs of conversion vary widely according to individual circumstances, arising from a combination of output reductions, new investments, information and experience gathering and changes in production costs. However, lack of access to organic prices during conversion and changes in eligibility for subsidies may be more important factors influencing the cost of conversion. Conversely, the Organic Farming Scheme in England (see DEFRA website) and similar schemes in

Scotland, Wales and Northern Ireland provide direct financial support to farms in conversion. Unlike most other EU countries, support was not available for continuing organic production, but maintenance payments have been introduced in England and Wales in 2003.

Model farm calculations prepared for DEFRA's Organic Conversion Information Service (Table 13.3) indicate that, depending on farm type, costs of conversion may be relatively low or non-existent for farms of average intensity (Lampkin, 2002). These models take account of prices for organic and conventional products in early 2002 and include the Organic Farming Scheme payments, as well as the flexibility built into the main arable and livestock support schemes for organic farmers. However, they are sensitive to the organic price assumptions, as the difference between organic and conventional prices can account for up to £800/ha in the specialist dairy and stockless arable organic results. Since 2001, conventional livestock prices have started to recover, and organic prices for some commodities have fallen, narrowing the gap between organic and conventional systems and reducing the financial attraction of crash conversions.

Role of the market and regulations

The profitability of organic farming is highly dependent on the prices received, and these are only possible as a result of the development of a distinct market for organic products. The first efforts to develop such a market began in the late 1960s and early 1970s, as a means to support the broader goals of organic farmers, and effectively to allow consumers willing to pay with a means to compensate farmers for internalizing external costs by restricting the use of certain inputs and production technologies. Since then the organic market has become a well-established, global phenomenon worth more than 20–25 billion Euro in retail sales value, with Europe, the United States and Japan the key consuming markets.

The very success of the organic market brings with it certain risks. First, that the market becomes an end in itself, not a means to an end, and that the original goals of organic farming are devalued, thereby leading to the possible loss of consumer confidence in the integrity of organic farming. Second, a market of this size clearly invites fraud, and there is a need for legislation to avoid this and to protect consumers and *bona fide* producers. As a consequence, organic farming is the only approach to sustainable agriculture (see above) that is enshrined in legislation in Europe (EU, 2001), in the United States (USDA, 2000) and at international level through the *Codex Alimentarius* agreement (FAO, 2001).

Under EC Regulation 2092/91 there has been a legal requirement since 1993 for all crop products sold as organic in the European Union to be certified. In August 2000, EC Regulation 1804/1999 came into full effect, extending the legal requirement for certification to organic livestock and livestock products. This legisla-

Table 13.3 Model farm estimates of financial changes during and after conversion (*Conv.*), by farm type, early 2002 prices (Lampkin, 2002).

Farm type	Size (eff. ha^2) (acres)	Annual whole farm gross margins less conversion-related fixed cost changes (£/eff. ha^2)			
		Conv.	Staged transition[1]	Crash transition[1]	Organic
Specialist dairy	60 (150)	1098	1088	1177	1336
Mainly dairy	145 (363)	782	889	990	1102
Stockless arable	210 (525)	564	756	726	882
Mainly arable	260 (650)	512	488	499	524
Lowland livestock	180 (450)	457	438	476	490
Upland livestock	83 (206)	259	336	343	314
Hill livestock	130 (325)	219	266	266	229

[1] Effective area (rough grazing adjusted).
[2] Five-year average.

tion and the relevant production standards based on it emphasize control of the production process, not the end product. DEFRA, advised by the Advisory Committee on Organic Standards (ACOS), which replaced the UK Register of Organic Food Standards (UKROFS) in late 2003, is the national authority responsible for implementing the EU legislation regarding organic production in the UK. DEFRA, supported by the UK Accreditation Service (UKAS), sets the UK baseline standards (ACOS, 2003) and is responsible for licensing organizations to carry out farm inspections and certification and for commissioning surveillance inspections to ensure procedures are followed correctly. The bodies licensed by UKROFS/DEFRA in 2003 to carry out producer inspections were Organic Farmers and Growers Ltd. (OFG), Scottish Organic Producers Association Ltd. (SOPA), the Organic Food Federation (OFF), Soil Association Certification Ltd. (SACert), Bio-dynamic Agricultural Association (BDAA/Demeter), Irish Organic Farmers and Growers Association Ltd. (IOFGA) and Organic Trust Ltd. (see Lampkin *et al.*, 2004, for further details and contact addresses).

Alternative business organizations

The emphasis on social justice goals and close links with consumers has led to a range of initiatives to develop new models of business activity as well as alternative production methods. There are close links between organic production and many Fairtrade initiatives designed to improve the returns to producers in developing countries, particularly with respect to tea, coffee, cocoa and banana producers (Maxted-Frost, 1997; Lockeretz and Geier, 2000; Fairtrade Foundation website). Community supported agriculture (CSA) is another increasingly popular approach, particularly with small producers in North America, Europe and Japan (see CSA website). The simplest models are based around box scheme subscriptions, whereby consumers commit to a fixed weekly subscription, sometimes paid in advance, and receive a box of vegetables in season in return. Variants on this theme include subscription gardening, where a fixed fee is paid to the grower to look after the consumer's plot, but the consumer can determine what is to be grown and can participate in the growing and harvesting of the crop. Other models include the consumer investing a capital stake in the holding and participating in the farm decision-

making process. Various alternative models with respect to worker participation can also be found, ranging from worker co-operatives to the large Sekem company in Egypt, which has developed a school, university, community hospital and other facilities for its workforce and their families on the back of a thriving organic food, fibre (cotton) and clothing business (Maxted-Frost, 1997; Sekem website).

Organic farming and society

The key goals of organic farming with respect to environmental protection, animal welfare, food security, safety and quality, human health and nutrition, resource use sustainability and social justice are ones for which the market mechanism does not normally provide an adequate financial return and are normally seen as public goods and services, of benefit to society as a whole rather than the individual. These goals are increasingly important to policy-makers too, leading to increasing interest in the potential of organic farming as a policy option. Organic farming has established a complex set of principles and practices that are believed to contribute to achieving these objectives, but that does not guarantee that the objectives are achieved, and much debate centres around the extent to which these objectives are achieved in practice (Tinker, 2000; House of Commons, 2001). It is not possible in a chapter such as this to provide a detailed and comprehensive assessment, so some 'heroic' generalizations will be needed and readers should consult the source material for further information.

Agricultural sustainability can be seen as a measure of the performance of different systems with respect to all these goals, as well as the financial viability and hence sustainability of individual farm businesses. However, the key factor determining the relative sustainability of different systems in such a multi-objective context is the weighting placed on the individual objectives by different parts of society. A high weighting on environmental factors may favour one approach, while a high weighting on yield will favour another. Thus it may prove impossible to come to a conclusive view on the relative merits of the different approaches to sustainable agriculture, such as organic farming, integrated crop management, agroforestry and permaculture.

Environmental impacts

The impact of organic farming on the environment has been reviewed most recently by Stolze *et al.* (2000), Greenwood (2000) and the Soil Association (2000). There is now a significant body of research indicating the beneficial effects of organic practices on soil structure, organic matter levels and biological activity, as well as plant, insect, bird and wild animal biodiversity.

However, differences can vary depending on farm type, the relative intensity of the conventional and organic systems compared, and the management ability and interest of the individual farmer, so that better performance is not necessarily guaranteed in all cases. As Stolze *et al.* (2000) point out, these benefits are clearer on a per unit land area basis, but the reduced yields from organic farming may mean that the benefits per unit food produced are not as great. There is an ongoing debate as to whether the reduced yields might require additional land currently not in production to be brought into production, with potential negative environmental consequences. However, this assumes current production structures will be retained, including the current level of feeding crops suitable for human consumption to livestock.

In terms of non-renewable resource use (and the related pollution risks/greenhouse gas emissions), several studies indicate that organic farming has the potential to reduce resource use and pollution, not only on a per unit land area basis, but also per unit food produced (ENOF, 1998), with significant implications for future global food security in the context of diminishing resources.

Food quality, nutrition and health

The impact of organic farming on food quality, nutrition and human health has been a core concern of organic farming since the research of McCarrison (1936). The issue has been subject to recent reviews (Williams *et al.* in Tinker, 2000; Soil Association, 2001). The evidence on food quality is less conclusive than that for environmental benefits, with some studies showing benefits with respect to increased valuable nutritional components (vitamins, minerals, trace elements, secondary metabolites) and reduced harmful components (nitrates, pesticide residues), while other studies show little or no differences, and some authors have raised the theoretical risk of increased levels of potentially harmful components such as *E. coli* 0157 and mycotoxins in organic foods, but with little evidence to substantiate this (FAO, 2000; FSA, 2000; Soil Association, 2001). Some animal studies have shown beneficial impacts on fertility and morbidity from organic diets.

Such differences as have been identified tend to be specific to particular crops or farming situations, so that it is difficult to generalize an overall benefit from organic food. However, it is clear from the focus of many agricultural research programmes that the way food is produced does affect its quality. Therefore it is reasonable to expect that quality differences, for better or for worse, could exist between organically and conventionally produced foods. There is a clear need for further research on this topic, which is now more likely to take place than in the past, as the resistance of governments to funding such research is waning.

Food security and developing countries

The relatively large crop yield reductions observed in the northern European context have led many to question whether organic farming is capable of meeting the food needs of a growing global population. This is a complex question, which has to take account of distribution as well as production issues, as well as the increasing demand for meat as incomes increase and the role of livestock production as a direct competitor with human food needs. In addition, the large yield reductions experienced in northern Europe are not reflected in other studies from countries where conventional production is less intensive (Lampkin and Padel, 1994). Pretty and Hine (2001) and FAO (2003) have demonstrated the potential for yields to be increased in resource-poor countries (where the ability to pay for external inputs, in particular agro-chemicals, is severely limited) through the adoption of ecological management principles. The experience of Cuba in pursuing organic farming as a key part of its food security strategy in the face of US economic sanctions is particularly relevant in this context (Pretty and Hine, 2001; Food First website).

Social impacts

Social impacts are perhaps the least considered aspect of organic farming, although social goals have long been part of the organic farming concept. The International Federation of Organic Agriculture Movements' standards (IFOAM website) include a section on social justice which covers workers' rights and expectations of appropriate working conditions, rewards for labour and educational opportunities. Further work is continuing at several levels on the integration of social issues in organic standards.

In a European context, basic rights are covered by national legislation, and have therefore not been a focus of organic farming standards and legislation, but there is a need to look critically at working conditions, employment and income levels on organic farms (Jansen, 2000; Offermann and Nieberg, 2000). In general terms, the case can be made that employment and incomes can be maintained or increased on organic farms, but even securing current farming businesses and existing employment and income levels might be beneficial for rural communities in the context of the dramatic structural changes currently taking place in conventional agriculture. There is clearly a question whether organic farms will not in the longer term be exposed to the same economic pressures for specialization and rationalization as conventional farms, and it may be that local marketing and processing initiatives are more important than production in terms of the rural development potential of organic farming (OMIARD website).

Organic farming and agricultural policy

More than 90% of the expansion in the organic land area in Europe up to 2002 took place in the 10 years since the implementation in 1993 of EC Regulation 2092/91, defining organic crop production, and the widespread application of policies to support conversion to and continued organic farming as part of the agri-environment programme (EC Regulation 2078/92, continued as part of the rural development regulation 1257/1999). The former provided a secure basis for the agri-food sector to respond to the rapidly increasing demand for organic food across Europe. The latter provided the financial basis to overcome perceived and real barriers to conversion.

The increasing role of EU policy support of this type during the 1990s has arisen because of a gradual convergence of policy goals with the underlying objectives of organic farming, including environmental protection, animal welfare, resource use sustainability, food quality and safety, financial viability and social justice. Organic farming is also perceived to contribute to reducing problems of overproduction and to rural development. In addition, organic farming offers three potential advantages over other, more targeted policy measures: it addresses many of these goals simultaneously, it utilizes the market mechanism to support these goals, and it has achieved global recognition.

Agri-environment support

The agri-environment measures came into effect in 1993, although the majority of organic support schemes under EC Regulation 2078/92 were not fully implemented by EU member states until 1996, and significant differences between the schemes implemented exist. Financial support for organic farming has continued to be provided, in many cases at increased levels, under the agri-environmental measures in the rural development regulation (EC Regulation 1257/1999). By 2001, agri-environmental support for organic farming in the EU amounted to 271 million Euro on 1.5 million ha, 16% of total agri-environment scheme expenditure and 8% of total agri-environment scheme supported land (unpublished European Commission data), the differing shares reflecting in part the widespread uptake of baseline programmes in some countries and the higher levels of per hectare payment for organic farming.

Mainstream commodity support

Like their conventional counterparts, organic farmers also qualify for the mainstream commodity support measures, including arable area payments and livestock headage payments, as well as support for capital investment and in less favoured areas where available. In most EU countries, the mainstream commodity support measures are seen as beneficial, at least for organic arable producers. Set-aside in particular is seen to have potential to support the fertility-building phase of organic rotations during conversion and on arable farms with little or no livestock.

Only in a few cases have significant adverse impacts of the mainstream measures on organic farmers been identified. In some cases, special provisions have been made to reduce these, for example flexibility with respect to use of clover in set-aside in the UK. There is a case that, since organic producers are producing significantly less output, and the organic market is undersupplied, then organic producers should not face compulsory set-aside. In 2001, the European Commission moved in this direction, permitting organic producers to use set-aside land for forage production.

The loss of eligibility for livestock headage premiums as a result of reduced stocking rates following conversion has been seen as potentially more problematic, but this has been mitigated by extensification payments and quota sales or leasing where applicable.

Several countries have made use of investment aids and national/regional measures to provide additional assistance, including special derogations for organic producers.

The June 2003 CAP Reform agreement provides for the decoupling of arable area and livestock headage payments to be provided in future (from 2005) as a single farm payment on an historical or regional basis. This will mean that producers converting in future will not face the same penalties with respect to support payments that previous converters have incurred. However, previous converters, unless they can make use of the special case provisions for participants in agri-environment schemes foreseen in the new regulations, may find themselves locked into lower levels of support in the long term.

Continued supplementary support for protein crops, and in several countries for beef production, is likely also to benefit organic producers. So will the specific provisions relating to organic farming in the agreement, including set-aside exemptions and specific supports envisaged under the new quality policy chapter of the rural development regulation. Within the UK, however, a key issue remains of whether it will be possible to resource these new measures and the agri-environmental and organic farming policy reforms under discussion in 2003/2004 for implementation in 2005/2006.

Rural development and structural measures

Organic farming is seen in many countries in Europe as having significant potential for rural development, in terms of its capacity to supply premium markets and thereby to support rural incomes and employment. Organic farming projects were favoured under the marketing and processing support and structural measures in the 1990s and this has continued under the new Rural Development Programme and structural measures under Agenda 2000. Some countries, for example Denmark, France, The Netherlands, Sweden, England and Wales, have developed integrated action plans for organic farming, which fully utilize the support available under these measures and aim to ensure a better balance between support for supply growth through the agri-environment programme and demand growth through market-focused measures.

Information programmes, including support for research, advice, training and demonstration farms as well as consumer information, are also seen in many countries as essential counterparts to the other programmes and have been supported at EU level through the Framework research programmes and national funding, as well as through the provision of specific training and advice under the Rural Development Programme.

Future potential

There is currently renewed debate about the potential for organic farming in Europe. The spread of bovine spongiform encephalopathy to other European countries and the outbreak of foot-and-mouth disease in early 2001 have led to many calls for a fundamental review of the future direction of agriculture, including an increased role for organic farming. Moves to develop a European Action Plan for organic farming have been supported by several agriculture ministers at an international conference held in Copenhagen in 2001 (MFAF, 2001), and these bore fruit with the publication of a plan by the European Commission in early 2004.

Several countries have set targets for organic farming to grow to 10 or 20% of total agriculture by 2005/2010. Although growth trends in individual countries have varied considerably, with periods of rapid expansion followed by periods of consolidation and occasionally decline, overall growth in Europe has been consistently around 20–25% per year for the last 10 years, i.e. exponential growth. Future growth may well be slower than this, but 5–10% of EU agriculture managed organically by 2005 and 15–20% by 2010 are conceivable. This would imply a several-fold increase in the size of the sector with significant implications for the provision of training, advice and other information to farmers, as well as for inspection and certification procedures. It also has implications for the resourcing of existing organic support schemes under the Rural Development Programme, as the cost could increase to more than 5 billion Euro annually. It is an open question whether policy-makers, farmers and consumers will respond to the challenge to make this sort of expansion possible.

References

ACOS (2003) Compendium *Standards for Organic Food Production*. UK Register of Organic Food Standards. DEFRA, London.

ADAS (2001) *Managing Livestock Manures*. Booklet series including organic farming. ADAS, Gleadthorpe.

Altieri, M. (1995) *Agroecology – The Scientific Basis of Alternative Agriculture*, 2nd edn. Intermediate Technology Publications, London.

Balfour, E.B. (1943) *The Living Soil*. Faber and Faber, London. Reprinted 1976 as *The Living Soil and the Haughley Experiment*. Universe Books, New York.

Blake, F. (1994) *Organic Farming and Growing*. Crowood Press, Swindon.

Boeringa, R. (ed.) (1980) *Alternative Methods of Agriculture. Agriculture and Environment*. Special Issue, 5. Elsevier, Amsterdam.

Conford, P. (2001) *The Origins of the Organic Movement*. Floris Books, Edinburgh.

Dabbert, S. (1994) Economics of conversion to organic farming – cross sectional analysis of survey data in Germany. In: *The Economics of Organic Farming – An International Perspective* (eds N.H. Lampkin and S. Padel), pp. 285–93. CAB International, Wallingford.

ENOF (1998) *Resource use in organic farming*. Proceedings of Conference of European Network of Organic Farming, Barcelona.

EU (2001) *Organic Farming – Guide to Community Rules*. Directorate-General for Agriculture, European Commission, Brussels. http://europa.eu.int/comm/agriculture/qual/organic/brochure/abio_en.pdf.

FAO (2000) *Food safety and quality as affected by organic farming*. Agenda Item 10.1, 22nd FAO Regional Conference for Europe. Food and Agriculture Organization, Rome. http://www.fao.org/docrep/meeting/X4983e.htm.

FAO (2001) *Guidelines for the Production, Processing, Labelling and Marketing of Organic/Biodynamic Foods*. Codex Alimentarius Commission, Food and Agriculture

Organization, Rome. http://www.fao.org/organicag/doc/glorganicfinal.doc.

FAO (2003) Organic Farming and Food Security Report. FAO, Rome.

Foelsch, D. (ed.) (1978) *The Ethology and Ethics of Animal Production*. Birkhauser, Basel.

Foster, C. and Lampkin, N.H. (2000) *European Organic Production Statistics 1993–1998*. University of Wales, Aberystwyth. http://www.organic.aber.ac.uk/library/European%20organic%20farming.pdf.

Fowler, S., Lampkin, N.H. and Midmore, P. (2000) *Organic Farm Incomes in England and Wales 1995/96–1997/98*. Report to MAFF. University of Wales, Aberystwyth. http://www.organic.aber.ac.uk/library/Organic%20Farm%20Incomes.pdf.

Fowler, S., Wynne-Jones, I. and Lampkin, N.H. (2001) *Organic Farm Incomes in England and Wales 1998/99*. Report to MAFF. University of Wales, Aberystwyth. http://www.organic.aber.ac.uk/library/Organic%20Farm%20Incomes%201998-99.pdf.

Francis, C.A., Flora, C.B. and King, L.D. (eds) (1990) *Sustainable Agriculture in Temperate Zones*. Wiley, New York.

FSA (2000) *Food Standards Agency View on Organic Foods*. Position paper. Food Standards Agency, London.

Greenwood, J.J.D. (2000) Biodiversity and environment. In: Tinker, P.B. (ed) *Shades of Green – A Review of UK Farming Systems*. Royal Agricultural Society of England, Stoneleigh, pp. 59–72.

Hamm, U., Gronefeld, F. and Halpin, D. (2002) *Analysis of the European Market for Organic Food*. University of Wales, Aberystwyth.

House of Commons (2001) *Organic Farming*. Second report of the House of Commons Agriculture Committee. Stationery Office, London.

Hovi, M., Roderick, S., Wassink, G. and Oakeley, R. (2000) *Compendium of Animal Health and Welfare in Organic Farming*. Veterinary Epidemiology and Economics Research Unit, University of Reading.

Howard, A. (1940) *An Agricultural Testament*. Oxford University Press, London.

IFOAM (2002) *Basic Standards of Organic Agriculture*. International Federation of Organic Agriculture Movements, Tholey-Theley.

ITC (2003) *Overview World Markets for Organic Food and Beverages*. International Trade Centre, Geneva.

Jansen, K. (2000) Labour, livelihoods, and the quality of life in organic agriculture. *Biological Agriculture and Horticulture*, **17**, 247–78.

King, F.H. (1911) *Farmers of Forty Centuries – Permanent Agriculture in China, Korea and Japan*. King, Madison.

Lampkin, N.H. (1990) *Organic Farming*. Farming Press, Ipswich.

Lampkin, N.H. (2002) *Organic Farming Conversion Models.* Unpublished report. University of Wales, Aberystwyth.

Lampkin, N.H. (2003) *Organic Farming Statistics.* University of Wales, Aberystwyth. http://www.organic.aber.ac.uk/stats.shtml.

Lampkin, N.H. and Padel, S. (eds) (1994) *The Economics of Organic Farming – An International Perspective.* CAB International, Wallingford.

Lampkin, N.H., Measures, M. and Padel, S. (eds) (2004) *2004 Organic Farm Management Handbook.* University of Wales, Aberystwyth.

LEAF (2001) *Handbook on Integrated Farm Management.* Linking Environment and Farming, Stoneleigh.

Lockeretz, W. and Geier, B. (eds) (2000) *Quality and communication for the organic market.* Proceedings of Conference of International Federation of Organic Agriculture Movements, Tholey-Theley.

Mäder, P., Pfiffner, L., Lützow, M.v., Fliessbach, A. and Munch, J.C. (1996). Soil ecology – the impact of organic and conventional agriculture on soil biota and its significance for soil fertility. In: *Fundamentals of Organic Agriculture* (ed. T. Ostergaard), pp. 24–46. Proceedings of Conference of International Federation of Organic Agriculture Movements, Tholey-Theley.

MAFF (1998) *Codes of Good Agricultural Practice for the Protection of Water, Soil and Air.* Ministry of Agriculture, Fisheries and Food (now DEFRA), London.

Maxted-Frost, T. (ed.) (1997) *Future agenda for organic trade.* Proceedings of Conference of the Soil Association, Bristol.

McCarrison. R. (1936) *Nutrition and Health.* McCarrison Society, London.

Merrill, M.C. (1983) Eco-agriculture – a review of its history and philosophy. *Biological Agriculture and Horticulture*, **1**, 181–210.

MFAF (2001) *Organic food and farming – towards partnership and action in Europe.* Proceedings of Conference of the Danish Ministry of Food, Agriculture and Fisheries, Copenhagen. www.fvm.dk/konferencer/organic_food_farming.

Mollison, B. (1990) *Permaculture – A Practical Guide for a Sustainable Future.* Island Press, Washington, D.C.

Newton, J. (1995) *Profitable Organic Farming.* Blackwell Science, Oxford.

Nix, J. (2003) *Farm Management Pocketbook.* Wye College, London.

Offermann, F. and Nieberg, H. (2000) *Economic Performance of Organic Farms in Europe.* Organic Farming in Europe – Economics and Policy. Vol. 5. University of Hohenheim, Stuttgart.

Padel, S. (2001) Conversion to organic farming – a typical example of the diffusion of an innovation? *Sociologia Ruralis*, **41**, 40–61.

Padel, S. and Lampkin, N.H. (1994) Conversion to organic farming – an overview. In: *The Economics of Organic Farming – An International Perspective* (eds N.H. Lampkin and S. Padel), pp. 295–313. CAB International, Wallingford.

Pretty, J. and Hine, R. (2001) *Reducing Food Poverty with Sustainable Agriculture – a Summary of New Evidence.* Final Report from the SAFE-World Research Project. University of Essex, Colchester.

Pretty, J.N., Brett, C., Gee, D., *et al.* (2000) An assessment of the total external costs of UK agriculture. *Agricultural Systems*, **65**, 113–36.

Reed, M. (2001) Fight the future! How the contemporary campaigns of the UK organic movement have arisen from their composting of the past. *Sociologia Ruralis*, **41**, 131–45.

Reganold, J.P., Palmer, A.S., Lockhart, J.C. and Macgregor, A.N. (1993) Soil quality and financial performance of biodynamic and conventional farms in New Zealand. *Science*, **260**, 344–9.

Rusch, H.-P. (1968) *Bodenfruchtbarkeit – eine Studie biologischen Denkens.* K.F. Haug, Heidelberg.

Sattler, F. and Wistinghausen, E.v. (1992) *Bio-Dynamic Farming Practice.* Bio-Dynamic Agricultural Association, Stourbridge.

Soil Association (2000) *The Biodiversity Benefits of Organic Farming.* Soil Association, Bristol.

Soil Association (2001) *Organic Farming, Food and Human Health – A Review of the Evidence.* Soil Association, Bristol.

Soil Association (2003) *Organic Food and Farming Report 2003.* Soil Association, Bristol.

Steiner, R. (1924) *Agriculture – A Course of Eight Lectures.* Rudolf Steiner Press/Bio-Dynamic Agricultural Association, London.

Stockdale, E.A., Lampkin, N.H., Hovi, M., *et al.* (2000) Agronomic and environmental implications of organic farming systems. *Advances in Agronomy*, **70**, 261–327.

Stolze, M., Piorr, A., Haering, A. and Dabbert, S. (2000) *The Environmental Impacts of Organic Farming in Europe.* Organic Farming in Europe: Economics and Policy. Vol. 6. University of Hohenheim, Stuttgart.

Tinker, P.B. (ed.) (2000) *Shades of Green – A Review of UK Farming Systems.* Royal Agricultural Society of England, Stoneleigh.

USDA (2000) *National Organic Program.* United States Department of Agriculture, Washington, DC. www.ams.usda.gov/nop/facts/index.htm.

Young, A. (1997) *Agro-Forestry for Soil Management.* CAB International, Wallingford.

Useful websites

www.adas.co.uk – ADAS.

www.defra.gov.uk/farm/organic – Department of Environment, Food and Rural Affairs (DEFRA).

www.efrc.com – Elm Farm Research Centre (EFRC).

www.eisfom.org – European Information System for Organic Markets (EISfOM).

http://europa.eu.int/comm/agriculture/qual/organic/index_en.htm – European Union.

www.fairtrade.org.uk – Fairtrade Foundation.

www.fao.org/organicag – Food and Agriculture Organization (FAO).

www.fibl.org – Swiss Research Institute for Organic Agriculture (FIBL).

www.foodfirst.org/cuba/index.html – Food First.

www.hdra.org.uk – Henry Doubleday Research Association (HDRA).

www.ifoam.org – International Federation of Organic Agriculture Movements (IFOAM).

www.irs.aber.ac.uk/OMIaRD – Organic Marketing Initiatives and Rural Development (OMIARD).

www.nal.usda.gov/afsic/csa – Community Supported Agriculture (CSA).

www.ncl.ac.uk/tcoa – Tesco Centre for Organic Agriculture.

www.organic.aber.ac.uk – Organic Centre Wales (OCW).

www.organic-europe.net – Organic Europe.

www.organic-research.com – CABI Organic Research.

www.organicts.com – Organic Trade Services.

www.sac.ac.uk – Scottish Agricultural Colleges (SAC).

www.sekem.com – Sekem, an initiative in Egypt.

www.soilassociation.org – Soil Association (SA).

www.veeru.reading.ac.uk/organic – Network for Animal Health and Welfare in Organic Agriculture (NAHWOA).

14

Farm woodland management

A.D. Carter and I. Willoughby

Introduction

Woodland covers 2 800 000 ha or about 11.5% of the total land area of Britain, of which farm woodlands make up about 350 000 ha. Less productive than many larger commercial forests, farm woodlands typically occur in small fragmented parcels and comprise a significant proportion of mixed broadleaved woods, especially in the lowlands of Britain, where they form important features in the landscape and may have considerable value for wildlife and game. Unfortunately, timber quality and productivity may be significantly reduced by a combination of factors, including difficult access, poor condition of the growing stock, small size and neglected or inappropriate management. Fortunately, both site conditions and the quality of the trees can be improved by the application of appropriate silvicultural techniques to create a quality timber resource with multiple benefits for the landscape, amenity, biodiversity and game. New farm woodland planting is also expanding, as surplus agricultural land becomes available and grant incentives encourage farmers to diversify into woodland enterprises.

Despite this clear potential, a number of perceived obstacles deter landowners from either managing existing woodlands or planting new ones, including a lack of knowledge of silviculture and how to market small volumes of timber. The most enduring financial obstacle is the long gestation period involved in forest investments, requiring a wait of at least 20 years for most crops before thinnings even begin to offset the early establishment costs. This 'income gap' can be partially offset by some of the following options:

- Revenue from harvesting an existing crop can be used to offset the cost of establishment.

- Grants are available to compensate landowners for planting trees and other forms of sustainable woodland management.
- Development of shorter rotation crops (e.g. coppice).
- Integration of forestry with other revenue-generating activities in the early life of the crop, such as game management, recreation or growing Christmas trees.
- Agroforestry systems that offer some potential to use the spaces between the trees in the early years for low-intensity grazing or other agricultural use.

Setting management objectives

While woodlands can be successfully managed for multiple benefits, competing objectives must be prioritized so that an appropriate silvicultural management plan can be formulated. For the majority of farm woodlands, if the primary objective of the owner is to maximize timber production, then this aim will have to be compromised to a certain extent to incorporate other objectives, such as game and wildlife habitat management, or to meet the requirements of landscape designations. Once the overall objectives are set, choice of specific silvicultural operations will depend on an assessment of the site conditions and the growing crop, together with a knowledge of the external environment, such as the market for timber produce, the wider policy framework and grant arrangements.

Timber production

Timber is normally sold on a price per unit volume basis, which is set according to the volume available,

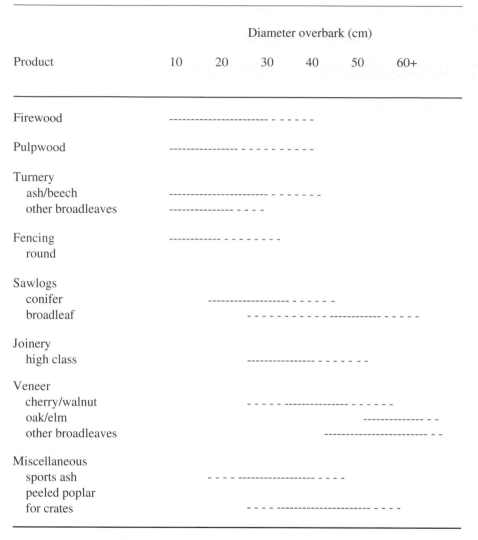

Figure 14.1 Diameter specifications for various end-uses (after Hibberd, 1988, Crown copyright, reproduced with permission of the Forestry Commission, 2001).

the state and location of the timber (standing, stacked in the forest or at the roadside), ease of access, distance to markets and the method of sale (auction, tender or private sale). The use of electronic marketing and auction systems is becoming increasingly popular. Each tree species will have recognized characteristics that must be met for particular markets, such as length, diameter (Figure 14.1), straightness and freedom from defects. Some markets are much more exacting than others (e.g. veneers), demanding higher prices, so that the harvested timber may be graded and separated into

different price bands according to quality. The difference between the market price for low and high quality is enormous. For example, oak veneers may fetch up to £300/m³, while poor quality oak may only be worth £10/m³ as firewood.

Small farm woodlands cannot compete with the large upland forests, which supply vast quantities of timber to bulk markets such as pulp processors and particle board manufacturers. The farm woodland owner is best advised to produce high quality timber, to take advantage of local specialized markets (e.g. craft, furniture, domestic fencing) and to add value wherever possible by processing on the farm. For further information on timber properties of farm woodland trees, see Brazier (1990).

Marketing timber from farm woodlands may require significant effort – seeking out and negotiating with potential buyers, completing the statutory paperwork required by the Forestry Commission, selecting and marking timber to be felled, drawing up contracts and organizing and supervising contractors are all time-consuming activities. The more of this work the woodland owner can carry out him/herself, the greater the eventual returns will be.

Social objectives

The idea of social forestry forms part of the concept of sustainable forest management. This was defined at Helsinki in 1993 to include 'social functions' and forms the basis of the UK Forestry Standard (Forestry Commission, 1998a), which underlies support for farm and other private woodlands and private woodlands through the forestry grant schemes. The England Forestry Strategy (Forestry Commission, 1999) emphasizes the need to 'increase awareness of the relevance of woodlands to all parts of society'.

Timber values have declined significantly in real terms over the last decade as a result of the availability of low-priced imported softwood, and the disappearance of some key markets due to increased use of recycled fibre. Other non-market or social objectives have therefore tended to increase in importance, and this is reflected in the conditions applied to the forestry grant schemes. For instance, specific supplements have been made available for woodland improvements that encourage public access. Woodlands provide marketable non-timber products, for example hunting and shooting; mushrooms and berries; horticultural moss and foliage. They also provide important environmental benefits, for instance in buffering surface waters from nitrogen and phosphorus pollution, and in conserving wildlife. Of particular importance here are the 'ancient' woodlands – woodlands that have never been converted to agriculture and still support wildlife species characteristic of the wild-wood. Woodland habitat covered most of Britain before the wild-wood was cleared by man, but now only around 570 000 ha

of woodland is classed as 'ancient' (Peterken, 1993). In addition, many important open habitats are associated with woodlands, for instance rides, clearings, wetlands and open water. Conserving and restoring existing woodland and establishing new woodland with conditions favourable for wildlife are therefore important objectives of woodland management and expansion. Recommendations for conservation measures are given in a later section.

Social objectives and benefits include those goods and services that are not readily marketable:

- recreation and access;
- health and well-being;
- education and learning;
- rural development;
- attractive landscape.

Recreation and access

Woodland access can be understood as benefiting either local users – where the woodland is part of the environment in which they live – or visitors, particularly urban-dwellers who may have limited access to wooded space near where they live. Community woodlands are particularly important for the quality of life of local users. Providing for use by tourists can attract inward investment and assist in rural development. In either case, well-managed woods where access is clearly welcomed, and where litter is quickly removed, offer a better experience and are less prone to abuse.

The grant schemes generally seek to enhance goods and services on behalf of the tax-paying public. However, 'the public' is made up of many different communities of place and communities of interest, so it is helpful to specify those communities that are expected to benefit when setting woodland objectives. For instance, provision for access by a local community in a situation where there is not much alternative access may be more important than an improvement to an existing right of way. Providing a link in a long-distance walking route may also be important in encouraging tourism.

Health and well-being

Woodlands and other natural spaces, depending on their location and on the access arrangements, offer opportunities for health-giving outdoor activity in a pleasant environment. There is also a body of research evidence on the likely benefits to psychological well-being when people are able to experience respite or recovery from stress. Psychological well-being has been linked with physical health, and both are influenced by physical activity and contact with natural environments (Tabbush and O'Brien, 2003).

Education and learning

Woodlands have traditionally been places for educating children and adults about 'nature', and this remains an important role for formal groups of school-children or informal family groups. Outdoor environments also offer opportunities for physical and mental challenge, and this is important in building self-esteem through the achievement of small tasks, and learning practical skills. In 'forest schools' children learn all their subjects in the woodland, and this idea is proving successful with all age groups, and with particular benefits for disaffected and autistic children. The educational role of woodlands is likely to be of increasing importance in a world dominated by sedentary activities based on computers at work and at play.

Rural development

Forestry for rural development is the first of the strategic priorities in the England Forestry Strategy. In rural areas a thriving economy can help combat social exclusion and poverty. Woodlands generate jobs both by providing raw materials and through service jobs in conservation, recreation and tourism. In this respect larger woodlands are more valuable than smaller ones, they are more likely to yield economically worthwhile quantities of timber and more likely to generate service jobs.

Attractive landscape

Woodlands can be a major component of beautiful landscapes, and the Forestry Commission has issued landscape design guidelines (Forestry Commission, 1995) which need to be borne in mind when designing new or managing existing woodlands. Enhancing landscape may also enhance property prices and business investment, and so have an indirect impact on rural

development (see section on 'Landscape design for farm woodlands').

Public involvement

One implication of sustainable forest management is the participation by stakeholders in making decisions that affect them. This avoids conflict and makes for better informed, more robust decisions. The first stage is usually some form of stakeholder analysis, in which local managers identify all those who will have an interest in or be affected by woodland management decisions. Decisions are then taken on the appropriate approach for each stakeholder group. Some will demand a much higher level of involvement than others. Statutory bodies will often require formal consultation, while it is usually best to develop an active dialogue with local neighbours. A 'toolbox' of methods (Hislop, 2004) can be used to develop a public involvement strategy. The tools chosen can range from posting notices at the beginning of footpaths giving non-controversial information, to holding 'citizens' juries' to make recommendations on complex and high-impact environmental proposals.

Creating new woodlands

The establishment of new farm woodlands gives the landowner a greater degree of control than for existing woodlands over the size, location and most appropriate type of silvicultural system. New planting is often concentrated on areas of lower productivity or agriculturally marginal sites, such as steep slopes or awkward corners of fields. Other sites might include extensions to existing woodland, linking fragmented areas of existing ancient semi-natural woodland, shelterbelts, or for amenity around farm buildings. Many farmers are now incorporating woodland planting with general conservation or landscape work around the farm.

Site investigation

Before any new woodland is planned, or any silvicultural operations take place, it is imperative to carry out a thorough site investigation to determine the nature of past land use and the presence of factors that may influence the final choice of planting design and establishment operations. These investigations can be broken down into five steps. The degree of importance of each step, or the detail to which it needs to take place, will be dependent on the nature of the site.

The level of detail can be divided into three stages:

(1) desk study;
(2) walkover/general site surveys;
(3) intrusive/detailed site surveys.

At each stage, the information collected should be used to develop a conceptual model of the site. This will help to determine the need to continue to the next, greater stage of detail. The five steps to consider at each level of detail are:

(1) Preliminary site investigation. This should include determining past and current land use, ownership, rights of way and other wayleaves, physical landform including the surrounding landscape, soil types, consulting the contaminated land register and identifying any risks or hazards to health or planning issues restricting planting.
(2) Archaeology survey. The site should be walked to identify any archaeological remains that may influence woodland design. The local authority archaeologist can give advice on the need for and scope of such a survey.
(3) Investigation into soil resource and physical conditions. This should include geomorphology, topography, suitability for tree planting (for example using the Forestry Commission ecological site classification system), soil depth, drainage, cultivation requirements to allow adequate tree rooting and, if necessary, a contaminated land survey.
(4) Ecology survey. A survey of key existing habitats and existing tree species should be carried out. The relationship with neighbouring habitat types should be examined. This process may identify areas that need to be protected, left unplanted, or might be enhanced by careful woodland design.
(5) Community liaison. As well as being a requirement of many grant aid schemes, in order to minimize possible future conflicts it makes good economic sense to consult with local communities that might be affected by the creation of any new woodland. Consultation should aim to determine current and possible future usage of the site, and the facilities or features of the site people would be interested in preserving or creating.

Further information on site investigation, in particular the detailed studies that are necessary when dealing with contaminated brownfield land, can be obtained from the Forestry Commission (2003).

Establishment

The effective establishment of trees is a key management operation in the life of the woodland. The cost of establishing a woodland crop can seem very high; however, neglected or poor establishment practice, in an effort to reduce costs, will lead to many problems that will be difficult (and even more costly) to remedy later. Careful planning and organization of species choice, site preparation, planting, weeding and tending will reduce the overall cost of establishment in the long run.

Site conditions

For new woodlands, one of the most important initial decisions is the correct choice of tree species and woodland design to match the site conditions and objectives of the landowner. Subsequent forest management will not be able to make up the lost ground as a result of a poorly adapted species growing in the wrong conditions. The following factors should be considered.

Climate and topography
While the mild climate of the UK is generally favourable for the growth of a wide range of trees, the choice of species for any particular area will be influenced by many factors. The most important to consider are:

- *Altitude:* increasing altitude results in reduced yields due to lower ambient temperature and shorter growing season (450 m above sea level is considered to be the economic tree line for timber production, although trees planted for other objectives will grow at most elevations present in the UK).
- *Exposure:* exposed sites have lower overall productivity and trees on the edge of woodlands will be shorter and more bushy with increased risk of windthrow.

- *Rainfall:* for the purposes of timber production, coniferous species with higher water requirements such as Douglas fir, larch, spruce, western hemlock and western red cedar grow better in the wetter north and west of Britain (rainfall 1000–1500 mm/year), while the drier eastern counties are more suited to species such as Corsican pine and Scots pine. In general, native broadleaved species tend to be less site sensitive in terms of rainfall requirements.
- *Length of growing season:* summer warmth is required for some species such as sweet chestnut and walnut, which restricts economic growth of these species to the southern counties.
- *Incidence of drought or frost:* the incidence of seasonal frosts, often exacerbated by local topography, may cause injury to young trees, especially at the beginning or end of the growing season. The risk of frost damage is greatest in hollows or valleys where cold air from higher ground is able to collect. Only frost-hardy species should be planted in these locations.

Soils and ground vegetation

While healthy forest growth is attainable on a wide range of soil types, tree planting tends to be concentrated on those soils less suitable for agriculture, including less fertile mineral soils, gleys, podzols and deep peats. Adequate root development is essential for sustained growth through anchorage and supply of water and nutrients. Where root growth is constricted by impeded drainage, an iron pan, a plough pan from previous agricultural cultivation, or by shallow or compacted soils, there is a greater risk of subsequent windthrow or moisture stress in drought periods. This may necessitate further cultivation or drainage work. A series of soil inspection pits should be dug pre-planting to check for soil depth, fertility, soil texture, drainage and the existence of hard pans. Further background information can be obtained through soil maps and land capability maps for forestry, available from the Soil Survey and Land Research Centre (SSLRC).

Ex-industrial brownfield land such as coalfields, landfills, quarries, etc. can be particularly problematic site types, as they require careful restoration to be able to support tree growth. The standard of restoration required is often far higher than that needed to support agricultural uses. For this reason, woodland establishment on brownfield sites should not normally be

Table 14.1 Assessment of site quality from vegetation present (reproduced with permission of the Forestry Commission).

Site quality	Indicator species
Good	Ash, beech, hazel, hornbeam, field maple, oak, bluebell (*Hyacinthoides non scriptus*), dog rose (*Rosa* spp.), primrose (*Primula vulgaris*), wild garlic (*Allium ursinum*), wild raspberry (*Rubus idaeus*), dog's mercury (*Mercurialis perennis*)
Moderate	Alder, bracken (*Pteridium aquilinum*), honeysuckle (*Lonicera periclymenum*), horsetails (*Equisetum* spp.), rhododendron (*Rhododendron ponticum*), soft rush (*Juncus effusus*)
Poor	Bilberry (*Vaccinium* spp.), cotton grass (*Eriophorum* spp.), deer grass (*Trichophorum caespitosum*), cross-leaved heath (*Erica tetralix*), heather (*Calluna vulgaris*), moor mat grass (*Nardus stricta*), purple moor grass (*Molinia caerulea*), sphagnum moss (*Sphagnum* spp.)

attempted unless there is an opportunity to influence restoration practices at the outset. Depending on the nature of the previous industrial use, sites are likely to require capping of waste, containment or buffering of toxic chemicals, landforming to encourage better drainage, and loose tipping of at least 1m of soil-forming material. Sites that are suspected to have been restored in the past should be carefully investigated before any planting takes place. See Moffat and McNeill (1994) for guidance on this subject. The Forestry Commission (Forest Research) and other independent advisors offer an advisory service on restoration.

Ground vegetation can be used as an indication of soil and site conditions as a guide to tree species selection (Table 14.1), although care is needed since the type and pattern of vegetation present can be greatly modified by previous land management (most agricultural sites will have been improved at some time in the past). The systems most commonly in use are based on the presence of particular indicator species which denote site fertility, soil pH and drainage conditions.

The Forestry Commission has developed a useful ecological site classification system as a guide to site assessment using a range of environmental indicators.

This helps the woodland manager to select those trees most suited to the site conditions, including those species 'native' to the local area. See Pyatt *et al.* (2001) for further information.

Species selection

The nature of the woodland site will impose certain restrictions on the choice of tree species for planting (Table 14.2). A fertile, sheltered site will enjoy a far longer list of possible tree species than an inhospitable one. Even in the most hostile locations, however, acceptable forest growth can be achieved by selecting species best adapted to the conditions. Following the site assessment, a list of suitable trees should be drawn up by relating the site conditions to their silvicultural characteristics. The final selection of species should be made on the basis of management objectives.

Tree species differ in their silvicultural characteristics according to their natural position within the woodland ecosystem. An important distinction relates to how tree species react to different light conditions. *Light demanders* tend to have thin crowns, light foliage and less dense timber and will grow rapidly in reasonably open situations, although in shade they are quickly suppressed and die. *Shade bearers* tend to have more layered crowns and heavier timber and will only grow slowly at first, although they are able to survive moderate shade. Where maximum volume production is a major objective, light-demanding species are clear winners. Other important silvicultural characteristics include climatic and soil requirements, timber quality, response to silvicultural treatment and sensitivity to drought, frost, wind, exposure, insect attack, browsing and disease.

In addition to appropriate species choice, consideration should be given to the origin or *provenance* of the planting stock, since, for the same species, the silvicultural characteristics of the trees may be slightly different for seed drawn from different geographical subpopulations. Many commercially grown exotic conifers, for example, are native to the western seaboard of North America and seed from the northern populations (e.g. Alaska) will be more resistant to winter cold, although slower growing than seed drawn from more southerly populations (e.g. California). Since the first introductions of these exotic conifers, seed may now be obtained from seed orchards (or provenances) in Britain, even though the origin remains unchanged. Both seed origin and provenance are recorded in tree planting catalogues which must be matched with the environmental conditions of the planting site for optimum growth. Further information is available from Lines (1987).

When attempting to create a new native woodland, the use of plants produced from seed from local sources may be important, particularly if there is a compelling history of genuinely local native origin for such sources.

Pure crops and mixtures

Pure crops are much simpler to manage than mixtures, although mixtures do offer other benefits if managed skilfully. For the creation of new native woodlands, mixtures of broadleaved species planted at a variety of spacings are usually required (see Rodwell and Patterson, 1994, for more details). Broadleaves can also be grown in mixture with conifers to provide an early economic return from the coniferous thinnings in order to offset the delayed returns from the longer rotation of slower-growing broadleaves. Appropriate mixtures may also be able to more fully utilize the site through differences in rooting depth and are far preferable on landscape, game and conservation grounds. Mixtures are often used where one species or *nurse* offers benefit to the main tree species, in the form of shelter against frost or cold, smothering weed growth, nutrition, support or suppression of side branches. Timely thinning is often required to prevent the nurse from outgrowing and suppressing the main tree species, so that care is needed in the choice of appropriate mixture. As a general rule, for broadleaf/conifer mixtures, the expected growth rate of the conifer should not be more than double that for the broadleaved species. Broadleaves should be established as blocks (9 or 25 to each block) within a conifer matrix, at an appropriate spacing to provide a final broadleaved crop. Alternatively line mixtures can be used, although these should be lines of at least three of each species to reduce the risk of suppression. Normal practice is to remove the adjacent lines of conifers to the broadleaves at the first thinning, followed by selective thinning thereafter. Greater care is required on hillside plantings to avoid regimented or geometric patterns of mixtures which are highly visible in the landscape.

Examples of suitably robust mixtures include:

Table 14.2 Species characteristics and site requirements.

Species	Recommended sites	Unsuitable conditions	Silvicultural characteristics	Timber quality
Scots pine (*Pinus sylvestris*)	Low rainfall areas on heather, poor gravel or sandy soils. Frost hardy.	Wet or soft ground, chalk, limestone, high rainfall moorlands.	Light demander, although slow growth rate and volume production. Useful as a nurse for hardwoods.	General-purpose softwood timber, good strength, takes preservative well.
Corsican pine (*Pinus nigra* ssp. *laricio*)	Low rainfall areas and elevations on sand and clay soils especially near the sea. Plant only in southern and eastern England.	High elevations, wet moorlands (increased risk of dieback from *Gremmeniella abietina*).	Higher volume production and better form than Scots pine, although more difficult to establish.	Timber similar to Scots pine but coarser and a little weaker.
Lodgepole pine (*Pinus contorta*)	Pioneer species on poor heaths, deep peats and sand dunes.	All but the poorest sites where no other tree will grow.	Coastal provenances give higher volume production but poorer form. Vulnerable to pine beauty moth in north Scotland.	Timber similar to Scots pine.
European larch (*Larix decidua*)	High rainfall areas on moist free-draining loams.	Dry, poorly drained and very exposed sites or frost hollows. Avoid areas with rainfall under 750 mm.	Deciduous conifer. Good nurse for hardwoods, although runs out of top growth quickly so thin from age 15–18 years. Susceptible to butt rot (*Heterobasidion annosum*).	Heavy and generally strong timber, best quality used for boat building. Heartwood is naturally durable, and makes a good farm timber for fencing, gates and other estate uses.
Japanese larch (*Larix kaempferi*)	Mild and wet regions, less exacting than European larch. Pioneer in uplands on heather and grass.	Dry, poorly drained and very exposed sites or frost hollows. Avoid areas with rainfall under 750 mm.	Corkscrews on very fertile sites.	See European larch.
Hybrid larch (*Larix × eurolepis*)	Hybrid larch is slightly hardier than European larch.	As for Japanese larch.	Higher yielding than Japanese larch. Only use first-generation seedlings.	See European larch.
Douglas fir (*Pseudotsuga menziesii*)	Sheltered valley slopes on well-drained and moderately fertile soils. Grows well in wetter regions.	Exposed or frosty sites, heather ground, wet, soft or shallow soils.	A high yielding species although susceptible to windblow on shallow soils.	Strong construction timber with a high weight : strength ratio.
Norway spruce (*Picea abies*)	Moist grassy or rushy sites, most reasonably fertile soils and fairly heavy clays.	All dry or exposed sites. More frost tolerant than sitka spruce.	Often grown in old woodland sites.	Good general-purpose timber, works and nails well. Stable in changing humidity conditions so suitable for building. Not suitable for preservative treatment or outdoor use.

Species	Site		Characteristics	Timber
Sitka spruce (*Picea sitchensis*)	Wet exposed uplands in the north and west of Britain. Thrives on peats and grasslands in high rainfall areas.	All sites liable to dry out and rainfall areas under 1000 mm.	Will withstand very exposed conditions, although avoid previous scrub land where there is risk of honey fungus (*Armillaria* spp.).	Timber superior to Norway spruce, but too coarse for joinery. Good pulpwood.
Western hemlock (*Tsuga heterophylla*)	Tolerant of high and low rainfall on acid mineral soils and better peats.	Slow to establish on heather and open ground without shelter.	Strong shade bearer, often grows best in mixture. Susceptible to butt rot (*Heterobasidion annosum*) and honey fungus on previous conifer sites.	General-purpose building timber and pulpwood.
Western red cedar (*Thuja plicata*)	High rainfall areas on moderately fertile soils in sheltered sites.	Exposed, poor and dry sites.	Shade bearing and narrow crown, useful in mixtures or nurse species.	Light-coloured heartwood which is very durable. Cladding used outside for greenhouses and sheds.
Lawsons cypress (*Chamaecyparis lawsoniana*)	Requirements not exacting but best on deep fertile soils, with moderate to high rainfall.	Dry infertile sites and heather ground. Avoid frosty, exposed and waterlogged sites.	Slow growing shade bearer for use in underplanting or in mixtures (e.g. with oak).	Heartwood reasonably durable for general-purpose uses. Small sizes used for fencing.
Grand fir (*Abies grandis*)	Well-drained moist deep soils.	All poor soils, dry, frosty or exposed sites.	High volume producer, useful for underplanting as moderately shade bearing.	Soft white timber of only moderate quality.
Noble fir (*Abies procera*)	Well-drained deep moist soils, tolerates acidity.	Poor dry soils.	Withstands exposure well. Tolerates drier sites than sitka spruce. A useful shelterbelt tree in Scotland.	Poor timber/quality exacerbated by high/taper.
Pedunculate oak (*Quercus robur*)	Deep clay loams with ample moisture, although will tolerate heavy clays.	Shallow, infertile or poorly drained soils on exposed sites.	Oak is a strong light demander and will not grow under shade. For top quality timber, close planting (at least 2500 plants/ha) is required for suppression of side branches. Bole shading by underplanting or pruning will prevent problems from epicormics. Oak grown on lighter soils may suffer from 'shake'. Windfirm.	Oak has strong, naturally durable heartwood, although the sapwood does require preservative treatment if used outside. Timber quality (and price) varies tremendously. Prime quality oak is scarce and is used for veneers, planking and furniture. Lower grades for beams, fencing, gates and other estate purposes. The branchwood and lowest grades may only be fit for firewood or pulp.

Cont.

Table 14.2 Continued.

Species	Recommended sites	Unsuitable conditions	Silvicultural characteristics	Timber quality
Sessile oak (*Quercus petraea*)	Best on deep porous brown earths, although will tolerate clay soils.	As above and wet clays or fluctuating water tables (risk of shake).		
Beech (*Fagus sylvatica*)	Light well-drained soils on chalk or limestone, deep sands and acid brown earths.	Cold, wet poorly drained soils, dry infertile sands.	Shade bearing, although may require a nurse (e.g. Scots pine) on exposed sites. An excellent amenity tree and suitable for underplanting. May suffer from severe squirrel damage, requiring close spacing for adequate selection.	Timber strong, fine, even texture which takes a stain well. Used for wide range of interior uses: furniture, kitchenware, turnery and flooring. Lower grades used for firewood or pulpwood. Not suitable for exterior uses.
Ash (*Fraxinus excelsior*)	Only thrives on moist, well-drained fertile soils, ideally deep calcareous loams in valley bottoms. Wild garlic or dog's mercury are good indicators.	Dry, shallow heavy clay or badly drained soils, heaths and moorlands. Avoid frost hollows and exposed sites.	Usually grown in mixture with beech, oak, cherry or larch. The stand needs to be thinned regularly to encourage large crowns.	Spring timber with high shock resistance. Top grade ash (annual rings 1.5–6 mm) is used for sports goods, tool handles and furniture. Lower grade timber makes excellent firewood.
Sweet chestnut (*Castanea sativa*)	Deep fertile soil in warm climate, ideally in southern England. Warm sunny acid sandy loam banks are ideal (pH 4.0–5.0).	Cold, wet, badly drained, exposed, frosty or infertile sites. Chalk, limestone or alkaline soils.	Often grown productively as coppice material used for estate products. Timber trees over 80 years are prone to shake on free-draining soils.	Sawn timber makes a strong substitute for oak. Coppice material used for cleft fencing, poles and stakes.
Sycamore (*Acer pseudoplatanus*)	Moderately fertile free-draining soil. Grows on exposed upland sites and resistant to frost.	Dry, shallow, ill-drained or heavy clay soils.	Moderate shade bearer and regenerates freely. Performs well on poor upland sites as windfirm. Prone to heavy squirrel damage.	White timber used in turnery and where in contact with food. Figured sycamore valuable for veneer and furniture.
Wild cherry (gean) (*Prunus avium*)	Deep fertile well-drained soil, especially over chalk.	Heavy soils, depressions and all infertile sites.	Grows best in small groups in mixture with oak, ash or beech. Not attacked by squirrels, but may be severely affected by bacterial canker and deer browsing. Regular thinning and pruning in summer for quality timber.	Decorative timber with rich reddish-brown heartwood used for veneers, furniture and paneling.
Common walnut (*Juglans regia*)	Sheltered south-facing slopes on deep, fertile, well-drained, medium texture soils. Optimum pH 6.0–7.0.	Only planted on carefully selected sites that meet requirements, avoiding frost hollows.	Best grown in open situation or in small groups, although pruning will be needed to improve form in July/August.	Valuable decorative dark brown timber, often dug out and taken to the sawmill with main roots as 'figuring' occurs in the base. Branches for firewood.

Species	Site requirements	Sites to avoid	Notes	Timber
Black walnut (*Juglans nigra*)	Chalk or limestone with more than 600 mm of overlying soil. Requires warm summers (ideally south and central England).			Timber dries red-brown, good for turnery, staining and polishing.
Common alder (*Alnus glutinosa*)	River and stream banks and wet marshy places in lowlands and uplands. Frost hardy.	All dry sites. For very acid sites prefer grey alder. Widespread incidence of *Phytophera* disease.	Good shelterbelt tree (windfirm) for use on nutrient-poor sites or in coastal locations, as resistant to salt spray. Mixes with birch and ash and is a good nurse for oak on wet clay soils. Fixes atmospheric nitrogen in root nodules. Grows rapidly for the first 25 years and coppices vigorously.	
Grey alder (*Alnus incana*)	Requires drier situation than common alder, but tolerates poorer sites.	Dry infertile situations, heathlands, shallow and thin chalky soils.	Widely planted for windbreaks, shelter and as a pioneer species for reclaiming derelict land. Extremely hardy.	Timber worthless.
Italian alder (*Alnus cordata*)	Drier chalk and limestone soils of southern England.	Avoid dry, thin, acid or infertile soils.	Pioneer species used for single row windbreaks.	Timber worthless.
Black poplar (*Populus robusta*)	Base-rich loamy soils in sheltered position with water table 1–1.5 m below the surface in summer.	Exposed, acid, dry or infertile sites.	Poplars often used for perimeter horticultural windbreaks.	Light white timber used for pallets, chip baskets and pulpwood.
Black hybrids (*Populus canadensis*)				
Balsam poplar (*Populus trichocarpa*)	Tolerates more acid soils than black hybrids and more suited to cooler, wetter parts of Britain.	As for black poplar.	Balsam poplars subject to bacterial canker; plant only resistant clones (Tabbush and Lonsdale, 1999). Preferred to black hybrids as they are fastigate and trim more easily. Hybrid *Populus txt* Clone 32 most commonly used. Other poplars include white and grey spp.	

Cont.

Table 14.2 *Continued.*

Species	Recommended sites	Unsuitable conditions	Silvicultural characteristics	Timber quality
Cricket bat willow (*Salix alba* var. *coerulea*)	Only margins of flowing streams or rivers with highly fertile soils. Useless elsewhere.		Planted as sets 10–12 m apart, with all side shoots removed on the bottom 3 m of the stem for knot free timber.	Cricket bats.
Silver birch (*Betula pendula*)	Brown earths, podzols, sands and gravels. Withstands frost and exposure.		Pioneer species with light crown. Regenerates easily on mineral soils, although not widely planted.	Grown in Scandinavia with improved cultivars for veneer. Low grade timber makes good firewood.
Downy birch (*Betula pubescens*)		Poorly drained heathlands and waterlogged conditions at higher elevations.	Used as soil improver and nurse for oak and beech.	Strong, fine textured timber not naturally durable.
Southern beech (*Nothofagus procera*) (*Nothofagus obliqua*)	Wide range of soils from heavy clay to deep sand.	Badly drained exposed or frosty sites.	Both species are fast growing. Prefer *N. procera* in the wetter west country, *N. obliqua* in the drier east. Light-demanding species, but start under a thin canopy, thin early in life. Frost damage leads to stem cankers.	Timber of both species similar to native beech, but with 20% less bending strength.
Norway maple (*Acer platanoides*)	Moist, deep, free rooting soils with high base status.	Will not thrive on infertile soils, although will survive on thin chalk soils. Avoid frost hollows and exposed sites.	Good amenity tree, suited for screens and mixed shelterbelts. Heavily damaged by squirrels.	Timber as for sycamore; flooring, furniture, turnery and veneer. Good firewood.
Hornbeam (*Carpinus betulus*)	Moist damp clays, chalk, limestone and acid brown earths. Frost hardy.	Thin, infertile, dry and very acid sites.	A substitute for beech on clay soils and where high frost resistance is required. Shade bearing and slow growing, therefore valuable as coppice understorey for shelterbelts or under oak to control epicormics.	Hard, heavy and tough timber giving a very smooth finish ('white beech' on the continent) used for carving and turnery. Makes very good firewood.

- oak, ash and cherry on moist, deep, fertile soils or clay over chalk;
- oak and European larch on free-draining and lighter soils;
- oak and Norway spruce on heavy acid clays;
- oak and western red cedar on free-draining soils;
- beech with western red cedar, Scots pine or Corsican pine;
- sweet chestnut and European larch;
- sitka spruce and European larch.

Site preparation

The aim of ground preparation is to create suitable planting conditions for rapid establishment. Where planting is to take place on the site of recently harvested woodland, consideration must be given to the treatment of brash from the former crop. Brash can be left on site to protect soil during harvesting, redistributed to facilitate easier planting and cultivation, burnt, mulched or chipped to possibly sell on as wood fuel. The actual method or combination of methods of disposal or retention chosen will depend on site factors, marketing opportunities and management objectives for the site. See Moffat *et al.* (2004) for further details.

Additional problems with restocking include the rewetting of the ground as soil moisture increases after felling, and compaction caused by heavy harvesting machinery.

On upland sites, the traditional method of ground preparation was by ploughing, using either a single furrow or double mouldboard, set at the appropriate planting distance. Ploughing is now rarely recommended because of the risk it poses to soil erosion and water quality.

Mounding or '*dolloping*' is now the preferred technique, especially on heavy, wet, compacted or organic soils and restock sites, using a machine to dig out dollops of soil and deposit them on the planting position. This causes far less site disturbance than ploughing and creates a similar raised mound of cultivated mineral soil, through which the trees are planted. On comparatively infertile sites this can almost eliminate the need to weed during the first growing season, as the vegetation is completely buried under the inverted soil. The main advantage, however, is through improved root development, as a result of increased soil aeration, drainage, soil temperature and the release of nutrients from the breakdown of organic matter.

An alternative to dolloping, where drainage is not required, is *scarifying*, which involves the removal of surface vegetation and harvesting residues by a mechanical scraper. Scarifying is the preferred technique where sites are to be established at least partly by natural regeneration. Patterson and Mason (1999) provide detailed cultivation prescriptions for different site types.

Site preparation on former agricultural land

Former agricultural land, particularly lowland arable or improved grassland, is likely to be more fertile than land available for commercial upland forestry and hence different establishment practices are required (Williamson, 1992; Willoughby and Moffat, 1996).

The initial preparation of the planting site will depend on the previous land use and soil conditions. Where sites are uncompacted and freely draining, surface cultivation is not always appropriate. Ploughing up former grassland is usually counterproductive, as the disturbance of the soil both exposes viable weed seeds and creates a bare seed bed for the rapid invasion of arable weeds which are difficult and expensive to control. On bare ground or arable stubble, the most suitable method of controlling weed growth on a site is to establish a low-productivity grass sward, or a wild flower meadow, prior to planting. Competing weed growth still needs to be controlled around trees, but the grass or wildflower/grass sward helps to suppress the growth and spread of noxious weeds such as ragwort (*Senecio jacobaea*) and thistles (*Cirsium arvense, C. vulgare*) on land that would otherwise lie unmanaged.

Heavy soils present particular problems due to shrinkage in summer and waterlogging in winter. In a dry summer, heavy land will shrink, leading to a network of tiny cracks. On recently planted land, this shrinkage is often concentrated in the bare strips along planting lines or around the circumference of trees that are spot weeded, leading to excessive cracking and loss of plants due to moisture loss from exposed roots. The poor structure and texture of heavy soils can pose problems at planting too, which should be delayed until the soil has a suitable moisture content in the autumn. If the soil is too hard and dry, it will be difficult to firm in the roots, while if the soil is too wet, the sides of the planting hole will be smeared, which may crack and expose the roots in a dry summer.

Some sites and soils may suffer from compaction as a result of poaching around cattle feed troughs or heavy machinery travelling along the headlands. The presence of a plough pan in the soil will impede the growth of trees and increase the risk of windthrow. Ripping or subsoiling of these areas is essential to improve soil structure, although trees should be planted to the side of the ripped lines due to the risk of soil cracking.

Drainage

The prime function of forest drains is to remove surface water and to prevent waterlogging, which, over pro-longed periods, will severely impede the growth of trees and may increase the risk of windthrow as root development is restricted. In the uplands, substantial areas of forest plantation suffer from excessive soil water, particularly on gleyed clays, peaty gleys and deep peats where the lateral movement of water is severely limited, requiring extensive open drainage systems. Open drains (60–90 cm deep) are cut to inter-cept surface water from above the plantation at a slight angle to the contour (no more than 2°), together with smaller collecting drains at 30–40 m spacing through the planting site. The environmental effects of increased levels of sediments in upland streams leaving afforested land have caused major concern in recent years, due to the impact on aquatic life. The Forestry Commission has produced useful guides on silvicul-tural practices that can help to prevent these effects (Forestry Commission, 1998b, 2004). In particular, drains should end 5–10 m from smaller streams and 15–30 m from main watercourses to reduce sediment loading.

Once fully established, trees draw water from the soil through transpiration which also helps to reduce soil waterlogging. Open drainage systems should be monitored regularly, although maintenance operations should be restricted to removing serious blockages and overflowing ditches. At harvesting, substantial damage to the drainage system can be avoided by working during drier periods and installing temporary culverts to remove surface water. Inevitably, some reinstatement work will be necessary, although the root channels formed by the previous crop will improve water move-ment, especially where these have broken through any surface pan or compacted layers.

Drainage work can be an expensive and time-consuming operation, normally involving the use of contractors with specialized tracked excavators. The majority of farm woodland sites are unlikely to merit anything other than localized drainage and the owner may be better advised to plant more tolerant species such as alder and willow or to incorporate those 'wet hollows' into areas of open ground within the woodland design.

Where planting is on former agricultural land with an existing below-ground field drainage system, owners should be aware that tree roots will inevitably block and break up these drains, reducing their effectiveness. Where the existing system also serves surrounding agricultural fields, the below-ground drains in the new woodland may need to be replaced with open drains. For further information, see the Forestry Commission (2004).

Protection from mammal damage

New plantations will require protection from grazing animals (rabbits, hares, deer and farm stock) while the young trees become established. The choice of protec-tion will depend primarily on the size and shape of the plantation and the density of planting. For sites smaller than 2–3 ha, individual protection by *tree shelters* is normally the cheapest method, while for larger areas, fencing becomes increasingly cost effective. Fencing may also be required on woodland sites bordering stock fields, roads or public places. Irregular-shaped and narrow belts of trees are the most expensive to fence, having a greater perimeter than square plantations.

The specification of the fence will depend on the type of animal present and the durability required. Rabbit fences should be at least 0.75 m high with 31-mm mesh wire netting, while for deer and stock a stronger fence is required to a height of 1.8–2.0 m. Spring steel wire is commonly chosen as a support for various grades of wire netting, since it will return to normal if acciden-tally stretched by an animal or fallen trees. If the fence is required for longer than 5 years, only treated wooden posts and struts should be used. The line of the fence should be chosen to avoid wet or shallow soils and snow hollows, which may allow entry of animals in winter. Where the fence cuts across badger runs, two-way gates should be provided. Further details of fence specifications are available in Mayle (1999) and Table 14.3.

Tree shelters are translucent open-ended tubes, placed over each tree, and have revolutionized the

Table 14.3 Protection methods for use against damaging mammals. After Hodge and Pepper (1998). Crown copyright, reproduced by permission of the Forestry Commission.

Mammal	Individual tree protection	Fencing	Direct control
Field voles (*Microtus agrestis*) (populations fluctuate and so first signs of extensive damage should trigger protection)	*Tree guards*: 200 mm tall split tubes, buried at least 5 mm into the soil. As trees grow, tubes open out and are easily collected. Tree shelters will not protect against voles unless staked firmly and buried 5 mm into soil. Plastic guards with aeration holes are ineffective. *Chemical repellents*: paint or spray Aaprotect on stem to 300 mm.	Vole guards may be required in fenced areas.	No viable options. Use of poisons is illegal. Good weed control will reduce the risk of damage.
Grey squirrels (*Sciurus carolinensis*)	*Chemical repellents*: paint or spray Aaprotect on stems to be protected.		*Poisoning*: use 0.02% warfarin/wheat bait presented in hoppers of specified dimension for tree protection between 15 March and 15 August in permitted areas. *Live trapping*: multi-capture traps are the preferred trap for woodland tree protection. A 4-day pre-bait period is required before traps are set. Set traps must be visited daily. *Control more effective when co-ordinated by a local Squirrel Management Group.*
Rabbits (*Oryctolagus cuniculus*)	*Tree guards*: 0.6 m tree shelters, split plastic tubes or plastic mesh guards (lateral growth may still be browsed); spiral guards. *Chemical repellents*: Aaprotect applied to dormant trees from mid November.	*Fencing*: 0.9 m; 18 gauge × 31 mm hexagonal mesh with bottom of netting turned out 150 mm towards the rabbits and turved.	*Shooting*: labour intensive and rarely effective. *Gassing*: phosphine gas (Phostoxin or Talunex) is used from November to March to fumigate burrow systems; hazardous to operators; requires properly trained and equipped personnel. *Live trapping*: box traps along fence-lines for large numbers, cage traps for small numbers. Traps must be visited daily.

Table 14.3 *Continued.*

Mammal	Individual tree protection	Fencing	Direct control
Hares (*Lepus capensis, Lepus timidus*)	*Tree guards:* 0.75 m tree shelters or plastic mesh guards. *Chemical repellents:* as for rabbits.	*Fencing:* 1.0 m. Use rabbit netting with a line wire 100 mm above netting.	*Shooting:* can be effective where damage is due to few individuals.
Deer Roe (*Capreolus capreolus*) Red (*Cervus elaphus*) Sika (*Cervus nippon*) Fallow (*Dama dama*) Muntjac (*Muntiacus reevesi*) Chinese water (*Hydropotes inermis*)	*Tree guards:* 1.2 m for roe and muntjac, 1.8 m for red, sika and fallow. Piling brash on coppice stools as a browsing deterrent is largely ineffective and provides ideal cover for rabbits and muntjac. *Chemical repellents:* as for rabbits.	*Fencing:* 1.8 m red, sika, fallow; 1.5 m roe, muntjac.	*Shooting:* set cull levels according to current population estimates, estimate of population growth and target density to ameliorate negative impacts. Optimal strategy may involve shooting and selective use of other deterrents for particularly vulnerable trees.

Fencing detail (within Deer row):

Species	Mesh size (mm)	Height (m)
Fallow	220 × 200	1.5
Roe	200 × 150	1.2 (< 2.5 ha)
		1.5 (> 2.5 ha)
Muntjac	100 × 100 but 75 × 75 preferred for humane reasons	1.5

Electric fencing: roe are not deterred by shocks given by currently available energizers. In recent (unpublished) trials, electric fences have provided an effective barrier against fallow but not against muntjac.

Control most effective when co-ordinated by a local Deer Management Group.

Mammal	Individual tree protection	Fencing	Direct control
Sheep and goats (*Ovis* spp. and *Capra hircus*)	*Tree guards:* 1.8 m (with regular access, two tall stout stakes needed for most breeds). Not reliable for goats.	*Fencing:* 1.5 m (goats) or 1.0 m (sheep) agricultural stock fence. *Electric fencing:* to recognized specification.	
Cattle and horses	Individual tree protection not viable other than for specimen trees.	A buffer zone is needed between fence and trees. *Fencing:* agricultural stock fence (without barbed wire for horses). *Electric fencing:* to recognized specification.	

establishment of broadleaved trees on sheltered lowland sites. They are also valuable for protecting naturally regenerated seedlings or small plantings (e.g. in hedgerows or enrichment schemes). The shelters act as 'tree greenhouses' providing the tree inside with a warm, moist microclimate. The results are more rapid initial growth and better root establishment, which increase tree survival, especially in dry summers. In addition, the shelters give protection against browsing animals and allow for the easier application of foliar-acting herbicides. Height growth can be doubled for some species (notably oak), although this effect is reduced once the tree emerges from the top of the shelter. The advantage of tree shelters is therefore to speed up the expensive process of tree establishment by faster growth and increased survival over the first 3–4 years. Following this, the shelter continues to give support to the tree, until it disintegrates at between 5 and 10 years. Tree shelters should not be used on exposed or waterlogged sites and any vegetation on the planting position should be screefed away. The base of the shelter will need to be pushed into the ground at least 5 cm to prevent air circulating up through the shelter and secured using a stout and preferably treated stake together with a special nylon tie. Different trees vary in their response to tree shelters (Table 14.4),

Table 14.4 Height increment improvement with tree shelters compared with mesh guards 3 years after planting (after Potter, 1991, Crown Copyright, reproduced by permission of the Forestry Commission, 2001).

Height improvement	Species
>100% increase	Oak, beech[1], walnut, lime, sycamore, field maple, birch, hawthorn
50–100% increase	Douglas fir[1], grand fir[1], sitka spruce[1], Norway spruce[1], Corsican pine[1], Japanese larch[1], Norway maple, alder, sweet chestnut, ash, holly, southern beech
<50% increase	Western hemlock[1], western red cedar[1], wild cherry, black walnut, hornbeam, horse chestnut, rowan, whitebeam

[1] Best results in shelters no bigger than 0.6 m. If deer pressure is high, fencing will be required. Beech should be free of fungal infection on receipt from the nursery before the use of tree shelters is considered.

which should only be used with appropriate species and using good transplants with a root collar of at least 6 mm. It is important to note that the use of tree shelters does not mean that weed control can be avoided (see section on weed control). As the primary reason for using a tree shelter should be for tree protection, the size chosen will depend on the browsing mammals present (see Hodge and Pepper, 1998, and Table 14.3).

Tree shelters have been the cause of great debate among foresters over the last 15 years, and they are by no means the best form of protection in all circumstances. Concerns have been raised over their stability in exposed conditions, the growth of trees after the shelter is removed, weeds in the shelter, the amount of maintenance they require, failure of shelters to break down, costs of removal and disposal, and not least the unattractive appearance of lines of tree shelters in the landscape. Some of these problems have been overcome, such as the use of different coloured tree shelters to blend in with the ground vegetation and better advice on the maintenance and removal of shelters.

The final choice of protection method should be based on a very careful preplanting assessment of the site, ensuring that any signs of potentially damaging browsing animals are taken seriously (Table 14.5). Where protection is required from domestic livestock (horses, cattle and sheep), tree shelters are not appropriate, and fencing should be used instead. Where the local population of browsing animals is particularly high (e.g. rabbits, roe deer), control of their numbers may be required whatever protection method is chosen, to prevent overwhelming damage to the growing trees.

Plant origin

The production of high quality timber depends on the selection of healthy planting stock grown from selected parents of superior quality from the correct choice of provenance. Under the Forest Reproductive Material Regulations, cuttings and plants of listed species (oak, beech, poplar and most conifers) intended for the production of timber may not be sold unless obtained from sources approved and registered by the Forestry Commission in Great Britain or other approved authorities elsewhere in the European Union. For unlisted species, seed from recommended sources is preferable to 'unknown' sources.

Table 14.5 Identification of browsing and bark damage to trees (after Hodge and Potter, 1998, Crown Copyright, reproduced by permission of the Forestry Commission, 2001).

Mammal	Time of year	Type of browsing damage	Type of bark damage
Bank vole	Winter and early spring.	Bud removal, especially pine on restock sites.	Short, irregular strips 5–10 mm wide with incisor marks 1 mm wide in pairs. Can climb to 4 m.
Field vole	All year, but greatest risk in winter.		Bark signs as bank vole but restricted to height of surrounding vegetation.
Rabbit	Winter and spring.	Sharp-angled clean cut to shoots 0.6 m above ground. Removed portion eaten.	Incisor marks 3–4 mm wide in pairs, running diagonally across the stem to 60 cm above ground.
Hare	Winter and spring.	As rabbits to 0.7 m above ground, but shoots often not eaten.	
Squirrel	April–July.		Incisor marks 1.5 mm wide in pairs, running parallel with stem on sycamore, beech, oak and pine trees 10–40 years old.
Deer	All year.	Ragged edge to damaged shoots (due to lack of teeth in front upper jaw) to 1.1 m above ground (roe deer and muntjac) or 1.8 m (fallow, red and sika deer. Fraying in March–May.	Vertical incisor marks (red, fallow and sika deer).
Sheep and goats	All year.	Coarse browsing of foliage to 1.5 m with uprooting of transplants.	Diagonal incisor marks to 1.5 m above ground.
Cattle and horses	All year.	Coarse browsing of foliage to 2.0 m (cattle) and 2.5 m (horses) with uprooting of transplants.	Diagonal incisor marks to 2.0 m (cattle) and 2.5 m (horses).

Planting stock

Planting stock should have a good balance between root and shoot and ideally conform to British Standard 3936 (Morgan, 1999). Large planting stock will tend to establish more quickly and hence require less weeding. However, on exposed sites, larger transplants are not appropriate as they may be more prone to desiccation. Whips (trees larger than 90 cm tall) and standards are more likely to suffer die-back or drought stress than well-balanced smaller stock, because their root system is poorly developed in relation to their transpiring foliage area.

Plants should be ordered well in advance from a forest tree nursery to ensure delivery and quantity required (names and addresses can be obtained from the Royal Forestry Society of England, Wales and Northern Ireland). Only healthy plants should be accepted, which are stout and well balanced, with plenty of fibrous moist roots and free of pests and diseases. A sample of plants should be sent for independent assessment of plant quality. The Forestry Commission (Forest Research) provides such a plant quality testing service, based upon root electrolyte leakage levels.

The following types of planting stock are commonly available.

- *Bare-rooted transplants* are widely available as forest trees, being reasonably priced and less bulky than containerized trees, although greater care is needed in planting and handling. Transplants are raised in a seedbed for one or two seasons and then lifted and planted out in transplant beds for a further year or two, which encourages the growth of a vigorous root system. Nursery catalogues may refer to these plants in a coded system (e.g. a 1 + 1 plant has spent one year in the seedbed and one year in the transplant bed). In some nursery systems, plants remain in the same seedbed but are sown at a lower initial density and are undercut by a reciprocating bar passed through the seedbed to promote bushy root growth, e.g. 1/2 u 1/2 (1/2 u 1/2 indicates a 1-year-old plant which has been undercut midway through the growing season).

- *Container-grown plants* retain a growing medium around the roots of the tree at planting. This may allow planting outside the growing season, and offers a reduced likelihood of mortality through poor handling. Plants are available in a range of containers and sizes, from reusable stiff plastic pots to degradable paper cells (Japanese paper pots) or flexible root trainers which produce small plants as 'plugs'. The plants are normally raised in polytunnels and hardened off outside prior to planting. Always be careful not to accept pot-bound trees, or very small, spindly specimens.
- *Cuttings and rooted sets* are used for poplars and willows which will grow from unrooted cuttings (20–25 by 1–2 cm diameter).

Time of planting

The planting season for bare-rooted transplants runs from late October until the end of March. The exact date is determined by the onset of dormancy, which is later in a mild season, together with the ground conditions, species, storage of plants and location. Planting is most successful when coupled with higher soil temperatures, although delaying planting until after budburst renders trees susceptible to decay. Budburst can be delayed, while extending the planting season, by protecting the trees in cold storage at a nursery before delivery. Broadleaves are best planted by late November to allow the roots to become settled-in over the winter and better able to withstand drought in the following season. Conifers are normally planted during early spring to avoid winter frost damage. Morgan (1999) gives detailed guidance on suitable planting dates for different species and locations.

Plant handling

Although disease, frost or drought may account for some losses of trees in the first year of planting, a major cause of death is the result of poor handling between the nursery and the planting site. Particular problems include the following:

- *Root desiccation* is a significant cause of tree mortality from the exposure of bare roots to drying winds, particularly while the planting hole is being prepared. Bare-rooted transplants should ideally be kept in co-extruded opaque plastic bags, which are white on the outside to reflect heat and black on the inside to keep the roots cool. Fertilizer bags are not suitable due to the presence of chemical residues which may scorch the plants. Trees are traditionally stored prior to planting by 'heeling in', whereby the trees are lined out in a shallow, slanted trench (30 cm deep) to cover the roots. However, care needs to be taken to avoid damage to roots. It is better to take delivery of plants immediately prior to planting, and if necessary store them temporarily in cool sheds or under dense canopies to reduce overheating and frost damage.
- *Physical damage* can be caused by poor handling between the nursery and the planting site, leading to broken, bruised or damaged shoots, roots and buds. Plants should never be dropped off the back of a trailer, stacked in piles or treated roughly.

Planting

Before the planting operation, all site preparations should be completed and the plants and equipment safely delivered to the site. Ideal planting weather is mild and wet, and trees should not be planted in severe frost or while the ground is waterlogged. Sites can be marked out with coloured pegs or canes, and the first planting line is laid out with sticks or twine, from which subsequent lines are measured. Experienced planters will judge the distance between plants by stepping or by the length of the planting tool.

Bare-rooted trees are normally planted by the *notch method*, which involves the cutting of a T- or L-shaped slit in the soil. Special planting spades with a strengthened handle and straight blade are preferable to garden spades. The first slit is levered open with the spade, allowing the tree to be inserted into the notch, until the old ground level mark (*soil collar*) of the tree coincides with the surface of the disturbed ground. The roots should be evenly distributed and the soil firmed down around the tree, which should remain vertical. For larger trees or standards, *pit planting* is the preferred technique. This involves the digging of a hole large enough to take the spread-out roots of the tree, the

Table 14.6 Examples of recommended spacing for different tree species.

Species and situation	Approximate square spacing (m)
Christmas trees	0.75–1.00
Oak/beech (for timber production)	1.8
Conifers and broadleaves	2.0
Poplar (pruned for timber)	8.0
Oak/beech restocking, when naturally occurring woody infill is anticipated	2.0
Broadleaf restocking, when naturally occurring woody infill is anticipated	2.5

bottom of which may be broken up to aid drainage. Well-rotted organic matter or a sprinkle of bone meal may be added, and with large specimens a stake should be driven in before inserting the tree.

On level and workable sites (free from large stones, compacted soil pans, stumps and ditches) mechanized planting by a tractor-mounted machine may be possible. This technique offers the advantages of quicker planting and may be more economical than hand planting on large sites with a simple planting design.

Spacing

The choice of spacing (Tables 14.6 and 14.7) between the trees will reflect the management objectives for the woodland. If good quality timber is required, it is important to obtain a high initial stocking density, which will lead to straighter and less branched trees, less need to prune, better returns from thinning and a greater number from which to select the final crop trees. However, regardless of whether timber production is an

Table 14.7 Tree requirements (number of trees required per hectare planted on the square).

Spacing (m)	Number of trees/ha	Spacing (m)	Number of trees/ha
0.75	17 778	2.75	1322
1.00	10 000	3.00	1111
1.25	6 400	4.00	625
1.50	4 444	4.50	494
1.75	3 625	5.00	400
2.00	2 500	6.00	278
2.25	1 975	8.00	156
2.50	1 600	10.00	100

objective, close spacing also allows more rapid establishment and lowers the risk of failure induced by other damaging factors, and reduces the length of time spent weeding, and hence often the amount herbicide used. Once a woodland environment has been created, respacing and thinning can take place if necessary to meet the owner's objectives for the stand. Wide spacing is cheaper initially, but lengthens the establishment period considerably. Where a grant is being paid to offset the costs of planting, a minimum average tree spacing or tree density per hectare will often be specified as part of the conditions of payment. This should not preclude planting at closer spacing for silvicultural reasons.

Beating up

Beating up involves the replacement of dead trees which are lost during the first year or two after planting. The level of tree losses that are considered acceptable will depend on the owner's objectives. Overall losses of greater than 10%, or failures occurring in groups and at wider tree spacings than recommended earlier, will require beating up. By using good quality plants and effective weed control for several years after the first planting, beat-ups should eventually be indistinguishable from the original crop.

Direct seeding

Direct seeding is a silvicultural system similar in some ways to agricultural cropping, whereby tree seed is sown by hand or machine into a prepared seedbed at the site intended for woodland creation. It has a number of potential advantages over conventional tree planting for new woodlands: it allows selection for better quality timber through higher stocking rates; it gives a more rapid establishment of a woodland environment; it can utilize farm-scale techniques and machinery for sowing and maintenance; it is also cheaper, and may offer a means of reducing herbicide inputs.

Woodland stands resulting from direct seeding tend to have a variety of spacings and randomly occurring open space, similar to that advocated for new native woodlands. Direct seeding has great potential therefore for establishing links between existing fragmented areas of ancient semi-natural woodland. The technique also appears promising for creating larger-scale woodlands, particularly where rapid establishment of a woodland cover is desirable, for example in community woodlands.

A major disadvantage of direct seeding is that it is technically much more challenging than conventional establishment, requiring a tight silvicultural prescription to be followed, particularly in the first year after sowing. Other disadvantages include unpredictability of seed germination and species and site limitations. As a technique it is also not particularly well suited to heavy-textured soils subject to winter waterlogging – generally such sites should be avoided. Direct seeding is not currently recommended for the establishment of any broadleaved or conifer species on restock sites where high numbers of seed-eating mammals are present.

Direct seeding may be worth considering for new broadleaved establishment, but only on good-quality lowland sites with lighter-textured soils, where mechanized access is possible, and providing ash, sycamore or oak form a dominant (75%) part of the species mix. Seed must be pretreated appropriately to overcome dormancy and sown in late autumn at a rate of at least 200 000 viable seeds/ha into a suitably prepared, fully cultivated and weed-free seedbed. Seedlings must be protected from browsing mammals by fencing. Control of all weed vegetation, particularly in the first year after sowing, is vital. This is best achieved through the use of selective herbicides. The use of an agricultural cover crop in conjunction with tree seeding is not recommended. Willoughby *et al.* (2004) provide detailed guidance and prescriptions for establishing and maintaining new broadleaved woodlands by direct seeding.

Fertilizer application

On fertile mineral soils, the use of fertilizer is unlikely to be economic as trees respond only weakly to nutrient supplements compared to most agricultural crops. In commercial upland plantations, fertilizer use is generally restricted to the establishment phase to improve initial growth rates of conifers until canopy closure, the decision to fertilize being based on soil and/or foliar analysis. After canopy closure, further applications are not normally necessary due to the shading of competing vegetation, improved nutrient cycling and the capture of atmospheric nutrients by the crowns of the trees. On restocked sites, the increased availability of nutrients from the breakdown of brash and litter normally precludes the use of fertilizers except for very poor heathland or moorland soils. Phosphate, potassium and nitrogen are normally applied as top dressings, using ground rock phosphate (50–75 kg P/ha), muriate of potash (75–100 kg K/ha) and prilled urea or ammonium nitrate (80–120 kg N/ha).

Examples of farm woodland sites where the use of fertilizer should be considered are:

- *Mineral soils* (iron pans, podzols, gley soils) may suffer from phosphate deficiency.
- *Peaty gleys and deep peats* may suffer from nitrogen, phosphate and potassium deficiency, particularly where the peat depth exceeds 45 cm.
- *Shallow soils on chalk and limestone* may suffer from nitrogen and potassium deficiency.
- *Derelict and restored soils* often suffer from nitrogen availability, in addition to other specific nutrient deficiencies or toxicity. Further soil analysis will normally be necessary.

Sitka spruce planted on heather-dominated heathland or moorland suffers from a well-documented 'heather check' which restricts growth through the limited availability of nitrogen and may delay canopy closure by up to 10 years. Although heather competition may be reduced by herbicide applications, additional inputs of nitrogen may be required. Alternatively, sitka spruce may be planted in mixture with a suitable provenance of Scots pine, lodgepole pine or Japanese larch, which appears to suppress the competing heather and increases the availability of nitrogen.

If a crop appears to be showing signs of nutrient deficiency once established, such as discoloration of leaves or needles, poor growth rate or early senescence, then samples of foliage can be sent to the Forestry Commission (Forest Research) for analysis and rapid diagnosis of the problem. Remedial applications of fertilizer may subsequently eliminate the deficiency.

Aftercare

Weeding

The most significant threat to the successful establishment of woodland is competition from weed growth, especially on drought-prone sites. Grasses and broadleaved weeds compete effectively with the growing trees for water, light and nutrients, reducing height growth and potentially increasing mortality. On fertile sites, the growth of rank weed vegetation may also cause physical damage by smothering or collapsing over the trees in the autumn. The cutting back of these weeds and grasses will reduce light competition and physical damage, although root competition will not be affected. In fact, mown grass exerts an even greater competition for soil moisture than unmown grass. The standard recommendation is that a *weed-free zone of at least 1 m* should be maintained around young trees until they are well established and above the height of competing vegetation. In reality, an initial target of 1.2 m is often required to actually achieve a 1.0-m weed-free zone in practice.

The most appropriate weed control treatment should be determined by regular inspection of the site prior to planting to reveal the type and abundance of weed species present, together with an assessment of the soil conditions and potential for weed invasion. The total removal of weeds from the site is rarely justified either by cost or on silvicultural grounds: in fact, there is some evidence that the retention of some weed growth may benefit the trees by providing shelter from frost or sun scorch. A totally weed-free plantation may also be prone to excessive soil erosion and has a substantially reduced wildlife value. Weed control therefore normally involves the selective spot treatment of competing vegetation around the trees or in strips along the planting lines, retaining weed growth in the central aisle. The timing of treatments will depend on the site, although the time of greatest potential moisture competition is early in the growing season (March–May) when tree roots are actively growing. Control should therefore aim to anticipate future weed growth before this becomes obvious on the ground. Current best practice involves preplanting treatments to establish a weed-free planting position, together with additional treatment during the growing season if necessary. Weeding is usually carried out once or twice a year, for 3–6 years after planting, depending on the site.

The main forms of weeding available are the use of herbicides and mulches. On less-fertile sites preplanting cultivation can be very effective, and on a very small scale, say around individual specimen trees, regular post-planting hoeing may be an option. Herbicides are the most commonly used method of post-planting weed control as they are the most cost-effective option in most cases; however, voluntary certification initiatives such as the UK Woodland Assurance Standard (UKWAS, 2000) call for managers to reduce the amount of synthetic pesticides that are used in woodlands. Herbicide use can be reduced through utilizing alternatives such as mulches, by following the good establishment practices outlined earlier, and through using larger, well-balanced planting stock established at closer initial spacing. Alternative silvicultural systems such as direct seeding may also require less overall herbicide use. Willoughby *et al.* (2004) provide detailed guidance on reducing pesticide use in woodlands.

Herbicides

Herbicides usually provide the most cost-effective form of weed control. When used in accordance with the manufacturer's recommendations, the environmental risks are minimized. A wide range of chemicals are available for use on sites classified by the UK Pesticides Safety Directorate of DEFRA in 'farm woodland' situations (i.e. sites that have recently been either arable or improved grassland), with a much narrower range available for unimproved 'forestry' sites. Most herbicides will only affect a limited number of species or may only be active at certain times of the year in particular weather conditions. It is therefore important to match the weed control spectrum of the herbicide with the weeds found on the site. Herbicides may be applied by tractor or ATV-mounted sprayers, although knapsack sprayers may be preferable in small or more irregular plantations.

Herbicides can be divided into two main groups, foliar acting and residual.

- *Foliar acting herbicides* are applied to the aerial parts of actively growing weeds, from where they are taken up through the stem and leaves. While some can be applied over the trees without harming them, few can be used during the growing period (May–July) without scorching the foliage or killing the tree completely.
- *Residual herbicides* are applied around the base of the tree and are most effective on moist, friable soils. The active chemical remains in the soil and is absorbed by the roots of weeds as growth commences. These herbicides are particularly useful for broadleaves susceptible to foliar herbicides, often

being applied in the dormant winter season to control weeds during the spring and early summer.

The need to reduce the cost of herbicide applications has led to the development of low-volume sprayers (weed wipers, spot guns or controlled drop application sprayers) which also reduce the environmental impact of these chemicals.

Current farm woodland practice involves the use of residual herbicides, such as propyzamide granules in the winter (slowly volatilizes in cold soil) followed by the use of foliar sprays in the summer (e.g. glyphosate) to cope with localized weed problems. Where larger woody weeds are present (e.g. rhododendron), a more concerted approach is often required. Alternatives include the cutting back of shrubs by mechanical means followed by herbicide treatment of the regrowth from the stumps, or uprooting of shrubs by machines followed by chemical control of the early regrowth.

The storage and use of all herbicides is now subject to strict environmental standards under COSHH (the Control of Substances Hazardous to Health Regulations 2002), and only fully approved contractors should be used for weed control operations. For further instruction on appropriate herbicides, storage, application and relevant legislation see Willoughby and Dewar (1995) and Willoughby and Clay (1996).

Mulches

The use of mulches is a technique normally only used in amenity or small-scale plantings since it tends to be expensive and time-consuming. More recently, interest in mulching has grown since it offers an alternative to the use of herbicides. Mulching involves the prevention of weed germination and growth by covering the ground around the tree with either organic material or artificial sheeting. Mulches also retain soil moisture close to the surface by reducing evaporation losses, which promotes root growth and nutrient cycling. The use of mulches is more commonplace in climates where soil moisture is in very short supply. Artificial mulches include specially formed black polythene sheets or mats (125 μm thickness) with ultra-violet inhibitors to slow degradation, although squares of old carpet may be equally as effective. The mulching material should be laid securely with pins or weighed down with soil or stones to prevent displacement by the wind and to deter voles from burrowing underneath. Successful organic mulches include straw, wood chips or crushed bark, although they may reduce tree growth by locking up available nitrogen as the mulch breaks down and are prone to being blown away by the wind. The mulch must be maintained for 3–4 years to prevent weed growth within 1.2 m of the tree to ensure good early growth.

Inter-row vegetation management

The inter-row vegetation between the weed-free spots or strips will grow unchecked if no weed control is carried out. On new farm woodlands, particularly those on arable or improved grassland sites, this may lead to invasion by persistent perennial species or tall growing weeds which fall over and smother the trees in the autumn. The development of rampant inter-row vegetation may also impede access by farm machinery. Occasional inter-row mowing (e.g. once or twice a year) can be undertaken in the first few years after planting. This will enable access throughout the crop and prevent weed species seeding into neighbouring fields. Machinery adapted to work within narrow rows may be required (e.g. mini-tractor or ATV-mounted equipment). As an alternative, the ground between the weed-free areas may be sown with a cover crop which has uses in addition to weed suppression (e.g. game cover or conservation). Inter-row crops should be chosen for low productivity, minimizing competition for moisture with the trees and be easily controlled by herbicide in the weed-free zone around each tree (e.g. some fescues). Kale has been successfully used as a game cover crop (the instant spinney); it provides a sheltered habitat for game birds for about 3 years, in addition to the physical support it lends to tree shelters. A hardy variety should be chosen, sown in rows 50–60 cm apart, in June–July following tree planting. Fencing to exclude farm stock is essential, although game birds do not combine well with rabbit-netted plantations unless an area inside the rabbit fence can be mown to provide the birds with a place to dry off and warm themselves in the sun. Where the new woodland is part of a project to increase the value of wildlife habitats on the farm, the inter-row area may be sown with a low-productivity grass and wild flower mixture, which should be chosen to match the underlying soil type. This type of mixture produces an attractive habitat which is particularly appropriate for the edges of the woodland and areas adjacent to footpaths and other public spaces.

Cleaning

This refers to the removal of woody plant growth which is either in competition with the tree crop or physically damaging. The operation normally involves the cutting back of unwanted broadleaved trees or regrowth from hardwood stumps and the severing of harmful climbers such as honeysuckle (*Lonicera periclymenum*) and old man's beard (*Clematis vitalba*). Methods of cleaning depend on the nature of the site, although the work generally involves the use of clearing saws to cut or girdle shoots. Chemical treatment is a cost-effective operation, using glyphosate, applied either to foliage or cut surfaces. Where individuals or groups of crop trees have failed, naturally regenerated broadleaves or regrowth from stumps should be retained in these areas to suppress the development of side branches on neighbouring crop trees.

Inspection racks

As the new plantation reaches canopy closure, it may become necessary to cut rackways through the thicket of trees in order to allow assessment of its silvicultural condition. These are normally made by cutting off the lower branches of two adjacent rows with a pruning saw, cut flush to the trunk.

Brashing

This involves the removal of dead and dying branches from crop trees up to a height of about 2 m, normally undertaken during the thicket or pole stage prior to first thinning, to improve access for marking and extraction. Brashing may also be carried out where access for game beaters and amenity is required or to release broadleaves when grown in mixture with conifers. The persistence and difficulty of removal of the lower branches varies with species and crop spacing. Shade-bearing trees tend to have the most long-lived branches (e.g. western hemlock), while larch tends to self-prune well. Spruce tends to have quite tough branches, which are difficult to remove, larch branches break off easily with a stick and pine is intermediate between the two. The use of a special curved brashing or pruning saw is preferable to a billhook, especially with larger branches where extra care is need to reduce the risk of damage. Complete brashing of plantations is rarely economic, so only sufficient trees are now brashed to allow for inspection and marking, concentrating on the final crop trees. Where plantations are line thinned, the rows to be removed are left unbrashed.

Pruning

Pruning involves the removal of side branches above the height of normal brashing to produce longer lengths of clean knot-free timber. The operation is restricted to selected final crop trees, in order to achieve the standards of quality required for high-grade timber (e.g. planking, joinery, beams and veneers). Pruning should be carried out before the side branches become larger than 5 cm, for ease of working and to reduce the incidence of disease entry. A sharp pruning saw or chisel is used with an extendable handle for high pruning. The branches are sawn not quite flush with the trunk, but slightly proud to avoid bark damage. High pruning will inevitably remove some 'live' branches, although there are no great harmful effects if the amount of crown removed ranges between 25 and 45%. Pruning regimes differ for conifers and broadleaves:

- *Conifers*: initial pruning to 4 m is carried out once the trees reach about 10 cm diameter or 20 years old. Further pruning to 5–6 m height is carried out once the trunk at this point reaches 10 cm diameter. The operation may be carried out at any time of year, although preferably during March–May for quicker healing. Pruning of conifers is not widely practised in larger commercial estates, although for smaller farm woodlands, if labour is available, it may be justified in order to increase the capital value of the woodland. An expanding market exists for pruned pine (Scots and Corsican), Douglas fir and European larch for 'boatskin' quality.
- *Broadleaves*: pruning begins early for broadleaves (5–10 years) to create a single main stem through the removal of competing shoots (*formative pruning*) and later by the removal of lower side branches from the trunk to about 5 m. The operation is particularly suitable for selected final crop trees grown for decorative veneers (e.g. oak, walnut, cherry and maple). Pruning should never be under-

taken during flushing (March–May), since resistance to infection is lower at this time. For most species, winter is the optimum time for pruning, although cherry should be pruned between June and August to reduce the risk from canker and silver leaf disease (*Chondrostereum purpureum*). See Kerr and Evans (1993) for further details on pruning broadleaved species.

- *Epicormic branches* arise from adventitious buds around pruning scars or from dormant buds on the stem, which are triggered into sprouting by increased levels of light (e.g. following thinning). These epicormic shoots remain semi-moribund in normal woodland conditions, although a further increase in light levels will enable the shoots to develop into larger branches, which will severely affect the quality of the timber. Only small knots will be tolerated for high quality sawnwood and may have to be completely absent for decorative veneers. Pedunculate oak, sweet chestnut, poplar and cricket-bat willow are the worst affected. The growth of epicormics can be restricted by avoiding sudden changes in the environment of the stand (e.g. through heavy thinning) or by pruning. If carried out annually, epicormics can be removed by rubbing. Classical silvicultural treatment of oak stands to reduce the incidence of epicormics involves growing an understorey of beech, hornbeam, Norway spruce or western red cedar to keep the boles of the oak trees shaded.

Managing existing woodlands

Silvicultural options for existing woodlands

Silvicultural options for existing woodlands can be classified into six groups as shown in Table 14.8. This divides up woodlands in terms of the silvicultural system, i.e. the structure of the woodland, and the method of regeneration, i.e. either planting trees raised in a nursery or using seed from trees already on the site to regenerate naturally. The main determinant of which system is used must be the objectives of management, as a woodland managed using clearfelling and replanting will be very different to one managed using a selection system and natural regeneration. However, there are other factors to consider; for example, the use of

shelterwood and selection systems requires good silvicultural skills and a knowledge of the site, tree species, vegetation and potential damaging mammals in the area. The following sections consider each of the main systems and then natural regeneration, which is more commonly used with shelterwood or selection systems.

Table 14.8 Silvicultural options for existing woodlands.

System	Method of regeneration	
	Planting	*Natural regeneration*
Clear felling	Option 1	Option 2
Shelterwood	Option 3	Option 4
Selection	Option 5	Option 6

Clear felling (options 1 and 2)

Clear felling involves the harvesting and removal of whole areas of the forest at once. It is probably the simplest and cheapest method of creating a uniform and even-aged area for restocking. Clear felling is usually combined with planting (option 1) because natural regeneration (option 2) is generally more difficult when the parent trees are removed from the site in one operation. Planting allows the forest manager to change species or to take advantage of improved planting stock.

The open nature of the clear fell creates substantially different environmental conditions to those under a selection or shelterwood system, which has a number of important implications. The increased levels of light, rapid breakdown of organic matter and the exposure of mineral soil may encourage the vigorous growth of weed species. Exposed soil may be prone to erosion and increased run-off, especially on sloping ground. The brash left from felling may provide a home for weevils and beetles which can threaten the establishment of future crops. The edges of neighbouring stands will be exposed to an increased risk of wind damage. In addition, the greater amounts of light may cause sun scorch on thin-barked species such as beech (*Fagus sylvatica*), and oak (*Quercus* spp.) trees may develop epicormic growth. On some sites, the large-scale clear felling of mature trees may be inadvisable due to amenity, landscape or biodiversity considerations.

Shelterwood systems (options 3 and 4)

A shelterwood system is characterized by a series of

partial fellings of the mature overstorey, which are timed to release the development of young trees which may have been planted (option 3) or developed through natural regeneration (option 4), or by a mixture of both regeneration options. The period over which this transition takes place is between 10 and 50 years, depending on the silvicultural characteristics of the young trees (i.e. light demanding or shade tolerant) and the success of regeneration. In common with clear felling, shelterwood systems produce a woodland with a reasonably simple structure; however, they avoid complete removal of the mature trees, which may be advantageous for landscape, amenity or conservation objectives. Different types of shelterwood systems exist, for example, the mature crop can be removed evenly over the area or in groups, again depending on the silvicultural characteristics of the young trees being regenerated.

Selection systems (options 5 and 6)

A selection system is characterized by felling and regeneration operations being intimately mixed throughout an area so that any one area of forest contains a mixture of regeneration, young trees, medium-sized trees and mature trees in a mosaic pattern. Selection systems are usually only realistic when combined with shade-tolerant species and natural regeneration (option 6) and generally produce conditions known as *continuous cover forestry* (Mason *et al.* 1999).

Natural regeneration

Natural regeneration may be more desirable than planting because it can conserve local genotypes, create more diverse woodlands and produce a wide choice of stems for selection. However, it can be unpredictable and sometimes difficult to achieve and is not always a 'free gift'. The main difficulties with natural regeneration include: the timing and quantity of seed production; predation of seed by squirrels and small mammals; the presence of competitive weed growth; and, perhaps most importantly, browsing of seedlings by mammals. To minimize these problems the best strategy is to develop advance regeneration, i.e. seedlings that are present beneath the canopy before regeneration fellings begin. Once advance regeneration has been secured

then fellings can encourage the growth and development of regeneration. The pattern of fellings is dictated to a large extent by the choice of silvicultural system (see above).

Where timber production is an objective of management, the parent crop must be of good quality, well suited to the site and capable of producing large quantities of viable seed. Seed production can be encouraged by good long-term management of the existing woodland to produce large crowned seed trees which are regularly spaced throughout the stand. Some species such as ash, birch and sycamore, for example, produce seed almost every year, but beech and oak bear large amounts of seed at irregular intervals of 4–5 years.

Operations to promote germination of seed should be timed to coincide with seed dispersal during good seed years. Although ground preparation may be required to incorporate humus layers and expose the mineral soil, it should be used with care as it can stimulate the growth of undesirable, competitive weed species that can prevent the establishment of tree seedlings. Retention of some overstorey will protect seedlings from frost and sun scorching, but it is unreliable as a method to control the growth of weeds, as the light levels required to allow seedling growth often allow weeds to grow vigorously as well. As weeds develop, remedial control may be required in the same way as for clear felling.

The young tree seedlings will require adequate protection from grazing by herbivores. This may be through culling to control populations, fencing to exclude animals or protection of individual seedlings with tree shelters. As the young trees become established, any remaining overstorey is gradually removed, increasing the amount of available light to the growing crop. Subsequent operations will depend on how the woodland develops, but can include the respacing of dense patches of regeneration and planting of understocked areas.

The creation of a fully stocked woodland through natural regeneration can prove a complex and consequently costly exercise. The key drawbacks for timber production include the reliance on seeding years rather than markets, the need for skilled management, overwhelming local populations of rabbits or deer, the possibility of adverse ground conditions [including brambles (*Rubus fruticosus*) or *Rhododendron ponticum*] or parent trees of inferior quality.

Planting

See earlier sections on creating new woodlands.

Coppice

Coppice is a traditional system of woodland management that relies on the ability of many trees to regrow from the stumps remaining after felling. As few conifers are capable of resprouting, the system is generally used only for broadleaved woodland and many native species such as ash, oak, hazel (*Corylus avellana*), willow, alder and field maple (*Acer campestre*) can be successfully managed as coppice.

New coppice shoots usually arise from dormant buds on the stump, but some species including beech can regenerate from newly produced adventitious buds. Other species such as cherry and some poplars sucker, producing shoots from roots. The ability of stumps to produce shoots depends on a variety of factors including species, age, diameter, site and vigour of the parent stump. Success is most likely for vigorous young stems less than 25 years old. As stump mortality increases with diameter and age, old or neglected coppice tends to respond less well to felling. Stumps should be cut in the dormant season between October and March using tools that produce a clean cut. Regularly cut stems can form large stools which, for species such as lime (*Tilia cordata*), field maple and hazel, can survive for many centuries.

Woodlands can be managed as simple coppice or coppice with standards. In simple coppice all stems in defined patches are clear felled on a regular cycle to produce a woodland comprising even-aged areas of coppice. The length of rotation varies with the species and type of product required. Short rotations of 5–7 years produce small-diameter, pliable material whereas longer rotations can be used for firewood, turnery and fencing. Coppice with standards is a system of management in which an understorey of simple coppice is grown with a partial overstorey of trees grown on longer rotations to produce large-diameter timber, which is often more readily marketable than other coppice products. This system is more difficult to manage than simple coppice as the number, size and distribution of standards, and the canopy shadow they cast, need to be controlled.

Initial regrowth of coppice shoots is usually much faster than seedlings, and therefore weed control is seldom necessary. However, adequate protection of new shoots from browsing during their first few growing seasons is necessary to ensure satisfactory regrowth. Care should be taken not to damage stools during harvesting operations. If stools die they can be replaced by planting robust nursery-grown stock, natural regeneration or layering shoots on existing stools.

Large areas of coppice woodland have been neglected for decades and recent management has often taken place for nature conservation and biodiversity objectives rather than timber production. At present, the quality of stools and harvestable material in many woodlands is poor and of little value. However, there are localized niche markets for coppice products, and good quality crops of hazel can command high prices. Coppice can be converted to high forest by reducing the number of coppice shoots to one for every stool. This stored coppice typically produces poor quality timber, as a result of basal sweep, shake (internal splitting) and a tendency to decay. For further information see Harmer and Howe (2003).

Short-rotation coppice for firewood and biomass

Firewood is traditionally a secondary product from the branchwood and low-quality timber of broadleaved woodland, especially coppice. As a fuel, only air-dried wood should be burnt, since wood with a high moisture content gives off little heat and causes tar deposits in stoves and flues. The best timbers for firewood include ash, beech, birch, hazel, holly, oak and old fruit trees, which may be cut and split or chipped by machine (Keighley, 1996).

High volume production can be secured over short rotations of coppice (2–5 years) from selected willow and poplar clones planted at high densities on agricultural land for energy generation or industrial cellulose. Recent trials suggest that average yields of around 8–9 oven-dry t/ha may be possible in the first 3-year rotation, which may increase in subsequent rotations. Field-scale production is underway in some parts of Britain where a regional market exists, as a result of government initiatives to increase the production of energy from renewable resources. High-yielding disease-resistant varieties of willow are planted as cuttings in a multiclonal mix into prepared agricultural land. During establishment and following harvest, regular herbicide applications are required to maintain rapid production of multistemmed stools. The aboveground biomass is mechanically harvested every 3–4 years in winter, followed by vigorous regrowth from the cut stools. Fungal infection, such as *Melampsora* rust, is a serious threat to some crops which may be offset by the use of carefully selected more resistant varieties. Further expansion of the biomass market will depend on the relative cost of fossil fuels and advances in energy generation technology. For further information see Tubby and Armstrong (2002).

Christmas trees

The production of Christmas trees can be successfully incorporated into woodland management by utilizing 'dead ground' under electricity pylons or other temporary spaces in newly established mixtures and along forest rides. However, the production of top quality, higher value trees usually requires intensive weeding, shearing and the control of potentially damaging insect pests. Such operations are best carried out as a specialist horticultural operation, on flat, fenced areas where access by the public or wildlife is prevented.

Traditionally, the most commonly grown species was Norway spruce, although more attractive and needle-retentive species, such as Noble fir (*Abies procera*) and Caucasian fir (*Abies nordmanniana*) are rapidly gaining ground. The objective is to produce healthy, well-shaped and bushy trees by planting at a close spacing of 1×1 m (10 000 stems/ha), with careful attention to weeding and inspection for the presence of insect damage. Shearing may be required in the fourth year to restrict the length of the leading shoot and to create a better shaped tree. The trees are normally harvested in a range of sizes to suit local markets, with the main crop being harvested at between 4 and 10 years.

Improvement of timber quality in unmanaged woodland

The poor timber quality of many farm woodlands generally owes more to inadequate or inconsistent past management than any inherent poor site fertility. For a woodland owner keen to improve the timber quality of neglected woodland, management will generally be based on the following options:

- improvement
- enrichment
- replacement.

However, it should be noted that many broadleaved woodlands that appear neglected from the point of view of timber production may be extremely valuable for the conservation of biodiversity. Given that management grants are available from the government to conserve such woodlands, the most appropriate action in most cases will not be to attempt to improve timber quality through enrichment or replacement at all.

Improvement

Improvement assumes that there are sufficient trees of 'final crop' potential to make the production of a timber crop worthwhile. Between 150 and 300 straight, defect-free trees of marketable species which are evenly spaced throughout the stand should be identified. From the pole stage onwards (10 m height), all competing trees and climbers are removed from around the final crop trees to allow complete crown freedom. Pruning is advisable to produce clean stems and to remove epicormic growth. The felling of the poorer quality matrix may help offset the cost of these silvicultural treatments.

Enrichment

Enrichment involves the planting of additional trees in a stand to increase the overall stocking. The degree of enrichment required depends on the quality of the existing crop. Where some utilizable trees remain, the focus of enrichment is usually existing gaps or bare ground. In the past, enrichment species were chosen for their rapid early growth and included species such as Norway maple, cherry, sycamore, western hemlock and western red cedar. However, in most cases within broadleaved woodlands, enrichment with tree species native to the locality is a better option. As for any new planting or natural regeneration, trees will need to be weeded and protected from damaging agents, and any overstorey carefully managed to avoid excessive shading.

Replacement

Replacement involves the partial or full removal of the

existing crop and its replacement by a new superior species, where the present stand is unlikely to produce a potential final crop. This is the most drastic option and is rarely advisable, due to the loss of many other woodland benefits such as landscape and wildlife habitat. Many former broadleaved woodlands have been converted to more productive coniferous plantations by this method since the 1940s with only mixed success and enormous losses to wildlife and landscape value.

Restoration of plantations in ancient woodland

Many woodlands that have been subject to enrichment practices in the past using conifers were originally ancient woodland sites. Restoration of these plantations to woodlands comprising native species is often regarded as an important objective of woodland management, as it will enhance the biodiversity value of the site. However, not all of these plantations are suitable for restoration and priority should be given to those where the impact will be greatest. The restoration methods used will vary with site and are likely to include a combination of silvicultural approaches, including continuous cover forestry, natural regeneration and enrichment with native species. Successful restoration will be a long and slow process. For further information see Thompson *et al.* (2003).

Forest protection

Fire

The chances of fire damage depend on the nature, size and location of the woodland. A key factor is the amount of inflammable material in the forest, either dead wood and brashings or dry undergrowth in young plantations before canopy closure. March to May is the most dangerous period, when a large amount of dead vegetation remains from the previous year, especially when combined with low humidity, high temperatures and windy conditions. Woodlands close to main roads or residential areas also tend to have a higher risk of fire outbreak. The risk of fires spreading can be reduced by cutting fire breaks, strips of at least 10 m width,

which are kept free of inflammable material by mowing or cultivation. Japanese larch or alder do not easily catch fire and can be used as fire belts 10–20 m wide, between compartments. Forest tracks and roads also act as internal fire barriers. Other precautions that should be taken include the following:

- Inform the local fire brigade of the location of woodland access points and fire-fighting equipment.
- Ensure adequate access to woodlands by removing fallen trees and repairing broken culverts.
- Maintain fire-fighting equipment at strategic points, to include spades, buckets and axes, and provide birch brooms and beaters at entrance points to woodlands.
- Assess the proximity to water and if necessary dig additional ponds.
- For high-risk sites (along footpaths and roads) clear inflammable material and brash young conifers early.
- Ensure that all staff and contractors observe the fire precautions and are trained in the fire drill.
- Insure against fire.

Wind

Prolonged exposure to high winds significantly reduces tree growth rates and in extreme conditions will lead to stem damage from abrasion by neighbouring trees. Where root development is restricted by waterlogging or an iron pan, woodlands are more susceptible to windthrow during autumn and winter gales since the shallow root plate gives less anchorage. In persistently windy regions, the risk of windthrow can be reduced by shortening crop rotation lengths and either avoiding line thinning or not thinning at all. Site drainage and cultivation can improve tree anchorage and root development. Some species are more susceptible than others; Douglas fir and larches are particularly prone to windthrow on shallow clay soils. Windthrown trees can be economically harvested, although the cost of harvesting fallen trees, and the danger involved, is much greater than if they were standing. Opening up of the woodland can lead to further windthrow along exposed unstable edges, so windthrown pockets should always be cleared back to windfirm edges or existing compartment boundaries to reduce this risk.

In order to predict the likely effect of wind on

the woodland, the Forestry Commission devised a Windthrow Hazard Classification (WHC) based on windiness, elevation and degree of exposure and soil conditions. The classification gave an indication of the height to which a tree crop can be grown before windthrow will become a limiting factor. The WHC system has now been superseded by the Forestry Commission's GALES computer software, which is a comprehensive management tool to assist with the planning of silvicultural operations on exposed sites. Further information on GALES can be obtained from the Forestry Commission Research Agency or see Quine (1995).

Insects

A large number of insects are dependent on trees for some part of their lifecycle, either for food, by siphoning off plant sap, eating leaves, needles, shoots, bark or sapwood, or for reproduction, using the trunk, stump or shoots for laying eggs or building brood chambers. The vast majority of these associations cause insufficient damage to be of concern, while some insect–tree relationships are mutually beneficial. The most damaging insects are associated with particular stages in the life of the tree crop, together with site and climatic conditions favouring a rapid increase in insect population size. There is some evidence in this respect that trees already under some environmental stress (e.g. from incorrect choice of species for site) are more prone to serious damage. The most susceptible stage in the life of the woodland is at establishment, particularly where this is adjacent to neighbouring mature trees of the same species or where stumps from the previous crop are still present, which may provide the initial source for an insect outbreak. While some insect pests are fairly general in their tastes, others are specific to particular trees or groups, conifers tending to be more susceptible than broadleaves.

The most important insects associated with particular stages in the life of the forest are as follows:

* *Establishment of all species on restock sites* may be prone to attack from large pine weevils (*Hylobius abietis*), which will breed and multiply in the stumps and brash of the previous crop, to emerge and feed on the newly planted trees, causing severe losses. Pre- or post-plant treatment with alpha cypermethrin or the incorporation of carbosulfan granules at planting give effective control. Alternatives include the use of tree shelters or delaying replanting for five or more years after felling. However, on more fertile sites, this latter option will result in a substantial invasion of weed vegetation that may be difficult and costly to control.
* From the *thicket stage* onwards, many tree species suffer varying degrees of defoliation from leaf and needle feeders. Most trees will recover from even quite severe defoliation, and control measures are normally only economic in large commercial woodlands. Specific examples of damage include pine looper moth (*Bupalus piniaria*) on Scots pine, pine beauty moth (*Panolis flammea*) on lodgepole pine and the oak leaf roller moth (*Tortrix viridana*).
* On *windblown sites* or where *felled timber* is available, a number of secondary pests (weevils and bark beetles) will breed and multiply under the bark of damaged, dead or felled timber. Where subsequent populations are high, the surrounding healthy crop may be damaged, spruce, larch and pine being the most severely affected. The pine shoot beetle (*Tomicus piniperda*) is a serious pest, requiring good forest hygiene to control outbreaks in susceptible crops. Felled timber should be either removed or debarked within 6 weeks, during the period March–August, to prevent broods of insects from being raised. The great spruce bark beetle (*Dendroctonus micans*) is another serious pest affecting sitka and Norway spruce. Its impact is controlled on a national scale by releases of a specific predatory wasp *Rhizophagus grandis*, and through the restriction of movement of timber within controlled areas of the country, currently Wales and neighbouring counties of England.

The identification and control of insect pests is a specialized activity, so that if insect damage is suspected, further advice should be sought from the Forestry Commission. Further information is available in Bevan (1987).

Diseases

There are a number of fungal and bacterial diseases of trees, which can lead to deterioration of the main stem, destruction of roots, shoots or cambium and ultimately death or windblow. As with insect damage, the severity of the disease infection is often exacerbated by both climatic conditions and environmental stress, wounding of trees from extraction damage or pruning being examples. Control measures are rarely justified, so correct planting choice using more resistant tree species is to be advised in susceptible conditions. The most important diseases to be aware of are discussed below.

Honey fungus (*Armillaria* spp.) can be a problem in old broadleaved woodlands, affecting both broadleaves and conifers. It spreads through the soil from infested wood, especially old broadleaved stumps, via a network of black strands (*rhizomorphs*) to infect young trees. Losses are sporadic and rarely affect a whole wood, although it may be wise to restock with a more resistant tree, such as Douglas fir or a broadleaved species.

The *stem and root rot* caused by *Heterobasidion annosum* is the most serious disease of coniferous woodland, leading to deterioration of the lower trunk and roots. Larch, spruce, western hemlock, western red cedar and pines (on former agricultural land or alkaline soils) are most susceptible, especially on the sites of previous coniferous crops. More resistant species include Douglas fir, grand fir and Corsican pine, or all conifers on peaty soils with a depth of peat of at least 15 cm. The risk of disease can be reduced by treating the freshly cut stump surfaces of felled trees with a concentrated coloured solution of urea within 15 minutes of felling. A biocontrol agent based on the fungus *Phlebiopsis gigantea* can be used as an alternative treatment on pine stumps.

Larch canker is caused by the fungus *Lachnellula willkommii* on European larch, favouring its replacement by hybrid larch. *Group dying of conifers* caused by the fungus *Rhizina undulata* spreads through litter from fire sites, and therefore fires should be excluded from the forest interior. *Beech bark disease* develops from the association of the scale insect *Cryptococcus fagisuga* and the fungus *Nectria coccinea*, forming a matt of white greasy wool on the main stem and branches. Infected trees should be removed during thinning operations.

Further information on identification and impacts of diseases can be found in Gregory (1998).

Mammal damage

Mammals can cause damage to trees and woodlands by browsing on foliage, shoots and buds or stripping and fraying bark. Heavy browsing may devastate unguarded transplants and saplings, natural regeneration and coppice regrowth. While light woodland grazing may benefit some wildlife, very high local populations of rabbits or deer will reduce the structural diversity of the woodland ecosystem with fewer shrubs and less abundant ground vegetation. Management should be based on identifying signs of damage (Table 14.5) assessing local population density and protecting vulnerable areas by fencing or other control methods. General measures to protect young trees are included in previous sections in this chapter on fencing and tree shelters. More specific measures to protect established trees are discussed below.

Grey squirrels damage pole-size trees by stripping bark between May and July, causing deformities in the main stem and increased forking. Beech, sycamore, ash and pine are particularly susceptible, but all trees can be damaged if population levels are high enough. The extent of damage can be reduced by controlling squirrels from April to July, by live cage trapping or using hoppers containing warfarin-poisoned bait. Warfarin is potentially hazardous to both people and wildlife, requiring training in its safe application. The use of warfarin is permitted only in approved areas, which effectively excludes much of Scotland, the Lake District and other areas where red squirrels may still exist. The designs of warfarin hoppers and guidelines for their use have become increasingly sophisticated over time in an attempt to reduce the risk of other species being poisoned. For maximum control the hoppers should be placed under large trees clear of vegetation at 200-m centres or 3–5 hoppers/ha. Research is continuing into alternatives to warfarin for the control of grey squirrels, including the use of immuno-contraceptives. Control by shooting is labour intensive, requires skilled staff and is rarely completely effective.

Deer will browse the tender growth of most tree species, especially broadleaves such as ash, cherry, willow, hazel and rowan, leaving a ragged end to the nibbled shoots and a clearly demarcated browse line throughout the woodland. Deer may strip bark from smooth-barked trees such as Norway spruce, larch, ash, willow and beech, leaving behind tell-tale broad teeth marks running up the peeled stem. Male deer mark out their territories and clean off velvet by fraying young saplings. Successful deer management depends on co-operation amongst neighbouring landowners, fencing around vulnerable areas and humane culling to ensure a healthy population. Design features to ensure safe and effective culling include the creation of sunny deer glades and wide grassy rides with strategically placed high seats. Shooting is strictly controlled by law (1995 Deer Act), although properly managed, it may generate substantial income from stalking and venison. Further information is available from the Deer Initiative (or Deer Commission for Scotland) or see Mayle (1999).

Managing woodlands for conservation of biodiversity

Landscape ecology

There is an increasing imperative to conserve biological diversity and ensure its future viability and integrity. In the UK, the conservation of biodiversity has historically been based on the protection of a series of small, isolated sites. However, continuing declines in biodiversity indicate the ineffectiveness of this approach. There is now an increasing recognition of the importance of including the surrounding wider countryside in future conservation strategies, and this has facilitated a more holistic view of biodiversity conservation at much larger spatial scales. The importance of reversing woodland fragmentation has been further reinforced by the UK Biodiversity Action Plan (Anonymous, 1995), which contains a number of specific woodland Habitat Action Plans and many Species Action Plans for species that live in or near the edge of semi-natural woodland. Almost 30% of the species in the UK Biodiversity Action Plan are associated with woodland in some way.

The England Forestry Strategy (Forestry Commission, 1999) points out that many of our ancient and semi-natural woods are fragments of historically more extensive woodland, and that their continued ecological viability is threatened. As a result, grants such as the Forestry Commission's JIGSAW Challenge fund are being piloted to attempt to encourage the *expansion* and *linkage* of semi-natural woodland, as a contribution to sustainable forest management. Due to the lack of statutory control in the wider countryside, management agreements, such as individual countries' forestry grant schemes and the JIGSAW Challenge, may be important

mechanisms for reversing habitat fragmentation in the future in the UK.

Site-specific management issues

The following section gives general guidance on the improvement of woodland for conservation of biodiversity. However, some woodlands will already have important conservation status or designation such as Sites of Special Scientific Interest (SSSI), National Nature Reserves (NNR) or ancient semi-natural woodland. These important sites will require a more specific management prescription, normally involving expert advice from either local or national wildlife organizations such as English Nature, Scottish Natural Heritage and the Countryside Council for Wales. The Forestry Commission has produced guidance on protection and enhancement of biodiverstiy in woodlands (Forestry Commission, 1990; Currie and Elliot, 1997; Ferris and Carter, 2000; Humphrey *et al.*, 2002; Thompson *et al.*, 2003), and can also be contacted to provide specific advice on woodland conservation issues.

- Native trees support large numbers of species (especially invertebrates) and by planting trees native to the area the local genetic stock is preserved. Natural regeneration of native trees should be encouraged wherever possible.
- In commercial coniferous woodlands, planting of a proportion of native broadleaves should take place in groups or along watercourses and woodland edges.
- Open space is an important woodland habitat with characteristic ground vegetation and associated wildlife. These areas can often be incorporated at planting by leaving bare ground around rock outcrops, stream sides and wetlands without reducing the overall productivity of the plantation (up to 20% open space is now accepted under the Woodland Grant Scheme). Woodland glades and rides (at least 5–10 m wide) can be managed by cutting back vegetation and mowing to create a graded profile.
- The ground vegetation which builds up following tree establishment supports a thriving population of small mammals and their predators. Following canopy closure, much of this vegetation is lost under

the dense shade of the thicket of trees, especially conifers. By thinning earlier (up to 5 years) and slightly heavier (up to 10% more volume removed) ground vegetation is more likely to survive.

- Structural diversity can be improved by planting an understorey of native shrubs, such as hazel, hawthorn, holly or juniper, particularly along woodland rides and edges.
- Harvesting creates the opportunity to increase the age diversity of uniform plantations by clearing smaller areas (0.2–0.5 ha) phased in over a longer period.
- Some mature trees should be retained beyond the normal rotation until physical maturity (on at least 1% of the forest area), to provide old-growth habitats and habitat continuity.
- Dead wood habitats are scarce in most commercial woodlands. Some deadwood should be left lying where it has fallen. Humphrey *et al.* (2002) provide specific target amounts for different woodland types. Forest hygiene is not thought to be as big an issue in the UK woodlands as it is, for example, in Scandinavia. Dead branchwood can be left in a stack as shelter for invertebrates, and some standing dead trees should be retained for hole-nesting birds, specialized invertebrates and epiphytes. In addition, nest boxes and bat boxes can be provided in a variety of designs.
- The timing of forest operations should be planned to avoid breeding seasons and the passage of felling and extraction machinery organized to avoid disturbance to sensitive areas such as streams and wetlands. Herbicide use should be minimized and extreme care taken to avoid contamination of watercourses and other habitats.

The landowner may find it useful to incorporate some of these improvements into a conservation management plan for the woodland. Following an appraisal of the wildlife value of the site, appropriate management prescriptions should be proposed to reflect the overall aims for the woodland.

Landscape design for farm woodlands

A woodland that is designed and managed using the principles of good landscape practice will enhance the surrounding landscape, as well as providing benefits for wildlife and public enjoyment. The Forestry Commission and other agencies (e.g. national parks and local authorities) are increasingly enforcing the adoption of a professional approach to forest landscape design through control of felling licences, forest strategies and grant schemes. Landscape design skills have become an essential part of the forester's toolkit. There is strong evidence that more attractively landscaped forests command greater prices in the market.

On level ground, the landscape sensitivity of the forest is less than for highly visible forests on steep slopes at higher elevations, or in designated areas of great landscape value. Good landscape practice therefore begins by assessing the nature of the surrounding landscape, which will dictate the most appropriate design solutions.

For *hedgerow landscapes* in the lowlands, the design of woodlands should mirror the often geometric shape of the surrounding hedgerow pattern, the shape of the woodland interlocking harmoniously with the surrounding fields.

For *open and upland landscapes*, the design of woodlands needs more careful consideration. In these areas the topography or landform will be the dominant force in the landscape so that any new woodland planting must reflect the scale and shape of the surroundings. Small woodlands may look out of scale and should be located close to existing woodlands or on lower slopes where a hedgerow pattern is more evident. The shape of the forest should follow natural boundaries and vegetation, rather than the contour or the edge of the fence. As a general rule, the edges of the forest should rise up the valleys and hollows and fall down the shoulders of hills and ridges.

The landscape of all forests can be improved by attention to the following measures:

- *Diversity* creates visual interest and enhances the landscape value of the forest. Different textures and colours can be created by using different species, such as broadleaves along watercourses or compartments of larch, which change from fresh green in summer to brown in winter. A change of species can reduce the scale of the forest and allow more light into recreational areas, although the main species should make up two-thirds of the composition for aesthetic reasons. Diversity can also be increased by the provision of open ground, exposing outcrops of rock and creating a mosaic of different ages of trees.

- *Forest boundaries* are particularly important at the skyline, where the forest should either completely cover the skyline or cross it at the lowest point, cutting diagonally across the main view or else curving gently over the skyline. The edges to the forest should be as natural as possible, varied in scale with the landscape by the use of irregular groups, different species or plant spacing and detailed shaping.
- *Forest operations* should be carried out in sympathy with the landscape. Planting should be organized to create open areas that follow the landform, and forest tracks should cut diagonally across the contour, rather than along it. The harvesting of the forest provides an excellent opportunity to improve the appearance of a previously poorly designed forest by introducing different species, a greater diversity of age structure and by reshaping the edges of compartments. Any such work will need to be carefully phased to avoid immediate drastic changes.

For larger woodlands, especially in sensitive locations, it will be essential to undertake a full landscape appraisal, possibly using professionals who will have access to specialist forest design software. For small-scale woodland design, photographs, acetates and overlays will suffice to show the effect of different forest operations, the phasing of harvesting and sites for new planting. For further information see the Forestry Commission (1995).

Forest management for non-timber end products

Shelter woods

The provision of correctly designed and located shelter woods can benefit both crop and livestock enterprises on the farm. Shelter woods also have the potential to contribute to secondary objectives, including the provision of game cover, facilities for recreation and occasional timber production. The most important design features are the height and porosity of the shelter wood, which determine the extent and degree of wind protection (Figure 14.2). The requirements of arable crops and livestock differ in this respect as follows:

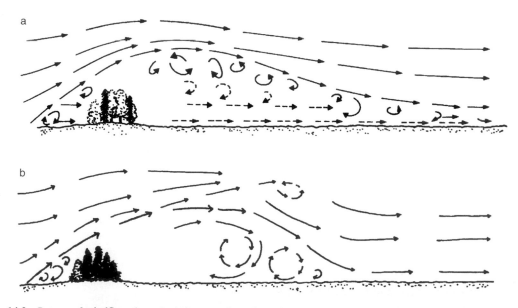

Figure 14.2 Pattern of windflow through shelter woods: **a** through a narrow and semi-permeable shelter wood: small rolling eddies create a long zone of shelter (15–30 times the height of the wood); **b** over an impermeable shelterbelt: large standing eddies and a short shelter zone (5–10 times the height of the wood).

- *Arable and grassland.* Shelterbelts within arable field crops or grassland will have the effect of reducing wind speed and moisture loss from the crop, leading to earlier harvests and greater productivity. Solid or impermeable windbreaks should be avoided, as they cause an upward deflection of the air stream, producing an area of low pressure to leeward of the barrier which results in intense turbulence more damaging than the original wind. More porous windbreaks (ideally 50% permeability) not only reduce turbulence but also reduce wind speed by up to 70% for a distance up to 30 times the height of the barrier. On sloping ground, the location of the windbreak should be carefully considered, as a trap is created for frosty air flowing down from higher up the hill. Single- or double-row windbreaks are normally used on horticultural and productive agricultural land, as wider windbreaks give no additional reduction in windspeed and may harbour pests such as pigeons or bullfinches and prevent normal drying, favouring fungal diseases such as grey mould (*Botrytis cinerea*). Shelterbelts may be inappropriate in some situations where they may shade neighbouring crops, harbour pests and vermin and interfere with machine access, or on heavy wet soils where the drying effect of the wind is reduced. Suitable species for field crop protection are balsam poplar hybrids (*Populus trichocarpa* × *P. balsamifera*), grey alder, or Italian alder on drier soils, all at 1- to 2-m spacings.
- *Upland livestock.* Shelter woods afford protection to stock from the cold driving winds, reducing heat loss and hence improving survival and cutting food consumption. A dense impermeable windbreak is most suitable, as animals tend to pack in tightly to the lee of the wood during winter gales. Dense shelter woods create a narrow belt of calmer conditions in their lee up to ten times the height of the trees and are also suitable for small areas such as lambing paddocks or farm buildings. Narrow belts of trees become draughty around the lower trunk with age and pose problems at maturity, as restocking will generally mean clear felling leading to a temporary loss of shelter. Neglected belts of conifers pose particular difficulties and require very gradual thinning to avoid catastrophic windthrow. Wider shelter woods should therefore be used (at least 20 m and 45 m if possible), which allow for the

use of a mixture of shrub and tree species to create a more densely graded profile. Wider shelter woods allow restocking to take place progressively without loss of protection and have a greater proportion of utilizable timber together with other benefits for wildlife and game. Livestock can be allowed access to the woodland in extreme weather for shelter and feeding on ground vegetation, although this should only be a temporary measure as long-term entry by livestock will lead to root damage, poaching and browsing of natural regeneration. A mixture of species is advisable, especially oak, ash, beech, pine and sycamore, with cherry, whitebeam, rowan and alder along the woodland margins. Sitka spruce and lodgepole pines are windfirm and useful on exposed ground in the west.

The height and porosity of the shelter wood is controlled by the width of the wood, species choice and silvicultural management. Draughts can be reduced by planting shrubs and coppicing tress on the woodland edge. Lighter canopy trees such as birch and oak encourage the growth of dense ground vegetation and the wood should be thinned regularly.

Other factors to consider in the design of shelter woods include the local climate, topography and surrounding landscape. Narrow lines of trees will not thrive in very exposed conditions and a landscape appraisal may be a condition of grant aid, especially for long straight shelterbelts in a predominately open landscape.

Pheasants

Farm woodland management for pheasants need not be in conflict with sound silvicultural practice. The value of woodlands can be substantially increased by incorporating some of the following specific measures to increase its potential for pheasants:

- Pheasants do not thrive in cold draughty conditions, so woodlands should be designed to give maximum protection to ground level using a mixture of shrubs and taller trees. The perimeter of the woodland may be planted with hedging species or Christmas trees, although temporary protection can be provided by strategically placed big-bales. Roosting trees should be retained on restocked sites and at the intersection of rides.

- The holding capacity of the wood can be increased by establishing ground cover shrubs, which provide additional shelter and food for the birds. Suitable species include butcher's broom (*Ruscus aculeatus*), rose of Sharon (*Hypericum calycinum*), dogwood (*Cornus alba*), elder (*Sambucus nigra*), flowering nutmeg (*Leycesteria formosa*), privet (*Ligustrum vulgare*), raspberry (*Rubus idaeus*) and snowberry (*Symphoricarpus albus*).

- A rich ground layer should be maintained by avoiding planting trees that cast a dense shade (beech, sycamore and close-spaced conifers), favouring trees with a lighter canopy (oak, ash, birch, cherry, rowan, larch and pine) and thinning at the appropriate time to increase the amount of light reaching the forest floor. Coppice is particularly suitable as it maintains a higher proportion of ground vegetation and a diversity of ages in the stand.

- Pheasants are woodland-edge birds, using the forest for winter cover and roosting. By planting a series of smaller and longer woodlands, the proportion of edge can be increased.

- In larger woodlands, open spaces and rides (30–50 m wide) should be provided to allow birds room to fly up and over the guns. Grassy glades maintained by annual mowing provide favourable conditions for young birds.

Further advice regarding woodland design for game is available from the Game Conservancy Trust.

Forest management for timber end products

Forest measurement

Forest measurement is usually required to provide an estimate of the quantity of timber for sale. In addition, forest inventory can provide the woodland owner with useful information for planning and forecasting future production. A periodic inventory will show the current state of the stand, its performance over time and any consequent deviations from the expected which may require treatment (e.g. restricted drainage). The amount of time invested in forest measurement is a balance between the degree of precision required and the costs involved. Generally more precise measures are used for more valuable timber in larger quantities, and sophisticated electronic equipment is now available for woodland surveys and volumetric analysis.

Forest measurement and inventory planning are described in two Forestry Commission publications, Edwards (1983) and Hamilton (1985). Timber is normally measured as solid volume, although other measures exist for specific purposes (green weight, dry weight, number of pieces). A number of conventions exist as described below.

Diameter

Diameter (cm) is normally measured using a girthing tape or electronic callipers around the circumference of the tree or log, rounded down to the nearest whole centimetre. *Diameter at breast height* (dbh) is measured at 1.3 m above ground level.

Length

Length (m) is measured by tape following the straight line distance along a log, rounded down to 0.1 m for lengths up to 10 m and 1 m for lengths over 10 m.

Height

Height (m) can be measured by an observer standing a known distance (at least 1.5 times the height of the tree) away from the base of the tree by either trigonometry or from direct readings from specifically designed instruments, such as the clinometer, altimeter or hypsometer. *Total height* is the vertical distance from the base to the uppermost tip of the standing tree. *Timber height* is the height to the lowest point on the main stem where the diameter is 7 cm overbark. In broadleaves, timber height may be the lowest point at which no main stem is distinguishable. The *top height* of the stand is the average total height of the 100 trees of largest diameter (dbh) per hectare and is used to estimate the yield class. A more practical method of estimating top height is given below for single-species, even-aged plantations or for the dominant species in mixtures:

1. Lay down random, 0.01-ha plots through the stand, the number of samples required depending on the size and variability of the crop.

Area of stand (ha)	Number of samples required	
	Uniform crop	Variable crop
0.5–2.0	6	8
2.0–10.0	8	12
Over 10.0	10	16

2. Record the total height of the single tree of largest dbh within 5.6 m of the plot centre.
3. Calculate the average height of the samples taken to give an estimate of the top height of the stand.

Basal area

Basal area (m²) is the cross-sectional area of a tree at 1.3 m above ground level (dbh). Basal area (m²) for an individual tree =

$$\frac{\pi \times dbh^2}{40\,000}$$

where *dbh* = diameter at breast height in cm.

The basal area of a stand (m²/ha) is the sum of the basal areas of all the trees in the stand. The basal area of a stand can be measured by laying down sample plots or using a relascope (or angle gauge) to take direct readings (see Edwards, 1983). Basal area is a useful value as it gives a better measure of the level of stocking than number of trees per hectare alone.

Volume

Volume (m³) is measured overbark to 7 cm diameter, or where no main stem is distinguishable, not including the volume of branchwood. The measurement of timber volume is always an estimate, due to the variable nature of trees and logs; in particular, the volume must be corrected to account for the taper of different species of tree. A number of useful tables have been produced in the aforementioned Forestry Commission booklets, which allow volume to be estimated by measuring the height and diameter of standing or felled timber. The following procedures are drawn from those publications.

Estimating the volume of a single standing tree
The volume of a standing tree can be estimated by correcting the total height (or timber height for broadleaves) by the use of a 'form factor' to account for the taper. The degree of taper increases in widely spaced stands and among open-grown broadleaves.

Appropriate form factors range from 0.5 for mature plantation conifers, 0.4 for open-grown conifers or close-spaced broadleaves, to 0.35 for younger trees. To calculate the volume using this method, first find the basal area at breast height (in m²) and multiply this by the total height (m) and the appropriate form factor.

Example
A mature plantation-grown Douglas fir with a total height of 22 m and dbh of 46 cm:

$$\text{Basal area} = \frac{\pi \times 46^2}{40\,000}$$

$$= 0.166\,\text{m}^2$$

$$\textit{Volume} = \text{basal area} \times \text{height in metres}$$
$$\times \text{form factor}$$
$$= 0.166 \times 22 \times 0.5$$
$$= 1.83\,\text{m}^3$$

An alternative method in Hamilton (1985) is based on the use of single tree tariff charts, by measuring dbh and total height.

Estimating the volume of a stand
The method of estimating stand volume will depend on the value of the timber and the method of sale. For moderate to high value crops the *tariff system* is appropriate. This system is based on the relationship between tree diameter and volume, which will vary depending on the tree species, age and site conditions. Tables have been produced for a wide range of conditions in British forestry, where this relationship is classified by tariff numbers from 1–60. By calculating the appropriate tariff number for any particular stand of trees, the volume of the stand is found by consulting the tariff tables in Edwards (1983) or Hamilton (1985).

Measuring the volume of felled timber
The volume of felled timber can be reasonably accurately calculated by measuring the diameter of the log at the mid point along its length (*mid-point diameter*) and using Huber's formula below:

$$\text{Volume (m}^3)$$
$$= \frac{\pi \times (\text{mid-point diameter})^2}{40\,000} \times \text{length of log}$$

where diameter is measured in cm and the length is measured in m.

Logs over 15 m should be measured in two or more sections. Volumes can also be calculated using the mid-point diameter tables in Edwards (1983) and Hamilton (1985). In addition, tables have been produced for sawlog and roundwood volumes on the basis of top diameter (normally underbark) and length. This assumes a standard taper (e.g. 1 : 120 for sawlogs) and gives a quicker but less accurate measure of log volumes.

Other measures

An estimation of timber in stacks is sometimes useful, although the volume of the stack must be converted to a solid volume by the use of conversion factors, which take account of log diameter, length, taper, straightness and method of stacking. Typical conversion factors range from 0.55–0.65 for broadleaves and 0.65–0.75 for conifers, although for short, straight billets the figure may be as high as 0.85.

Some timber merchants dealing in quality broad-leaved timber may still refer to the traditional measurements of hoppus and cubic feet. One hoppus foot $= 0.036 \, \text{m}^3 = 1.273 \, \text{ft}^3$.

The yield class system

The annual increase in timber volume for even-aged stands of trees is termed the *current annual increment* (CAI, $\text{m}^3\text{ha}^{-1}\text{year}^{-1}$), which rises rapidly in the early life of the plantation and falls gradually as growth rates decline in old age. The *mean annual increment* (MAI, $\text{m}^3\text{ha}^{-1}\text{year}^{-1}$), is the average annual yield up to any particular age, which reaches a maximum at the ideal rotation age for maximum volume production. The maximum MAI also defines the *local yield class* (LYC, $\text{m}^3\text{ha}^{-1}\text{year}^{-1}$) of the stand, which is a measure of the productivity of the trees when grown under a specific management regime for those site conditions (Fig. 14.3).

The *general yield class* (GYC) of an even-aged plantation can be estimated without detailed volume measurements, since for a defined management regime a good correlation exists between top height, species, age and local yield class. By estimating the top height of the stand, the general yield class for that species can be

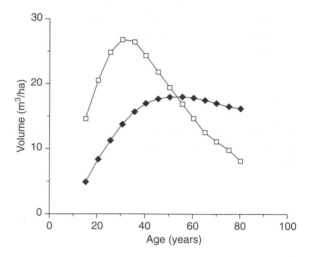

Figure 14.3   Volume increment in an even-aged stand (sitka spruce GYC 18: line with open squares shows CAI; line with solid diamonds shows MAI) (after Edwards and Christie, 1981, Crown Copyright, reproduced with permission of the Forestry Commission, 2001).

found by reference to the general yield class curves for that particular species given in Edwards and Christie (1981) or Rollinson (1999). General yield class is usually sufficient for management purposes. However, the more accurate local yield class can be assessed using direct measurement by sampling cumulative volume or cumulative basal area of the stand before the point of first thinning (Edwards and Christie, 1981). Alternatively, it can often be found more simply by observing the value of maximum MAI shown in a relevant table in Edwards and Christie (1981), the table being chosen because it accurately reflects the species, general yield class and actual management regime applied to the stand in question. Production class is sometimes used to describe the difference between general yield class and actual local yield class for a site. Production class A reflects a local yield class of $2 \, \text{m}^3\text{ha}^{-1}$ year^{-1} higher than general yield class, B no difference, and C $2 \, \text{m}^3\text{ha}^{-1}$ year^{-1} less than general yield class. For example, suppose that observations of top height in a sitka spruce stand show it to have a general yield class of 20. However, the yield class curves in Edwards and Christie (1981) are based on a standard management regime involving an initial spacing of 1.7 × 1.7 m and intermediate thinning. Suppose, however, that the stand in question had an initial spacing of 3.0

× 3.0 m. Fortunately a 3-m initial spacing table exists in Edwards and Christie (1981), and for general yield class 20 this shows a maximum MAI of 18.1, implying a local yield class of 18, or a general yield class of 20 with production class C. It should be understood that these tables give estimated future volume production on the basis of normal silvicultural treatments and full stocking. A reduction of 15% from the gross total area of the plantation is normally made for forest tracks, rides and stacking areas, although this should be increased where other unstocked ground exists.

Thinning

The gaps between tree canopies in a young plantation quickly close, as the crowns of individual plants coalesce. Trees compete for light, moisture and nutrients and eventually some trees become suppressed or overtopped by their more dominant neighbours. The growth of these suppressed trees is arrested and they become susceptible to disease and decay. This natural process of self-thinning reduces the numbers of stems in a woodland by some 50–70% of the original planting density at harvest.

Thinning, the artificial removal of some trees, does not increase the total yield from woodland. However, it does allow the forest manager to remove trees of poor quality, to selectively release the better trees, and to increase diameter growth and hence value of selected final crop trees. The effective utilizable yield and stand quality is therefore increased. Where the products of thinning are marketable, thinning gives a useful financial return throughout the life of a plantation.

The major determinants of timber quality are:

- *Sufficient length of clean stem or bole*. This is achieved by a high initial planting density, with a consequent rapid canopy closure. The competition from neighbouring trees suppresses the development of side branches and promotes apical growth (especially in broadleaves). In more open grown conditions, high pruning of selected final crop trees (to 5 m) may be necessary.
- *Large girth*. This is achieved by reducing competition from neighbouring trees to allow for the full development of the canopy of the selected trees.
- *Evenness of growth*. For most conifers, the faster the rate of growth, the lower the timber density and

strength, while for broadleaves, evenness of growth may be more important for some markets, such as veneers. Oak, by contrast, is stronger the faster the rate of growth (greater proportion of denser late wood).

The date of first thinning depends on timber species, site productivity and spacing, normally within the range of 18–30 years old or about 8 m in height. Faster growing trees on good sites are thinned earliest.

While the overall objective may be to maximize the financial return from the woodland, the first thinning usually yields very little utilizable timber, although its effect on the subsequent crop value may be considerable. Over the life of the crop, 80–90% of the trees could be removed as thinnings, without affecting the productive capacity of the stand. The amount removed at each thinning and the method of selection are determined by the thinning intensity, thinning cycle and thinning regime.

Thinning intensity

This is the rate at which volume is removed from the stand, often termed the *annual thinning yield*. The higher the intensity, the greater the amount of timber removed and consequently the larger the amount of growing space left around the remaining trees. Higher intensities may produce larger diameter trees, and consequently in some cases a greater financial return. The maximum intensity that can be maintained without a loss in overall volume production is termed the *marginal thinning intensity*. For a wide range of species and conditions, this point approximates to an annual rate of removal of 70% of the maximum MAI. For example, a stand of local yield class 20 will have an annual thinning yield of $14 \, m^3 \, ha^{-1} \, year^{-1}$, although this rate of removal only holds for a stand within the normal thinning period, which varies according to species. See Rollinson (1999) for further information.

Thinning cycle

This is the number of years between each thinning, which is determined by the silvicultural requirements of the timber species, the thinning volume generated and the prevailing economic climate. Longer cycles allow for a greater volume out-turn, but may reduce volume production and increase the risk from wind-blow. Where small amounts of thinnings are used on the farm, it may be preferable to carry out light thinnings at more frequent intervals, with distinct silvicultural advantages, since the trees are released more gradually. Typical thinning cycles range between 3 and 6 years for conifers and up to 10 years for some slow-growing broadleaves.

Thinning yield

This is the actual volume removed at any one thinning, that is, the thinning intensity multiplied by the thinning cycle. For most commercial operations about 50–60 m^3/ha is the minimum amount of timber that can profitably be extracted. For example, a stand of Douglas fir (local yield class 24) might have a thinning cycle of 3 years, which enables about 50 m^3/ha to be removed at each thinning (annual thinning yield 16.8 m^3/ha).

For even-aged stands of trees that are fully stocked it is possible to determine when thinning is required according to published thinning tables. Where stocking or management differs from normal (i.e. many farm woodlands), the point of thinning may have to be judged by visual inspection. A more accurate method is to estimate the basal area of the stand and to compare this with threshold basal area tables for fully stocked stands (see Table 14.9). If the basal area is equal to or more than the figure in the table, the stand is ready for thinning. The amount of timber removed is also best controlled by the volume or basal area of the thinnings, on the basis of marked plots (see Rollinson, 1999).

Table 14.9 Before-thinning basal areas for fully stocked even-aged stands (basal areas in m^2/ha) (after Rollinson, 1999, Crown Copyright, reproduced by permission of the Forestry Commission, 2001).

Species		Top height (m)										
		10	12	14	16	18	20	22	24	26	28	30
Scots pine		26	26	27	30	32	35	38	40	43	46	—
Corsican pine		34	34	33	33	33	34	35	36	37	39	—
Lodgepole pine		33	31	31	30	30	31	31	32	33	34	—
Sitka spruce		33	34	34	35	35	36	37	38	39	40	42
Norway spruce		33	33	34	35	36	38	40	42	44	46	49
European larch		23	22	22	22	23	24	25	27	28	30	—
Japanese/hybrid larch		22	22	23	23	24	24	25	27	28	29	—
Douglas fir		28	28	28	29	30	31	32	34	35	37	40
Western hemlock		32	34	35	36	36	36	37	38	38	39	40
Western red cedar		—	49	50	51	53	55	57	60	63	66	70
Grand fir		—	39	39	39	39	39	39	40	41	43	45
Noble fir		—	45	46	46	47	48	49	51	52	54	—
Oak	Local yield class											
	4	24	24	23	23	24	24	—	—	—	—	—
	6	—	26	25	24	24	25	25	25	25	—	—
	8	—	27	25	24	24	24	25	26	26	26	—
Beech	4	—	22	23	25	27	30					
and sweet chestnut	6	—	24	25	25	27	29	31	33	36		
	8	—	—	27	27	27	28	29	31	33	35	37
	10	—	—	28	28	27	27	28	29	31	33	35
Sycamore,	4	—	17	17	18	21	—	—	—	—	—	—
ash, alder and birch	6	—	17	18	19	22	25	—	—	—	—	—
	8	—	17	18	20	22	25	28	—	—	—	—
	10	—	18	19	20	23	26	30	33	—	—	—
	12	—	19	20	21	24	27	31	35	—	—	—

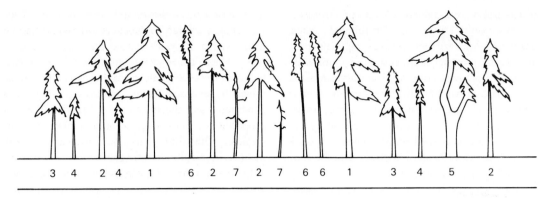

Figure 14.4 Classification of types of tree likely to be found in the crop. *1* Dominants; *2* co-dominants; *3* sub-dominants; *4* suppressed trees; *5* wolf trees; *6* whips; *7* leaning, dead or dying trees.

Thinning type (selective thinning)

Thinning type determines the type of tree in the stand that is removed by thinning (see Fig. 14.4). It is normal practice to identify the final crop trees at an early stage, although it may be prudent to choose a larger number than required to allow for further selection at a later date. Thinning should aim to remove trees that are forked, spiral barked, coarse branched or leaning and any that look unhealthy or spindly. While marking for thinning, canopy and crowns must be continually observed so that excessive gaps are not produced after thinning. A *low thinning* removes trees that are currently in the lower levels of the canopy (sub-dominants and suppressed trees), while a *crown thinning* is directed to releasing the crowns of selected final crop trees, which may mean felling competing dominants. An *intermediate thinning* is the most common form of thinning, and comprises a low thinning to remove suppressed and sub-dominant trees, together with opening up of the canopy to release better dominant trees and create a more uniform stand.

A flexible approach, based on crown thinning, is likely to be required when transforming woodlands to continuous cover systems. In selection systems, the aim should be to encourage a full range of sizes and ages of trees.

Systematic thinning

The first thinning differs from most other thinnings in that very little, if any, financial return is likely, due to the small size of trees and low volumes produced. In order to reduce the cost of this operation a number of systematic thinning systems can be used. All involve the removal of trees according to a predetermined pattern, irrespective of the merits of individual trees, and are often carried out by mechanical harvesters. Extraction and marking costs may be reduced greatly by the adoption of systematic methods of thinning, particularly in early thinnings, as trees felled in selectively thinned stands are easily caught up in the dense thicket. *Line thinning* is a systematic form of thinning where trees are removed in lines along the planting rows or in more sophisticated systems as a series of interconnected lines (e.g. chevron thinning). The most common systems in practice are based on removing one planting row in four or five, although for close spacings (less than 2 m), two adjacent rows may be removed to allow for the passage of extraction machinery. The benefit of line thinning is restricted to those trees neighbouring the line removed; for 1 : 5 or 1 : 6 row thinning it may be advisable to carry out an additional selective thinning in between the rows. The removal of rows wider than 5 m should not be contemplated, due to the loss in volume production and especially on sites prone to windthrow.

No-thin systems and delayed thinning

Where the danger of windthrow is particularly high, especially in the extreme north and west of Britain, a no-thinning system with a reduced rotation age may be preferable. However, although total volume of production may not be reduced, the volume of sawlogs is usually significantly lower than for thinned stands. Other approaches on very exposed sites include the use of chemical line thinning and self-thinning mixtures, which may be more cost effective than harvesting and have a lower risk of windthrow. Elsewhere, the decline in the real value of small roundwood in recent years has led to a reappraisal of the economic advantages of thinning on large commercial forest estates. On some poorer, windfirm sites (local yield class <10), the time of first thinning may be delayed by up to 10 years, allowing for a greater out-turn of volume at a larger average diameter. For small farm woodlands, however, thinning is usually advisable on silvicultural grounds, in addition to the advantages of providing a more open canopy for wildlife, game and amenity.

Where previous thinning has been neglected, this may lead to a large proportion of weak or spindly trees. It is important at this stage not to correct the past neglect by a single heavy thinning, as this may open up the canopy to a greater risk of wind or snow damage, epicormic growth or sun scorch. Some species will recover better than others: for example, crown expansion is most rapid in oak, although ash will recover only weakly. In some older stands a decision has to be taken whether or not sufficient trees will respond to thinning, especially if there are already signs of windthrow or crown dieback. Any thinning that may be carried out should be restricted to 5% of the total basal area.

Harvesting

In conventional even-aged clear fell and replant systems, where timber production is a primary objective, a time comes for the harvesting of the mature final crop trees. The age of the trees at this point or *rotation length* is set according to the management objectives of the owner. For some markets, the rotation length is relatively fixed, according to specific size and diameter requirements, such as for sawlogs or veneers. The point of maximum rate of growth (maximum MAI) can be a useful indicator of an economic rotation length, although it takes no account of the different prices obtained for different sizes of timber. Many commercial growers prefer the concept of *financial rotation*, whereby the maximum income from timber is the overriding objective for the site. In this respect the length

of rotation can be drastically affected by the discount rate chosen, which in practice forecasts a harvesting date some 5–10 years before the point of maximum MAI. These concepts are useful in forest planning, although the exact date of felling will often be determined by external factors, such as the local market conditions, and the biodiversity benefits of leaving a proportion of trees to grow on to biological maturity (typically 200–300 years) as opposed to economic maturity (typically 40–60 years).

In recent years, the efficiency of forest-scale harvesting has been greatly improved by the use of specialized felling, processing and extraction machinery for use in large commercial forests. The high capital expense of this machinery largely precludes its use in smaller farm woodlands, so that harvesting is likely to involve the use of chainsaws, tractor-mounted harvesters and smaller extraction machinery. A range of farm-scale machinery is becoming more widely available, including equipment that can be mounted on all-terrain vehicles.

The harvesting (or thinning) operation itself requires careful planning, to determine the course and direction of felling and the location of extraction routes and stacking areas. Chainsaws and other forest machinery are extremely dangerous, and woodland owners contemplating a DIY approach to woodland operations should undertake appropriate training in all relevant aspects before embarking on any harvesting themselves. The choice of harvesting system is a choice for the forest manager, and is determined by the nature of the terrain, the silvicultural system, the product specification, the ease of access, degree of mechanization available and landscape considerations. Where simple product mixes are required (e.g. sawlogs and pulp), the trees are normally cut to size in the wood and extracted separately (the *shortwood system*). For a more complex range of sizes and lengths, trees are extracted whole from the forest to a cleared area, where the timber is cut to size (the *tree length system*). The costs of felling and clearing by contractor will depend on the terrain and size of the timber cut, although it should be indicated whether the price includes the treatment of stumps with urea (against *Heterobasidion annosum*) or the treatment of branchwood and other harvesting residues (brash).

Control of tree felling

Woodland owners *must* apply to the Forestry Commission for permission to fell any trees. Exceptions to this requirement include a small entitlement of up to $5\,m^3$ every 3 months for own use, small-diameter trees (under 10 cm dbh for thinning, under 15 cm dbh for multi-stemmed trees, under 8 cm dbh for other felling) and for dead or dangerous trees. Felling is controlled by a Felling Licence from the Forestry Commission usually through the Woodland Grant Scheme and will be issued only if certain conditions are fulfilled, such as a commitment to replanting. Special permission may also be required for sites within areas designated as Conservation Areas, Sites of Special Scientific Interest (SSSIs) or where the trees are protected by a Tree Preservation Order (TPO) imposed by the local planning authority. Owners failing to obtain all necessary permission risk prosecution and fines.

Forest investment

In financial terms, the forest plantation represents a long-term investment, where the owner faces a patient wait before revenue from the later thinnings and final harvest offset the early establishment expenditure. In order to compare forest investment with other alternatives, a discounted cashflow may be calculated, using an appropriate interest rate (net of inflation), to give a *net present value* (NPV) for the proposed woodland. The return from different rotations and species can be compared easily, since all future revenues and expenditure are discounted back to year 0. The *internal rate of return* (IRR) is calculated as the interest rate at which NPV = 0. The choice of discount rate is critical to the effect on NPV, since the higher the discount rate chosen, the lower the value of future returns, especially for longer rotation crops, such as oak. A figure of 3–5% is normally appropriate for most commercial woodlands, although lower rates may be acceptable where other less tangible benefits, such as landscape and wildlife, are considered important.

For a more immediate assessment of the financial impact of woodland planting, a *partial budget* or a *cash flow forecast* are useful tools. A partial budget involves a review of all the potential costs of planting set against

any income over the first 5–8 years in the life of the new woodland. The costs of planting and aftercare should include proceeds foregone from land taken out of agricultural production and income should include grant aid, the sale of machinery or livestock and other cost savings to the enterprise. Fixed and variable costs should be included in the calculations. The woodland owner will then at least be aware of the short-term financial implications of a proposed woodland venture.

Integration of agriculture and forestry

An increasing interest in genuinely integrated farm forestry systems is apparent in some parts of the UK, where tight agricultural margins are forcing farmers to look at all the resources at their disposal. Integrated farming/forestry systems have the potential to provide many benefits. For example, labour can be deployed in the woodlands at less busy times of the year, marginal land may be more productive growing trees than being farmed and home-grown timber can be used on the farm or sold at local markets. A farmer or farm staff trained in woodland operations may be able to generate income from contracting work on neighbouring farms. Woodlands planted with a shelter objective can be highly beneficial to crops and livestock and income may be generated from letting out a game shoot. Timber produced can be used on the farm, for example as fencing, or in wood-fired heating systems.

The potential for intimate integration of agriculture and tree production – *agroforestry* – continues to be assessed by researchers in the UK. Agroforestry involves the establishment of widely spaced trees with grazing or arable crops in between the rows. The productivity of the agricultural component will depend on the tree spacing and resulting point of canopy closure. The timber objective is to produce a main stem of at least 40–50 cm dbh over a short rotation of 40–50 years, combined with pruning to remove side branches. Suitable tree species include ash, cherry, walnut, poplar, Douglas fir and hybrid larch. Economic models have shown that such systems may prove financially viable, particularly in marginal areas, although further field-testing is required. See Hislop and Claridge (2000) for further information.

Grants

Forestry Commission grants

The Woodland Grant Scheme (WGS) (England and Wales) and the Scottish Forestry Grant Scheme (Scotland) aim to encourage the management of existing woodland together with the expansion of private forestry in a way that achieves a reasonable balance between the needs of the environment, increasing timber production, providing rural employment and enhancing the landscape, amenity and wildlife conservation. The scheme can provide a substantial contribution towards the costs of woodland establishment and management, and a professional woodland adviser will be able to assist owners to make the most of the grants on offer.

Considerable national and local variations in the WGS now exist with different Forest Conservancies targeting funds towards the type of woodland considered most valuable in their country or region. This more discretionary approach involves the assessment of proposed woodland plans against defined criteria, with grants only being awarded to applications that score above a minimum standard. For specially designated areas, additional 'Challenge Funds' have been procured for woodland planting and are distributed by a competitive bidding process (e.g. the New National Forest in the Midlands and the Grampian Forest in north-east Scotland).

A number of general provisions must normally be met in order to obtain WGS funding. These include the maintenance of a broadleaved component of the woodland, protection of statutory designations (e.g. ancient monuments, SSSIs, rights of way), appropriate management of ancient woodland and compliance with various Forestry Commission guidelines such as landscape design, watercourse management, fire protection and nature conservation. Major afforestation projects may need to go through a formal Environmental Impact Assessment process. The assessment will usually include a description of the proposal, an environmental statement and details of how consultation is to take place.

Grants are available for restocking, new planting or natural regeneration on areas larger than 0.25 ha or 15 m wide, although smaller areas may be accepted by agreement with the Forestry Commission. Additional supplements may be available for planting on arable land or improved grassland (Better Land Contribution) or creating woodlands incorporating public access within 5 miles of selected urban areas (Community Woodland Contribution). Woodland Improvement Grants may also be available to contribute towards the cost of forest management where there is well-defined provision of public benefits such as access or high-quality wildlife habitats. The use of long-term management plans is currently being piloted with a 20-year time scale and all associated permissions and grants agreed in advance.

In some cases enhanced grants may be available in conjunction with other grant-aiding organizations.

The Farm Woodland Premium Scheme (Departments of Agriculture)

The Farm Woodland Premium Scheme (FWPS) aims to encourage the planting of woods on farmland by providing the farmer with annual payments to compensate for the loss of agricultural revenue for up to 15 years after planting. The scheme is weighted towards better quality land in lowland areas and runs in conjunction with the WGS:

- farmland that has been under arable cultivation for at least 3 years prior to the WGS application;
- grassland that has been improved for at least 3 years prior to the WGS application, the sward comprising at least 50% ryegrass (*Lolium* spp.), cocksfoot (*Dactylis glomerata*), timothy (*Phleum pratense*) or white clover (*Trifolium repens*).

Information packs about the WGS and FWPS are available from the Forestry Commission. Both the Forestry Commission and Department of Agriculture grant schemes in England and Wales will be extensively revised in 2004/2005.

Other sources of grant assistance

Grants are available for approved amenity planting from a range of sources such as the Countryside Agency, Countryside Council for Wales and Scottish Natural Heritage operating through the relevant local authority. For woodlands of high wildlife conservation value, assistance may be available from English Nature (or equivalent organization), National Park Authorities or through charitable organizations such as the Woodland Trust. Additional help may also be available for sites within other schemes such as Countryside Stewardship or Environmentally Sensitive Areas.

Further information on these and other local schemes can be obtained from the regional office of the Forestry Commission, professional woodland advisers, your local Farming and Wildlife Advisory Group (FWAG) officer or the Woodlands Advisor of the local authority.

References

Anonymous (1995) *Biodiversity, the UK Steering Group Report.* HMSO, London.

Bevan, D. (1987) *Forest Insects.* Forestry Commission Handbook 1. HMSO, London.

Brazier, J.D. (1990) *The Timbers of Farm Woodland Trees.* Forestry Commission Bulletin 90. HMSO, London.

Currie, F. and Elliot, E. (1997) *Forests and Birds: A Guide to Managing Forests for Rare Birds.* RSPB, Sandy, Bedfordshire.

Edwards, P.N. (1983) *Timber Measurement.* Forestry Commission Booklet 49. HMSO, London.

Edwards, P.N. and Christie, J.M. (1981) *Yield Models for Forest Management.* Forestry Commission Booklet 48. Forestry Commission, Edinburgh.

Ferris, R. and Carter, C. (2000) *Managing Rides, Roadsides and Edge Habitats in Lowland Forests.* Forestry Commission Bulletin 123. Forestry Commission, Edinburgh.

Forestry Commission (1990) *Forest Nature Conservation Guidelines.* Forestry Commission, Edinburgh.

Forestry Commission (1995) *Forest Landscape Design Guidelines*, 2nd edition. Forestry Commission, Edinburgh.

Forestry Commission (1998a) *The UK Forestry Standard.* Forestry Commission, Edinburgh.

Forestry Commission (1998b) *Forests and Soil Conservation Guidelines.* Forestry Commission, Edinburgh.

Forestry Commission (1999) *England Forestry Strategy: A New Focus for England's Woodlands.* Forestry Commission, Edinburgh.

Forestry Commission (2003) *The Brief for Stage 2 – Part A, Exploratory Contamination and Technical Feasibility Survey on Potential Community Woodland Sites.* Internal report prepared on behalf of the Forestry Commission and the Land Regeneration Unit by Forest Research, Farnham, Surrey.

Forestry Commission (2004) *Forests and Water Guidelines*, 4th edition. Forestry Commission, Edinburgh.

Gregory, S.C. (1998) *Diseases and Disorders of Forest Trees.* Forestry Commission Field Book 16. HMSO, London.

Hamilton, G.J. (1985) *Forest Mensuration Handbook.* Forestry Commission Booklet 39. HMSO, London.

Harmer, R. and Howe, J. (2003) *The Silviculture and Management of Coppice Woodlands.* Forestry Commission, Edinburgh.

Hibberd, B.G. (1988) *Farm Woodland Practice.* Forestry Commission Handbook 3. HMSO, London.

Hislop, A.M. and Claridge, J.N. (2000) *Agroforestry in the UK.* Forestry Commission Bulletin 122. HMSO, London.

Hislop, M. (2004) *A Toolbox for Public Involvement in Forest Planning.* Forestry Commission, Edinburgh.

Hodge, S. and Pepper, H. (1998) *The Prevention of Mammal Damage to Trees in Woodland.* Forestry Commission Practice Note 3. Forestry Commission, Edinburgh.

Humphrey, J.W., Stevenson, A. and Swailes, J. (2002) *Life in the Deadwood: A Guide to Managing Deadwood in Forestry Commission Forests.* Forest Enterprise Living Forest Series. Forestry Commission, Edinburgh.

Keighley, G. (1996) *Wood as Fuel. A Guide to Burning Wood Efficiently.* Forestry Commission, Edinburgh.

Kerr, G. and Evans, J. (1993) *Growing Broadleaves for Timber.* Forestry Commission Handbook 9. HMSO, London.

Lines, A. (1987) *Choice of Seed Origins for the Main Forest Species in Britain.* Forestry Commission Bulletin 66. HMSO, London.

Mason, B., Kerr, G. and Simpson, J. (1999) *What Is Continuous Cover Forestry?* Forestry Commission Information Note 29. Forestry Commission, Edinburgh.

Mayle, B. (1999) *Managing Deer in the Countryside.* Forestry Commission Practice Note 6. Forestry Commission, Edinburgh.

Moffat, A. and McNeill, J. (1994) *Reclaiming Disturbed Land for Forestry.* Forestry Commission Bulletin 110. HMSO, London.

Moffat, A., Jones, B.M. and Mason, B. (2004) *Brash: Its Importance and Management.* Forestry Commission Information Note. Forestry Commission, Edinburgh.

Morgan, J. (1999) *Forest Tree Seedlings.* Forestry Commission Bulletin 121. Forestry Commission, Edinburgh.

Patterson, D.B. and Mason, W.L. (1999) *Cultivation of Soils for Forestry.* Forestry Commission Bulletin 119. HMSO, London.

Peterken, G. (1993) *Woodland Conservation and Management*, 2nd edition. Chapman and Hall, London.

Potter, M.J. (1991) *Treeshelters.* Forestry Commission Handbook 7. HMSO, London.

Pyatt, G., Ray, D. and Fletcher, J. (2001) *An Ecological Site Classification for Forestry in Great Britain.* Forestry Commission Bulletin 124. Forestry Commission, Edinburgh.

Quine, C. (1995) *Forests and Wind.* Forestry Commission Bulletin 114. HMSO, London.

Rodwell, J.S. and Patterson, G. (1994) *Creating New Native Woodlands*. Forestry Commission Bulletin 101. HMSO, London.

Rollinson, T.J.D. (1999) *Thinning Control*. Forestry Commission Field Book 2. HMSO, London.

Tabbush, P. and Lonsdale, D.L. (1999) *Approved Poplar Varieties*. Forestry Commission Information Note 21. Forestry Commission, Edinburgh.

Tabbush, P. and O'Brien, E. (2003) *Health and Well-being: Trees, Woodlands and Natural Spaces*. Forestry Commission, Edinburgh.

Thompson, R., Humphrey, J., Harmer, R. and Ferris, R. (2003) *Restoration of Native Woodland on Ancient Woodland Sites*. Forestry Commission Practice Guide. Forestry Commission, Edinburgh.

Tubby, I. and Armstrong, A. (2002) *Establishment of Short Rotation Coppice*. Forestry Commission Practice Note 7 (revised). Forestry Commission, Edinburgh.

UKWAS (2000) *Certification Standard for the UK Woodland Assurance Scheme*. UK Woodland Assurance Standard support unit, Forestry Commission, Edinburgh.

Williamson, D.R. (1992) *Establishing Farm Woodlands*. Forestry Commission Handbook 8. HMSO, London.

Willoughby, I. and Clay, D.V. (1996) *Herbicides for Farm Woodlands and Short Rotation Coppice*. Forestry Commission Field Book 14. HMSO, London.

Willoughby, I. and Dewar, J. (1995) *The Use of Herbicides in the Forest*, 4th edition. Forestry Commission Field Book 8. HMSO, London.

Willoughby, I. and Moffat, A. (1996) *Cultivation of Lowland Sites for New Woodland Establishment*. Forestry Commission Research Information Note 288. Forestry Commission, Edinburgh.

Willoughby, I., Evans, H., Gibbs, J., *et al.* (2004) *Reducing Pesticide Use in Forestry: A Decision Guide*. Forestry Commission Practice Guide. Forestry Commission, Edinburgh.

Willoughby, I., Jinks, R., Gosling, P. and Kerr, G. (2004) *Creating New Broadleaved Woodland by Direct Seeding*. Forestry Commission Practice Guide. Forestry Commission, Edinburgh.

Further reading

Evans, J. (1984) *Silviculture of Broadleaved Woodland*. Forestry Commission Bulletin 62. HMSO, London.

Ferris-Kaan, R. (1995) *Managing Forests for Biodiversity*. Forestry Commission Technical Paper 8. Forestry Commission, Edinburgh.

Hart, C.E. (1991) *Practical Forestry for the Agent and Surveyor*. Alan Sutton, Stroud.

James, N.D.G. (1989) *The Forester's Companion*, 4th edition. Blackwell, Oxford.

Kerr, G., Mason, B., Boswell, R. and Pommerening, A. (2002) *Monitoring the Transformation to Continuous Cover Management*. Forestry Commission Information Note 45. Forestry Commission, Edinburgh.

Lucas, O.W.R. (1991) *The Design of Forest Landscapes*. Oxford University Press, Oxford.

MAFF (1993) *Farm Woodlands – Practical Guide*. HMSO, London.

Mason, B. and Kerr, G. (2001) *Transforming Even-Aged Conifer Stands to Continuous Cover Management*. Forestry Commission Information Note 40. Forestry Commission, Edinburgh.

Matthews, J.D. (1989) *Silvicultural Systems*. Clarendon Press, Oxford.

McCall, I. (1988) *Woodlands for Pheasants*. Game Conservancy Trust, Fordingbridge.

Robertson, P.A. (1992) *Woodland Management for Pheasants*. Forestry Commission Bulletin 106. HMSO, London.

Savill, P.S. (1991) *The Silviculture of Trees Used in British Forestry*. CAB International, Wallingford.

Savill, P.S. (1997) *Plantation Silviculture in Europe*. Oxford University Press, Oxford.

Taylor, C.M.A. (1991) *Forest Fertilisation in Britain*. Forestry Commission Bulletin 95. HMSO, London.

Yorke, D.M.B. (1998) *Continuous Cover Silviculture: An Alternative to Clear Felling. A Practical Guide to Transformation of Even-Aged Plantations to Uneven-Aged Continuous Cover*. Mark Yorke, Tyddyn Bach, Llanegryn, Tywyn, Gwynedd LL37 9UF.

Useful websites

The following sites are good starting places for UK forestry and woodland information:

www.british-trees.com – general information and useful links.

www.forestry.gov.uk – Forestry Commission website with details of grant schemes, market reports and research information.

www.woodnet.org.uk – information for woodland owners on timber sales and contractors.

Useful addresses and sources of advice

Arboricultural Association
Ampfield House, Ampfield, Nr Romsey, Hampshire SO51 9PA. Tel.: 01794 368717. www.trees.org.uk.

British Christmas Tree Growers' Association
18 Cluny Place, Edinburgh EH10 4RL. Tel.: 0131 447 0499. www.bctga.org.uk.

British Standards Institute
389 Chiswick High Road, London W4 4AL. Tel.: 020 8996 9000. www.bsi.org.uk.

Coed Cymru
Ladywell House, Newtown, Powys SY16 1RD. Tel.: 01686 26799. www.coedcymru.org.uk.

Countryside Agency
John Dower House, Crescent Place, Cheltenham GL50 3RA. Tel.: 01242 533311. www.countryside.gov.uk.

Countryside Council for Wales
Maes-y-Ffynnon, Penrhosgarnedd, Bangor, Gwynedd LL57 2DW. Tel.: 0845 1306229. www.ccw.gov.uk.

Deer Commission
Knowsley, 82 Fairfield Road, Inverness IV3 5LH. Tel.: 01463 231751. www.dcs.gov.uk.

Deer Initiative (England)
c/o Great Eastern House, Tenison Road, Cambridge CB1 2DU. Tel.: 01223 314546.

Deer Initiative (Wales)
c/o Victoria Terrace, Aberystwyth SY23 2DQ. Tel.: 01970 625866.

English Nature
Northminster House, Peterborough PE1 1UA. Tel.: 01733 455000. www.english-nature.org.uk.

Farming and Wildlife Advisory Group
National Agricultural Centre, Stoneleigh, Kenilworth, Warwickshire CV8 2RX. Tel.: 01246 696699. www.fwag.org.uk.

Forestry and Timber Association
5 Dublin Street Lane South, Edinburgh EH1 3PX. Tel.: 0131 5387111. www.forestryandtimber.org.

Forestry Commission (main headquarters)
231 Corstorphine Road, Edinburgh EH12 7AT. Tel.: 0131 334 0303. www.forestry.gov.uk.

For Forestry Commission local conservancy offices consult your phone directory or the Forestry Commission website.

Forest Research (publications, research, advice, plant testing; www.forestresearch.gov.uk)
Northern Research Station, Roslin, Midlothian EH25 9SY. Tel.: 0131 445 2176.
Forest Research, Alice Holt Lodge, Wrecclesham, Farnham, Surrey GU10 4LH. Tel.: 01420 22255.

Game Conservancy Trust
Fordingbridge, Hampshire SP6 1EF. Tel.: 01425 652381. www.game-conservancy.org.uk.

Institute of Chartered Foresters
7a St Colme Street, Edinburgh EH3 6AA. Tel.: 0131 225 2705. www.charteredforesters.org.

National Small Woods Association
The Cabins, Malehurst Estate, Minsterley, Shropshire SY5 0EQ. Tel.: 01743 792644. www.woodnet.org.uk/nswa/.

Royal Forestry Society of England, Wales and Northern Ireland
102 High Street, Tring, Hertfordshire HP23 4AH. Tel.: 01442 822028. www.rfs.org.uk.

Royal Scottish Forestry Society
Hagg-on-Esk, Canonbie, Dumfriesshire DG14 0XE. Tel.: 01387 371518. www.foresters.org/rsfs.

Scottish Natural Heritage
12 Hope Terrace, Edinburgh EH9 2AS. Tel.: 0131 446 2277. www.snh.org.uk.

Soil Survey and Land Research Centre (SSLRC) National Soil Resources Institute, Cranfield University, Silsoe, Bedfordshire, MK4S 4DT. Tel.: 01525 863000. www.silsoe.cranfield.ac.uk.

Tree Council
51 Catherine Place, London SW1E 6DY. Tel.: 020 7828 9928. www.treecouncil.org.uk.

Woodland Trust
Autumn Park, Dysart Road, Grantham, Lincolnshire NG31 6LL. Tel.: 01476 581111. www.woodland-trust.org.uk.

15

Game bird management

S. Tapper

Shooting and the rise of game management

The first useable firearms for sport were the muzzle-loading fowling pieces of the sixteenth century. They were mainly used against large targets like flocks of ducks or waders. The sporting limitations of these fowling pieces were imposed by the roughly made hail shot, which gave poor ballistics, and the match or wheel lock firing mechanism with its inherent delay between pulling the trigger and detonation of the charge. This meant that the quarry had to be close by and standing still. There was no question of shooting a flying bird – it had to be stalked. One approach was to hide behind a grazing 'stalking' horse to get within range. By the eighteenth century shooting gamebirds was becoming a popular activity and writers like Hawker extolled the pleasures of this new form of country pursuit. The method was to use pointing dogs such as setters or pointers to find birds and then spaniels to flush them in front of the guns for shooting. Shooting in this era was much closer to what we commonly refer to as rough shooting today. The skills of finding and stalking the quarry were paramount and shooting grouse in the Scottish Highlands was more akin to a hunting safari than the social event of the late nineteenth and twentieth centuries (Longrigg, 1977).

Modern driven game shooting owes much to the progressive development of firearms in the nineteenth century. London gun-makers like Manton drove forward a succession of innovations which reached the apogee of development in the classic English double-barrelled shotgun of the 1920s produced by gun-makers such as Purdey or Holland & Holland (Pollard, 1923). The cartridge by then had a percussion cap and the lead pellets were smooth, spherical and manufactured in shot towers. Combined with the breech loading gun and smokeless gun-powder, invented in the late nineteenth century, sportsmen were able to shoot flying birds in rapid succession, honing their skill of quick-fire marksmanship on fast, high and jinking gamebirds driven over their heads by a cordon of beaters. This evolution was in part driven by fashion and it is said much came from the continental influence of Prince Albert and then later from Edward Prince of Wales. This royal interest meant that game shooting became highly fashionable and the winter shooting weekends on country estates became a key feature of the upper-class Edwardian calendar (Ruffer, 1989). However, it was not just fashion that drove these events but other factors like access too. Of key influence was mid-nineteenth century development of the railway. Prior to the railway, much of the countryside (especially the uplands) tended to be too remote for city dwellers and country sports were the preserve of the local squires. With the railway, a London-based gentlemen could take his family to his Highland lodge for the 3-month shooting season with little or no discomfort (Eden, 1979).

All this required money which was generated in the first place by increasing prosperity of landowners who were able to secure higher rents from their tenants following the agricultural revolution of the eighteenth century, and later also by industry and commerce across the empire which enabled wealthy businessmen to buy up country properties (especially during periods of agricultural recession) for purely sporting purposes. This transfer of capital from city to country has been and continues to be an important component in the development of the countryside.

On the Edwardian sporting estate, the numbers of game shot were important not only as a measure of an individual's shooting prowess but also as a measure of the sporting value of an estate or grouse moor. Game-books in which the details of the day's shooting bag were recorded became as important as a company ledger, and the value of sporting properties depended more on the size of the average bag than on the acreage. Today these gamebooks coincidentally provide a valuable ecological record of changes in game and other species numbers (Tapper, 1992).

From this desire to shoot game in quantity came the need to breed them in abundance. This task was assigned to the gamekeeper. Although gamekeeping as a profession goes back at least to the eighteenth century, their main role was to discourage poaching – indeed by the early nineteenth century this had become practically a rural war. In 1823 a third of all prisoners in English gaols had been convicted under the Game Acts (Longrigg, 1977). However, later, in a post-enclosure countryside, with a farming system run on organic lines, the main issue for the late Victorian gamekeeper was the war on game predators or so-called vermin.

By the end of the Edwardian era, driven game shooting had reached its pinnacle of fashion. There were over 25 000 full-time gamekeepers in employment and their influence in some counties such as Norfolk and Hampshire must have been such that practically all of the rural land was subject to their regime (Figure 15.1). This not only helped to secure abundant stocks of wild grey partridge and pheasant but also contributed to, if not caused, the disappearance of some birds of prey like buzzard, red kite and hen harrier (Moore, 1956; Brown,

1976; Watson, 1977), as well as, probably, small carnivores like the polecat and pine marten (Langley and Yalden, 1977).

A more benign influence, however, was on the character of the landscape. Country estates were laid out with game coverts so that they held pheasants and these coverts were arranged to facilitate successive drives on pheasant shooting days (Figure 15.2). The imprint of this era remains on much of the landscape and, even though most of the country houses have gone, the woodland often remains. Likewise, even though there are now far fewer full-time gamekeepers (Table 15.1), some of the predatory species have yet to recover their former range.

Records in estate gamebooks allow us to compare shooting 100 years ago with today. Table 15.2 shows a comparison of the average bag per unit area of important game species from two eras of the twentieth century. Bearing in mind that the number of sporting properties today is much less than in 1900, so the total head of game shot in Britain will be lower, but the number of game shot (per unit area) on lowland estates today is comparable with the Edwardian era. On the upland grouse moors, however, gamebags are about half what they were.

While the Edwardians seemed to have shot a wider range of game, today game shooting is dominated by two game species: pheasant in the lowlands and red

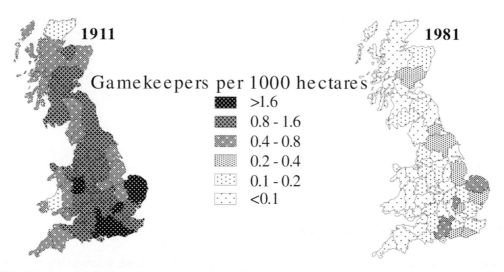

Figure 15.1 Number of full-time gamekeepers employed in Britain, expressed in per unit area in each county (data from national censuses; after Tapper, 1992).

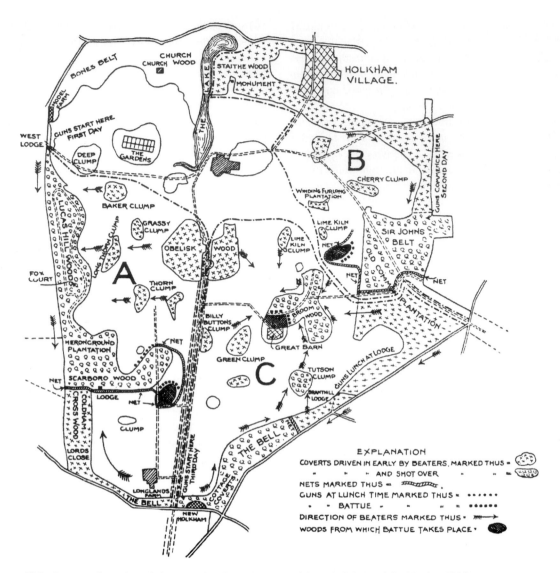

Figure 15.2   Layout of woods and pheasant shooting drives at Holkham Hall in Norfolk (Napier, 1903).

grouse in the uplands. The management of these two gamebirds influences large tracts of country and we consider this in more detail below.

The principles of game management

Before looking at details of management for red grouse and pheasants there are some important principles that underlie the management of any wildlife stock that is harvested. It applies not only to game, but also to the commercial exploitation of fish stocks and fur bearers. A hundred years ago much of this theory was unknown and management was driven empirically and by good stockmanship. However, even in the Edwardian era scientific analysis was brought to bear on grouse populations (Lovat, 1911).

The science of game management has run parallel with the evolving science of population ecology and the two were pioneered by scientists of the 1930s and

Table 15.1 Changes in the number of full-time game-keepers employed in Britain (data from the national censuses; see Tapper 1992).

Census year	Gamekeepers in England and Wales	Gamekeepers in Scotland
1871	12 429	
1881	12 633	4246
1891	13 814	4589
1901	16 677	5367
1911	17 148	5908
1921	9 367	3908
1931	10 706	4050
1951	3 776	615
1981[1]	1 790*	720*

[1] 1981 data is an estimate based on a 0.5% sample.

Table 15.2 Head of game shot during two eras of the twentieth century. Numbers are head of game shot per 100 ha of ground (from Tapper, 1992).

Species	Edwardians 1900–1909	Today 1980–1989
Lowland estates		
Pheasant	14.80	48.60
Grey partridge	27.50	5.50
Redleg partridge	0.82	9.60
Woodcock	0.50	2.10
Common snipe	0.45	0.14
Brown hare	15.80	9.00
Rabbit	36.50	12.40
Total	**96.37**	**87.34**
Upland estates		
Red grouse	58.59	26.23
Black grouse	0.59	0.03
Capercaillie	0.23	0.06
Mountain hare	2.03	1.08
Total	**61.44**	**27.40**

1940s like Charles Elton in Oxford (Elton, 1942) and Aldo Leopold in Wisconsin (Leopold, 1933). In population ecology animals are viewed as numbers subject to change brought about by births (natality) and deaths (mortality). Birth and death rates are looked at in relation to changes imposed by man; thus shooting increases mortality, but providing a better food supply may improve the birth rate.

Animal populations are often stable over time and have reproduction and death rates that exactly balance each other. This happens because natural resources (e.g. food supply) are limited and as these resources are used up, competition for them intensifies. This tends to increase mortality (*density-dependent mortality*) and reduce fecundity (*density-dependent natality*). This density dependence maintains the population around its equilibrium level. Density dependence is a key concept in population ecology.

If a stable population is subjected to regular shooting its numbers will be reduced, but this reduction will free-up natural resources which in turn either lower the natural mortality rate or increase the birth rate. Thus a regularly shot population can also be stable, but at a lower level than otherwise would be the case if it was not shot.

As the proportion of the population killed by shooting is increased, the level at which the population stabilizes becomes lower. With a lower population the number of breeding individuals is reduced – even though they may be breeding at a faster rate. Thus there are two opposing tendencies in operation, a shrinking

breeding stock and rising productivity. Together these determine the number that can be killed sustainably at a given level of shooting.

Clearly the maximum number that can be shot each year will be achieved when the largest number are breeding at the fastest possible rate. This is termed the *maximum sustainable yield* and is most elegantly presented as one of the points on the curve in Figure 15.3. Because of vagaries in ecological systems, culling and harvesting strategies are usually set at a rate somewhat lower than the maximum sustainable yield – this is the optimum sustainable yield.

When populations are subject to game management some additional factors also apply. Game managers (gamekeepers and others) will try both to enhance the productivity, by providing better breeding habitat and more food, and to reduce the natural mortality due to predators and disease (Figure 15.4). This works best with the resident game animals since a gamekeeper can protect his local population year round and because the economic benefits of a higher bag are also local.

An additional concept, which is built on the above, is the idea that steps taken to improve the conditions for game species also benefit a range of other animals and plants that have similar requirements. In this way game management can aim to support wildlife conservation overall. It is a concept referred to as *conserva-*

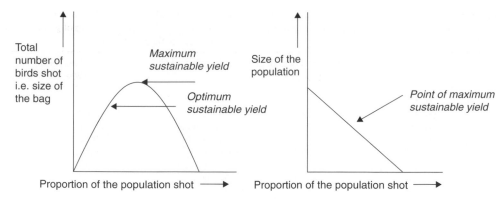

Figure 15.3 **Left** Graph showing relationship between the average annual bag and proportion of the population that is shot each year. On the left side of the curve, shooting a higher proportion of the population produces a higher bag. However, on the right side, beyond the point of maximum sustainable yield, increasing the proportion of the population shot leads to diminishing returns as the breeding stock is reduced further. Note, however, that shooting beyond the maximum sustainable yield leads to low bags and low stocks – it does not necessarily lead to declining stocks, although it may. **Right** Graph of the same population but showing the relationship of the breeding stock to the proportion that is shot. The relationship is not always linear and the maximum sustainable yield not always at the mid point. A population can remain stable at all points along this line.

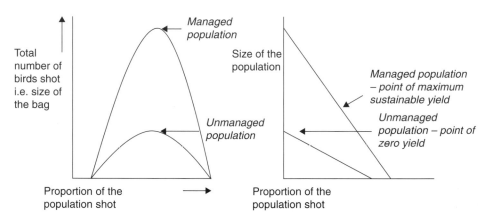

Figure 15.4 **Left** Graph showing how the maximum sustainable yield of gamebirds differs between managed and unmanaged populations. Increased productivity and reduced (non-shooting) mortality mean that yields of managed populations are much higher than unmanaged ones. **Right** Further, in many instances, the managed bird populations are maintained at much higher densities – even after shooting – than are unmanaged ones with no shooting at all.

tion through wise use. Some examples are heather moor being retained because of grouse shooting, grouse management benefiting upland waders, and woodland that is managed for pheasants being more diverse than woodland managed for forestry. We consider this in more detail later.

However, game management can have a negative effect on biodiversity. Most notable is the impact on predators, especially rare carnivores and birds of prey.

Thus shooting and game management, to qualify as sustainable in the modern context, must resolve this in a way that ensures the conservation of these species too.

Grouse management

The red grouse (*Lagopus lagopus scoticus*) is a gamebird unique to the British Isles and is a sub-species of

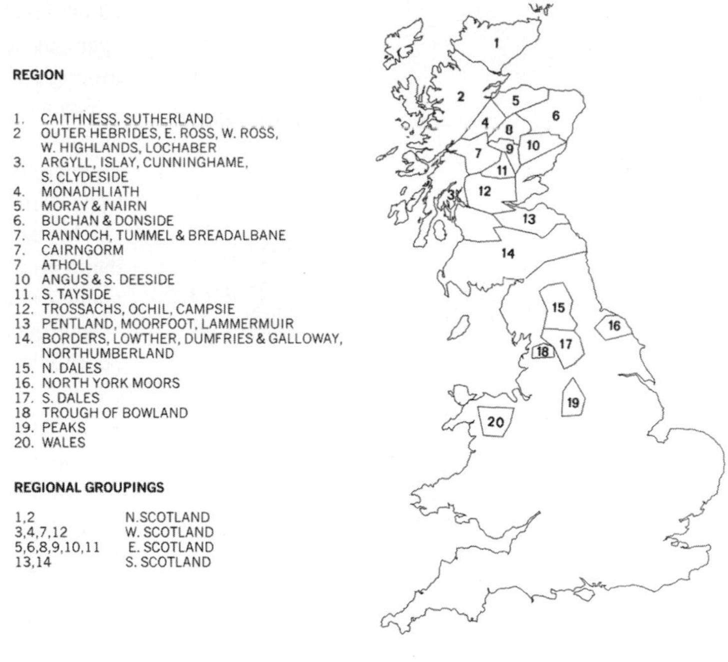

REGION

1. CAITHNESS, SUTHERLAND
2. OUTER HEBRIDES, E. ROSS, W. ROSS,
   W. HIGHLANDS, LOCHABER
3. ARGYLL, ISLAY, CUNNINGHAME,
   S. CLYDESIDE
4. MONADHLIATH
5. MORAY & NAIRN
6. BUCHAN & DONSIDE
7. RANNOCH, TUMMEL & BREADALBANE
7. CAIRNGORM
7. ATHOLL
10. ANGUS & S. DEESIDE
11. S. TAYSIDE
12. TROSSACHS, OCHIL, CAMPSIE
13. PENTLAND, MOORFOOT, LAMMERMUIR
14. BORDERS, LOWTHER, DUMFRIES & GALLOWAY,
    NORTHUMBERLAND
15. N. DALES
16. NORTH YORK MOORS
17. S. DALES
18. TROUGH OF BOWLAND
19. PEAKS
20. WALES

REGIONAL GROUPINGS

1,2	N.SCOTLAND
3,4,7,12	W. SCOTLAND
5,6,8,9,10,11	E. SCOTLAND
13,14	S. SCOTLAND

Figure 15.5 Main grouse shooting areas of Britain (Hudson, 1992).

the European willow grouse. It frequents open moorland where it feeds mainly on young heather shoots and other plants.

Grouse moors, where red grouse are managed for shooting, are bare, treeless, upland areas dominated by extensive coverage of heather (*Calluna vulgaris*), mixed to a lesser degree with other dwarf shrubs and grasses. The most obvious landscape feature of many grouse moors is the checkerboard appearance of the moor which results from the rotational patch burning of the heather. There are some 459 grouse moors in Britain, with the principal ones in the Pennines, the North York Moors, the southern uplands and the Scottish Highlands. Grouse shooting has mostly disappeared in Wales and South West England. Figure 15.5 shows the main areas of grouse production and Table 15.3 shows the estimated extent of these areas and how the average annual bag of 450 000 grouse was distributed between these regions in the 1980s. Although grouse moors are more common and more extensive in Scotland, English moors on the Pennines are more productive and have densities of grouse typically some three times as high as those in Scotland (Figure 15.6).

Grouse shooting is not a profitable business and most moors run at a net loss underwritten by their owners. Some moors will let days shooting to recoup some of the costs. In Scotland in 2000, moors that let some of

Table 15.3 Estimated extent of managed grouse moor (ha) in Britain by region with the estimated average annual bag during the 1980s (from Hudson, 1992).

Region	Grouse moor area	Annual bag
N. Highlands	525 800	15 000
N.W. Highlands	938 300	10 000
Argyll	448 300	1 500
Monadhialth	290 400	10 000
Moray and Nairn	96 700	12 000
Buchan and Donside	63 600	15 000
Rannoch	133 400	9 000
Cairngorm	119 800	25 000
Atholl	66 700	10 000
Angus and S. Deeside	192 500	40 000
S. Tayside	198 000	40 000
Trossachs	32 300	3 000
Highland total	**993 300**	**190 500**
Pentlands, Moorfoots, and Lammermuirs	86 300	35 000
Borders	318 300	25 000
S. Uplands total	**404 600**	**60 000**
Scotland total	**1 397 900**	**250 500**
N. Dales	112 000	90 000
N. York Moors	47 700	25 000
S. Dales	50 300	55 000
Bowland	17 400	9 000
Peak District	43 000	20 000
N. England	**270 400**	**199 000**
Wales	**8 000**	**1 000**
Britain	**16 763**	**450 000**

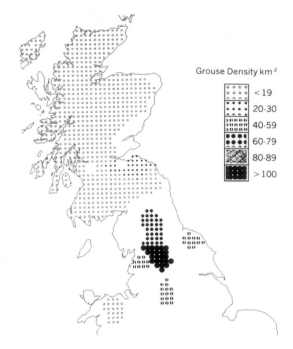

Grouse Density km²

° ° ° °	< 19
• • • •	20·30
(grid pattern)	40·59
● ● ● ●	60·79
(hatched)	80·89
■ ■ ■ ■	> 100

Figure 15.6 Average numbers of grouse shot (density/km²) across grouse moors in Britain (Hudson, 1992).

their shooting expended £49 000 per annum, with an average revenue of £35 000. Expenses on moors that had no let days were lower, but their net costs were higher. In Scotland, 459 grouse shooting estates employed some 630 full-time jobs and paid about £9.3 million in wages (Fraser of Allander, 2001). In England and Wales, the Moorland Association (www.moorlandassociation.org.uk) estimates that its members employ 279 full-time jobs and spend £4.4 million in wages.

As well as the disappearance of stocks of grouse on Exmoor and Dartmoor, historically grouse stocks have declined substantially in Scotland over the last century, but especially since the mid-1970s. Grouse shooting has virtually stopped in Wales. Only in northern England have grouse numbers been maintained (Figure 15.7). Where grouse shooting stops, grouse stocks not only decline but may disappear entirely. This is evident in Ireland, Wales and the west of England (Table 15.4).

Red grouse are a territorial species, with males establishing their first territories in the autumn (September to October). They display by showing inflated red wattles above their eyes, and giving song flights and dawn calls. Females pair up with these males. Over winter, territorial activity subsides but resumes again to

a high level in spring (March to May). It then subsides again during the summer months while females are nesting and rearing a brood.

Female grouse produce a single clutch of 6–11 eggs around the end of April laid in tall heather, and following incubation of 30 days the young brood is taken by the female to forage on the ground. Initially young chicks feed on insects such as craneflies especially on damp areas of moor (Park *et al.*, 2001), but later after about 10 days they feed mainly on plant material.

The managing of a grouse population for shooting depends mainly on the following:

- the effectiveness of the predator control. Predators take grouse at all times of the year, but they are particularly vulnerable during the summer breeding months;
- the degree to which the habitat is maintained through correct burning and grazing;
- the degree to which stocks are kept free from diseases like louping ill and parasites, especially the nematode worm *Trichostrongylus tenuis*.

Predator control

Gamekeepers aim to keep grouse moors free of grouse predators throughout most of the year. Foxes are important predators of grouse in all seasons but especially during the grouse breeding period when they kill nesting hen birds, thus destroying the potential for a whole brood. Gamekeepers control foxes by: (1) rifle shooting – often at night using a spot-lamp and rabbit call to attract them within range; (2) using snares – these catch foxes alive which are then killed by the gamekeeper when the snares are checked. Snares are often set in rushy areas around a buried carcase bait or midden; and (3) the use of dogs. Terriers are used to flush foxes from dens and earths. Previously in some districts packs of hounds, especially in combination with shooting with shotguns (gun-packs), used to contribute to fox control but this is now forbidden under the Hunting Act 2004.

Carrion crows systematically search out and destroy the nests of many birds in the spring and a gamekeeper will ensure that none is nesting on his moor. Apart from shooting, he will mainly use a combination of large cage traps and small movable Larsen cage traps for catching crows.

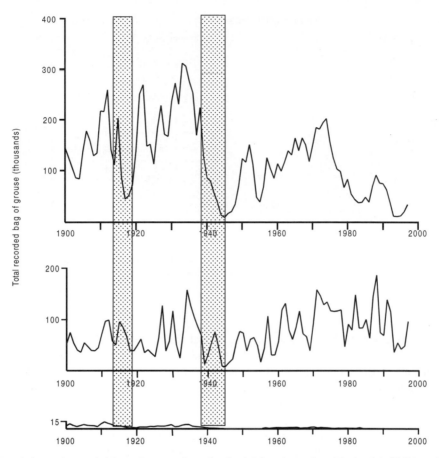

Figure 15.7 Trends in total recorded bag of grouse from Scotland (**above**), northern England (**middle**) and Wales (**below**) since the start of the twentieth century; war years are indicated by bars. (Data from the National Gamebag Census [(Game Conservancy Trust, 2003).]

Table 15.4 The decrease in numbers of 10 km² where grouse were recorded present from 1988–1991 compared to 1968–1972 (from Galbraith and Tucker, 2000).

Country	Percentage loss of range between surveys
Ireland	−66
Wales	−27
South West England	−46
North of England and Scotland	−11

Stoats can be important predators of grouse, especially when they have young broods, and spring traps set in tunnels are the main means of their control.

Birds of prey, particularly the peregrine falcon and the hen harrier, also take grouse and, when the latter breed in high numbers, they make take sufficient to make grouse management impossible (Figure 15.8). Given that birds of prey are protected, the post-1970s, recovery of populations of these larger raptors is of increasing concern to grouse moor owners. Research conducted as part of the Joint Raptor Study (Redpath and Thirgood, 1997) has demonstrated the significance of this raptor predation. On one moor (Langholm) breeding numbers of hen harriers built up rapidly following complete protection, and drove down grouse stocks to such an extent that all shooting was stopped.

Heather management

The flora of most moorland has been derived from the ground vegetation left in the uplands after forest clear-

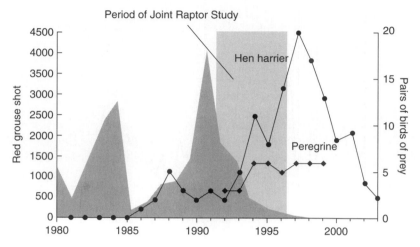

Figure 15.8 Changes in numbers of breeding hen harriers and peregrine falcons in relation to grouse shooting bags on Langholm moor before and after the Joint Raptor Study (Game Conservancy Trust, 2003).

ance was followed by extensive burning and grazing. Parts of the uplands were also cultivated in Neolithic times. Thus managing these areas for grouse is recent compared to the thousands of years during which the principal land use would have been livestock.

Today, grouse moors remain multi-purpose with upland farmers and gamekeepers both having a hand in managing the moors. The quality of the heather sward and grouse production depend on an appropriate grazing regime, and a good pattern of burning which produces a patchwork of different aged heather – often classed as pioneer, building, mature and degenerate heather.

In Scotland, grazing is by sheep, cattle and deer, but in England and Wales, it is mainly sheep. Since the late 1960s overgrazing has been a significant problem because English sheep numbers have more that doubled over the last 50 years and more of this stock has become concentrated in the uplands. The management of these sheep flocks, with less emphasis on shepherding and more concentrated winter feeding (fothering) along the moorland edge adjacent to roads and tracks, has led to heather being eaten out in many places and replaced with coarse, less palatable grasses. Heather cover has disappeared from some hillsides and retreated up the hill on others. Overgrazing can be recognized along the lower edge of the moor as heather plants become small and stunted until they disappear beneath coarse grasses

and the heather/grass borderline retreats uphill. Over the moor itself young heather plants which should be erect become prostrate and nibbled to the ground, and older plants take on a topiary or drumstick effect, rather than being bushy. On typical dry heather moor, if in excess of 40% of the growing tips have been removed over winter, the sward has been overgrazed (Hudson and Newborn, 1995).

Heather is burnt for two main reasons: first, to provide a varied age structure of plants across the moor so that the grouse have ready access to relatively long heather for nesting adjacent to short heather for feeding. Second, the repeated burning removes dead wood and litter, and releases nutrient, locked up in this, back into the soil. The burning season is 1 October to 15 April – or later with restrictions. On Sites of Special Scientific Interest (SSSIs) additional restrictions may be imposed. Most burning takes place in the early autumn and in the spring when conditions are dry. It is usual to burn heather on a 6- to 10-year cycle depending on local conditions – but some areas of slow growth may be left much longer. Steep slopes and boggy areas are avoided, because of erosion risks in the first case and damage to the flora in the second. The usual approach is for a three-man team to burn a strip about 20–30 m wide across the side of a slope. The aim is to have a cool fire that removes the leaf litter and heather, but does not burn into the peat or kill the plant root-stock. After a burn new shoots will appear from the root-stock and new heather plants germinate from seedlings – fire encourages heather seed to germinate.

Disease and parasitism

Unlike some gamebirds, red grouse are not subject to a wide variety of diseases and parasites, but two important ones have a significant impact on their numbers. The first of these is the gut parasite *Trichostrongylus tenuis*. This is a small nematode worm that can infest the blind gut (caeca) of red grouse. The lifecycle of the worm is a simple one with no intermediate hosts. The nematode lays eggs whilst in the grouse and these pass out with the grouse faeces. These eggs then hatch and, after a period of quiescence, the microscopic nematode larvae make their way up heather plants until they end up on the tips of the shoots. Grouse then ingest these larvae as they feed on the heather. Most grouse have some of these parasites, but some birds can acquire worm burdens of over 10 000 in their guts. When grouse become heavily infested it affects their survival, the number of eggs that they lay and subsequent survival of their brood. The average number of worms carried by grouse varies between years and is partially affected by the weather which affects the survival of the larvae during the main periods when they infect grouse (late summer and late winter). In addition, the average worm burden in grouse is driven by the number of grouse around the previous year and the number of worms that they were carrying. Thus worms tend to build up in a grouse stock over several years to a point at which grouse survival and productivity is poor. This causes the grouse stock to decline or crash. These intermittent declines in stock cause many populations to cycle. This is very evident in the bag records from some moors (Figure 15.9), where the population crashes every three or four years. The length of cycle seems to depend partly on climate, from southern Pennine moors peaking on average every 4 years to the northern highlands where numbers peak roughly every 8 years. Nevertheless, the length of individual cycles is variable and so predicting peak and trough years is not possible. This has a bearing on the management of grouse shooting and imposes considerable difficulty for a landowner especially when he is relying on letting some of the shooting to recoup the costs of running the moor.

To combat this, gamekeepers first try to ensure that not too large a grouse stock is left after the shooting season which may host a lot of parasites. Second, in recent years they have started using medications to reduce the worm burdens of the grouse stock. The keeper may monitor the worm burdens of shot grouse during the shooting season by doing routine worm counts and then either catching up grouse and dosing them with an anthelmintic in the winter or leaving out medicated grit for the birds in late winter or early spring. Direct dosing is accomplished by catching birds individually at night with a spotlight and net and then squirting a dose of anthelmintic down their throats. Medicated grit is a quartz grit that has been coated in medication and that is added to the small piles of grit that are left out for grouse on many moors. Trials of both these techniques have been shown to be effective under experimental conditions.

The other health problem that some grouse stocks are subject to is louping ill. This is a tick-borne viral infection which kills young grouse especially. The presence of louping ill on a moor is maintained by upland herbivores such as red deer and sheep which both sustain the tick population and amplify the virus. Small mammals and birds which may also host ticks do not become viraemic and therefore do not sustain the disease. Control of the disease therefore depends mainly on the vaccination and dipping of hill sheep where they are the primary reservoir, and reducing habitat conditions such as bracken and deep litter in areas where red deer are the reservoir.

Pheasant management

The pheasant (*Phasianus colchicus*) is the most popular gamebird in Britain. It is easy to manage, and as well suited to a small farm shoot as it is to a prestigious country estate. Although wild pheasants require the same care with habitat and protection from predators as other wild gamebirds, their numbers can be supplemented by hand-reared birds released onto the shoot prior to the shooting season.

The pheasant is not a native bird but is Asian. In the Caucasus between the Caspian and Black Seas we find pheasants conforming to the 'old English pattern', with green heads and a brown to purplish body plumage. Further east in central Asia is the 'mongoloid' group with birds that have white neck collars, paler wings and

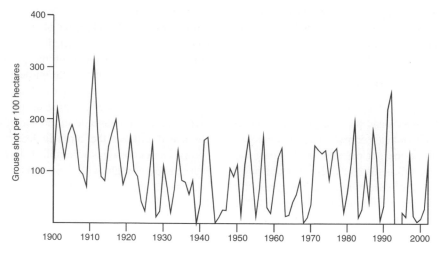

Figure 15.9   Annual changes in numbers of grouse shot on a Pennine moor, showing the cyclic nature of grouse stocks. [Data from the National Gamebag Census (Game Conservancy Trust, 2003).]

a greenish tinge to their body plumage. Even further east in China we find a large group of sub-species that tend to have grey backs. There is also an additional species, *P. versicolor*, which is a dark green pheasant, found in Japan.

It is not entirely clear when the first pheasants arrived in Britain, but the fact that pheasants are depicted in Romano-British mosaics suggests to some that they were introduced then. However, this is by no means certain and the Norman period seems more probable (Hill and Robertson, 1988). Most of these early pheasants were probably derived from Mediterranean stock which in turn had come from birds east of the Black Sea. Only later during the eighteenth and nineteenth centuries were other races like the Chinese and Japanese green introduced.

In Britain the pheasant is primarily a bird of the woodland edge even though it can be found scattered across arable farmland. On most areas where there is pheasant shooting, measures are taken to improve the breeding success of the wild birds, but the annual production is augmented by hand-reared stock.

Managing wild pheasants

The first management principle is to provide woodland-edge habitat. Pheasants do not tend to venture deep into woods (Figure 15.10) so large blocks should be inter-

sected by wide rides – 70 m in mature woods. For new woods, planting in a long spinney gives more edge and is preferable to a square block. The amount of woodland edge is directly related to the numbers of cock pheasants that establish territories each spring. The maximum density of breeding pheasant territories is achieved where woodland and farmland are mixed in a 30:70 ratio.

Pheasants are polygamous birds and in March and April some males dominate and take up large territories which attract a harem of up to a dozen hen birds. The males have large red wattles that can cover the entire face when inflated; they display by crowing with a drumming wing-beat and a 'kok-kok' call. In May hens lay a clutch of normally between 8 and 13 eggs. Woodland, rough grass, cereal and hedgerow are typical nesting habitats. Incubation lasts 25 days and the hen spends 23 hours a day on the nest. In the event of a nesting failure (predation, disturbance or weather) the hen will remate and lay a repeat clutch. Later, repeat clutches are more often found in growing cereal crops.

After hatching the hen takes the brood in search of food along the margins of crops, hedgerow and woodland. Gamebird chicks depend on insects for proper development (Table 15.5). When insects are in short supply because of cold weather or the application of pesticides, chick mortality increases as the hen forages more widely in search of better feeding areas. Under

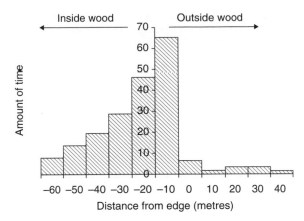

Figure 15.10 Amount of time pheasants spend along woodland edges. Time estimated by radio-tracking birds (after Robertson, 1992).

Table 15.5 Typical diet of pheasant chicks in their first 10 days. Under poor feeding conditions the proportion of insects in the diet will be smaller (after the Game Conservancy Trust, 1996).

Food	Proportion in diet (%)
Aphids	11
Hoppers	11
True bugs	22
Sawfly larvae	23
Caterpillars	12
Other	21

modern farm conditions, where pesticides are used routinely, the absence of weeds in crops leads to a low abundance of invertebrates and chick survival is often poor (between 20 and 40%) and certainly lower than would have been the case under old-fashioned farming systems. This has been well documented in grey partridges and loss of insect biodiversity has been the main cause of the disappearance of this species from many arable districts. After 10 weeks pheasant chicks become independent of the hen.

Productivity of wild pheasants can be increased by the control of hen and nest predators in the spring and early summer, and by improving the availability of farmland insects by modifying the crop spraying regimes along field margins. These practices have been tested with field experiments on wild grey partridge populations and the results are applicable to wild pheasants. Table 15.6 shows the results of a predator control experiment on two partridge populations subject to alternate predator control/no predator control regimes (Tapper *et al.*, 1996). Table 15.7 shows the improvements on grey partridge chick survival that can be achieved by reduced spray regimes on field margins (Rands, 1985).

Predator control for farmland gamebirds needs to be concentrated in a spring effort which coincides with the nesting period. Foxes and carrion crows in particular cannot be allowed to breed or hunt over areas where wild pheasants are being protected. Fox control is mainly done by rifle shooting on open areas and by snaring in areas of cover. Cage traps, especially Larsen traps, are the principal method of corvid control.

Reduced-spray field margins – conservation headlands (grant aided under some agri-environment schemes) – involve not spraying the outer 6 m of wheat and barley with herbicides and insecticides during the spring and early summer. The effect is that the crop in this 6-m zone has a weed ground flora which supports high insect numbers, improving chick survival.

Modern farmland, after harvest and cultivation, provides little food for gamebirds and it is usual to provide pheasants with wheat from feed hoppers. The usual design is a small (25-litre) metal drum attached to a post with a slot underneath through which a pheasant can peck out wheat. Since pheasants are territorial most of the time, feeders need to be scattered around all parts of the farm.

By using these methods, modest numbers of wild pheasants can be produced under modern farming regimes. Figure 15.11 shows the autumn numbers of wild pheasants produced from the Game Conservancy Trust's Allerton Project farm in Leicestershire over a 9-year period. This stock produced an annual bag of between 120 and 220 birds per year for about 45 man days of shooting annually.

Hand-reared pheasants

Largely because of poor wild production, pheasant shooting has come to rely mostly on birds that are hand reared and released. Pheasants are relatively easy to rear and post-war improvements in the poultry industry were simply applied to pheasants. Pheasants and partridges were hand reared by Edwardian gamekeep-

Table 15.6 Summary of grey partridge production from two experimental areas on Salisbury Plain: Collingbourne was subject to predator control between 1985 and 1987 and Milston between 1988 and 1990. Figures are averages over 3 years. Brood sizes tend to reflect chick survival, but the number of young per breeding pair also reflects losses of nests (data from Tapper *et al.*, 1996).

	Mean brood size	Young per pair	Autumn stock per 100 ha	Subsequent pairs per 100 ha
Collingbourne				
With predator control	6.32	3.9	59.16	10.46
Without predator control	5.35	1.57	25.06	5.08
Milston				
With predator control	7.04	4.95	59.34	9.74
Without predator control	4.73	3.98	32.31	7.99

Table 15.7 Effect on grey partridge chick survival of a reduced crop spraying regime on cereal field edges (headlands). The cereal fields in three areas (beats X, Y and Z) of a large mixed farm divided into those with normally sprayed fields and those with a reduced spraying regime. Broods sizes were significantly larger on the fields with reduced spray headland (from Rands 1985).

Area	Broods on field with normally sprayed headland (number of broods in parentheses)	Broods on field with reduced spray headland (number of broods in parentheses)
Beat X	2.35 (20)	5.14 (14)
Beat Y	1.75 (4)	10.33 (3)
Beat Z	2.27 (15)	6.83 (12)

ers, but it was very labour intensive as pheasant eggs had to be incubated under bantam hens and a protein-rich diet for the chicks had to be prepared by hand. In modern systems, eggs from pheasant laying pens are put into incubators. After hatching, chicks are moved to brooder houses (Figure 15.12), and then pens on a rearing field before being released onto the shoot. Ready-prepared pelleted feeds are available for all stages of growth and a single-handed gamekeeper can raise up to 10 000 birds with the right equipment. However, many shoots now prefer not to rear their own birds but buy them in as young birds (poults) ready for release onto the shoot.

Releasing pheasants for shooting is a process of acclimatization that begins in July when birds are put into a release pen usually situated at an appropriate central woodland site on the shoot. The pen is open topped so that the birds, once fully fledged, can fly out and can walk back in through predator-proof ground level pop holes through the netting. The pen, made of wire netting at least 2-m high which is buried along the bottom edge and supported by stout posts, often sur-

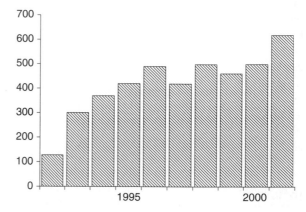

Figure 15.11 Autumn numbers of wild pheasants on a 325-ha arable farm in Leicestershire, following systematic game management in 1993 (Stoate and Leake, 2002).

rounds a whole or part of a wood, and should contain shrubs for roosting as well as feeders and watering points for the birds. Release pens should be spacious enough to allow for 16 m² per bird released.

Near the release pen and at other strategic points on

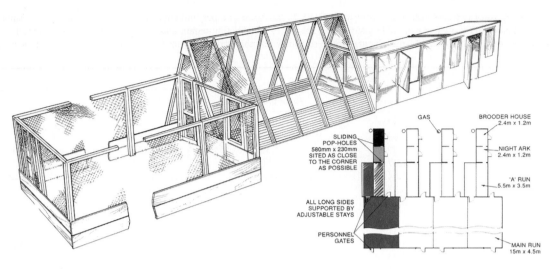

Figure 15.12   Layout of a brooder house system for 1000 pheasants (reproduced from Game Conservancy Ltd., 1994).

the shoot, game crops like kale, millet, maize or quinoa are grown. These crops provide cover and food for the pheasants and will be in places where the birds need to be concentrated and can be flushed on shooting days.

The cost of pheasant shooting has remained low (Table 15.8) over the last two decades. As a result it has increased in popularity but become dependent on hand-reared birds (Figure 15.13). Accompanying this has been a tendency on some shoots to increase the size of the bag and run the enterprise on commercial lines. This can lead to intensive releasing programmes with a high frequency of winter shooting. Increasing concern about excessive releasing and shooting of pheasants has led to the drawing up of *The Code of Good Shooting Practice* to encourage higher sporting, animal welfare and conservation standards (Anonymous, 2003).

Influence on countryside biodiversity

There are three main areas where game management impinges on conservation: (1) the retention or creation of habitat that would otherwise have been used solely for intensive forestry or farming; (2) supplementary feeding of game which also supports other species; and (3) predator control which reduces the abundance and the diversity of predators, but may also serve to protect vulnerable species of conservation value.

Habitat retention and creation

In the later half of the twentieth century, farming and forestry were subsidized through government research, subsidy and tax breaks on the basis that they were strategically important to an island nation which, in war time, could be cut off from world trade. Conservation interests (government agencies as well as privately funded bodies) have tried to mitigate this by designation (nature reserves and SSSIs), as well as through habitat creation – especially lately with agri-environment schemes.

Game management has certainly moderated the influence of subsidized farming and forestry even though it has not been powerful enough to halt or reverse trends in loss of habitat. One habitat of international importance is heathland and this is now subject to a national biodiversity Habitat Action Plan. Heathland is a secondary climax vegetation derived from forest clearance, fire and grazing on nutrient-poor soils. It was once widespread in Europe, but in most countries its extent has contracted. Only in Britain do significant areas remain, mainly as grouse moors (Gimingham, 1981). Much of this upland has now been designated as SSSIs, but grouse moors to a large extent were responsible for protecting this habitat before designation. Robertson *et al.* (2001) demonstrated that in Scotland heather moorland has been retained to a much greater extent where grouse management has

Table 15.8 Cost of running a pheasant shoot in 2002/2003 (expressed in £ per bird shot). Owners' shoots are where the shoot is run on a landowner's own ground and rented shoots are where the shooting tenant and the landowner are different. Premium costs are the costs of the top 25% in each category. (Previously unpublished data from Game Conservancy Ltd. Advisory Services.)

Inputs	Owners' shoots		Rented shoots	
	Average costs	Premium costs	Average costs	Premium costs
Rent	0.52	0	1.43	2.91
Rates	0.09	0	0.13	0
Keepering	7.99	1.33	4.72	1.92
Equipment	1.27	0.58	1.12	1.07
Restocking	3.64	4.08	5.28	5.05
Post-release feed	3.33	2.20	2.53	1.87
Game crops	0.34	0.22	1.94	2.64
Beaters	1.47	1.11	1.74	1.40
Total costs	**18.41**	**9.28**	**17.53**	**15.18**
Receipts	**0.98**	**0.35**	**0.29**	**0.25**
Net cost	**17.43**	**8.93**	**17.24**	**14.93**

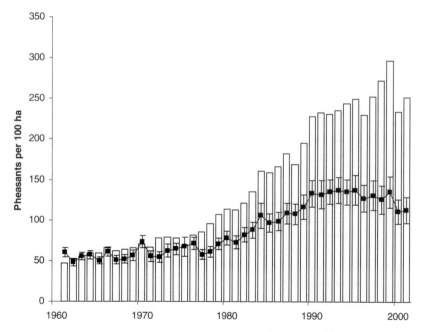

Figure 15.13 Increase in bag of pheasants. [Data from the National Gamebag Census (Game Conservancy Trust, 2003).]

continued, than on moors where grouse shooting has ceased (Figure 15.14).

In lowland districts habitat retention and creation for game has mainly been through the creation of small woods for pheasants. This has been supported by government-funded schemes which partly underwrite the costs of planting and managing small woods. The motivation of landowners taking up these grants is often difficult to assess, but one study (Cox *et al.*, 1996) found that farmers who released pheasants for shooting were three times as likely to plant new woods, four times as likely to plant shrubs, seven times as likely to manage rides and nine times as likely to coppice woodland as those farmers not interested in pheasants.

Improving woods for pheasants involves increasing the amount of edge by planting in lengths rather than blocks, cutting wide rides within the woods and sky-lighting (removing the occasional canopy tree) to allow pheasants to fly out on shooting days. Robertson *et al.* (1988) found that woodlands managed for pheasants had significantly more species and numbers of butter-flies than either commercial forest blocks or closed-canopy broad-leaved woods (Figure 15.15).

Food supply

It is customary to provide supplementary food for both pheasants and partridges, with a network of feed grain hoppers distributed across the shoot, as well as patches

of special game crops planted for both cover and seed production. Feeders for pheasants provide spilt grain for seed-eating passerines like yellow hammers in late winter when other natural food supplies are low (Stoate and Leake, 2002), and game crops and wild-bird cover crops, grown under set-aside rules, are used by a wide variety of species (Stoate *et al.*, 2003). A mix of plants within a game crop provides the best conditions for a range of songbirds. Kale, for example, with its large leaves, gives a moist humid environment, making ideal cover and feeding habitat for blackbird, song thrush and dunnock, whereas a good seed-bearing plant like quinoa provides food for goldfinch, redpoll, chaffinch and tree sparrow (Stoate and Leake, 2002).

Predator control

Predator control is certainly the most contentious area of game management and, in Britain, as in North America, has been subject to large swings in scientific thought and public policy. Up to the middle of the twen-tieth century the control of certain predators was con-sidered as axiomatic both for the protection of livestock and for the production of game. It was not an invention of nineteenth century gamekeeping, as the many have claimed, but actually a development from much older practices, as many churchwarden accounts reveal. Predator control was also early conservation policy in North America (Dunlap, 1988; Budiansky, 1995), with wolf and bear control being a notable feature of the man-agement of Yellowstone National Park in its early days.

However, some leading ecologists of the 1960s challenged the idea that predation was important in

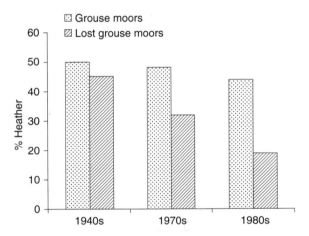

Figure 15.14 Retention of heather on Scottish grouse moors compared to areas where grouse shooting has been abandoned (data from Robertson *et al.*, 2001).

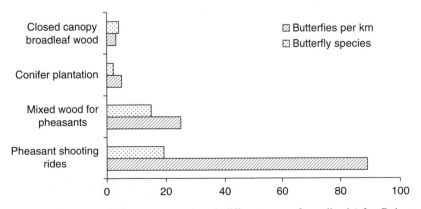

Figure 15.15 Butterfly numbers counted along transects though different types of woodland (after Robertson *et al.*, 1988).

regulating animal numbers (Jenkins *et al.*, 1964), thus suggesting that predator control by gamekeepers was of little or no consequence. The basis of this idea is *compensatory mortality*. This proposes that prey tend to produce more young each year than can be sustained by the habitat so if they are not eaten by predators they would die anyway. However, field experiments with predator control have shown that this is not always the case and that predator control significantly increases both autumn and spring stocks of wild gamebirds (Marcstrom *et al.*, 1988; Tapper *et al.*, 1996).

The effect of predator control programmes on non-game species is less well understood and experimental studies currently underway have yet to be completed. However, it is known that non-native mink have largely been responsible for the disappearance of the water vole (Strachan and Jefferies, 1990), and the introduction of hedgehogs into the Hebrides is responsible for the significant reduction in wader numbers there. It does appear that ground-nesting waders seem to be a group that may substantially benefit from gamekeepers' predator control programmes. Tharme *et al.* (2001) in a survey of upland moors found that some species of wader were up to five times more abundant on grouse moors than on other moorland and this was mostly attributed to predator control. The effect was not universal, and snipe, for instance, were equally common on both types of ground (Figure 15.16). On farmland, crow and magpie control to help pheasants may also improve the nesting success of some species such as chaffinch, blackbird and song thrush, which have poor nesting success, compared to others like yellowhammer and whitethroat which have better nest survival (Table 15.9).

Taken as a whole, improving the habitat, improving the food supply and reducing predation pressure certainly appear to bring about a substantial improvement in breeding numbers of farmland songbirds which have been declining nationally for several decades (Figure 15.17).

Given that Victorian gamekeeping contributed to the disappearance of some predators, already referred to, finding a sensible strategy for predation control is perhaps the most significant challenge for game management and indeed for conservation (Reynolds and Tapper, 1996). Accommodating the presence of a diverse array of predators is certainly a requirement if game management is to qualify as *conservation through wise use*, and it is likely that an acceptance of lower game bags and predator management instead of predator control is likely to be the way forward if the latter is to be accepted by conservation agencies.

References

Anonymous (2003) *The Code of Good Shooting Practice.* BASC, Marford Mill, Rossett.

Brown, L. (1976) *British Birds of Prey.* New Naturalist Series. Collins, London.

Budiansky, S. (1995) *Nature's Keepers. The New Science of Nature Management.* Weidenfeld and Nicholson, London.

Cox, G., Watkins, C. and Winter, M. (1996) *Game Management in England: Implications for Public Access, the Rural Economy and the Environment.* Countryside and Commu-

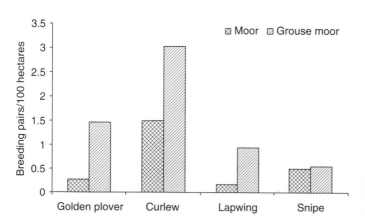

Figure 15.16 Numbers of waders breeding on upland moors, areas with and without grouse management compared (data from Tharme *et al.*, 2001).

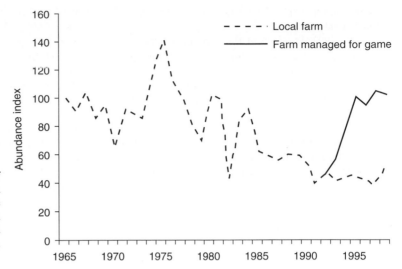

Figure 15.17 Changes in numbers of nationally declining songbirds on a Leicestershire farm managed for game compared with another nearby farm where songbirds have been monitored over the long term (Stoate and Leake, 2002).

Table 15.9 Probability of daily nest failure for six species of passerine birds on farmland (from Stoate and Leake, 2002).

Species	Probability of nest failure per day
Chaffinch	0.08
Blackbird	0.08
Song thrush	0.06
Dunnock	0.04
Yellowhammer	0.02
Whitethroat	0.02

nity Research Unit, Cheltenham and Gloucester College, Gloucester.

Dunlap, T.R. (1988) *Saving America's Wildlife*. University Press, Princeton.

Eden, R. (1979) *Going to the Moors*. John Murray, London.

Elton, C. (1942) *Voles, Mice and Lemmings: Problems in Population Dymanics*. Oxford University Press, Oxford.

Fraser of Allander (2001) *An Economic Study of Scottish Grouse Moors: An Update (2001)*. Report by the Fraser of Allander Institute for research on the Scottish economy, published for the Game Conservancy Scottish Research Trust, Fordingbridge.

Galbraith, C.A. and Tucker, C. (2000) *Report of the UK Raptor Working Group*. Joint Nature Conservation Committee, Peterborough.

Game Conservancy Ltd. (1994) *Gamebird Rearing*. Game Conservancy Ltd., Fordingbridge.

Game Conservancy Trust (1996) *Special Report on Pheasants*. Game Conservancy Trust, Fordingbridge.

Game Conservancy Trust (2003) *Review of 2002*. Game Conservancy Trust, Fordingbridge.

Gimingham, C.H. (1981) Conservation: European heathlands. In: *Heathlands and Related Shrublands of the World* (ed. R.L. Sprecht). Elsevier, Amsterdam.

Hill, D. and Robertson, P. (1988) *The Pheasant. Ecology, Management and Conservation*. BSP Professional Books, Oxford.

Hudson, P.J. (1992) *Grouse in Space and Time: The Population Biology of a Managed Gamebird*. Game Conservancy Trust, Fordingbridge.

Hudson, P. and Newborn, D. (1995) *A Manual of Red Grouse and Moorland Management*. Game Conservancy Trust, Fordingbridge.

Jenkins, D., Watson, A. and Miller, G.R. (1964) Predation and red grouse populations. *Journal of Applied Ecology*, **1**, 183–95.

Langley, P.J.W. and Yalden, D.W. (1977) The decline of the rarer carnivores in Great Britain during the nineteenth century. *Mammal Review*, **7**, 95–116.

Leopold, A. (1933) *Game Management*. Scribner's, New York.

Longrigg, R. (1977) *The English Squire and his Sport*. Michael Joseph, London.

Lovat, L. (1911) *The Grouse in Health and in Disease*. Smith, Elder and Co., London.

Marcstrom, V., Kenward, R.E. and Engren, E. (1988) The impact of predation on boreal tetronids during vole cycles: an experimental study. *Journal of Animal Ecology*, **57**, 859–72.

Moore, N.W. (1956) Rabbits, buzzards and hares. Two studies

on the indirect effects of myxomatosis. *La Terre et la Vie*, **103**, 220–25.

Napier, A. (1903) Pheasants at Holkham. In: *Shooting. The 'Country Life' Library of Sport* (ed. H.G. Hutchinson). Country Life, London.

Park, K.J., Robertson, P.A., Campbell, S.T., *et al.* (2001) The role of invertebrates in the diet, growth and survival of red grouse (*Lagopus lagopus scoticus*) chicks. *Journal of Zoology, London*, **254**, 137–45.

Pollard, H.B.C. (1923) *Shot-guns: Their History and Development*. Pittman and Sons, London.

Rands, M.R.W. (1985) Pesticide use on cereals and the survival of grey partridge chicks: a field experiment. *Journal of Applied Ecology*, **22**, 49–54.

Redpath, S. and Thirgood, S. (1997) *Birds of Prey and Red Grouse. Report of the Joint Raptor Study*. Stationery Office, London.

Reynolds, J.C. and Tapper, S.C. (1996) Control of mammalian predators in game management and conservation. *Mammal Review*, **26**, 127–56.

Robertson, P.A. (1992) *Woodland Management for Pheasants*. Forestry Commission Bulletin 106. HMSO, London.

Robertson, P.A., Woodburn, M.I.A. and Hill, D.A. (1988) The effects of woodland management for pheasants on the abundance of butterflies in Dorset. *Biological Conservation*, **45**, 159–67.

Robertson, P.A., Park, K.J. and Barton, A.F. (2001) Loss of heather *Calluna vulgaris* moorland in the Scottish Uplands:

the role of red grouse *Lagopus lagopus scoticus* management. *Wildlife Biology*, **7** (1), 11–16.

Ruffer, J.G. (1989) *The Big Shots: Edwardian Shooting Parties*. Quiller Press, London.

Stoate, C. and Leake, A. (2002) *Where the Birds Sing. The Allerton Project: 10 Years of Conservation on Farmland*. Game Conservancy Trust and Allerton Research Education Trust, Fordingbridge and Loddington.

Stoate, C., Szczur, J. and Aebischer, N.J. (2003) Winter use of wild bird cover crops by passerines on farmland in northeast England. *Bird Study*, **50**, 15–21.

Strachan, R. and Jefferies, D.J. (1990) *The Water Vole Arvicola terrestris in Britain, 1989–1990: Its Distribution and Changing Status*. The Vincent Wildlife Trust, London.

Tapper, S.C. (1992) *Game Heritage*. Game Conservancy Trust, Fordingbridge.

Tapper, S.C., Potts, G.R. and Brockless, M.H. (1996) The effect of an experimental reduction in predation pressure on the breeding success and population density of grey partridges *Perdix perdix*. *Journal of Applied Ecology*, **33**, 965–78.

Tharme, A.P., Green, R.E., Baines, D. and Bainbridge, I.P. (2001) The effect of management for sport shooting of red grouse on the density of breeding birds on heather-dominated uplands. *Journal of Applied Ecology*, **38**, 439–57.

Watson, D. (1977) *The Hen Harrier*. T. and A.D. Poyser, Berkhamsted.

16

Equine

J. Houghton Brown

The British horse industry

Development of the industry

Horses have been significant throughout British history: for power, draught, military, transport and agriculture. They have been gradually superseded, particularly during the first half of the twentieth century, so that by the late 1940s the number of horses had declined markedly. Then, in the second half of that century, the growth of equine sports and recreational pursuits caused the number of horses to grow steadily: horses and related activities now form an important aspect of the British rural economy.

The British horse industry, due to the small scale of its individual units and its disparate nature, has suffered from lack of recognition and reliable data. In the late 1990s, it began to restructure and reorganize to reflect the needs of a modern industry. The relationship between horse and rider gives unique pleasure and is now more widely available and appreciated; however, some horse people find it difficult to reconcile the concepts of horses and industry or business.

Considering terminology, the term 'horsey-culture' should not be used. The word 'hippology' means the study of horses; 'equine' means pertaining to horses and 'equestrian' means pertaining to horsemanship.

Structure of the industry

In summary, the British horse industry:

- gross output is £3.4 billion, making it the second largest economic activity in the countryside to farming;
- utilizes 500 000 ha of farmland for equine pasture plus the production of fodder and concentrates;

- directly employs about 100 000 people and as many again indirectly; it also gives work to many more outside the industry;
- is based on an estimated 600 000–900 000 horses and ponies;
- has 2.4 million riders, of which 80% ride for pleasure; 5 million actively interested.

The industry can be considered in four sectors – racing, sports, recreation and services – as shown in Figure 16.1. These four sectors are represented by the British Horse Industry Confederation (BHIC) which is responsible for lobbying the Government on behalf of the equine industry; BHIC also represents the UK horse industry in the European Union. The structure of BHIC is shown in Figure 16.2. In 2003, the Department of Environment, Food and Rural Affairs (DEFRA, previously MAFF), together with BHIC, started to survey the UK equine industry to quantify its size and economic importance to the UK.

The principal activities of the British horse industry are shown in Table 16.1 and are discussed in the following sections.

The racing sector

The richest and most organized sector of the British horse industry is both glamorous and tough. In 2002, 5.5 million people went racing and prize money was in excess of £84 million. Each year in the UK there are, on average, 14 000 horses in training and approximately 9500 racehorse owners. These horses will be trained by one of the 1000 or so racehorse trainers or permit holders. The professional trainer manages and trains horses on behalf of the owner, whilst the permit holders are amateur trainers who train their own horses. In recent years the racing sector generated about £500

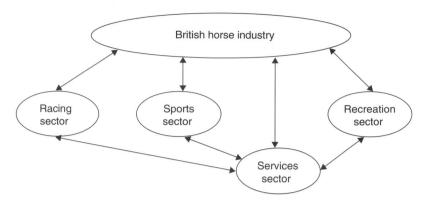

Figure 16.1 The four sectors of the British horse industry.

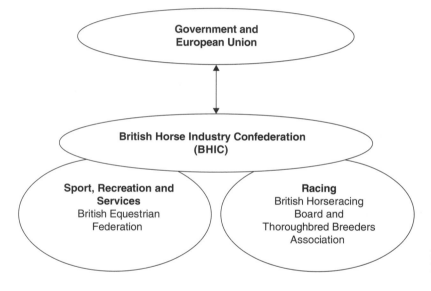

Figure 16.2 The British Horse Industry Confederation.

Table 16.1 Activities of the British horse industry.

Racing	Sports	Recreation	Services
Breeding	Breeding	Breeding	Vets, farriers, etc.
Flat	Show jumping	Hacking	Saddlers and tack
National Hunt	Eventing	Trekking and trails	Clothing and kit
Point-to-point	Dressage	Riding schools	Stables and arenas
Arab	Endurance	Equestrian centres	Hay and feeds
Harness	Vaulting	Livery yards	Bedding
Sales	Trials driving	Individuals (largest group in the industry)	Media
Racecourses	Reining (western)		Vehicles
Betting	Hunting		Insurance
	Hunter trials and team chases		Dealers
	Polo and ball sports		Show centres
	Showing		Colleges
			Working horses
			Welfare organizations

million a year through taxation and betting duty. It supports about 60 000 jobs overall plus some 40 000 in the betting industry which generates 70% of its business from horseracing.

The British Horseracing Board is the governing body for horseracing in the UK. The Jockey Club is responsible for the integrity of racing and for the licensing of all those who work in trainers' yards or as officials on racecourses. The racehorse breeding industry, represented by the Thoroughbred Breeders Association (TBA), generates £310 million in sales and export revenue annually, with a workforce exceeding 11 000 people. In addition to Thoroughbred racing, separate organizations conduct Arab racing and harness racing, which includes trotting and pacing.

The names of all Thoroughbreds are held in the *General Stud Book*, kept by Weatherbys, together with the details of their parentage and any progeny. Thoroughbred breeders select a suitable mare or stallion from the *Stud Book*. Covering fees can be expensive. The resultant foals are sold at the sales, then sent to trainers to race on the flat as 2-year-olds or to begin a jump racing career at four or more years old. There is a high wastage due to physical setbacks and practical problems, but the successful flat race animals will go to stud: the unsuccessful are sold off.

The sports sector

The sports sector can be divided into four main categories: international sports disciplines, ball sports, field sports and showing.

The international sports disciplines are show jumping, eventing, dressage, trials driving, vaulting, endurance and reining (western riding). Ball sports include polo, horseball and polocrosse. Field sports cover all forms of hunting (250 000 regularly hunted in 2003), although in 2004 the UK Government banned hunting with dogs. Showing includes classes for all types and breeds of horse and pony.

The British Equestrian Federation (BEF) is the governing body of the British international equine sports disciplines and has access to Government funding. The BEF also represents the recreation and services sectors at BHIC. BEF has 12 member bodies and represents some 200 000 riders. Its associations and members are shown in Figure 16.3.

The recreation sector

About 2 million people in Britain ride just for pleasure. The British Horse Society (BHS) has 57 000 members, and a further 35 000 belong to those riding clubs that are affiliated to the BHS. The BHS is a charity which promotes access, safety, welfare, quality riding instruction and provides insurance benefits to members. The Pony Club is a major youth organization with some 40 000 members. With the aid of 14 000 volunteers, Riding for the Disabled (RDA) enables 25 000 disabled people to benefit from and enjoy ponies either riding or driving. Horse sports for 'pleasure' riders are also provided at a lower level by many local clubs and centres.

The service sector

The service sector covers businesses that provide supplementary, ancillary or complementary services to horses, riders and the equine industry; it includes both professionals, such as vets and farriers, and trades, such as feed merchants and retailers. The size and scope of this sector is indicated in Table 16.2. These traders are generally members of the British Equestrian Trade Association (BETA) which sets codes of conduct, holds international trade fairs, lobbies on behalf of this sector and conducts surveys of the equestrian industry.

Training, education and careers

The equine industry includes people at many levels from professional, such as veterinary surgeon, to trainee stable lad or lass. Employment opportunities lie in a diverse array of establishments from livery yard to Thoroughbred racing stable and feed merchant to insurance office.

The agricultural colleges have been well placed to take advantage of this increasing demand for professional horsemen and women to work in an industry that requires more specialists and a higher skill level. There are numerous training courses which cover all aspects of equine science and management. However, the best career preparation for practical work with horses is based on a mix of academic qualifications together with extended periods of work in the industry. BHS qualifications are highly esteemed. Further information concerning careers in the equine industry can be obtained from the Sector Skills Council for the Environmental

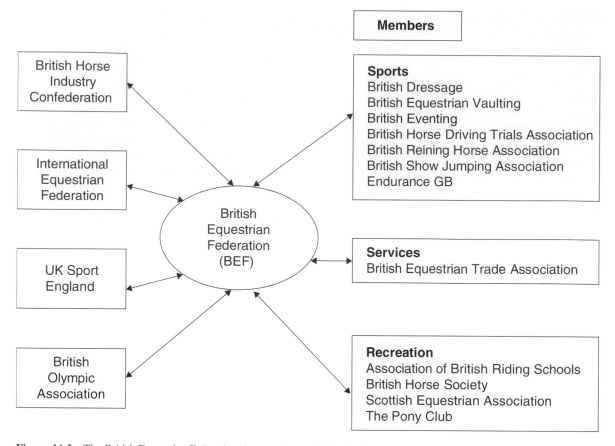

Figure 16.3   The British Equestrian Federation, its members and principal contacts.

and Land-based Sector (LANTRA), from colleges and from the BHS.

Enterprise and profitability

There are three principal consumers in the equestrian industry: horses, riders/owners and followers (e.g. race-goers, spectators and family supporters). These consumers support industries in trade, products and services. Thus, there are many business opportunities to provide one or more of a broad range of services. A range of business opportunities are discussed in the section 'Equine business enterprises', some of which can form complementary businesses.

Commercial considerations

The initial reasons for the development or taking on of a new business venture may include:

- development or expansion of a business;
- diversification;
- utilization of current resources; or, simply,
- wanting to work with horses.

It is necessary to consider the above objectively and to undertake a business plan before commencement. Warwickshire College has researched equine management data and its findings are published in the *Equine Business Guide*, which is updated every few years (Bacon, 2002).

Table 16.2 Size of the service sector (examples of activities).

Activity	Approximate number of businesses
Farriers	2000
Feed merchants	1600
Holiday and trekking centres	600
Livery and accommodation yards	1000
Manufacturing and distribution	1200
Organizations and clubs	1100
Principal breeders, auctioneers, bloodstock agents, dealers	2400
Racing including trotting	1500
Retailing including saddleries, mail order	1600
Riding schools and freelance instructors	1600
Services and direct supplies	3000
Shows and events	1300
Trade	4050
Veterinary practices	800

Table 16.3 An indication of the likely annual net profit per horse.

Business type	Annual net profit per horse (£)	Notes
Livery (full)	600–1400	
Race training	1300–2800	Needs Jockey Club licence
Riding school	900–2400	Needs Riding Establishment licence
Stud	100–700	

The *Equine Business Guide* is an essential business tool as it includes business performance, costs and prices, labour, taxation, finance, and so on. Its findings are published in the gross margin format which is widely used in the land-based industries. It enables a comparison to be drawn between different types of equine business based on a per horse margin. Thus different enterprises of similar size which may be under consideration can be compared. A study of the results of surveys of businesses indicates that a likely annual net profit per horse will vary from £100 to £2800; an essence of this is shown in Table 16.3. Further considerations on the management of a successful business can be found in the book *Horse Business Management* (Houghton Brown, 2001, also published by Blackwell Publishing).

Equine business enterprises

Breeding non-Thoroughbreds

The individual breed societies support the owners and breeders of a wide variety of horses and ponies. Previously, fragmentation and lack of coordination impeded UK breeding programmes for sports horses. There is now better synchrony between the global breed societies. Also, supported by accurate records, unique identification of animals and a strong incentive to use superior stock, the BEF is promoting a more professional approach. The introduction of performance grading for stallions and mares together with techniques such as artificial insemination have increased the quality of the breeding stock. Cloning remains a future possibility for sports horses (but not for racehorses due to their strict regulation).

Dealing yards

The success of a dealing yard relies on the proprietor's ability to match successfully the requirements of the client to the appropriate animal in stock. Profit is made on speedy turnover. Dealers had a poor image in the past, but customer care is now better. See also the following section on 'Livery yards' and the 'Equine law' section on 'Buying or selling a horse'.

Feed and bedding sales

The sales of feed and bedding are generally separate businesses, but they may supplement enterprises such as a livery yard, riding school or other retail outlet. Retailers will stock a range of feeds and supplements. A Feed Code of Practice has been developed by BETA.

Livery yards

A livery yard cares for horses and ponies that do not belong to the proprietor; the proprietor contracts with the owner to provide a certain level of care and management. Prices vary according to the level of care and management provided.

The top level of care is full livery. This is a complete care and management package. It will include stabling, exercise, feeding and care. Full livery may also include some specialist services such as care of hunters, breaking or reschooling. At the other end of the scale, DIY livery means that care and management are provided by the horse owner, but the livery yard provides grazing and/or stable accommodation.

Contracts should define who the legal keeper (carer) is and the yard owner's right to call a vet at the horse owner's expense. Apart from the voluntary BHS Approval Scheme, there is no current specific law that applies to the standards of care and management of a yard, but a Code of Practice for Livery Yards is available from the BHS. Proposals have been made to license all livery yards. Yards that offer riding tuition may require a Riding Establishment licence (see the legal section on the Riding Establishments Acts).

Trekking centres

Trekking centres are often found in areas of natural beauty. They provide a unique holiday experience for those who wish to ride. The trekking season normally lasts from Easter to October although some centres stay open throughout the year, supplementing their income by providing riding lessons and livery services. A trekking centre provides horses and ponies for hire on supervised rides or day treks. Trail rides that last several days over a circular route have become increasingly popular.

Clients are generally accommodated on site or locally and come with a wide range of riding abilities. Young females dominate the client base, but trekking is popular with school groups and youth clubs, so centres will require sufficient ponies and horses to cope with variation in size and ability. A trekking centre has to hold a Riding Establishment licence. There are specialist qualifications for trekking staff.

Those managing a trekking centre must consider the regulations contained in the following legislation:

- Activity Centres (Young Persons Safety) Act 1995.
- Health and Safety at Work Act 1974.
- Horse Safety Act 1990.
- Package Travel, Package Holidays and Package Tours Regulations 1992 (as amended).
- The Riding Establishments Acts 1964 and 1970.
- The Welfare of Animals Acts 1911–1988.

Riding schools

There are over 2000 riding establishments, all of which have to be licensed under the Riding Establishments Acts 1964 and 1970. Establishments may also be approved by the BHS or the Association of British Riding Schools (ABRS). Over one-third of riders are under 15, and particular attention should be given to all aspects of the regulations covering riding establishments. A survey in 2002 by the ABRS of their members reported that overall business is thriving; 95% of their riding schools had experienced an increase in the demand for their services in recent years and 80% of proprietors were confident about their future. Riding lessons generate about £500 million in revenue each year.

A riding school will draw people, mainly children, from the locality. Although riding schools have weekend peaks in activity, mid-week is encouraged through classes for particular groups such as schools, mothers or returning riders. Some offer additional services such as livery, schooling, dealing, tack and clothing sales and refreshments.

The Riding Establishments Act 1970 specifies that the proprietor shall hold public liability insurance. BHS or ABRS approval is significant as the Local Authority Riding Establishment licence does not give any indication of the quality of teaching offered. Both the BHS and ABRS provide teaching qualifications.

Saddlers and tack shops

Saddlers are retail outlets that sell saddles and related equipment. Today, it is unlikely that a saddler will manufacture saddles, but he/she may offer a saddle fitting and repair service. Additional income is generated through the sale of other products such as tack, rugs, riding clothing, horse health products and other requisites. Membership of an organization such as the BETA or professional body of Master Saddlers gives a competitive advantage.

A tack shop sells equipment and other accessories for horses and their riders. Additional items such as clothing, books, feed and feed supplements and giftware may also be found here. Most tack shops will carry a range of both formal and leisure riding wear.

A Code of Conduct for members involved in retail or supply is available from BETA.

Special events and shows

There are many types of event held to promote or encourage interest in a business, such as an open day, show day, competition, specialist training day or demonstration day. Considerable work is involved in putting on an event, which will bring both tangible and fringe benefits to the business, and the objectives of undertaking a project of this kind should be clear from inception. Consideration should be given to traffic, noise, neighbours, marketing, sponsorship, and so on.

Premises

Site and suitability

Often there is no choice of premises as the business is already established. Critical and impartial evaluation is essential when considering the siting of the business and its suitability for the proposed purpose. The following should be considered:

- location – accessibility to services and resources, customer base, restrictions on use, neighbours;
- competition – competitors, market niche, market sector;
- product offered – suitability of product(s) offered, future development potential;
- physical qualities – free-draining land, quality of pasture, buildings, availability of hacking and exercising.

Planning permission

Although planning permission may already exist for a permitted development, confirmation that there are no special conditions attached to it should be sought; building regulations approval may be needed before any structural work is undertaken. Horse activities are generally not agricultural, so farm buildings will require planning permission for change of use. In 1990 the Town and Country Planning Act (TCPA 1990) came into force, followed in 1991 by the Planning and Compensation Act which made fundamental amendments to any previous legislation. The 1995 Town and Country Planning General Permitted Development Order (GPDO 1995) set out a list of developments on agricultural land that do not require planning permission.

All the above legislation is complemented by a number of government circulars and Planning Policy Guidance Notes (PPGs) (see DEFRA website).

Yard design

Horses and ponies need a calm environment with comfortable bedding, adequate food, water and regular exercise. To achieve this any yard should be designed with the following in mind:

- aesthetic appeal;
- economic use of resources;
- efficiency and effectiveness;
- equine welfare;
- safety.

The efficiency of the yard can be enhanced by:

- the location of tack and feed rooms;
- siting of hay barn and muck heap;
- waste-free and pest-free feed storage.

Further discussion on yard layout can be found in *Horse Business Management* (Houghton Brown, 2001, Blackwell Publishing).

Stable design

Stabling should not only look good but also be appropriate and fit for the purpose for which it is to be used. Small ponies or youngsters will do well in a large barn in small single-sex groups.

The current preference for American barn dwellings reflects their ease of use. Although less fashionable, stalls, especially for mares, can be most economical in terms of cost, bedding and labour, but the horses must be exercised every day.

Horses should be provided with a clean, dry area which allows them to stand up or lie down comfortably at all times. The design and its materials should allow thorough cleaning and disinfection from time to time. Boxes are generally 3.7 × 3.7 m for horses and 3.0 × 3.0 m for ponies. Foaling boxes are larger, usually about 5.0 × 5.0 m.

The design should incorporate plenty of headroom and good ventilation. The use of air inlets and stale air out-takes high in the building, in combination with window and door openings, will normally provide adequate ventilation without creating draughts. Stables should be safe, without sharp edges or protrusions and with inbuilt fire safety. Natural light sources should be well utilized and additional lighting should enable the horses to be properly observed. Electric installations must be protected from chewing by horses and rodents.

Feeding via externally filled mangers is more efficient. Similarly, it is preferable to have a hayrack that is refillable from the outside. Hayracks that are placed lower on the walls are easier to fill and prevent eye irritation. Automatic drinkers also save labour and water so soon recover their cost.

The floor should be designed and constructed to provide good traction, proper drainage and comfort and to prevent injury.

Arenas and facilities

The modern equestrian establishment benefits by providing all or some of the facilities listed below. These can provide economic alternatives to the employment of additional staff or transportation of horses.

- arenas;
- free-draining turnout paddock or yard – turnout during wet weather and in winter;
- lunge ring – 20-m ring for lunging or free schooling;
- horse walkers – for fittening, limbering up or cooling down;
- cross-county fences – need not be too high for training purposes, best arranged like a show jumping course for supervision and ground economy;
- cantering track (gallop) – for faster fitness training.

For most riding businesses their main work will take place in an arena. Arenas can either be indoors or, more commonly, outdoors. For schooling a horse an area of 20×60 m is ideal. A compromise is to have an outdoor arena together with a covered area about $20\,m^2$ which provides exercise shelter in very bad weather and is also useful for breaking and lunging.

Yard management

Records

Efficient management requires that a business keeps the appropriate records on its own behalf and on behalf of others, e.g. the Inland Revenue. These records can be classified into the following areas:

- accounts and bookkeeping (including asset management and insurance);
- client accounts (livery charges, professional treatments, etc.);
- estate (including overall health and safety);
- livestock;
- marketing and event management;
- personnel.

Diaries are an essential management tool, for planning and monitoring daily, weekly, monthly and annual jobs or events. In future years they become a reference for what was done and its success.

Quality and standards

Under the Health, Safety and Welfare Acts 1974–1993, an establishment needs to maintain safe and good practices. An organization may be judged by its outward appearances and the efficiency and effectiveness of its staff. Setting and maintaining quality standards provides staff and management with a framework within which to work. This is important for all staff, but especially for the inexperienced or younger employees. Rushed and slip-shod or makeshift methods and practices lead to accidents and anxious horses. See also the 'Equine law' section on negligence.

Fire procedures

Fire prevention is part of good management and to minimize fire risks the location of fodder, bedding and flammable materials should be given careful thought. The stable yard should be a no-smoking area. Fire-fighting equipment should be located in highly visible places and regularly checked according to manufacturers' instructions. Regular inspections of electrical and heating systems should be carried out by an approved person. The fire drill should be clearly visible and all staff should be trained and practised in the procedure.

Work planning

European working time regulations restrict employees over 18 to working no more than 48 hours a week on a regular basis and this is likely to reduce the expectations put upon present staff in some yards. The majority of racing yards now work a basic 40-hour week and it is likely that the non-Thoroughbred industry will drift in this direction. The organization of sound routines and rotas will establish an effective system for work planning.

Daily routine

Having relatively small stomachs, horses become stressed by irregular, over-large meals or if food is unavailable for long periods of time. The ideal routine is centred on a regular daily timetable which balances the needs of the horse and the staff. Routines will depend on the type of business and the ratio of staff to horses to be cared for. The daily routine ensures that stock is fed and watered and checked for signs of ill health on a regular basis, especially first thing in the morning and last thing at night.

Tack and equine clothing

Tack includes the saddle and the bridle. Other items of tack and harness depend on the horse and the work expected of it. The resultant combinations of bit, bridle and saddle give an infinite choice, so simplicity using tried and well-proven designs is generally best. Tack should fit well, be in good repair, be cleaned regularly and be routinely inspected for wear and safety.

Apart from tack, stabled horses will need clothing (rugs, blankets, sheets) for both winter and summer. Care must be given in the fitting to avoid tight pressure on the spine and to ensure they stay put without chafing.

Bedding

Bedding is essential to the horses' thriftiness in that it not only encourages them to rest and urinate, keeps them warm, and also prevents concussive injuries from hard floors.

The ideal bedding is economical, dry and readily available, disposable and not harmful if eaten. Commonly straw or wood shavings are used. Other options are shredded paper, peat or hemp. 'Mucking out' at morning stables normally removes all damp and soiled litter. Alternatively, a partial deep litter system requires

thorough cleansing of the box at periodic intervals. Rubber matting may be installed and used with a little straw. All systems work best if droppings are removed throughout the day.

Disposal of manure and other noxious substances

Muck heaps or trailers should be sited near enough to the stables to be within easy reach but far enough away so that flies and smells do not pervade the work areas. Care is required to prevent soiled waste or effluent contaminating the yard or water courses.

EU regulations now prohibit burning of waste material or manure. Although local farmers or market gardeners may be willing to dispose of the waste, the heap will take many months to decompose sufficiently to be useful. Bunkers may be needed, ideally filled by a tipping trailer from above, and the yard trailer will need a fixed ramp so that wheelbarrows can be emptied into it.

Grassland care

Pasture for horses

Unlike cows, horses prefer a short sward. Their grazing patterns vary according to their breed characteristics and the work required of them. Thoroughbreds thrive less well in winter at grass, whereas native ponies cope much better. Quality pasture can provide both nutrition and exercise throughout the year, but ponies in particular may become laminitic on fast-growing grass.

During late autumn and winter, supplementary feeding is usually required. Stock may need to be removed from the land to prevent poaching or restricted to one paddock which will be salvaged after winter.

The best type of grazing is found in well-managed, established leys which provide a variety of grasses and herbs. Older pastures tend to stand up better to poaching during wet weather. Good horse pasture combines productive grasses with those designed to give a good bottom to the sward. An excellent seed mix is one based on a late-flowering perennial ryegrass to which has been added some creeping red bent and crested dog-tail to improve the turf. White clover, so long as it represents no more than 5% of the total, will not harm horses and will boost herbage growth.

Paddock management

The essence of good equine pasture management is to:

- extend the grazing period and reduce poaching;
- maintain a sward suitable for equine grazing;
- prevent over- or under-grazing (roughs and lawns);
- reduce infestation by intestinal parasites.

An annual plan will incorporate any spraying, fertilizing, harrowing, rolling, topping and grazing. A rule of thumb is that 1 ha of good pasture will provide two horses or four ponies with sufficient grazing throughout the year. Worm control is a major consideration; cross- or mixed grazing, 'poo-picking' or vacuum cleaning and harrowing in hot, arid conditions all have their place. Neglected paddocks can be enhanced by:

- improvement in boundaries (fencing, gates, etc.);
- improvement of drainage;
- improvement of soil structure, texture, acidity;
- installation of a water supply;
- reducing weed infestation (give special attention to ragwort and topping grasses in rough areas);
- regular collection of dung.

Ragwort was responsible for some 500 equine deaths in 2001; it remains toxic in both its fresh and dry states. Under the Equine Welfare (Ragwort Control) Bill there is a duty of responsibility imposed on the occupier of the land or, in the case of public land, the local authority, to effectively control any ragwort infestation and prevent its spread onto adjoining land. Ragwort is usually hand pulled in June and July (gloves should be worn) or dug up using a special fork. The weed should be removed for rotting in a restricted heap. Spraying at the rosette stage is an option.

Fencing and shelters

Fencing must not permit animals to gain access to the public highways or adjoining land. Horses kept in studs benefit from a double line of fencing over 2 m apart to avoid contact. Commonly used methods of fencing are:

- post and rail with anti-chew wire;
- synthetic rails (various styles and brands);
- high-tensile wire or equine mesh fence;
- electric fencing with a broad specialist tape.

Barbed wire, wire netting and pig wire are not recommended.

Horses in paddocks should be given ample protection from bad or hot weather and flies. Protection can be afforded by a purpose-built open-fronted shelter, or by shelter belts, trees, hedges or walls.

Equine fundamentals

Equine breeds

The horse (*Equus caballus*) is in the family Equidae; other species in the genus are asses and zebras. *Equus* has existed for about 5 million years and horses have been domesticated for over 6000 years. Early equine populations were endemic in locations around the world and research through gene mapping has shown that equine domestication occurred independently in different places. Science has discovered that there were several distinct horse populations, which has led to the diversity in breed and types we see today. According to *Mason's World Dictionary of Livestock, Breeds, Types and Varieties* (Porter, 2002), in 2002 there were 705 named horse and pony breeds, types and varieties. In the UK there are a variety of native ponies, draught and light horses and sports horses suitable for a multiplicity of uses. These breeds and their details are outlined in Table 16.4.

The reference to blood type in their description refers to temperament not blood heat. Evolving in the desert, the hot-blooded Arab is fine boned, thin skinned, speedy and sensitive. The docile, draught horses of the northern hemisphere are cold bloods with their strong bodies and stamina. Warm bloods have parentage from both hot bloods and cold bloods which has given rise to sports horses.

Structure of the horse

Key characteristics of an equine include:

- a large hind gut or fermentation chamber containing huge quantities of micro-organisms to break down cellulose and fibre;
- a strong back to carry the weight of the large gut slung beneath;
- acute senses;
- an elongated jaw with large cutting and grinding plates; and
- long legs with light feet and powerful muscles to enable them to be fleet of foot.

Table 16.4 Summary of British breeds and their uses.

Breed	Colour variations	Usual size in hands	Use
Native ponies			
Connemara	Grey, bay, dun, brown, chestnut (rare)	14.2	
Dales	Black, dark brown or bay	14.2	
Dartmoor	Bay, brown or black	12.2	
Exmoor	Dun, brown or bay	12–12.3	Hacking,
Fell	Black, dark brown or bay	14	hunting,
Highland	Dun, grey, bay and black	14.2	competing,
New Forest	Any colour except piebald or skewbald	12–14.2	recreation,
Shetland	Black, chestnut, brown, piebald, skewbald	10.2	and
Welsh Mountain (A)	Any colour except piebald or skewbald	12	showing
Welsh Pony (B)	Any colour except piebald or skewbald	13.2	
Welsh Pony Cob (C)	Any colour except piebald or skewbald	13.2	
Welsh Cob (D)	Any colour except piebald or skewbald	15.2	
Draught horses			
Clydesdale	Bay, roan with white legs	16–17	Draught
Irish Draught	Bay, chestnut or grey	15.3–17	Hunting
Percheron	Grey or black (originally French)	15.2–17	Draught
Shire	Bay, chestnut, grey or brown	17+	Draught
Suffolk Punch	Chestnut	16–16.3	Draught
Light horses			
Arab and Anglo-Arab	Bay, brown, chestnut or grey	15–16.3	Endurance/racing
Cleveland Bay	Bay with black points	16.2	Breeding
Hackney (horse)	Any solid colour	14–15.3	Harness
Hackney (pony)	Bay, black, brown or chestnut	12.2–14	Harness
Thoroughbred	Any colour except piebald or skewbald	16–17	Racing
Sports horses			
Cross breeds*	Any colour including piebald and skewbald	16–17	Sport
Warmbloods	Varies with breed	16–17	Sport

* Note: cobs, hunters and hacks are types not breeds.

The skeleton consists of two parts: the axial skeleton (from the skull to the tail vertebrae including the rib cage) and the appendicular skeleton (which are the bones of the four limbs). If the animal possesses a strong, well-balanced, skeletal structure and good muscle formation, it is described as having good conformation. The activity that the animal is required to perform will influence its conformation assessment.

Horse health

Each animal is individual; any variation to their normal patterns of behaviour or routine may indicate a problem, and so should be investigated further. The normal and average ranges of equine temperature, pulse

Table 16.5 TPR – vital signs of a healthy adult horse or pony at rest.

Sign	Average	Range
Temperature (rectal)	38°C (100.5°F)	+ or −0.5°C (1°F)
Pulse	32 beats/minute	25–40 beats/minute
Respiration	12 breaths/minute	+ or −2 breaths/minute

and respiration (TPR) are shown in Table 16.5. A healthy animal is one that is alert and bright both physically and mentally as shown in Figure 16.4. It should have a good appetite and be quick to respond to stimuli. Horses kept in paddocks should be inspected at least once a day, whilst those kept in stables and yards at

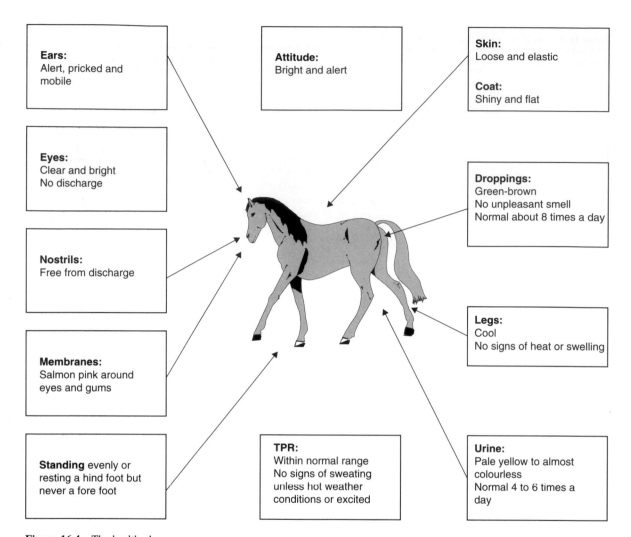

Figure 16.4   The healthy horse.

least twice a day. Mares that are due to foal should also be inspected twice a day, but as the expected delivery date draws near this frequency should be increased.

Horses should be inspected for:

- availability of food and water;
- body condition (see Table 16.6);
- ill-fitting head collars and/or rugs;
- signs of ill health;
- signs of injury.

The sick horse

The owner or the keeper of a horse is responsible for its welfare and maintaining the well-being of the animal in his or her care. Table 16.7 summarizes the signs of good health and possible problems. Ideally, provisions should be made for isolation stables on any yard, either for new horses arriving or those that are sick. Horses are commonly vaccinated against equine influenza and tetanus.

Table 16.6 Body condition scoring of horses and ponies.

Condition score[1]	Description	Hind quarters	Back and ribs	Neck
0	Very poor (emaciated)	• Angular, skin tight • Very sunken rump • Deep cavity under tail • No subcutaneous fat detectable	• Skin tight over ribs • Very prominent and sharp backbones • Skin tight over rib cage	• Marked ewe neck (shaped like a sheep's neck) • Narrow and slack at base
1	Poor (thin)	• Well-defined pelvis and croup • Sunken rump but skin supple • Poverty lines apparent • Deep cavity under tail	• Ribs easily visible • Prominent backbone with skin sunken on either side	• Ewe neck • Narrow and slack at base • Some fat on shoulder
2	Moderate	• Rump flat either side of backbone • Croup well defined, some fat • Slight cavity under tail	• Ribs just visible • Backbone covered, but spines can be felt	• Narrow but firm
3	Good (well covered)	• Covered by fat and rounded • No gutter • Pelvis easily felt	• Ribs just covered and easily felt • No gutter along back • Backbone well covered, but spines can be felt	• No crest (except for stallions) • Firm neck
4	Fat	• Gutter to root of tail • Pelvis covered by soft fat	• Backbone and ribs well covered – need pressure to feel	• Slight crest even in mares • Wide and firm • Fat deposits on shoulders
5	Very fat (obese)	• Deep gutter to root of tail • Skin distended • Pelvis buried, cannot be felt	• Ribs buried, cannot be felt • Deep gutter along back • Back broad but flat	• Marked crest • Very wide and firm • Fold of fat • Shoulders bulge with fat deposits

[1] To obtain the condition score, first score the hind quarters, then adjust this score up or down by 0.5 if it differs by 1 or more points from the back or neck score. This system is based on the work of Carroll and Huntington (1988).

Behaviour

Horses are naturally gregarious and enjoy the company of others. They live in a herd, communicating by subtle body language and vocally. Within the herd society there is always a pecking order of which all members are aware.

In the natural state, horses lead free-ranging lives, travelling many kilometres for fresh grazing and water. In domestication, their environment is limited, and their movements are restricted; they are constrained within an uncharacteristic routine.

Horses kept at grass, living out for the majority of the time, seem more content and demonstrate fewer symptoms of stress, which may be expressed as stereotypic behaviours, commonly known as vices, such as cribbing, weaving and box walking.

Complementary medicine and alternative therapies

Today there are a wealth of alternative and complementary therapies available to horse owners. These include such services as ultrasound, physiotherapy, massage, aromatherapy, hydrotherapy, homeopathy, and so on.

Some animal therapies are regulated by their own governing body, such as the Association of Chartered Physiotherapists in Animal Therapy (ACPAT), but some other therapists are less qualified. It should be noted that any alternative medication or therapy should only be administered after prior consultation with or under the direction of a veterinary surgeon.

Basic dental care, such as examination of teeth and rasping, may be undertaken by non-veterinarians. However, recent changes to the Veterinary Surgeons

Table 16.7 Signs of good health or possible problems.

Sign	Good health	Poor health
Appetite	• Eats up normal ration	• Leaves good food
Attitude	• Normal responsive self	• Lethargic and miserable
Behaviour	• Stable or field appears normal	• Signs of disturbance in stable
	• No abnormal behaviour	• Signs of discomfort: rolling, pacing, kicking at
	• Alert, comfortable, at ease	belly, etc.
Coat and skin	• Smooth, lying flat	• Sweating unexpectedly. Hair stark, staring, rough
	• Glossy. Mobile and loose	patches. Evidence of dried-on sweat
	• When pinched quickly goes flat against the skin	
Dung	• As usual. Passes droppings easily about 8–10 times a day, may break on impact with ground	• Droppings hard, smelly, slimy or absent (droppings consistency and colour vary with diet)
Ears	• Alert, mobile	• Unresponsive
Eyes	• Clear, bright	• Discharge. Unresponsive
Feet	• Stands and moves distributing weight evenly (may rest a hind leg)	• Smell foul
		• Tender
		• Limps (lame)
Legs	• Cool, firm, smooth	• Swollen, painful
		• One warmer
		• Pointing a leg/foot
Membranes	• Eyelids and gums appear normal colour (pink)	• Paler or different in colour (yellowish or bluish)
Nostrils	• Nostrils clear	• Discharging. Normal only after first fast work
TPR	• Within normal limits	• One or more 'normal signs' being abnormal (see Table 16.5)
Thirst	• 20–70 litres according to size, work, climate, etc.	• Abnormal thirst
Urine	• Pale yellow to clear, passed 4–6 times a day	• Changed in frequency, consistency or colour
Weight	• Appropriate condition (score 2–3)	• See Table 16.6

Act 1966 aim to ensure that dental procedures are carried out by a qualified equine dental practitioner who may also undertake minor dental surgery. This will include extraction of loose or wolf teeth, dental hook removal, techniques requiring the use of dental shears, inertia hooks and powered dental equipment. Any procedure that necessitates an incision to be made requires consultation with and supervision by a veterinary surgeon.

Training

Early training work has several phases. First comes the early handling from the day of birth until the horse or pony is sufficiently strong and well grown to start in light work. This stage is shortest in the flat racehorse, which begins work as a 2-year-old, and longest in tall, docile and late developing horses which may do little

before they are four. During this stage calm consistency is crucial; requests to the horse must be fair and clear and always obeyed promptly. The horse learns to be safe, obedient and pleasant with its handler and groom.

The second stage generally occurs at 3 years old and requires skilled expertise. The horse is first taught to be obedient and well behaved on a lunge. Then the aim is for the horse to feel neither alarmed nor uncomfortable nor unbalanced when it has a rider on its back. This breaking process takes several weeks of quiet patient work.

In the third stage the horse becomes stronger and more confident with its rider. The horse is developed as an equine athlete. It learns to jump, first on the lunge and then with a rider. If it is to be a driving horse it is now broken to harness and to pulling and then to shafts and so to slowing the vehicle. The well-developed, obedient, confident happy riding or driving horse or pony becomes, in due time, ready to learn its career role and to enjoy being a partner in the performance of its work or sport.

An individual training programme, unique to a particular horse, will allow it to achieve its potential with the minimum of fatigue, which reduces the likelihood of injury. The basis of all training programmes is a slow and steady increase in work load, building up strength, stamina and ability. It takes into account baseline fitness, the age and soundness of the horse and prior training. This is supported and sustained by an individual nutrition plan which takes into account the type and size of the horse and its work.

There are many training methods and programmes. Variety will keep the horse interested and keen. Apart from riding, exercise may be taken in hand, on the lunge or using a treadmill or horse walker.

Travelling

Today's horse industry stock is moved by road or air. These movements are covered by the Welfare of Animals (Transport) Order 1997.

Experience, skill, preparation and care taken during loading, travelling and unloading will minimize risk. The driver is responsible for the welfare and well-being of the horse during road transportation. The journey should be carried out smoothly, expeditiously, with appropriate rest breaks and with the animal suitably clothed for travelling. The requirements for trailers, equipment and vehicles and loading densities are contained in the EU directive 95/29/EC.

Euthanasia

Horses are considered 'aged' once they have reached their teens, but occasionally achieve 30 years or so, their working life ending in their late teens. At some point in time, for nearly every horse, a decision needs to be made concerning when to put the horse down; this decision may be taken in consultation with the vet. Consents from the owner and the insurers are wise before proceeding.

A horse can be euthanased by shooting or by lethal injection. Shooting can be performed by a veterinary surgeon, knackerman or hunt kennel man. Lethal injection may only be performed by a vet, but disposal options are then reduced. Burial is permitted only in restricted circumstances, otherwise the carcase is disposed of via the knackers yard, hunt kennel, incinerator or rendering plant. Owners deciding to send the horse for human consumption can find a few suitably registered slaughter houses, providing the horse's passport status allows this.

Feeding and nutrition

Rules for feeding and watering

Whereas the wild horse can generally meet its own needs, the domesticated horse relies on its keeper to provide all its dietary needs. There should be sufficient nutrients, minerals, vitamins and water for planned activity plus maintenance of appropriate body weight.

Traditional diets for horses include hay, oats and bran, but feed companies now produce a diversity of specific feeds to meet the needs of all types of horse and performance requirements. The 'rules' of feeding and watering are explained in any good horse and stable management book, such as *Horse and Stable Management* (Houghton Brown *et al.*, 2003, Blackwell Publishing), but essentially these are as follows:

- Ensure the diet contains plenty of fibrous material; reducing fibre will impair gut function.
- Feed according to age, condition, temperament and work required.
- Feed little and often (concentrates two to four times a day for stabled horse).
- Feed must be free from moulds or taints.
- Keep to a routine.
- Make any changes to diet gradually.
- Supply *ad libitum* fresh, clean water.

Rationing

Food may represent a substantial part of the budget, but food has significant impact on the horse's ability to do the work asked of it. Rationing relies on an understanding of the nutritional requirement of the individual horse and the nutritional value of the foodstuffs, taking into account economy and availability. Only good quality feeds should be fed and in sufficient quantities to enable the horse to perform well and flourish. All feed should be stored and handled in a manner that will prevent waste, damage, decay or spoiling and ensure that its nutritional value remains as high as possible.

There are two main categories of food:

- forages – grass, hay, haylage, feed straw (used mainly for equine maintenance);
- concentrates – straight or compound feed (used for performance and to supplement forages in winter).

Forages provide the bulk element to the diet which allows the food to be digested properly; reducing this element below 50% of the total daily intake will require extra skill to avoid problems. Concentrates are starch-rich and may be fed either as straight grains (e.g. oats or barley) or as compound feeds (cubes or mixes). Straight grains will not supply all the nutritional needs, but feed balancers are available. Cereals can be processed to make them more digestible and together with the addition of minerals and vitamins are formulated to make a variety of compound feeds. Compound feeds are convenient, of a consistent quality and created for a particular purpose. They do not normally require further additives or supplements.

The basic rules of the simplest rationing system are as follows:

- Feed according to weight. Appetite is about 2.5% of body weight if fed on hay and concentrates. Table 16.8 shows the likely range.
- The ratio of forage:concentrates by weight is generally as follows:

Resting	100:0
Light work	75:25 (hacking)
Medium work	60:40 (hunting, novice competing)

| Hard work | 50:50 (more strenuous or higher levels) |

- Vary the ration according to observation. Consider the horse's ideal weight, its temperament, performance and environment.
- Select appropriate concentrates.

A formula, devised by the author of this chapter, for calculating energy requirements measured in megajoules of digestible energy per day can be found in many horse management books.

Feeding practice

A horse's body condition should be observed and the amount of feed and exercise varied accordingly. It is recommended that most ponies or horses should be maintained between condition score 2 and 3 (see Table 16.6) where score 0 is very poor and score 5 is very fat.

The stabled horse

Feed rations are often measured in scoops, but it is important to know the weight of a scoop of each feed type, so that the ration may be correctly formulated. Initially, hay should also be weighed. Horses in stables are usually offered water in a bucket or via automatic drinkers; the latter save both water and labour.

The feed room board will list the horses and ponies by name and show quantities of ingredients at each feed. For light work 'horse and pony' cubes suffice, but for more demanding work 'competition' cubes are suggested. As an alternative to cubes, concentrates are supplied as 'mixes', which look nicer but cost more.

Table 16.8 Estimate of the ration required for different sized horses and ponies.

Height (in hands)[1]	Height (in cm)	Body weight (kg)	Appetite (in kg) (2.5% of body weight)
11	111.8	120–260	3.0–6.5
12	121.9	230–290	5.75–7.25
13	132.0	290–350	7.25–8.75
14	142.2	350–420	8.75–10.5
15	152.4	420–520	10.5–13.0
16	162.6	500–600	12.5–15.0
17	172.7	600–725	15.0–18.1

[1] Traditionally horses and ponies are measured in hands, where 1 hand = 4 inches or 10.16cm.

The quantity and quality of the hay will clearly affect the overall ration; haylage is also suitable. Silage can be used and is economical, but special instruction is required concerning feeding, storage and use.

The horse at grass

Later in the year, grass needs supplementing with hay or feed straw and for work a horse will also need concentrates. Supplementary feeds should be placed in appropriate containers and sited to minimize squabbling and considering safety of horse and handler.

Whilst it is possible to use suitable streams or ponds for watering it is by and large not recommended. It is better to provide mains water via a piped supply to a safely designed trough.

Welfare

Welfare standards

All welfare is based on the 'Five Freedoms' for animals set out by the Farm Animal Welfare Council. It states that all animals have a right not to suffer hunger and thirst, discomfort or pain caused by injury or disease. Animals should also be able to display any natural behaviour without fear or distress. Cruelty is sometimes hard to define and considerations tend to be emotive rather than objective; therefore the Five Freedoms form an important framework for equine welfare.

The International Equestrian Federation (FEI) promotes ethical considerations for competition horses, which can be summarized as follows:

*   In all equestrian sports equine welfare is paramount.
*   Competition riders and drivers must be suitably fit and competent.
*   Education and research should aid equine welfare.
*   High standards of horse management, safety and transport must be maintained.
*   National rules must reflect these ethics.
*   Riding, driving and training must use acceptable and non-abusive techniques and practices.
*   The well-being of the animal takes precedence over the demands and interests of its connections, including riders, owners, trainers, teams and sponsors, and any other commercial reasons.

Organizations

The leading equine welfare organizations work together under the National Equine Welfare Council (NEWC). Some organizations are sanctuaries, but the best of these also give grants for welfare research. The Blue Cross and the BHS have field officers. The International League for the Protection of Horses (ILPH) is active in issues such as the international transport of horses for slaughter. Further information is contained in the *Equine Industry Welfare Guidelines Compendium for Horses, Ponies, and Donkeys* (ADAS Consulting Ltd., 2002).

Equine law

The Animal Acts

Much welfare law covers pet animals or farm animals; generally horses are neither; the majority of UK horses and ponies are kept for leisure or sport. Under British law, the horse remains a non-agricultural animal, unlike in many EU countries. However, horses used in farming are subject to the provisions of the Welfare of Farmed Animals (England) Regulations (2000) (with similar regulations for Northern Ireland, Scotland and Wales). These regulations are the application of EU directive 98/58/EC (general directive on farmed animal welfare). Corresponding regulations are required to provide the minimum standard of welfare for all horses and ponies in the UK irrespective of the purpose for which they are kept.

An overhaul of all the existing legislation was begun in 2002. DEFRA sought the opinions of a number of equine organizations to ascertain which areas of the law required changing or modernizing to bring it up to date. Proposals so far include:

*   amendments on the laws of abandonment and tethering;
*   amendments to reduce indiscriminate breeding and the overproduction of poor quality, unsaleable stock;
*   a new offence of 'likely to cause unnecessary suffering';
*   national disqualification from keeping animals of anyone convicted of animal cruelty;

- increasing the powers of the police to access premises where they believe animal welfare laws are being breached;
- redefinition of the terms cruelty and neglect;
- regulation and licensing of livery yards.

In 2003, the Law Lords clarified the Animal Act 1971 in that a horse owner is now usually liable for all damages should his or her horse or pony escape and cause an accident. Thus it is prudent for all equine owners to take out at least third party insurance.

Riding Establishments Acts

Any establishment that hires out horses or ponies is included in the Riding Establishments Acts 1964 and 1970 and must be licensed. Such establishments will be inspected annually by the local authority, with an assessment of horse welfare, premises and tack by an approved veterinary inspector.

Horse passports

Under the European Commission Decision 2000/68/EC, from February 2005, all equines in the UK are required to have a passport identifying the animal; exceptions are free-running ponies on Dartmoor and in the New Forest. The passport identifies those animals that have been treated with certain medicines in order to eliminate them from entering the human food chain; it also aids disease control.

It is the responsibility of the horse owner to obtain a passport for any animal aged 6 months or over that he or she possesses. The passports are issued by an authorized passport issuing organization (PIO), a list of which can be found on the DEFRA website. Each animal gets a Unique Equine Life Number (UELN), which shows country, issuing organization and individual animal number. The information is held on the National Equine Database (NED).

The passport should accompany the animal when it is:

- moved into or out of Great Britain;
- moved to new premises due to change of keeper;
- moved to other premises for competition purposes;
- moved to other premises for the purpose of breeding;
- moved to a slaughterhouse for slaughter;

- moved on any other occasion specified by the recognized organization.

Veterinary and farriery

All vets come under the rules of the Royal College of Veterinary Surgeons, but specialist equine vets may also be members of the British Equine Veterinary Association (BEVA). Likewise all farriers are registered with the Farriers Registration Council.

The Veterinary Surgeons Act 1966 states that no individual is allowed to make a diagnosis or treat an animal, except in rendering emergency first aid, unless registered as a vet. It is recognized that owners, or a person in the employment of the owner, may treat minor ailments and disorders without reference to a veterinary surgeon.

Similarly, the Farriers (Registration) Act 1975, amended in 1977, prohibits any individual, other than a farrier registered with the Farriers Registration Council, from carrying out work in connection with the foot of a horse or pony immediately prior to receiving or fitting a shoe.

Buying or selling a horse

The law concerning buying and selling horses and ponies is, as for other items, the law of contract. Particular points to note are as follows:

- Caveat emptor is a common law rule meaning 'let the buyer beware', so before purchasing carefully investigate.
- Sold 'subject to vet' means there is not yet a contract or any obligation in law.
- The horse must not be falsely described.
- If bought from a dealer, the horse must be of 'satisfactory quality' and 'suitable for the purpose' described.
- The horse should normally be vetted as 'suitable for the purpose' (vets do not claim it is 'sound' for legal reasons), this contract being between the vet and the purchaser.

Negligence

Negligence is a tort in law and so is between two people. If a person fails to take proper care and in consequence someone is injured, damages may be awarded. Liability cannot be excluded and it is important to show that proper care has always been taken. Riding is a risk sport and it is important that staff and clients clearly understand the risks involved. Proper risk assessments must be made and safety procedures documented and adhered to. Consider the relevant legislation including:

- Control of Substances Hazardous to Health 1989 (COSHH);
- Health, Safety and Welfare Acts 1974–1993;
- Reporting of Injuries, Diseases and Dangerous Occurrences Regulations 1995 (RIDDOR).

Warwickshire College provides a number of distance learning programmes known as 'Equistudy' courses, which cover these areas specifically aimed at the equine industry (The Business School Series, 2003).

Acknowledgement

I am grateful to Sandra Tyrrell for her time and effort in assisting with the preparation of this chapter; particularly I thank her for providing research and design skills, helping to bring together the information in a concise form.

References

ADAS Consulting Ltd. (2002) *Equine Industry Welfare Guidelines Compendium for Horses, Ponies, and Donkeys*. DEFRA, London.

Bacon, R. (2002) *Equine Business Guide*. Warwickshire College, Leamington Spa.

Carroll, C.L. and Huntington, P.J. (1988) Body condition scoring and weight. Estimation in horses. *Equine Veterinary Journal*, **20**, 41–5.

Houghton Brown, J. (2001*) Horse Business Management*, 3rd edition. Blackwell Science, Oxford.

Houghton Brown, J., Pilliner, S. and Davies, Z. (2003) *Horse and Stable Management*, 4th edition. Blackwell, Oxford.

Porter, V. (2002) *Mason's World Dictionary of Livestock, Breeds, Types and Varieties*, 5th Edition. CABI, Oxford.

The Business School Series (2003) *Equistudy Business School*. Warwickshire College, Moreton Morrell.

Further reading

Auty, I. (ed.) (1998) *The BHS Complete Manual of Stable Management*. Kenilworth Press, Buckingham.

Bacon, R. (1996) *Horses and Money: How to Manage an Equine Business*. Blackwell Science, Oxford.

British Equestrian Trade Association (annually) *The British Equestrian Directory*. BETA, Wetherby.

British Horse Society (2001) *Guide to Horse Care and Welfare*. British Horse Society, Kenilworth, and the Home of Rest for Horses, Princes Risborough.

Hastie, P.S. (2001) *The BHS Veterinary Manual*. Kenilworth Press, Buckingham.

Pilliner, S. (1999) *Horse Nutrition and Feeding*, 2nd edition. Blackwell Science, Oxford.

Pilliner, S. and Davies, Z. (1996) *Equine Science, Health and Performance*. Blackwell Science, Oxford.

Soffe, R.J. (ed.) (2003) *The Agricultural Notebook*, 20th edition. Blackwell Science, Oxford.

Useful websites

www.abrs.org – Association of British Riding Schools.
www.bef.co.uk – British Equestrian Federation.
www.beta-uk.org – British Equestrian Trade Association.
www.beva.org.uk – British Equine Veterinary Association.
www.bhb.co.uk – British Horse Racing Board.
www.bhs.org.uk – British Horse Society.
www.defra.gov.uk – Department of Environment, Food and Rural Affairs.
www.farrier-reg.gov.uk – Farriers Registration Council.
www.lantra.co.uk – Sector Skills Council for the Environmental and Land-based Sector (careers in land-based industries).
www.newc.co.uk – National Equine Welfare Council.
www.thoroughbredbreedersassociation.co.uk – Thoroughbred Breeders Association.

17

Amenity fisheries and aquaculture

P. Haughton and D. Horsely

Introduction

Aquaculture involves the farming and marketing of aquatic organisms, including fish, molluscs, crustaceans and seaweed (Figure 17.1 presents industry growth trends). The activity requires intervention during the cultivation process, to a higher or lesser degree, depending on production method and objectives, with the overall goal being to increase stock yield in a suitably efficient manner. Although categorically similar, aquaculture operations can be broadly divided; they may market a product intended for (1) direct human consumption or (2) enhancement and/or recreational purposes. Unless otherwise stated, the term 'aquaculture' in this chapter will focus on the marketing of products directed for human consumption. Special attention is given to aquaculture for amenity purposes in the section on 'Freshwater amenity/ recreational fisheries'.

Traditional capture production is typically performed to fulfil the demand for high quality, nutritious food – much in the same way as aquaculture is doing now. However, due to the depletion of global fisheries stocks aquaculture is today the only alternative for making a significant contribution to the world's aquatic food supply. It is predicted that the world's growing population will require an extra 30 million t of fish for food by 2010, for which aquaculture will be the primary source (Poulter, 2001). Consequently, in the same way that agriculture has replaced the gathering of wild plants and livestock breeding has replaced hunting, aquaculture is supplementing fishing in order to save endangered sea stocks and promote sustainability. Aquaculture also has an important role in the alleviation of poverty in the world – growth of aquaculture in developing countries is well above the compounded

global average and there are no indications to suggest this will change (FAO, 2002).

Commercial aquaculture trends and production statistics

Global trends

Table 17.1 shows the development of world fish, mollusc and crustacean production during the period 1995–1999. Following a decline to 78.3 million t in 1998 (during the most significant El Niño year), total marine capture fisheries production increased to 86 million t in 2000, close to the historical maximum recorded for 1997 (FAO, 2002). As fishing pressure continues to increase, the number of underexploited fishery resources is decreasing, the number of fully exploited stocks is remaining relatively stable and the number of overexploited, depleted and recovering stocks is increasing slightly. Of particular concern is the failure of haddock (*Melanogrammus aeglefinus*), redfish (*Centroberyx affinis*) and Atlantic cod (*Gadus morhua*) stocks to respond to the drastic management measures adopted in the Northwest Atlantic. Another source of concern is the rapid increase in fishing pressure on some of the deep-water resources, since many of these stocks are slow-growing animals highly vulnerable to depletion, particularly when the stocks' distribution and dynamics are largely unknown. There is a high risk that, in the absence of effective fishery management regimes, these stocks may also be depleted long before much is known about their populations. As a consequence of the Common Fisheries Policy (CFP) failing to achieve sustainable harvests, fisheries in a growing number of ocean areas are now coming under

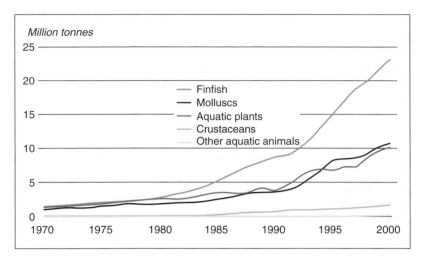

Figure 17.1 Trend of world aquaculture production by major species groups (FAO, 2002).

Table 17.1 Evolution of world fish, mollusc and crustacean production from 1995–1999 (FAO, 2002).

Production (millions of tonnes)	Marine					Inland					Marine + inland				
	1995	1996	1997	1998	1999	1995	1996	1997	1998	1999	1995	1996	1997	1998	1999
Fisheries	84.3	86.0	86.1	78.3	84.1	7.2	7.4	7.5	8.0	8.2	91.5	93.4	93.6	86.3	92.3
Aquaculture	10.5	10.9	11.2	12.1	13.1	14.1	16.0	17.6	18.7	19.8	24.6	26.9	28.8	30.8	32.9
Total	94.8	96.9	97.3	90.4	97.2	21.3	23.4	25.1	26.7	28.0	116.1	120.3	122.4	117.1	125.2

the purview of regional fisheries management organizations (RFMOs).

According to FAO statistics, aquaculture's contribution to global supplies of fish, crustaceans and molluscs continues to grow, increasing from 3.9% of total production by weight in 1970 to 27.3% in 2000 (Figure 17.2 illustrates this growth trend). Worldwide, aquaculture is growing more rapidly than all other animal-producing sectors – an average compounded rate of 9.2% per annum since 1970 compared with only 1.4% for capture fisheries and 2.8% for terrestrial farmed meat. In 2000, reported total aquaculture production was 45.7 million t by weight, 71% of which came from China. More than half of the total world aquaculture production was finfish, with growth of the major species groups continuing to be rapid with no apparent slowdown in production to date. With the exception of marine shrimp, in 2000 the bulk of aquaculture pro-

duction in developing countries comprised omnivorous/herbivorous fish or filter-feeding species, in contrast to developed countries in which 74% of finfish production was from carnivorous species.

European trends

Highlighting fish species cultured in Europe, the production of Atlantic salmon (*Salmo salar*), rainbow trout (*Oncorhynchus mykiss*), gilthead seabream (*Sparus aurata*), European seabass (*Dicentrarchus labrax*) and turbot (*Scophthalmus maximus*) is of greatest significance. The European seabass and the gilthead seabream are the two species that have characterized the development of marine aquaculture in the last two decades (Figure 17.3 presents EU production figures), with Greece being the largest producer, accounting for 60% of the European Union's production (Anonymous,

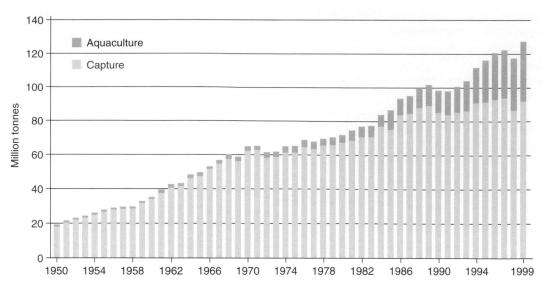

Figure 17.2 World capture and aquaculture production (FAO, 2002).

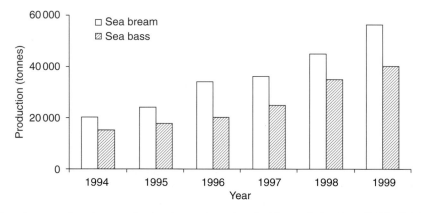

Figure 17.3 EU sea bream and sea bass production figures (reproduced with permission from FEAP, 2001).

2001). The Greeks almost exclusively farm these two species, and, until recent developments in hatchery production aided by EC grants, imported all juvenile fish from other countries. The number of hatchery-produced fry is now, however, far greater than the supply coming from the wild. European production of turbot is currently approaching the amount of the original wild landing – approximately 5500t in 2001, with the majority of this coming from France and Spain.

In recent years, a clear trend can be traced in European fish culture towards species diversification, a decline in common species market value and increased consumer interest in the more exotic or unusual fish species, which can generally demand high market price. Additionally, much research has been performed on reducing juvenile production costs of existing marine food fish species for which wild stocks have been largely depleted, the prime difficulty being that almost all marine fish larvae require live food for efficient growth. Key examples include the Atlantic cod, Atlantic halibut (*Hippoglossus hippoglossus*) and haddock.

Shellfish cultivation within Europe is almost solely concerned with bivalve molluscs. More than half the total shellfish landings can be attributed to aquaculture production; however, within the UK, cultivated outputs are less than one-quarter of the total shellfish production in volume terms (Lake, 2000). The prime aquaculture species in the shellfish sector is the European mussel (*Mytilus edulis*). Figures for 1997 show that mussel production within the EU exceeds 500 000 t, of which the UK produced only 2%.

UK trends

Salmon farming has been the most important economic development in Scotland's Highlands and Islands over the past 30 years. The industry has become the third largest of its kind in the world, being worth £300 million a year and employing around 6500 people (direct, indirect and induced), with 70% of them living in remote rural areas along the west coast of Scotland. From around 800 t of Scottish salmon being produced in a dozen or so marine sites in 1980, there are now in excess of 350 outlets producing 130 000 t of fish a year; around 98% of the salmon consumed in the UK. Rainbow trout are second to salmon in the UK aquaculture market, with the majority of production occurring throughout southern England and Wessex (Table 17.2). UK production has slowly increased over the past 5 years to around 17 000 t per annum. Although this

is not set to rise significantly, the proportion of 'value-added' products marketed is rising as the market becomes more competitive and producers seek to increase gross margins. It is worth noting the high proportion of restocking and ongrowing production sites from the data in Table 17.2 – reasons for this will be discussed in the section on 'Freshwater amenity/ recreational fisheries'.

Since the ongrowing stage of the salmon production cycle is almost identical to that of cod, much of the necessary technology, equipment, infrastructure and experience are already in place. Cod farming hence appears to be a good adjunct to existing salmon farming operations. While only small amounts have so far been produced in the UK (circa 2000 t in 2002), production in the next 10 years is forecast to increase dramatically. The British Marine Finfish Association (BMFA) has a cautious and realistic target of 25 000 t per annum of farmed cod by 2010 – some 7–8% of the current UK market. Figure 17.4 highlights how the prices for cod have been steadily on the increase over recent years. However, market prices for new aquaculture species are frequently not elastic; minor increases in supply generally lead to large decreases in prices. Nonetheless, should wild stocks fail to recover, as it seems likely, prices for farmed cod could remain strong.

Research into the farming of Atlantic halibut has reached a stage that strongly indicates such farming will become commercially and technically viable.

Table 17.2   1999 Rainbow trout production in 1999. Data from Environment Agency Region for England and Wales (Dunn, 2000. Crown copyright. Reproduced with permission of CEFAS, Lowestoft).

Environment Agency area	Number of sites					Production		
	No production	Table production	Restocking/ ongrowing production	Both (table and restocking)	Total number of sites	Table (t)	Restock/ ongrowing (t)	Fry (thousands)
Anglian	1	1	6	6	14	32	438	1 081
North West	0	1	9	9	19	200	45	83
Northumbria	1	1	3	0	5	0	158	0
Midlands	0	1	12	2	15	19	201	130
Southern	2	8	12	6	28	1893	174	238
South West	4	6	7	17	34	611	671	770
Thames	0	4	10	5	19	410	306	1 130
Welsh	2	3	21	10	35	361	270	3 380
Wessex	3	13	19	8	43	2349	687	9 497
Yorkshire	2	5	9	10	26	836	273	8 800
Totals	15	43	108	73	238	6710	3224	25 109

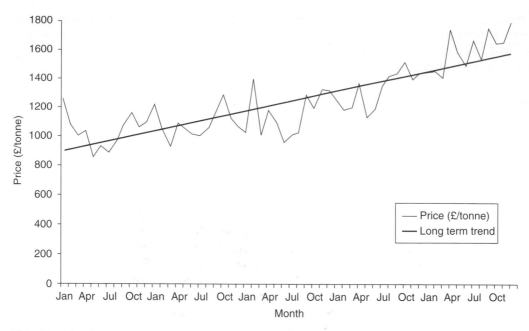

Figure 17.4   Monthly price per tonne of cod landed into Scotland 1995–1999 (Scottish Sea Fisheries Statistics 1999, SERAD, with permission of HMSO).

Securing live, wild female broodfish has been a limiting factor for the UK hatchery operations, four of which now exist, whilst for juveniles the whole developmental period is complex and long compared to other marine species, meaning that operating costs are high. However, with high demand due to the premium consumer image, market values of around £10/kg are common, making it a very attractive species for fish farming. It is unlikely that major volumes will be produced in the foreseeable future such that the market will be saturated and price adversely impacted upon. Current thinking is that a sustainable level of farmed halibut price at existing market demands is likely to be £5–6/kg for a 3- to 4-kg fish, with an acceptable production target being 6000–10 000 t per annum (Slaski, 2000).

Despite the wide range of shellfish landed from the wild, only eight molluscan species are commercially produced in the UK. Total shellfish production, by weight, continues to be dominated by mussels (Table 17.3). The bulk are produced in Wales (64%) and East Anglia (29%), with the remainder coming mainly from South West England. Farmed production of native

oysters (*Ostrea edulis*) and manila clams (*Tapes semidecussata*) in 2001 increased slightly compared to 2000 (Table 17.4). There has been a 28% drop in Pacific oyster (*Crassostrea gigas*) production, and farmed cockle production also decreased compared with 2000. A small quantity of palourdes was produced in 2001, and a few tonnes of hard shell clams were fished at one farm site, following 4 years with no recorded production. It appears unlikely that any 'new' species will revolutionize outputs from the UK industry – more emphasis is being placed on building outputs of the existing species, the market demand and on the need to establish the inherent production process constraints.

Production methods

As technical knowledge has improved and production technology has advanced, focus has been placed on producing fish in the most cost-efficient manner, without any compromise on product quality. Table 17.5 highlights annual production data for the salmon farming

Table 17.3 Production (in tonnes) of farmed shellfish in the UK in 2001 (Dunn and Laing, 2002. Crown copyright. Reproduced with permission of CEFAS, Lowestoft).

	Scotland	England	Wales	Northern Ireland	UK total
Pacific oysters	279	209	16	322	826
Native (flat) oysters	8	117	0	20	145
Scallops	28	0	–	–	28
Queens	47	–	–	–	47
Mussels	2988	4799	8568	977	17 332
Clams	–	34	–	2	36
Cockles	–	105	–	–	105
Estimated value (£ million)	4.0	2.5	3.0	1.6	11.1

Table 17.4 Farmed shellfish (table) production (in tonnes) in England and Wales 1996–2001 (Dunn and Laing, 2002. Crown copyright. Reproduced with permission of CEFAS, Lowestoft).

	Native oysters	Pacific oysters	Mussels	Manila clams	Hard clams	Palourdes	Cockles
1996	111.4	584	7618	11.5	0.3	0	0
1997	68	401	11 684	32	0	0	0
1998	106	330	9 295	19	0	12	43
1999	93	386	8 009	17	0	12	43
2000	115	313	11 224	25	0	3	147
2001	127	225	13 367	29	4	1	105

Table 17.5 UK salmon production and staff employed 1989–1998 (staff employed figures solely relate to direct salmon production).

	1989	1990	1991	1992	1993	1994	1995	1996	1997	1998
Production (t)	28 553	32 351	40 593	36 101	48 691	64 066	70 060	83 121	99 197	110 784
Staff employed										
Full-time	1 102	1 165	1 014	985	976	1 003	1 104	1 150	1 088	1 117
Part-time	316	326	272	275	248	242	251	241	207	192
Total staff	1 418	1 491	1 286	1 260	1 224	1 245	1 355	1 391	1 295	1 309
Mean productivity (ts/person)	20.1	21.7	31.6	28.7	39.8	51.4	51.7	59.8	76.6	84.6

industry over an 11-year period and illustrates how the manpower per unit of production has decreased substantially. Production technology is continually taking steps forward, and with some reductions occurring in product market value, all aspects of production are now being critically analysed.

Water recirculation systems

The constraints of freshwater use in aquaculture have been well recognized at a global level, and have been widely addressed at the practical local level, in planning new developments, managing water use in existing operations and, in an increasing number of cases, re-engineering production units to improve efficiency of water use (Muir, 2003). These changes are not just driven by a greater appreciation and concern for responsible water use, but by the very real cost pressures associated with water abstraction and waste discharge.

Fish rearing in land-based systems with water recirculation aims at producing fish with a minimum of water consumption and pollution impact on the environment. Furthermore, recirculation gives the possibility to control essential production parameters such as temperature and oxygen concentration. Recirculation has proven to be a successful technique within eel farming, and during the past 10 years several other species are now cultivated using recirculated water (Olsen, 2003). The high initial capital and the operational costs have generally limited their development to anything but highly intensive operations, or where the product is of particularly high market value. However, it has been suggested that by applying lower degrees of water recirculation, enabling reductions to be made in both investment and operational costs, the technology can be implemented in more extensive operations.

Cage farming

So-called cages – net pens with fixed tarpaulin bottoms – are commonly used as culture facilities for many of the principal aquaculture species, representing the favoured culture method for almost all the UK salmon ongrowing operations. Floating cage structures have generally been rectangular, with platforms made up of two walkways joined by the main crossbeams, with many of the more recent designs working on modular principles, i.e. once a four-sided collar is constructed, further sections are added on to it to form subsequent collars. Currently, however, the most commonly used design is the polar circle. These are made of high-density polyethene (HDP) plastic and contain no parts of wood or metal that can be destroyed by saltwater or sea worms. In recent years the trend has been for larger sized cages. Diameters of 70 m are common with depths of 15 m; hence at maximum stocking densities of 15 kg/m³ almost 60 t of fish can be held within each. The principal reason for such large-sized cages is economies of scale; however, as cage size and stock number increase, so does intra-stock size variance. It is essential therefore that careful feed management be practised with such large-sized cages.

Cages are now commonly used as culture facilities for ongrowing halibut. Full-grown halibut are typically deep-water fish, so shade netting on the top of the pen can help to protect the halibut from sunlight and appears to improve feeding response. Cod are also grown in cage net pens; however, it has been found that individuals can 'pick' their way through the net and escape. It is hence important that cod are well fed to prevent them from doing this, and to avoid cannibalism amongst stocks.

Feeding

Feeding in aquaculture has traditionally been performed by hand, but new technologies have now been implemented on larger-scale sites due to the labour intensity of hand feeding, and to negate the problem of fish being fed when appetite is not at its largest. Modern automatic feeders are now highly sophisticated, electronically controlled systems with extensive use in the salmon industry. Centralized air-blown feeders are capable of feeding numerous cages at doses predetermined by the operator, being fully controllable from the feeding system barge. Further developments include sensors at the base of cages; as pellets begin to fall onto the sensor the stock is known to have approached satiation, and hence the feeding system can be programmed to directly respond to appetite. This allows for reductions in feed wastage, having significant economic and environmental benefits. Manual cyclone blowers are also commonly used, but generally on smaller cage sites. Although these may cause quite high levels of pellet breakage, the initial investment is low, and when used in combination with submerged cameras offers an excellent alternative to the fully automated systems.

Nutritional developments

Fishmeal and fish oil replacement

Modern fish feed is characterized by high protein and energy (fat) content and produced using a high-tech extrusion process. Traditionally, fish feed has depended to a large extent on marine raw materials – fishmeal and fish oil – as they provide high amounts of digestible protein and fat. However, as global aquaculture production continues to expand, production of fishmeal and fish oil remains at best static (Figure 17.5) while their price continues to escalate. Furthermore, certain environmental pollutants can pose a threat as they bioaccumulate in the marine food chain and also end up in fishmeal and fish oil (see section on 'Food safety'). Recent development of feeds in the already established salmonid culture has therefore focused upon finding

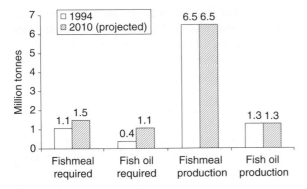

Figure 17.5 Requirements of intensive and semi-intensive aquaculture for fishmeal and fish oil (Roberts and Talbot, 1997, with permission of CEFAS).

cost-effective protein and fat sources that can partially substitute for expensive marine raw materials of inconsistent quality.

Research on alternative protein sources has concentrated on the use of plant products due to their high availability and relatively inexpensive cost. Numerous plant forms have been shown to be suitable for replacing fishmeal in practical salmonid diets, primarily various soy products, rapeseed, maize gluten meal, wheat gluten meal and sunflower meal. Figure 17.6 presents information on the relative availability of some of these. The use of plant-derived materials as fish feed ingredients for salmonids is, however, often limited by the presence of indigestible carbohydrates and a wide variety of antinutritional factors (ANFs). Common processing techniques like dry and wet heating, extracting with water and addition of feed supplements have been widely and successfully used to reduce the concentration of ANFs in plant feeds, but such techniques add

cost to the raw material, making them less competitive fishmeal substitutes.

Future work on finding alternative protein sources will inevitably continue. The range of plant products available for use in fish feeds is undoubtedly large; hence by incorporating numerous products, each with their own particular nutritional qualities, a more balanced and nutritionally complete feed may be produced. The formulation of such diets requires extensive research and should be reviewed. Further work is being performed to find other potential protein sources, or variations thereof, to replace fishmeal. It is unlikely that a single analogue will be found to totally replace fishmeal; however, more substantial replacement of fishmeal (without reducing performance) is an aim that at some point must be achieved if aquaculture growth is to remain sustainable.

Larval nutrition

The emphasis for developments in larval nutrition has been on the replacement of live foods with artificial micro-diets. The production of live food represents a significant proportion of hatchery operating costs, and, together with providing live food of suitable nutritional quality, this has been the major bottleneck in mariculture to date. Artificial micro-diets using special, high-grade fishmeals and agglomeration technology have now been successfully applied in research experiments, but only in a limited number of commercially important species. Stimulating the production of digestive enzymes is reported to be the main limitation surrounding their use. An alternative area of interest to producing artificial diets is the use of copepods. Copepods are the natural food of marine fish, and hence feeding these may overcome the nutritional deficits of

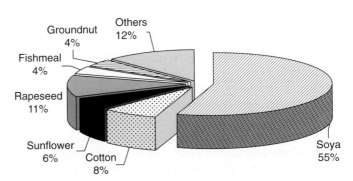

Figure 17.6 Availability (as a percentage of global production) of protein raw materials suitable for fish feed.

Artemia and rotifer. The technology to produce them on a large scale and the problems associated with production have, however, so far restricted their application.

New aquaculture species

Fish farmers in Scotland are familiar with using high-energy feeds for producing salmon, but there are a number of key differences in the nutritional requirements of cod, namely the requirement for very low oil levels in the feed. Cod is a lean fish, with less than 2% fat in the muscle. The main storage facility for surplus fat is therefore the liver, which can comprise up to 23% of the total body weight. Cod liver oil is of course a potentially valuable resource, rich in vitamins A and D and high in polyunsaturated fatty acids; however, in order to optimize fillet yield, cod need low oil diets to avoid excessive processing loss. The scenario is similar for haddock, whereas halibut seem to respond best to higher levels of dietary fat. Interestingly, cod liver oil, which is in itself a high-value commodity, may be a potential by-product for cod farming providing the chemical composition is identical to that from wild fish.

Genetic developments

Selective breeding

The science of applied selective breeding and genetics has contributed greatly to the steadily increasing productivity of terrestrial agriculture. This has not been true for aquaculture, where only 1% of aquaculture production is based on improved stocks (Johannsson, 2003).

Artificial selection is favourable in fish because: (1) females are highly fecund, (2) fertilization is external and (3) the variance in economic traits is large. Presently, the most significant traits being selected for include growth, disease resistance, size at sexual maturity and carcase quality (body shape, fillet fat content and pigment retention). Many of the current selective breeding programmes incorporate the latest genetic 'fingerprinting' technology within a traditional pedigree family breeding system. By incorporating microsatellite analysis (MSA), best-performing fish on customer sea sites can be easily traced back to their broodstock family held at the breeding unit. Full sib-

lings are then used to provide the optimum crosses between unrelated individuals to maximize the improvements of the desired commercial trait. There is no doubt that selective breeding does offer enormous potential; the problem so far has been resourcing it.

Transgenics

Transgenics involves transferring one or more genes from one animal to another to improve production characteristics. Perhaps the most appealing application of transgenics to aquaculture is the production of fish capable of digesting and utilizing dietary ingredients more efficiently than is naturally possible. Considering the requirement for aquaculture to reduce its dependency on marine resources, such a development would be a massive boon to the industry. In comparison to lengthy selective breeding programmes, the use of transgenics can significantly increase the degree of control over what traits the progeny are to express; selective breeding often requires high heritability values. Other potential applications of transgenics to aquaculture include the production of growth-enhanced fish and the development of fish resistant to economically significant diseases.

Currently there are many federal, state and local laws, rules and regulations governing research and commercial exploitation. The key problem facing commercial application is overcoming bad perception on ethical grounds, welfare issues and proving their environmental safety should stock be allowed to escape. Two key environmental issues must be answered to confirm their safety: (1) What is the probability of a transgene spreading from escapees into a natural population through interbreeding? (2) What is the potential for competitive displacement of wild populations due to altered traits of transgenic organisms?

The use of transgenics offers the potential for significant production gains; however, until further work is performed to confirm its safety, both environmentally and as a food source, the use of transgenic fish will be strictly restricted to research projects only.

Sex/chromosome manipulation

During early embryology fish embryos are phenotypically neither male nor female, in that they do not possess ovaries, testes or other characteristics associated with the reproductive system, i.e. the embryo is totipotent. To direct the phenotype, fish fry are typically fed diets containing either androgens or oestrogens or are raised in baths containing the hormone, allowing for the production of monosex populations. This can prevent early recruitment and stunted population growth, negate differential growth between sexes and delay sexual maturity, allowing for greater flesh quality at harvest.

Due to the perception of food fish being directly exposed to sex-reversing hormones, hormonal sex control is always taken a step further using genetics. In trout and salmon, for example, it is favourable to produce female fish since they are less aggressive than males, and female flesh quality is greater at the point harvest. By sex-reversing genotypic XX females to phenotypic males, then subsequently reproducing these with normal XX females, offspring produced are 100% XX female.

The technology of chromosome manipulation is now widely applied in salmonid culture, the aim being to delay sexual maturity and cause sterility within all production fish. When combined with sex control this can produce monosex sterile populations, negating any possibility of escapee stockfish outbreeding with those from the wild. There has been much discussion surrounding the potential for the use of all-female triploid brown trout (*Salmo trutta*) for stocking to avoid genetic impact upon native stocks. All-female triploids appear to satisfy the above requirement and have all the beneficial attributes of normal diploid fish. However, it remains necessary to further examine the performance of stocked all-female triploids with respect to possible interactions with wild fish.

Safety and welfare issues

Food safety

The enormous attention to food safety generated by food safety scares in the last 10 years, such as BSE, *Escherichia coli*, salmonella, listeria, polychlorinated biphenyl (PCB), heavy metals, etc., has permeated all aspects of food safety legislation in the EU (e.g. Elliot and Young, 2002). The Food Standards Agency (FSA) has raised concern surrounding mercury contamination levels in some of the increasingly popular fish species,

such as shark, swordfish and in fresh tuna. Whilst mercury residues discharged into the sea have been a known risk for many years, causing a similar concern to that regarding dioxin and PCB concentrations increasing in wild stocks being used for fishmeal production, any possible risks associated with eating fish, or fish oil, would have to be balanced against its well-documented and important nutritional benefits. New data are being examined by scientists to determine the severity of the mercury problem, which is likely to result in recommending maximum intake limits for the affected fish species. New technology is needed to remove the contaminants from marine feed ingredients in order to produce safe feed and aquaculture foods that meet EU legislation. Substantial focus is to be expected on the area of improving feed hygiene and removing undesirable substances.

We have also seen additional magnitudes of safety factors being applied by the EU scientific committees to products previously licensed for use in fish. The use of canthaxanthin in salmon feeds, an artificial carotenoid required to give the flesh its typically orange hue, has came under scrutiny by the committees due to the ill-effects that have been observed through use of sun-tanning pills. Although evidence suggests that for this to be of significance when eating salmon approximately 10 kg of fish would need to be eaten daily, the Food and Drug Administration (FDA) is to reduce the maximum inclusion level from 80 to 25 mg/kg feed. Since colour is an important sensory assessment of 'quality' by consumers, fish feed manufacturers will be required to replace a portion of pigment with the more expensive, naturally occurring carotenoid astaxanthin, increasing feed cost and putting further pressure on the fish farmer.

Fish welfare

Mirroring public concerns about fish health is the closely interrelated subject of fish welfare. The argument as to whether fish feel pain has gone back and forth for many years and is increasingly becoming a topic of public concern. Whilst the experience and practices of successfully rearing farmed fish means that the industry now has a better knowledge of good welfare, retailers are determined to prevent public criticism of industry products. This has led to considerable research into slaughter methods that improve fish welfare without compromising product quality. However, in the present absence of definitive indicators of discomfort and distress in freshwater species, including the correlation with stress hormones, ideologies as to what is a humane slaughter method and what is not are generally only based on perception. The Standing Committee of the European Convention for the Protection of Animals kept for Farming Purposes turned its attention to farmed fish some 4 years ago and started drafting conditions of fish welfare, although it still has a long way to go before completion. It is assumed that focus on welfare issues will strongly increase in the future.

Bodies and organizations

The political position of European aquaculture comes under the Common Fisheries Policy (CFP), which has recently been the subject of a Communication on *A Strategy for the Sustainable Development of European Aquaculture* (available from www.ieep.org.uk/pdffiles/publications/cfpbriefings) by the Commission – the first time that the subject has been treated in depth at this level. Within the recommendations and application of the Commission's Strategy for Sustainable Aquaculture, broadly welcomed by the production sector, the CFP will remain the sector's guideline for the coming years. For aquaculture's contribution and potential to be fulfilled by this Policy, the sector must assure its economic viability while guaranteeing food safety to the consumer and addressing all of the issues that comprise sustainability, for which a wide range of actions have been proposed.

The Environment Agency has a duty to secure the proper use of water resources in England and Wales. It monitors water in the environment, and issues abstraction licences to regulate who can take water from the environment. These specify the amount of water someone can take from a location over a period of time.

As the seabed and most of the foreshore are part of the Crown Estate, developers of fish farms have to apply to the Crown Estate Commissioners for a lease of the area that the fish farm will occupy. Consent for the installation of fish farming equipment in sea areas has to be obtained from the Marine Directorate of the Department of Transport – the main concern being the effects of navigation generally, and particularly listed anchorages and restricted channels. The relevant plan-

ning authorities are notified by the Crown Estate of fish farming applications in sea areas to enable them to indicate any onshore constraints and general views on the proposals. Planning permission will usually be necessary for the onshore development.

Freshwater amenity/recreational fisheries

Recreational fisheries

There is a lack of relevant data on recreational fisheries from many European countries, or many of the available data are not comparable due to different interpretations of 'recreational fisheries' between countries. For the purposes of reporting catch statistics, the FAO (1997) defined recreational fisheries as 'Fisheries conducted by individuals primarily for sport but with a possible secondary objective of capturing fish for domestic consumption but not for onward sale' – the implication being that the activity is not undertaken for the purpose of commercial gain by the angler. Thus recreational fisheries involve both subsistence fishing, where the catch is consumed, and leisure fishing, where the fish are returned live to the water. As a generalization within UK recreational fisheries, subsistence fishing can be linked with game angling, and leisure fishing linked with coarse angling. Since many game fisheries, and game anglers alike, now practise a 'catch and release' policy, this generalization may be somewhat inaccurate. Sea angling can involve subsistence and/or leisure fishing; however, the sea in itself is not a recreational fishery. Figure 17.7 gives an indication of the popularity of each of the forms of angling, taken during a National Rivers Authority (NRA) survey in 1994. Following on from this, Figure 17.8 shows where the majority of these anglers are located.

The broad pattern within the coarse angling community has been a decline in interest in river and canal fishing and increased interest in fishing newly developed intensively managed fisheries. It is generally accepted that the canal match fishing circuit is stable or slowly declining, whereas the number of pleasure anglers regularly fishing canals has declined significantly; there appears to be less interest in acquiring the skill necessary to catch large numbers of small fish, and less desire to catch small fish *per se* (Ellis, 1998). Carp

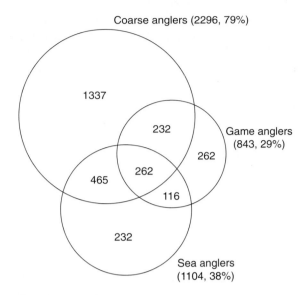

Coarse anglers (2296, 79%)

1337

232

Game anglers
(843, 29%)

262

262

465

116

232

Sea anglers
(1104, 38%)

Figure 17.7 Number of anglers (thousands) practising each type of angling in England and Wales (reproduced with permission from Environment Agency, 1994).

fishing is the most rapidly growing sector of angling in the UK, reflected by the five-fold increase in the development of carp-only fisheries in the last 15 years. The majority of modern anglers expect to catch these large, hard-fighting fish almost every visit, with little effort on their behalf, and it is apparent that the angler is prepared to pay for this. Market-leading intensively managed fishery complexes in the UK include the Gold Valley/Willow Park complex in Hampshire, Old Bury Hill Lakes in Surrey and Castle Ashby in Northamptonshire.

Angling

The *Public Attitudes to Angling* research commissioned by the Environment Agency (EA) revealed there were circa 3.9 million people who had fished at least once in the last 2 years (2000–2001) – some 9% of the population over the age of 12 years (Simpson and Mawle, 2001). A further 8% were 'very' or 'quite' interested in learning how to fish, giving angling the potential to involve 7.4 million participants – more than any other pastime, by a long shot! For angling to surpass walking and swimming as the favoured pastime, however, those people interested in learning must be attracted into doing so. Knight (2002, *Anglers Survey*, unpublished)

highlights how the Joint Angling Governing Bodies (JAGBs) now have a 4-year development plan that aims to achieve this, encompassing policies for juniors, women, ethnic minorities and the disadvantaged. It contains the blueprint for a Coach Licensing System, designed to professionalize angling and bring it into line with other major participation sports, and contains the requirement for a Child Protection Policy – deemed essential to keep the sport credible. Increasing awareness that facilities exist where those interested can come and learn to fish, and increasing the number of these facilities, is a must for the development plan to succeed, along with the financial investment required to do so.

In lieu of the relative stability in angler numbers, the area and/or the number of waters managed for angling purposes is rapidly increasing, possibly due to the increasing diversification in the use of lakes (Pinter and Wolos, 1998). Only 10 and 45% of inland waters less or more than 1 ha respectively are currently used for fishing in the UK, the main constraint in development being land ownership and access. Nevertheless, this statistic indicates striking potential for growth in the UK fisheries industry, particularly so if the further 8% of the UK population interested in angling can be encouraged to join in! In 1997 the collective annual expenditure on countryside sports in the UK was estimated to be £3.86 billion; angling contributed a massive £3.26 billion towards this (Cobham Resource Consultants, 1997). Of particular interest is a point made in the recent 2001 report prepared for the National Assembly for Wales by Nautilus Consultants: 'sea sport fishing in Wales alone generates expenditure of £28 million per annum – much of it in tourism – for species that commercially are worth just £3 million' (Broughton, 2002). The economic benefits of increasing and diversifying angling participation are clear; it will be interesting to see how it develops over the coming years.

Amenity aquaculture

The tremendous growth in UK carp waters has led to the inevitable increase in demand for carp to stock with. Although the production of carp for recreational purposes is not new – the small carp market could be classed as fully exploited – the potential for growth in *large* carp production is significant. Over the past decade, as angler demand for larger fish has grown, a

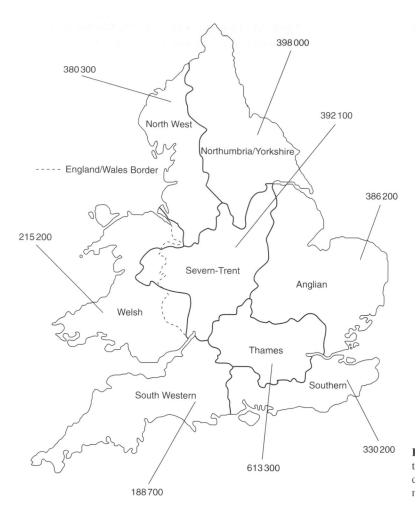

380 300

398 000

392 100

North West

Northumbria/Yorkshire

- - - - England/Wales Border

386 200

215 200

Severn-Trent

Anglian

Welsh

Thames

South Western

Southern

330 200

613 300

188 700

Figure 17.8 Distribution of anglers throughout England and Wales (reproduced with permission from the Environment Agency, 1994).

proportion of fishery managers have sought to stock these larger fish to gain 'the competitive advantage' over the fishery next door. A lack of consistent supply from legitimate UK suppliers led to a substantial increase in illegally imported carp from the European mainland, fish in these parts being much more abundant than in the UK. Furthermore, each importation of carp has the potential for transfer of spring viraemia in carp (SVC) – dozens of instances have been documented where the introduction of these fish has resulted in the complete eradication of native stocks. Environment Agency figures show that in 2000 approximately 300 000 carp were legally stocked in UK waters, with an approximate value of £7 million. This figure does not include illegal importation and movements, which may account for half as many stockings again. Given the

recent government crackdown on illegal importations, demand in the UK for large, English-bred carp is rising. Aquaculture will be the predominant supply of this.

A relatively new and rapidly increasing development in recreational fishing is the availability of put-and-take fisheries. There has been a tendency to promote this type of fishery as a commercial venture, not just in the UK but in all countries, not only making use of the substantial aquaculture facilities available but also creating new ones (FAO, 1998). It is quite possible that these stocked game fisheries provide up to 20% of the UK market for rainbow trout. Additionally, put-and-take fisheries provide anglers with easy access throughout the UK to trout stillwaters, offering relatively cheap fishing on almost any size of water the angler requires.

Table 17.6   Current and forecast workforce and business numbers with fishery management, 1999–2005 (reproduced with permission from Lantra, 2000).

	Current estimate					Annual forecast		Change 1999–2005
	1999	2000	2001	2002	2003	2004	2005	
Businesses	3100	3100	3200	3200	3300	3400	3400	+300
Workforce	5900	6000	6100	6200	6300	6400	6500	+600

Management

With at least 20 million people in the countries of the European Inland Fisheries Advisory Commission (EIFAC) participating in recreational fishing, there is a definite need for greater understanding of the value and management requirements of recreational fisheries. National Agencies of EIFAC member countries are now aiming to integrate sustainable, long-term approaches to aquatic resource management, which ensures full consultation between various interest groups and provides for the maintenance of biodiversity, protection and enhancement of fish stock, animal welfare issues and optimal socio-economic benefits to society (FAO, 1998). Table 17.6 gives an indication of the forecast workforce and business numbers within the fishery management sector from 1999–2005.

The ornamental industry

Retail sales of ornamental fish and associated dry goods amount to between £200 and £300 million per annum. Coldwater specimens and associated materials contribute very significantly to this figure. Some 2000 retail outlets sell live fish, over 100 businesses import and wholesale ornamental fish, while there are several hundred manufacturers servicing the needs of the industry and hobby. As many as 10 000 people may be employed directly or indirectly by the industry in the UK. All of the sales made by the industry ultimately rely on continued confidence in the quality of the fish offered to the hobby by the industry.

The majority of coldwater ornamental fish are imported. During 1997 approximately 6 million coldwater ornamental fish were imported into the UK via Heathrow from four countries – Israel, Japan, China (including Hong Kong) and the USA. Coldwater orna-

mentals are also imported from other countries such as Thailand, Malaysia and Singapore. In total, more than 12 million coldwater ornamental fish are imported from all sources annually. Home production of coldwater ornamentals is a small but growing source of supply.

In 1997 the UK accounted for 37% of imports into the EU of coldwater ornamental fish from Israel, China and Japan. While interest in the coldwater fish-keeping hobby has historically been strong in the UK, it is also growing in popularity in the remainder of the EU.

References

Anonymous (2001) *Fish Farming Expands to Meet Demand*. Available at http://www.greece.gr/business/food anddrinkindustry/fishfarmingexpandstomeetdemand.stm (accessed 17 Jan 2002).

Broughton, B. (2002) *Economic Benefits of Angling*. Report prepared for the National Assembly for Wales by Nautilus Consultants.

Cobham Resource Consultants (1997) *Countryside Sports – Their Economic and Conservation Significance*. Proceedings of Standing Conference on Countryside Sports, College of Estate Management, Whiteknights, Reading.

Dunn, P. (2000) 1999 Survey of Trout Production in England and Wales. *Trout News*, **30**, 5–7.

Dunn, P. and Laing, I. (2002) Shellfish production in the UK in 2001 – England and Wales. *Shellfish News*, **14**, 39–43.

Elliot, V. and Young, R. (2002) Fish 'a danger' to pregnant women and children. *The Times*, 11 May.

Ellis, J.W. (1998) Trends in recreational angling: the British Waterways experience. In: FAO (1998) *Recreational Fisheries: Social, Economic and Management Aspects* (FAO), pp. 63–9. Blackwell Science, Oxford.

FAO (1997) *Inland Fisheries*. Technical Guidelines for Responsible Fisheries no. 6, p. 36. FAO, Rome.

FAO (1998) *Recreational Fisheries: Social, Economic and Management Aspects*. Blackwell Science, Oxford.

FAO (2002) *The State of World Fisheries and Aquaculture Part 1 – World Review of Fisheries and Aquaculture. Fisheries Resources: Trends in Production, Utilisation and Trade.* Available from http://fao.org/sof/sofia/index_en.htm.

FEAP (2001) *Marine Fish Production – Bass and Bream.* Available at http://www.feap.org/breamprod.html (accessed 4 Feb 2002).

Johannsson, V. (2003) Benefits of Selection and Hatchery Technology. In: Abstracts of PROFET, *Freshwater Aquaculture and the Use of Water*, Billund, Denmark. Available at www.aquamedia.org.

Lake, N. (2000) Future of UK shellfish production. In: Abstracts of 3rd Annual *C-Mar/PESCA Aquaculture Workshop*, Portaferry, 7–8 Sept.

Lantra (2000) *Skills Foresight 2000 – Identifying Skills Needs in the Land-Based Sector.* Lantra, Coventry.

Muir, J. (2003) The use of freshwater in aquaculture – challenges and prospects. In: Abstracts of PROFET, *Freshwater Aquaculture and the Use of Water*, Billund, Denmark. Available at www.aquamedia.org.

National Rivers Authority (1994) *National Angling Survey* 1994. Fisheries Technical Report 5. Environment Agency, Bristol.

Olsen, B.H. (2003) Recirculation and sustainable development – intensive systems. In: Abstracts of PROFET, *Freshwater Aquaculture and the Use of Water*, Billund, Denmark. Available at www.aquamedia.org.

Pinter, K. and Wolos, A. (1998) Summary report of the symposium topic session on the current status and trends in recreational fisheries. In: *Recreational Fisheries: Social, Economic and Management Aspects* (FAO), pp. 1–4. Blackwell Science, Oxford.

Poulter, S. (2001) Salmon poison alert. *Daily Mail*, 4 Jan.

Roberts, J. and Talbot, C. (1997) Will trout be forced to become vegetarian? *CEFAS*, **25**, 16–19.

Simpson, D. and Mawle, G. (2001) *Public Attitudes to Angling.* Technical Report W2-060. Environment Agency R&D Dissemination Centre, Wiltshire.

Slaski, R. (2000) Marine fin fish farming – focus on cod. In: Abstracts of 3rd Annual *C-Mar/PESCA Aquaculture Workshop*, Portaferry, 7–8 Sept.

Further reading

Arthur, G. (1999) *The Atlantic Halibut – A Potential Species for Fish Farming in Shetland.* Fisheries information note, no. 2. North Atlantic Fisheries College, Scalloway Shetland.

FEAP (2000) *AGM – Press Release.* Available at http://www.feap.org/press_releases.html (accessed 4 Feb 2002).

Walden, J. (2000) *The Atlantic Cod – a Potential Species for Fish Farming in Shetland.* Fisheries information note, no. 3. North Atlantic Fisheries College, Scalloway, Shetland.

18

Tourism

P.R. Brunt

What is tourism?

Within literature relating to tourism and across various national and international agencies, there has long been debate over the precise definitions of tourism and issues relating to tourism. In general terms, it is accepted that there are four groupings that participate in and are affected by the tourism industry:

(1)   the tourist;
(2)   the business providing tourist goods and services;
(3)   the government of the host community/area;
(4)   the host community.

Lowry (1994) suggests that many people believe that tourism is a service industry that takes care of visitors when they are away from home. Some restrict the definition of tourism by the number of miles away from home, overnight stays in paid accommodation, or travel for the purpose of pleasure or leisure. Others think that travel and tourism should not even be referred to as an industry. Gunn (1994, p. 4) believes that tourism 'encompasses all travel with the exception of commuting', and that it is more than just a service industry. McIntosh and Goeldner (1986, p. ix) say that 'tourism can be defined as the science, art, and business of attracting and transporting visitors, accommodating them, and graciously catering to their needs and wants'. They also introduce the notion that tourism is interactive in that they believe that 'tourism may be defined as the sum of the phenomena and relationships arising from the interaction of tourists, business suppliers, host governments, and host communities in the process of attracting and hosting these tourists and other visitors' (p. 4). Accepting then that tourism relates to activities of persons travelling to and staying in places outside their usual environment (for not more than one con-

secutive year) for leisure, business and other purposes, this broad definition can be seen to have numerous subcategories.

There are also some elementary forms of tourism when considering particular destination areas or a country. These include:

- domestic tourism – relating to the activities of residents of a given area travelling only within that area, but outside their usual environment;
- inbound tourism – the activities of non-residents travelling to and within a given area that is outside their usual environment;
- outbound tourism – the activities of residents of a given area travelling to and staying in places outside that area (and outside their usual environment).

When referring to a country, we can think of:

- internal tourism, being made up of 'domestic tourism' and 'inbound tourism';
- national tourism as 'domestic tourism' and 'outbound tourism';
- international tourism as 'inbound tourism' and 'outbound tourism'.

Although beyond the scope of this chapter, it is worth noting that also within the literature there are additional tourism categories. These include, for example, ethnic, cultural, historical, environmental, recreational and business tourism.

Benefits of tourism

The benefits of tourism are numerous, and they can be thought of in several ways, as follows.

Personal value

- Millions of people regard their holidays as some of the most important aspects of their lives. This may well be when families get together or to re-establish relationships with friends and relatives.
- The holiday can be thought of as a means of reducing tension and is therefore vital during times of personal stress and national crisis.
- Visiting natural and rural areas provides opportunities for private contemplation, unique experiences that are unattainable elsewhere.

Increased knowledge and concern

- Travel improves knowledge of other people, areas and natural resources which may lead to a greater understanding of and concern for these.
- Travel is a means of raising the level of human experience and achievements in areas such as education, research and artistic activity.

Economic benefits

- Every country now views tourism as an important factor in national prosperity and in increasing employment opportunities.
- Spending by tourists can be 'multiplied' within an area. For example, the money paid in accommodation expenses is used by the hotelier to purchase foods locally, laundry services, stationery, etc. Hence there is, potentially, a 'knock-on' effect that multiplies the value of the injection of money from tourists.
- Tourism is characterized by a large number of very small businesses that support and are supplementary to the industry. As tourism flourishes so do they.

Promotion and protection of natural and cultural areas

- The creation and preservation of wildlife reserves and parks as well as national monuments and other cultural resources are often encouraged when tourism begins to be a force in the society.

- Nature-based and rural tourism has the potential to provide unique, regional experiences that create an awareness of the special value of the areas visited.
- When properly managed, nature-based tourism is a way of offering low-impact experiences that bring benefits to tourists, locals and to the resource itself through efforts to preserve its quality.

Measuring the value of tourism

In many ways the value of tourism is thought of in terms of the economic benefits generated. These are measured by investigating the expenditures made by tourists before, during and after their travel. The World Travel and Tourism Council (WTTC) (2003a) sees travel and tourism comprising basically five groups of activities:

(1) transportation: e.g. airlines, ferries, trains, rental cars, taxis;
(2) accommodation: e.g. bed and breakfast, guest houses, hotels, motels, caravan sites, cruise ships;
(3) catering/retail: clothing/footwear, restaurants, bars, souvenirs, luggage;
(4) recreation: e.g. arts and music festivals, historic sites, museums, zoos, gardens, wildlife parks, beaches, sporting events;
(5) travel-related services: e.g. credit cards, currency exchange, tour operators, travel agents, travel insurance, travellers cheques.

Within this list it can be seen that some activities (e.g. airlines and hotels) can be thought of as being wholly associated with the tourism industry, while in others (e.g. restaurants and sporting events) only a portion of their total business is related to tourism. In short, although an activity may be defined as being part of the tourism industry, tourism may account for only a portion of that activity's total economic contribution. It is these types of issue that make measuring the economic value of tourism very difficult. Measuring this contribution, is, however, important because in such a large industry it is essential to know what is going on. So how large is the industry and how reliable are the statistics?

Various organizations measure the value and extent of tourism; among the most respected is the WTTC who put together its measures of tourism through extensive

Table 18.1 The value of tourism at different geographic levels (WTTC, 2003a, 2003b, 2003c).

	World	*Europe*	*UK*
Number of jobs in the industry directly employed	67 441 100 2.6% of total employed	7 388 500 4.4% of total employed	1 061 470 3.6% of total employed
Number of jobs directly and indirectly employed	194 562 000 7.6% of total employed	20 678 000 12.4% of total employed	2 953 330 10.0% of total employed
Direct impact in US$	$1280.4 billion or 3.7% of total GDP	$385.6 billion or 4.1% of total GDP	$65.9 billion (£42.7 billion) or 4% of total GDP

Table 18.2 Internal tourism trips in England by destination, 2001. (Reproduced with permission from United Kingdom Tourism Survey, cited in Star UK, 2003.)

Destination type	*Millions of trips*	*%*
Countryside/village	29.6	22
Seaside	26.6	20
Large city/large town	49.2	38
Small town	25.3	19
Not stated	1.2	1

research on consumers and businesses within the industry. Some of the WTTC's key findings are shown in Table 18.1. These figures show just how large tourism really is and why it is referred to as the largest global industry. What is more is that the WTTC (2003a) expects growth within the industry to be 2.9% in 2003 and 4.6% annual growth between 2004 and 2013. In England alone some 131.9 million trips were made by UK residents in 2001 (internal tourism) and 19.3 million by overseas residents, the majority of which were for holidays or to visit friends and relatives (Star UK, 2003). The UKTS records staying trips of one night or more to a variety of different types of location made by UK residents for social, recreational or business purposes. These are shown in Table 18.2.

It can be seen, therefore, that a particular feature of tourism, among others, within the UK is in the number of trips made to the countryside and its villages. The figures from 2001 represented a decline from the year 2000 due to the restrictions in terms of recreational travel caused by the foot-and-mouth epidemic. Another particular feature of tourism within the countryside is that much of it is dominated by people who

stay overnight (70%) as opposed to with friends or relatives (22%) or for business purposes (8%). Compared with visits to other locations, the countryside and its villages represent a place where people go on holiday. This is referred to as a sub-category of tourism – rural tourism – and the next section investigates this more fully.

What is rural tourism?

First some facts. According to the Countryside Agency (2003):

- defining rural areas as including settlements with a population under 10 000, 9.3 million people live in rural areas of the UK (1 in 5).
- 1.3 billion day visits were made to the countryside in 1996.
- 144 million domestic tourism nights were spent in the countryside.
- Visitors to the countryside spent £9 billion in 1994.

According to Lane (1994), at first glance, defining rural tourism seems easy – rural tourism is tourism that takes place in the countryside. However, as most writers point out, once you start delving into the subject such a straightforward definition is unsatisfactory. However, trying to develop a better definition may not be that useful either, because of the complexity of what should or should not be included. Lane discusses these problems in depth, but we should consider the following points:

- What is a rural area anyway? There is considerable literature which attempts to define this, but there has yet to be a universally accepted definition.

Table 18.3 Activities carried out while on trips to the English countryside, 2001. (Reproduced with permission from United Kingdom Tourism Survey 2001, cited in Star UK, 2003.)

Countryside activity	Millions of trips	%
Short walk (up to 2 miles)	15.9	54
Long walk (over 2 miles)	7.4	25
Visiting heritage sites	7.1	24
Field study/nature study/bird or wildlife watching	5.6	19
Swimming	5.0	17
Visiting artistic or heritage exhibits, e.g. museum	4.7	16
Watching performing arts	3.1	11
Cycling	2.5	9
Visiting theme park/activity park	2.4	8
Golf	1.3	4
Fishing	1.2	4
Watching sporting event	0.9	3
Visiting traditional regional music event	0.8	3

- Not all tourism in rural areas is rural in nature. Theme parks and large accommodation complexes could be argued to be more akin to urban or resort-based tourism that just happens to be located in rural areas. The degree of 'rurality' is an interesting concept, possibly emotive at times. For example, where does Centre Parcs or Alton Towers fit?

- Different countries have different forms of rural tourism. Here, we often make an immediate connection with, say, farm tourism or farm accommodation. Certainly this form of tourism is popular in the UK and places like Germany or Austria, but rather uncommon in the USA or Canada. In France, for instance, *gîte* holidays are an important component of rural tourism.

- Rural areas are dynamic not static. This perhaps goes without saying, but the influence of technology, desire for quality of life and the like may mean that many choose to live in rural areas or retire there. As such there is a blurring between the characteristics of those living in rural, suburban and urban areas, with second homeowners not fitting a particular category.

- Rural tourism activity – when you begin to consider tourist practices, recognition of definitional difficulties is confirmed. What activities are in and what are out? Conceivable rural tourism could include nature holidays, eco-tourism, walking, climbing, riding, adventure, sports, health tourism, hunting, angling, educational travel, arts and heritage. However,

when you consider each of these separately there are clear difficulties in whether they are solely 'rural' activities.

Table 18.3 shows a selection of activities that could conceivably be thought of as being included within those associated with rural tourism. However, among these there may be some (e.g. countryside walks) that are easy to see being associated with rural tourism, while others (e.g. visiting a theme park) that are not.

Clearly, the issue of defining rural tourism is not straightforward, as what might constitute 'rural' and what might constitute 'tourism' both have difficulties in terms of definition. In order to move towards something workable, each of these will be looked at in turn.

It is not the purpose here to delve deeply into what is rural; however, some facts worth noting include:

- The national criteria for what is a rural settlement varies. In the UK it would appear that towns with more than 10 000 inhabitants are excluded (Countryside Agency, 2003), whereas in Denmark and Norway, rural settlements are agglomerations (collections) of less than 200 inhabitants. Despite the differences, it is fair to say that we are dealing with small settlements and low population densities.

- Rural areas and land use – most writers agree that agriculture, forestry and wilderness areas characterize rural areas.

- Traditional social structure – this is a problematic area, but many consider there is something to be

Table 18.4 Contrasting features of rural tourism and urban/resort tourism (Lane, 1994).

Urban/resort tourism	Rural tourism
Little open space	Much open space
Settlements over 10 000 inhabitants	Settlements under 10 000
Densely populated	Sparsely populated
Built environment	Natural environment
Many indoor activities	Many outdoor activities
Large establishments	Small establishments
Nationally/internationally owned firms	Locally owned businesses
No farm/forestry involvement	Some farm/forestry involvement
Tourism interests self-supporting	Tourism supports other interests
Many visitors	Fewer visitors
Cosmopolitan in atmosphere	Local in atmosphere
Modern buildings	Often older buildings
Broad marketing	Niche marketing

said about rural societies (close-knit, family farming businesses, integrated with the environment, etc.).

Based on a consideration of these issues, Lane (1994) argues that rural tourism is, therefore:

- located in rural areas;
- functionally rural (built upon features of rural areas being small-scale, open spaces, 'traditional', etc.);
- rural in scale (i.e. small);
- traditional in character (slow growing, connected with local families);
- of many different kinds (representing the complex pattern of the rural environment and economy).

Lane further contrasts rural tourism with urban or resort tourism that is more typical of densely populated areas and international in style and ownership. Table 18.4 displays some of these contrasting features and Table 18.5 lists some activities and accommodation types that are likely to be rural in focus. There may well, however, be a continuum between rural and urban features of tourism. For example, while Table 18.4 suggests 'black and white' definitions, some tourist practices being either specifically urban or rural, there could be a third type that might be either. Few areas will display all of the urban or rural characteristics and some will be in the process of possibly developing into larger resorts. Similarly the activities and accommodation types in Table 18.5 may also exist in urban or semi-urban areas.

Moving away from problems of definition and typology, it can be said that rural areas, despite their diver-

Table 18.5 Rural holiday activities and accommodation (Lane, 1994).

Typically rural holiday activities	Rural accommodation
Walking	Country hotels/guest houses
Climbing	Farmhouse bed and breakfast
Canoeing/rafting	Touring caravans
Cross-country skiing	Static caravans
Horse riding	Woodland lodges
Mountain biking/cycling tours	Stone tent/barn
River/canal fishing	Holiday cottages
Rural festivals	Camping
Hunting	
Birdwatching/nature study	
Hunting	

sity around the world, often face a common challenge of economic regeneration and employment creation especially for young people. Agriculture and other traditional sectors can no longer be relied upon for employment and growth. Tourism has emerged as one of the leading service sector drivers of wealth and job creation. It is a catalyst for other sectors and is responsible directly and indirectly for 10% of jobs and investment in the global economy (World Travel and Tourism Council, 1998).

Many rural areas, especially in Europe, already benefit from tourism. However, often there is only limited implementation of integrated development strategies, which incorporate the sector in order to optimize its economic and social contribution. Often there are gaps between authorities responsible for economic development and the tourism industry, particularly in rural areas (Butler, 1998). These problems are sometimes made more acute because rural tourism operations are typically small scale and often have difficulty in attracting adequate investment.

There is, however, evidence of increasing market appeal, reflecting growing consumer interests in the countryside, cultural tradition and concern for the environment (English Tourist Board and Employment Department Group, 1991). As such, the tourism industry may be particularly well suited as an agent for rural development because it can:

- create employment across the local economy with a 'flow-through' effect into agriculture, construction and other local activities;
- help to stem the migration of certain groups from rural to urban areas;
- offer employment opportunities for young people and women to help alleviate rural poverty;
- provide an opportunity for the unemployed to enter the labour market with training;
- encourage small and medium-sized enterprises to be rooted within the local community;
- stimulate local food production, crafts, community pride, heritage and nature conservation;
- help to sustain local services and facilities and thereby improve the quality of life.

However, the history of tourism development makes it clear that adequate planning procedures are essential if a quality tourism product is to be developed and maintained. It can be argued, therefore, that the role of rural areas has become increasingly recognized as having importance in the local, regional and national development of countries. Within the development and diversification of rural areas, tourism has been portrayed as an opportunity for rural areas to not only improve their economic viability but also to strengthen their identities and social networks (Gannon, 1994). As such, educational programmes designed to develop students' professional and managerial skills to higher levels will, ultimately, provide the necessary expertise for the rural tourism industry.

Rural tourism issues

One of the main issues associated with rural tourism is the extent to which it contributes to economic and social growth. It could be argued that the overriding purpose of all tourism development is the potential for economic and social development in destination areas. This includes rural tourism, which, although a relatively small sector of the total world market, makes a significant contribution to rural economies (Sharpley and Sharpley, 1997). Like the tourism industry in general, rural tourism undoubtedly has both positive and negative aspects in terms of its contribution to economic and social development. Davidson (1993) classifies the influences that tourism makes, many of which are relevant to rural tourism. For example, tourism increases foreign exchange and international trade, growth in invisible imports, improvements to the balance of payments, capital investments, foreign investments and employment. The multiplier effect mentioned at the outset of this chapter increases this economic impact widely. The downside of this, in rural areas particularly, comes from the potential for overdependence of income and employment on tourism. The industry is subject to many external forces that can have considerable impact locally (e.g. the foot-and-mouth disease outbreak). This and other themes are referred to by Page and Getz (1997) who suggest that income 'leakages' (where money generated by tourism is not 'multiplied' within the generating area but its benefits leak elsewhere) and labour issues (high proportion of poorly paid part-time seasonal jobs) highlight that tourism may not always be worthy of the praise it is given.

In social terms, Davidson (1993) also suggests there are numerous influences on the host population caused by tourism. Of benefit may be the changing attitudes of host populations, capitalizing on cultural differences, social awareness, increased opportunities and education. Similarly, though more unwelcome, social influences of tourism may include issues surrounding the authenticity of the tourism product (overcommercialization), the hostility of local people towards tourists, and dwindling local customs and traditions. The hostility towards tourists is aptly illustrated by Doxey's (1976) *Index of Irritation*. This model suggests four stages of increasing irritation felt by the host population as tourism grows in their local area. From an initial welcome (the euphoria phase), visitors are taken for granted and the relationships become more formal (apathy) until a saturation point is achieved and hosts have misgivings about tourism (annoyance). The final stage is 'antagonism', where irritations are openly expressed. In addition, Sharpley and Sharpley (1997) refer to several negative impacts of tourism on the social life of rural areas; crime, congestion, accessibility and housing being the core issues for rural tourism. Table 18.6 summarizes these benefits and costs of rural tourism.

What is clear from the discussion of rural tourism issues up to this point is that a relationship exists between all the demands on the countryside, and as greater and more diverse demands are placed on the countryside, the nature of that relationship changes. In particular, as the contribution of agriculture to employment and income generation declines, tourism is increasingly viewed as an available and justifiable economic and social activity in rural areas. In the UK, for example, less than 2% of the population controls and farms more than 70% of the land (Shaw and Williams, 1994) thus explaining why tourism in rural communities is now actively promoted as a new 'cash crop'. Often rural tourism has been referred to as the 'panacea' for the economic problems facing rural areas. However, Burns and Holden (1995) indicate that there are particular problems, highlighting the following issues with rural tourism. It:

- increases the demand for, and cost of, public services, such as refuse collection, medical services and the police;
- incurs development costs of attractions, facilities and infrastructure improvements;
- may create jobs that are part-time and seasonal. Furthermore, local people may neither wish nor possess the relevant skills to respond to the opportunities offered by tourism. As a consequence jobs go to 'outsiders' rather than benefiting local employment;
- frequently leads to increases in the price of land, property and some goods and services;
- may result in local communities becoming overdependent on a single industry, the success of which may be beyond the control of the local community.

As rural communities become more dependent on tourism as a source of income and employment, it is possible that, in the extreme, they will lose their local identity and become objects of curiosity to visitors. Furthermore, economic dependence itself can have a serious social impact; the greater the dependence on the income from visitors, the greater will be the hardship if the number of visitors decline (Davidson 1993).

While the benefits of income and employment could be said to be 'potential' rather than 'guaranteed', there are other potential benefits suggested in Table 18.6. Many historical buildings, for instance, depend on income from visitors for their upkeep. Tourism can also bring a new lease of life to old or redundant buildings

Table 18.6   Benefits and costs of rural tourism. (Adapted from Lane (1994) and Sharpley and Sharpley (1997).)

Benefits	Costs
Enhanced living standards	Pressure on agricultural production
Source of alternative employment	Inmigrants may benefit from higher paid jobs while locals experience poorly paid seasonal jobs
Creation of new markets for traditional and new products	Decline of other products in less demand
Creation of new tourism attractions – for locals' use too	Overuse of existing attractions
Earns foreign exchange	Leakage of exchange to intermediaries outside the local area
Increases GDP directly and indirectly via multiplier effects	Brings greater external control to the area
Stems rural depopulation	Increased congestion and crime levels
Contributes to conservation of monuments and buildings	Degradation of monuments and buildings
Funds new infrastructure – water, roads, etc.	Saturates existing infrastructure
Reduces strain on urban areas	Increases hostility towards tourists
Income from other unused land/improved landscape	Destroys landscape and leads to non-integrated tourism complexes
Protection of natural environment	Destruction of natural environment

in the countryside. Farm buildings can be converted to stone tents (bunk-barns) for walkers or other forms of accommodation. Similarly other rural buildings such as mills can be converted into craft centres. Organizations such as the Trevithick Trust in Cornwall have restored a variety of industrial buildings such as engine houses, some in rural areas, in an attempt to generate revenue from rural tourists.

It can be seen that rural tourism can be a valuable tool to improve the economic and social development of rural areas, and, as Lane (1994) suggests, rural tourism has been a major growth sector of both domestic and inbound tourism. Despite potential pitfalls, rural tourism does offer a significant source of employment and income that cannot be provided by other industries. In many countries rural tourism is closely linked with agriculture policies and is promoted as a means of conserving rural social and cultural environments.

Farm tourism

As Busby and Rendle (2000, p. 635) state, 'Farm tourism is not a new phenomenon'. Farming has suffered repeated difficulties and pressures in recent years and incomes from traditional food production have declined dramatically. This has forced many out of the industry and many others to reconsider their business options. One option has been to diversify into tourism as it offers some potential for farm businesses – though it must be remembered that it may not be suitable for all farms. Around 20% of farmers already earn over a quarter of their income from non-agricultural sources (www.ruraltourism.org.uk)

Farm tourism incorporates any provision of facilities for tourists on working farms (Hill and Busby, 2002, p. 459), which can be subdivided into three categories, accommodation, activities and the day visitor (Davies and Gilbert, 1992), with the provision of accommodation being the largest sector. Farms can draw on various existing parts of their business to provide a variety of tourism-related activities and most local planning authorities now take a positive approach to farm diversification proposals.

Farm accommodation

Farms offer considerable flexibility in the type of accommodation they can provide. From bed-and-breakfast-serviced and self-catering accommodation to camping and caravanning, farms often have a variety of built infrastructure that can be adapted to tourism uses. The surrounding farmland is a natural choice for those seeking a relaxing break and a base to explore the local area. Farms also have an appeal as a base for short breaks in the countryside. However, substantial investment may be required in conversion or building costs. Table 18.7 provides some of the criteria for success in and major costs of farm accommodation. Below, the case study of Devon Farms demonstrates how successful farm accommodation can be.

Case study 1: Devon Farms

Devon Farms is a co-operative of working farms that provide bed and breakfast or self-catering accommodation. There are approximately 130 members who farm all parts of Devon, from Exmoor in the north of the county to Dartmoor in the west, the South Hams and to Axminster in the east. Self-catering accommodation typically consists of one to four or five 'units' located on the farm. These may have once been farm workers' cottages, an annex to the farmhouse or farm buildings that have been converted. Refurbishment to a high standard is a requirement. The majority will sleep at least four and prices vary according to size, standard and throughout the season (for example, £150 per week in low season to £500 in the height of the season). Bed and breakfast accommodation is located within the farmstead in the main. There are often two or three letting rooms costing around £20–25 per person per night, with a further £10–12 for some who additionally offer an evening meal.

The co-operative itself began in 1988, when eight separate area groups within Devon decided to come together in the hope of improving their marketing opportunities. It was felt that marketing farm accommodation within Devon as a county brand could lead to greater success for all concerned. Today the eight groups remain and each provides one representative for every ten members who serves on an overall executive committee. The committee votes in its own officers, to include a chair, vice chair, marketing officer, treasurer, secretary and other posts with responsibility for projects, the website and brochure editing. The committee meets six times per year, throughout the winter. The

Table 18.7 Criteria for success in and major costs of farm accommodation.

Criteria for success	Major cost implications
Farmhouse bed and breakfast and holiday cottages	
• Location – close to tourist route in area of scenic quality, (e.g. National Park)	• Meals
• Attractively decorated rooms, scrupulously clean, in building of architectural charm/merit	• Heating
• Some grading – highly commended for overseas market	• Marketing
• Safe car parking close by	• Refurbishment of rooms and linen (redecoration every 2–3 years, high-quality furniture and decoration)
• Full central heating (especially if open all year) plus real fire somewhere, to 'play' with	• Capital costs of major conversions are considerable, but value has been put into property and value of fixed dwellings will be in line with local property prices
• On working farm, though not too noisy, smelly and dangerous, friendly dogs, cats, sheep, cattle in view	
• Friendly proprietor welcoming guests to his/her home can be vital (enhances repeat visit potential), provision of a generous breakfast	
• Adequate marketing, e.g. membership of successful co-operative	
Touring and static caravans	
• High-quality grounds, plenty of space between units, in traditional holiday location	• Infrastructure – electricity and water, landscaping
• For commercial lets, some grading by associations such as Camping and Caravan Club, AA, RAC, Tourist Board, etc.	• Purchase and interest charges on new caravans
	• Annual refurbishment
• Marketing – logos, letterheads, display advertising, commercial guides	• Site labour – fairly low (maintenance, grass cutting, reception)
• Easy access to site and car parking by caravan	• Waste removal
• Provision of facilities, electricity, TV, water, rubbish disposal	• Marketing – at least 10% of revenue, more if letting agent used (–25%)
• Close to principal holiday touring route	
• Good signposting	

group has developed a centralized vacancy helpline facility. This enables potential guests to ring a single telephone number to check which members have vacancies rather than having to ring each one. This facility helps to keep potential bookings within the farming sector.

To become a member of Devon Farms, applicants must achieve criteria that are set out and vetted by the local group – the requirement being that the accommodation is of a high standard and must be based on a working farm. In recent years this definition has come to include some that have purchased 'hobby' farms, where, perhaps, tourism might be seen as the main business. To some extent this may have introduced conflict with more traditional farms, but such new members

have also brought with them skills that have been of benefit to the group as a whole.

The main aim of the group is to increase the occupancy of its members. This is principally achieved through the production of a brochure, paid for out of members' subscriptions. The brochure is free and is distributed to Tourist Information Centres and libraries. In addition, a nationwide organic vegetable distributor, a brochure-stand company and overseas tourism exhibitions ensure that the brochure achieves wide coverage. Around 60 000 brochures are produced annually. In recent years, the production of a website has enhanced business, especially from overseas visitors (Germany, The Netherlands and the USA being the main markets). Individual members have their own sites linked to this, and the site itself is maintained and developed by one of the members.

Other aims of the group include generating funds, maintaining the profile of the group as a whole and trying to better serve the group's members. Funds are sought from appropriate sources that become available on a periodic basis. For example, a source of additional funding to enhance overseas marketing was success-fully won via Devon County Council. Members col-lected data in the form of a questionnaire from their overseas visitors and a consultancy firm analysed the data, which showed the type of experience such cus-tomers were seeking. Maintaining a public relations presence is also of high importance. As individual small-scale operators it would be all too easy for this type of provider to be 'overlooked' by the regional tourist board, whose membership has a greater propor-tion of hoteliers and larger concerns. The group works hard to ensure that this does not occur and achieves a fair share of grants and other funds when they become available. This was apparent in terms of additional monies made available following the foot-and-mouth epidemic in 2001. The experience of the epidemic was a sobering lesson on how farming and rural tourism is linked, particularly in the way that so many responsible rural tourists made the choice of keeping away so as not to increase the spread of the disease.

The co-operative also attempts to make improve-ments for its members. One example of this is 'peer group training', where experienced members assist others. This has also attracted funding from the county council to help members improve their computer skills. The theme of developing their business also extends to the provision of 'add-ons' such as farm walks, which have been created with assistance from a local envi-ronmental group, the Devon Wildlife Trust. Another outcome that partially resulted from the foot-and-mouth disease outbreak and also as a direct consequence of farming's downturn is the promotion of locally pro-duced food. In the South West, as elsewhere, this has become a high-profile tourism issue and many rural pubs and hotels display notices telling their guests which farms they source their meat from. With the Devon Farms businesses, home-produced produce is now a major selling point, and farming leaders are hoping that such a strategy will help to save many of the smaller family farms.

Overall, Devon Farms can be considered a classic example of rural tourism. Although, at an individual level, member's businesses are small, together as a co-operative there are clear benefits of working together, leading to improved occupancy levels and generally enhancing business opportunities. Within Devon there is considerable competition from accommodation in coastal locations, so to remain competitive Devon Farms must continually evolve. This includes embrac-ing new technology and keeping up to date with visitor requirements through market research. However, at the heart of the visitor experience is a high standard of accommodation and a personal service that provides a unique glimpse into a rural way of life. This, added to good homegrown food and plenty of space for children to play, ensures that this form of rural tourism remains successful.

Farm attractions and activities for the day visitor

In addition to accommodation, diversification on farms into tourism can extend to the provision of attractions and activities for day visitors. Farm attractions can be based on an existing aspect of the farm business or work-related skill such as animal husbandry or spe-ciality food production. Examples include vineyards, a rare breeds display or the demonstration of products such as cheese making. Exploiting the space and inter-est of farmland can include farm trails, riding stables and pony trekking, clay pigeon shooting, pick your own fruit, local walks and cycling. Opportunities can include cross-age-group activities such as riding or quad biking.

Clearly, funding diversification such as this may often be beyond the scope of some farms. However, numerous measures exist to assist in the development, especially where the farms are located in deprived areas. The following case study describes one such opportunity for farmers in this context.

Case study 2: European Objective One Programme for Cornwall and the Isles of Scilly Rural Tourism Improvement Fund

The aim of the fund (www.swtourism.co.uk) is to improve the quality and range of facilities available for visitors to Cornwall and the Isles of Scilly. To qualify for an Objective One grant under the Rural Tourism Improvement Fund, the applicant must present a suitable project for diversification. However, there are also some basic requirements. For example, all applicants must:

- be in Cornwall or the Isles of Scilly,
- have an Agricultural Holding Number

and comply with one of the following:

- have at least 2 ha (5 acres) of land (excluding dwellings, gardens, landscaped areas and such like for permanent caravan parks, etc.) in agriculture, grassland or woodland setting;
- have an annual turnover from agriculture of £1500 or more;
- be a member of Cornish Farm Holidays or a Cartwheel brand user.

If the project proposal complies with this then a variety of funds are available, as follows.

Business Plan Development Fund

This grant is designed to help farmers put together an application for a grant, and is funded at up to 80% of a maximum total cost of £1500 (total possible grant = £1200). The grant is discretionary for projects costing more than £30 000, and a short application form has to be obtained from the regional tourist board (who operate a list of approved consultants to assist farmers to correctly apply for grants).

Rural Tourism Improvement Fund

This is a capital grant scheme to help improve the quality and range of facilities that will attract more visitors to farm enterprises, especially out of season. The types of projects that are eligible include:

- upgrading facilities;
- quality improvements;
- additional facilities to encourage greater tourism outside the main season;
- projects that meet a proven identified niche market, e.g. activity holidays.

New accommodation may be eligible if a clear case can be made by the applicant of a new or special niche market, i.e. bringing guests into the area for an activity or interest or focusing on out-of-season bookings.

Additional accommodation for an existing tourism business may be eligible where:

- the applicant can show that he or she is diversifying from his/her existing market to a new market; this must operate outside as well as inside the main season;
- the work eligible for funding is capital investment, such as building and groundwork, e.g. landscaping and infrastructure.

However, this fund does not extend to the purchase of furniture, furnishings, vehicles and equipment or for repairs and renewals. This particular programme ran for 3 years to April 2004, and grants were offered at up to 50% of the eligible capital costs, to a maximum of £60 000.

Conclusions

As Page *et al.* (2001, p. 364) state, 'it is clear that rural areas are an integral part of the modern tourism experience. However, rural areas need to be understood to ensure that appropriate forms of tourism are developed'. This chapter has discussed a range of issues associated with tourism, its value and more specifically rural tourism, within a UK context. It was noted that there is no universally accepted definition of rural tourism, and the term is sometimes confused with 'farm' tourism, 'green' tourism or 'agritourism' and the like. Part of this difficulty lies in the different interpretations of what constitutes the term 'rural', in that while definitions exist in relation to, say, population density, in the context of tourism, culturally significant aspects emerge to complicate matters. For instance, it is often thought that rural tourism should represent a leisure experience based on a simpler lifestyle, away from the rat-race, in pleasant countryside surroundings. Whatever the definition, it is clear that rural tourism is popular in terms of the number of visitors and the range of activities they undertake. Despite this, the literature suggests that rural tourism should be small-scale, traditional in character, slow to grow, controlled by local people and representative of the rural environment, culture, heritage and socio-economy.

The growth of rural tourism is thought to have come about as a result of a restructuring of the agricultural sector, diminishing rural industry and, in some locations, the out-migration of young people from rural areas. However, over the same time period tourists' demands have changed. There has been an increased emphasis on the countryside as demand has moved away from traditional 'mass' products towards more individual tourist preferences. Similarly, tourists increasingly require greater flexibility and a closer interaction with traditional cultures, and yearn for the rural characteristics of tranquillity and genuineness. It is perhaps more the case that rural tourists have varied motivations, whether a desire for ecological uniqueness, a specific adventure opportunity or certain cultural attractions, or simply the peace and quiet of the countryside. Whatever the motivation, tourists staying in the countryside demand personalized hospitality of high quality.

From a supply side, there is an emphasis on the development of products that attract, satisfy and retain the potential of repeat business. The Devon Farms case study is a good example of this ethos. Here visitors are often accommodated on the farm homestead and opportunities are provided for the observation of, and perhaps participation in, farm activities. Of crucial importance is the personal interaction and close host–guest relationships in addition to high-quality accommodation. For international visitors especially, farm stays are seen to present the chance to meet local people and enjoy what is perceived as an 'authentic' experience. The importance of generating an appropriate experience for tourists is reflected in the opportunities afforded to rural tourism businesses in some areas to gain funding for suitable improvements. The Cornwall case study is just one of numerous opportunities that occur in different locations at various times.

While these issues present a 'rosy' picture of rural tourism, the chapter has also shown that a variety of impacts result from rural tourism. Among these are impacts upon agriculture. There are both positive and negative sides to this. These range from the creation of new markets for traditional products to severe pressure on the rural infrastructure.

In conclusion, rural tourism, at best, can be visualized as small-scale, decentralized, locally owned and environmentally, culturally and socially sustainable. While some might argue that such sentiments are idealistic and not achievable, this can be countered if it is felt that the aim of tourism is about elevating the quality of life of host communities.

References

Burns, P. and Holden, A. (1995) *Tourism: A New Perspective*. Prentice Hall, Hemel Hempstead.

Busby, G. and Rendle, S. (2000) The transition from tourism in farms to farm tourism. *Tourism Management*, **21** (4), 635–42.

Butler, R. (1998) Rural recreation and tourism. In: *The Geography of Rural Change* (ed. B. Ibery), pp. 211–32. Longman, London.

Countryside Agency (2003) *The State of the Countryside 2003, Environment and Recreation*. Available at www.countrysideagency.gov.

Davidson, R. (1993) *Tourism*. Pitman, London.

Davies, E.T. and Gilbert, D.C. (1992) A case study of the development of farm tourism in Wales. *Tourism Management*, **13** (1), 56–63.

Doxey, G.V. (1976) When enough's enough: the natives are restless in old Niagara. *Heritage Canada*, **2**, 26–7.

English Tourist Board and Employment Department Group (1991) *Tourism and the Environment: Maintaining the Balance*. English Tourist Board, London.

Gannon, A. (1994) Rural tourism as a factor in rural community. Economic development for economies in transition. *Journal of Sustainable Tourism*, **2** (1), 51–60.

Gunn, C.A. (1994) *Tourism Planning*, 3rd edition. Taylor and Francis, Washington.

Hill, R. and Busby, G. (2002) An inspector calls: farm accommodation providers' attitudes to quality assurance schemes in the county of Devon. *International Journal of Tourism Research*, **4** (6), 459–78.

Lane, B. (1994) What is rural tourism? *Journal of Sustainable Tourism*, **2** (1), 7–21.

Lowry, L.L. (1994) What is travel and tourism and is there a difference between them? A continuing discussion. *New England Journal of Travel and Tourism*, **5**, 28–9.

McIntosh, R.W. and Goeldner, C.R. (1986). *Tourism Principles, Practices, Philosophies*. Wiley, New York.

Page, S.J. and Getz, D. (1997) *The Business of Rural Tourism: International Perspective*. Thomson, London.

Page, S.J., Brunt, P., Busby, G. and Connell, J. (2001) *Tourism: A Modern Synthesis*. Thomson, London.

Sharpley, R. and Sharpley, J. (1997) *Rural Tourism: An Introduction*. Thomson, London.

Shaw, G. and Williams, A. (1994) *Critical Issues in Tourism*. Blackwell, Oxford.

Star UK (2003) *Statistics on Tourism Research*. Available at www.staruk.org.uk.

World Travel and Tourism Council (1998) *Econett Discussion*. Available at www.wttc.org.

World Travel and Tourism Council (2003a) *World Travel and Tourism*. Available at www.wttc.org/measure/pdf/world.

World Travel and Tourism Council (2003b) *European Union Travel and Tourism*. Available at www.wttc.org/measure/pdf/european.

World Travel and Tourism Council (2003c) *United Kingdom Travel and Tourism*. Available at www.wttc.org/measure/pdf/uk.

Further reading

Krippendorf, J. (1987) *The Holiday Makers*. Butterworth Heinemann, Oxford.

Page, S.J., Brunt, P., Busby, G. and Connell, J. (2001) *Tourism: A Modern Synthesis*. Thomson, London.

Useful websites

www.abtanet.com – Association of British Travel Agents.
www.bha-online.org.uk – British Hospitality Association.
www.bitoa.co.uk – British Incoming Tour Operators Association.
www.englishtourism.org.uk – English Tourism Council.
www.etoa.org – European Tour Operators Association.
www.staruk.com – Tourism Statistics.
www.statistics.gov.uk – Government Statistics.
www.tourism.wales.gov.uk – Wales Tourist Board.
www.tourist-offices.org.uk – Association of National Tourist Offices.
www.ttgweb.com – Travel Trade Gazette.
www.un.org – World Tourism Organization.
www.ustoa.com – United States Tour Operators Association.
www.visitbritain.com – British Tourist Authority.
www.visitscotland.net – Scottish Tourist Board.
www.wttc.org – World Travel and Tourism Council.

19

Marketing management

R. Soffe

Introduction

Traditionally, marketing in rural areas meant taking products to market. More recently, marketing has often been perceived simply as how to sell or advertise a product. Today marketing encompasses the ability to influence policy making and decision making as much as promoting products and services.

Marketing management encompasses promotions to price setting; public relations to product branding. Marketing management may offer small and micro businesses in rural locations an important route to success. Additionally marketing management is increasingly being used by well-organized communities, organizations and groups to influence policy.

Marketing is an ongoing process involving understanding the business environment, defining markets, setting objectives and monitoring implementation. As a business grows, businesses should be capable of refining their operations to changes in the marketplace. Developing a marketing strategy will help a business to build an understanding of the various parts of the business and the market environment. A simple explanation of marketing may be 'a process for meeting customer needs and achieving business goals efficiently'.

In theory, marketing may be viewed as a step by step process undertaken by a business:

- analysing markets and internal capabilities;
- matching these to an audience seeking a specific product or service;
- targeting through the audience's choice of medium;
- achieving company objectives, be that sales, profitability, audience penetration or membership take-up.

Marketing influences every facet of a business operation. This even includes conversations between employees, their attitudes and beliefs, and sales personnel on the road talking to customers, how they sound, how they sell and what they are selling. It includes the manner in which a telephone is answered, corporate literature, emails, websites, and how these are perceived by the stakeholders of the business.

Marketing is not just about TV, radio and newspaper advertising or sponsorship of football matches. Marketing is about researching, analysing, agreeing and implementing courses of action to achieve objectives. If objectives are achieved a marketer must then think about how the business is going to achieve the next stage of objectives. It is a constant and evolving process of agreeing what data are needed, analysing those data and making intelligent decisions based on individual and group deliberation. Promotion and advertising seen on the television is merely the tip of the iceberg.

Definitions of marketing

There are many definitions of marketing. The Chartered Institute of Marketing definition is:

> 'Marketing is the management process responsible for identifying, anticipating and satisfying customer requirements profitably.'

The American Association of Marketing refer to marketing as:

> '. . . the process of planning and executing the conception, pricing, promotion and distribution of ideas, goods, services to create exchanges that satisfy individual and organisational goals.'

US Marketer Phillip Kotler suggests:

> 'The marketing concept holds that the key to achieving organisational goals lies in determining the needs

and wants of target markets and delivering the desired satisfaction more efficiently and effectively than the competition.'

Finally, Dibb *et al.* (1997) list their priorities of marketing as:

- satisfying customers;
- identifying/maximizing marketing opportunities;
- targeting the 'right' customers;
- facilitating exchange relationships;
- staying ahead in dynamic environments;
- endeavouring to beat or pre-empt competitors;
- utilizing resources/assets effectively;
- increasing market share; and
- enhancing profitability.

These priorities are not for everyone. A firm's marketing priorities will heavily depend on the company's objectives. For example, the RNLI charity may not seek to be profitable, but may seek to be an efficient and cost-effective operation whilst saving lives. Alternatively, Virgin's goal may be to innovate and make a profit from high-risk ventures.

The marketing information system (MIS)

The key to successful marketing is information. A marketing information system (MIS) is a process for managing and analysing information, from both internal and external environments, in order to produce management data. Businesses that develop an understanding of MIS can then develop a competitive advantage over the competition. As Kotler (2003) states:

'A MIS system consists of people, equipment, and procedures to gather, sort, analyse, evaluate, and distribute needed, timely, and accurate information to marketing decision makers.'

Having knowledge of customers' habits, attitudes, the newspapers they may read or number of times they hear a radio advert aids a marketer to make decisions on pricing, packaging, promotional activity, strategy and plans. The internal environment includes all strategic business units: marketing, finance, production and human resources. The external environment seeks to understand the wider issues affecting the company's environment: customers, competition, industry and markets. These include political and legal issues; economics and stability; social and cultural issues; and technology.

Understanding the information a firm needs in order to make decisions is vital to the establishment of a successful system. Reviewing information against requirements will ensure that the right data are available and in the correct format.

The following four elements provide a useful framework for information gathering (see also Figure 19.1):

(1) Internal records process: the collection of internal reports on results, for example, from quarterly sales, inventory and sales forecasts, and production costs.

(2) Marketing research process: the methodology. This is the process of collecting, analysing and reporting information, which is transferable and easy to adopt by managers.

(3) Marketing intelligence process: this sub-category is about understanding the every-day environment and scanning for information, which may lead to the next development or competitive advantage. Data sources can be books, newspapers, mystery shoppers, distributors, retailers and customer feedback.

(4) Marketing decision support process: the process of converting data through modelling techniques into usable information. Well-known examples include the Boston Consultancy Group matrix, value chain analysis and profit lifecycle.

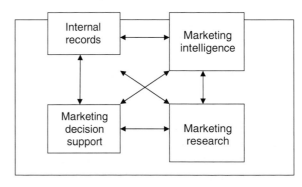

Figure 19.1   A marketing information system (adapted from Kotler, 2003).

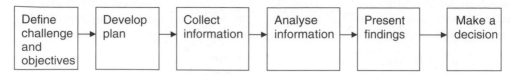

Figure 19.2   The market research process.

A major reason for setting up a MIS is to identify opportunities and potential threats. After careful analysis a company can decide which opportunities to pursue and which threats to prepare for (Figure 19.2).

Marketing strategy

The strategic marketing plan is a framework to guide planning (Figure 19.3). The marketing strategy outlines the target segments and the tactics to be employed whilst making efficient use of company resources to meet objectives.

Developing a marketing strategy has three basic elements:

(1) Strategic analysis: what position does the company occupy, who are the customers and competition, and what skills and abilities does the company possess?
(2) Strategic intent: formalization, evaluation and choice of strategy. Establishment of objectives and strategies for each element of the marketing mix.
(3) Strategic implementation: action and implementation of plans and control.

It also incorporates a mission and vision statement:

• Mission – corporate-level purpose in line with major stakeholder expectations.
• Vision – strategic choice to achieve future aspirations.

A marketer should develop skills to analyse and research the environment. Scanning is one skill for reviewing the market without being specific. Focusing on key information as the search continues is crucial to using analysis techniques. The objective is to build an impression of the company's position and understand its strengths, weaknesses, opportunities and threats (SWOT).

Figure 19.3 Strategic marketing plan (adapted from Drummond and Ensor, 2001).

External analysis

By analysing the external environment a company can begin to understand what threats exist and which opportunities are developing in the market place. This information is useful in aligning internal operations and resources to external environments (Figure 19.4).

There are two external environments:

(1) Macro environment – the environment beyond the control of the company: political and legal issues;

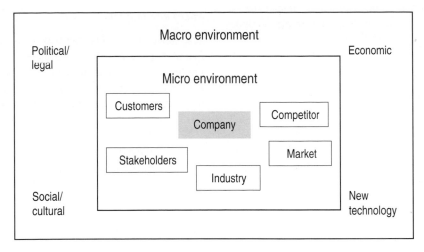

Figure 19.4   Strategic analysis (adapted from Dibb *et al.*, 1997).

economics and stability; social and cultural issues; and technology.

(2)   Micro environment – concerns competitors, customers, stakeholders and channels.

The things that a company has a direct relationship with in the immediate environment.

Macro environment

The macro environment reviews the broader environment. Usually, macro forces are beyond a company's control. A standard model commonly used to assess the environment is PEST analysis; other acronyms include STEP or STEEPLE. PEST stands for political/legal, economic, social/cultural and technology. The idea is to identify key factors in the environment that may have an effect on the company. Potential opportunities and threats may emerge as a result of this type of analysis.

Political/legal forces

Political, legal and regulatory requirements from local councils to government decisions to the international arena, for example the European Parliament, have an influence on all businesses. These forces mould the potential prosperity and dynamics or challenges a market may operate under. Factors may include taxation, employment law, foreign trade, trading laws, monopoly control, gambling, licensing and age limits.

It is the role of a marketing manager or researcher to identify the key factors at a political, legal and regulatory level that may have a bearing on the business, in the present and the future.

Emerging examples include:

- new rules regulating the use of emails;
- the implications of new European directives legally enforcing motor vehicle manufacturers to be made more accountable for the recycling and collection of disused vehicles (2005).

Economic forces

These forces relate to global and domestic economic conditions, for example buying power, exchange rates, economic stability, inflation and supply and demand. Companies need to be aware of any changes in the economic market place. Favourable economic conditions may present opportunities or potential threats in areas such as:

- employment;
- interest rates;
- investment grants;
- GDP;
- imports and exports.

Social/cultural and green forces

Social, cultural and green forces relate to the structures and dynamics within society as a whole: cultural beliefs, behaviour, religion, community responsibility and environmental concerns. Recent examples include:

- quality of life;
- demographics;
- attitudes;
- cultural expectation;
- lifestyle;
- leisure time;
- image.

Today's governments are increasingly seeking to improve our quality of life and this involves environmental policies. This is driven by groups of individuals who are increasingly concerned about ethical business practices, conservation and environmental protection. A few examples are given below:

- GM crops;
- child labour in less developed countries;
- environmental protection;
- sustainable developments;
- greenhouse gases.

Technology

Technology plays a pivotal role in the delivery of timely and effective services and products. Keeping apace with technology can prove expensive. It is important to understand market needs in order to have in place the correct technology resources and to sustain a position. Equally, it is important to identify future changes in technology and assess the potential impact these may have on the business and the customer.

Micro environment

The objective of this analysis is to identify critical factors or patterns in the immediate environment affecting the industry, market, competition and customers. It is important to understand that more than one scenario may emerge from the analysis. All that is required at this stage is to collect information for processing.

Industry analysis

To gain a good understanding of the industry served, it is wise to assess the dynamics of the market place. A useful tool is Porter's 'five forces analysis' model for identifying balances of power within a market (Porter, 1998; Figure 19.5).

(1) *Suppliers*: the supply chain may be controlled by many or few. Demand for products or services from few suppliers will place power with the suppliers. Many suppliers will reduce the overall power of a supplier because buyers have more choice for driving down prices. For example, UK rail has three train carriage suppliers who rent carriages to the UK franchised rail network. Some network operators believe these carriage supply companies control too much of the market and therefore control what prices the hire companies are charging.

(2) *Buyers*: when there are few buyers or control of large market share rests with a few, suppliers will have to consider their pricing strategies. An example of buyer power is the UK's supermarkets, forcing farm gate prices down and insisting

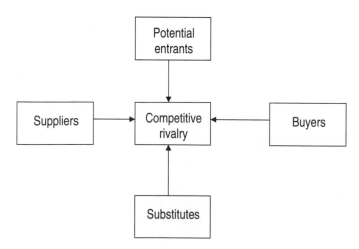

Figure 19.5 Porter's 'five forces analysis'.

on traceability. Another example is Amazon.com. From independent publishers Amazon insists on a large, some say 45%, discount on all book order prices. They can then pass this benefit to customers, who find Amazon a low-cost centre for book purchasing.

(3) *Potential entrants*: new entrants face barriers in any given market. Barriers can be in the form of monetary barriers, such as cost of set-up and cost of machinery; regulations controlling the supply and distribution of products and services; licences controlling who sells what and when, as found with 3G mobile; and distribution channels. Competitors may control supply chains and try to force another company out. For instance, Coca Cola, it is reported, persuaded MacDonald's not to stock Iron-Bru in Scottish outlets.

(4) *Substitutes*: new brands, new products or new uses of an existing product are examples of substitutes. Other examples include new products displacing older products; new products or services removing a previous need; and direct substitution of a product or service for another. Customers may query 'Where else can my buying power be used?'. The internet has had a marked effect on shopping patterns and behaviour. Consumers can now buy direct from the manufacturer, bypassing traditional distributor and retail outlets.

(5) *Competitive rivalry*: this concerns the intensity of competition, size, number of competitors, ease of

differentiation and competitive advantage. Other factors may include the stage of industry life-cycle: is it emerging, nurturing, maturing or in decline?

Porter's 'five forces' model is a useful tool to highlight key factors that affect an industry and ultimately the market. When comparing the key external factors, found within the PEST analysis, certain patterns may be identifiable. Managers should ask themselves how the information would affect the company's direction, taking into account wider environmental forces. Will the factors have a direct or indirect effect? What action can a company take to improve prosperity? Which areas of the market dominate the competitive nature of the industry? Who are the competitors?

Competitor analysis

Competition comes in many forms. Porter's 'five forces' is a useful model to identify a number of these competitors. However, many seemingly like-for-like competitors can be split into strategic groups. For example, looking at the airline industry we can divide the main players into three strategic groups – low-cost, medium-cost and high-cost operator (Figure 19.6).

This form of analysis allows a company to identify true competitors and concentrate resources and effort on a few competitors rather than taking a generic view. Once a company recognizes its competitors, it can then conduct a detailed analysis of the competition, considering:

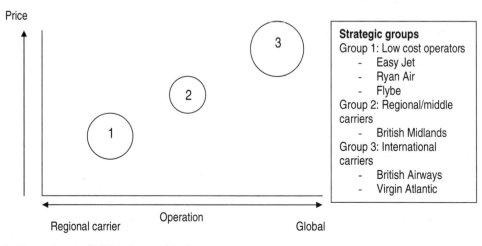

Figure 19.6 Competitor analysis/product positioning.

- competitor objectives in each definable market;
- past strategies;
- future strategies;
- the type of defensive or offensive strategies it may employ towards the competition;
- position in the market place;
- strengths and weaknesses;
- competitive advantage;
- orientation;
- marketing activity – price, promotion, product, place of purchase;
- its capabilities in terms of management and marketing,
- innovation, NPD and finance.

When analysing competitors it is easy to miss other players, for example:

- small firms that have been overlooked could pose a risk in the form of new entrants;
- a large corporation moving in to capitalize on market prosperity; and
- international companies. In 2004, ten new countries joined the European Union trading block. These included Poland, the Czech Republic and Hungary. These countries present potential new competitors, particularly in the field of manufacturing and engineering. Many of these countries have a low-cost, highly educated workforce.

Some companies design a competitive intelligence system, or CIS, to aid competitor data collection and analysis. A useful model to refer to is profit impact of marketing strategy (PIMS) market analysis.

On completion of PEST, industry and competitor analyses, a more in-depth review of the market can take place. According to Drummond and Ensor (2001), market analysis would normally include the following:

- *Actual and potential markets:* growth of the market, number of existing and potential sales in the form of forecasts, sustainable markets and accessible *trends* (and understanding the reason and the background for a trend). This may present new opportunities in the market or signify a contraction in the market size.
- *Customers and customer segments*: analysing customer segments, identifying the customers and their buying habits, is important to understanding how they buy, when they buy and what they are looking

for. Attitudes, motivation and the process by which the buyer makes a decision to purchase heavily influence these factors. All this information will help a company to align its product offerings, service or image to the needs of the customer.
- *Distribution channels*: the channels through which customers can buy; the process of distributing the product or service; and customer access to services or even the decision makers who make up a business buying team. This information is important if a company wants to move products.

Customers

Understanding the customer is one of the most important issues for a business. It is vital to analyse customer groupings and from which group customers come. How can you differentiate large numbers of customers in order to identify potential customers? Back in the 1950s all customers were viewed as a homogenous group. Today's customers often have:

- more complexity;
- very specific needs;
- been better educated; and
- increased levels of spending power.

A customer can be viewed as

- a consumer on the street;
- a business buyer in a company; or
- a stakeholder who has a vested interest in the company.

The most common approach to identifying the customer is through *segmentation*. The process of segmenting is fundamental to marketing. It is a waste of time and resources to target large and unidentifiable markets. Careful segmentation helps to highlight potential groups and provides an understanding of how to communicate with them more effectively.

Marketers often find that some segments are more profitable than others, and by concentrating on these segments and not others, a niche or several niche markets are identified. By applying segmentation techniques the marketing team will better understand the customer in a particular segment and become aware of changes in the market place and the circumstances of the group. This would normally be underpinned by ongoing market research. Relationship marketing and keeping close to the customer are all vital as markets change.

Many different parameters are used to segment a particular market; examples include:

- age, sex, family, lifecycle, job type – segmentation by demographics;
- cities, towns, villages, regions, countries – segmentation by geographics;
- lifestyle, attitudes, personalities – segmentation by psychographics;
- usage of product (heavy or light users), uses of the product – segmentation by behaviour patterns.

Successful segmentation starts by establishing criteria for similar customer groups to be identified. For example:

- How do buyers respond to different marketing tactics?
- Do they respond in a specific way?
- Is the segment identifiable?
- Is the customer group large enough?
- Is the segment sustainable?

To aid segmentation, many marketing books will present the same type of factors that managers can use to identify customer segments. Consumer segmentation can be divided into three main categories (Drummond and Ensor, 2001):

(1) Profile variables – these include:
  - demographics – age, gender, lifecycle;
  - socio-economics – refers to customers' backgrounds in terms of occupation, education, where they live and their income. For example, newspapers will quote ABC1 or C2DE social groups. This directly relates to the potential to become a customer, spending power and disposable income;
  - geographic segmentation – for example, geographic and social geographic. Where you live will identify someone as affluent, middle class or working class (using old stereotypes). A company may decide to target certain postcode areas for specific marketing activities.
(2) Behaviour variables:
  - benefit sought;
  - usage;
  - purchase occasion;
  - purchasing process.

(3) Psychographics variables:
  - lifestyle;
  - attitudes;
  - personalities.

Business/industry segmentation and stakeholder analysis

Businesses can be segmented by the characteristics of the organization. These can be:

- industry sector;
- size of business;
- location;
- end use application.

Additionally, businesses can be segmented by their decision-making unit (DMU), including:

- the structure of the DMU;
- the decision-making process;
- the buying process;
- the key criteria for reaching a decision;
- personal characteristics of the decision makers.

A number of models can be used to identify the main criteria for segmenting the business environment. These include the Webster-Wind framework, the Sheth framework and the nested approach by the Bonoma and Shapiro model (1983). The nested approach was developed to highlight the major factors for segmenting organizational markets. The nested approach involves a series of steps, moving from wider macro-level factors to personal micro factors until a business profile can be developed (Figure 19.7 and Table 19.1).

On completion of the business analysis potential, patterns may emerge. Depending on research objectives, the information will either present future opportunities or confirm that the business profile matches current strategies.

Internal (company) analysis

The internal analysis involves every resource of the business, from staff and finance to computers and processes. Positioning a business successfully requires a thorough understanding of the firm's capabilities in the form of assets and competencies plus an understanding of the business's competitive advantage. Therefore it is important to understand what strengths

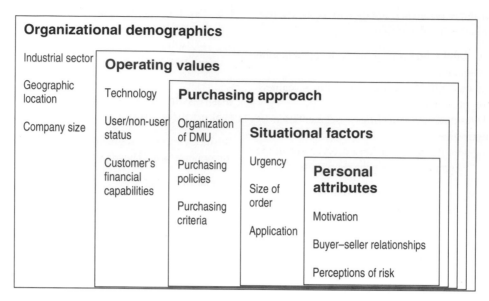

Figure 19.7   Business segmenting (Bonoma and Shapiro, 1983).

and weaknesses a firm possesses. Good auditing tools for analysing information include the balanced score card approach, value chain analysis and product life-cycle curve. The key with auditing tools is knowing which one is the most appropriate.

Assets

Assets are the disposal capital that a company possesses. This includes financial and non-financial, tangible and intangible assets. Staff knowledge and skills are an asset to a firm. Others may include lean manufacturing processes, technology and the ordering system. Example areas to review are:

- Financial.
- Physical – buildings, locations, infrastructure.
- Operations – production, manufacturing and transport.
- People – skills, abilities, knowledge, human resources.
- Legal – licences, patent protection, copyright.
- Systems – processes in place or being developed.
- Marketing:
  - distributors and channels: analysing external relationships with distributors and channels will assist a company to understand:

  - the location of infrastructure, such as buying centres, geographic locations and access points to products and service;
  - level of influence over the distribution chain;
  - level of complexity – cost, skills base, simple or complex.
- Alliances and partnerships – this may lead to accessing efficient distribution, access to other markets and technology and exclusivity through legal protection.
- Customers – what assets do customers perceive to be important? For example:
  - brand and image;
  - build quality or service delivery;
  - environmental stance;
  - origin of manufacturer.
- Internal – this can cover processes, communications, internal relationships and attitudes to change.

Competencies

Competence is the ability to maximize the use of assets through the effective and efficient use of company skills.

Table 19.1 Business segmenting: the nested approach.

Organizational demographics
- Industrial sector – e.g. wireless telecommunications, information technology, manufacturing and marketing.
- Geographic location:
  - local, regional, national, international;
  - proximity, transport routes, logistics and channels;
  - number of sites.
- Company size – number of employees and turnover.

Operating values
- Technology – a firm's orientation towards technology:
  - communications (telephone, internet, email, mobile, radio);
  - use of information technology and software;
  - manufacturing, production and operations;
  - logistics and transport.
- User/non-user status – to understand if a company uses or does not use a product or service. Companies can be divided into user and non-user groups:
  - heavy user;
  - occasional user;
  - light user;
  - non-user;
- Customer's financial capabilities:
  - buying power;
  - credit reference;
  - payment;
  - securities;
  - history.

Purchasing approach
- Organization of decision-making unit (DMU):
  - who is involved with decision making?
  - what are the individuals' roles?
    - initiator
    - user
    - influencer
    - decider
    - purchaser.
- Purchasing policies
  - what is the format for decision making?
  - supplier selection;
  - reason for purchase;
  - value;
  - decision-making process;
  - environmental and cultural stance;
  - regulations.
- Purchasing criteria
  - what criteria do they use to make decisions?
  - prior conditions;
  - accreditation.

Situational factors
- Urgency – level of urgency.
- Size of order – value, purchasing pattern, quantities and volume.
- Application – use.

Personal attributes
These relate to the individual at a personal level. Sales people try to win trust from buyers and therefore create a buyer–seller relationship. It is the sales person's job to understand the buyer and feed this information back to marketing. This approach creates a one-to-one relationship.

Competitive advantage

Competitive advantage is the critical advantage, in assets and competencies, over other rival firms in the market.

Marketing function

The marketing mix is made up of seven key marketing platforms known as the 7Ps. These are discussed below.

(1) *Product*: anything that can be offered to a market for attention, acquisition, use or consumption that might satisfy a want or need. It includes physical objects, services, persons, organizations and ideas.

(2) *Price*: the amount of money charged for a product or service, or the sum of the values that consumers exchange for the benefits of having or using the product or service. Unlike products, services cannot be tried and tested before purchase, they are intangible in nature. Pricing strategies need to consider three critical factors:
   • As services are difficult to evaluate before purchase, price may act as an indicator of perceived quality.
   • Price controls demand. Matching supply and demand in a service is critical as a service cannot be stored.
   • A key segmentation variable for services is price sensitivity. There is often a lack of creative pricing in services.

(3) *Promotion*: activities that communicate the product or service and its merits to target customers and persuade them to buy. Promotions include every type of platform to communicate a message, including word of mouth, television, newspaper stories and demonstrations. It is important to have a promotions strategy that targets each of the identifiable customer groups, including internal employees, distributors and end-users.

For example, to promote a branded service, the communication must deliver a message that can influence perception:
   • *Distinctiveness*: it immediately identifies the service provider and differentiates it from other providers.

   • *Relevance*: it communicates the nature of the service and the service benefit.
   • *Memorability*: it is easily understood and remembered.
   • *Flexibility*: it not only expresses the service organization's current business but also is broad enough to cover foreseeable new ventures.

(4) *Place*: all the company activities that make the product or service available to target customers. These include distribution channels, retail outlets, locations and so on.

(5) *People*: personnel are critical to influencing customer perceptions. Perceptions include service quality, image, training, knowledge and attitudes. It is important to constantly monitor and improve employee skills and abilities, to deliver customer satisfaction through products or services. Consistency in quality and avoiding variability in employee performance will ultimately lead to a better quality of service provided.

(6) *Physical evidence*: buildings, sites, the tangible aspects of a business offering.

(7) *Process*: the procedures, mechanism and flow of activities by which a service or product is developed and delivered. Process decisions radically affect how a product/service is delivered.

SWOT analysis

SWOT refers to strengths, weaknesses, opportunities and threats. The completion of a SWOT analysis is usually used to summarize all research, by dividing the main findings under each sub-category.

Strengths and weaknesses can be found with the internal analysis:

• *Strengths* – can be processes, advantages, skills or employees that a company possesses. These strengths will be the foundation of the company's competitive advantage.
• *Weaknesses* – having an understanding of weaknesses allows a company to consider options either to protect the weakness or do something about it.

Opportunities and threats are found in the external audit, from macro and micro environments:

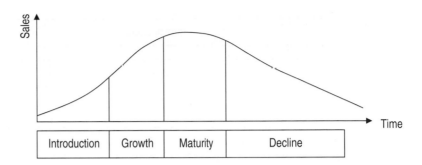

Figure 19.8 Product lifecycle.

- *Opportunities* – areas that a company can exploit, plan for or develop.
- *Threats* – actual and potential threats. Knowing of a threat in advance will assist a company to plan a counter to that threat or develop a means to lessen the threat.

Auditing tools to aid marketing

To audit external and internal environments it is important to know what data to collect and the model to use with the information. Auditing tools are used to assess past, current and future strategies.

The product lifecycle curve

The product lifecycle curve is a useful model for the assessment of:

- product markets;
- brands;
- customers (who, when and how many) and timing;
- the marketing mix;
- industries and markets.

Each stage from introduction to decline can be assessed (Figure 19.8).

- *The introduction stage* identifies common factors relating to introducing new ideas to the market, the right marketing mix and strategies to launch ideas and begin new relationships. Such factors may include competition, customers and distribution.
- *The growth stage* is about rapid developments, in terms of increasing sales and competition, penetration, product/service developments and price strategies.

- *The maturity stage* signifies plateaux in sales, customers and market potential. It is important for companies to have other ideas in order to extend the brand or product usage to gain new market ideas or customers.
- *The decline stage* is the stage when markets are diminishing, and so are customers and competition. Here companies need to decide if they are going to leave the market or push for new developments.

The product lifecycle curve can only be a guide. It helps to visualize aspects of the lifecycle in operation, as it aids future decision making, plans for new ideas and aligns corporate capabilities with developing opportunities.

Value chain analysis

Value chain analysis is a model used to assess the internal capabilities of a company, analysing functional areas in order to review assets and competencies (Figure 19.9). Primary activities include:

- Inbound logistics – actions associated, from receiving and storing to internal distribution.
- Operations – day-to-day functional activities involving converting raw material into customer products.
- Outbound logistics – external delivery, storing and transporting of customer products.
- Marketing and sales – activities associated with customer acquisitions.
- Service – pre- and post-sales support, warranties, assistance and training.

Secondary activities include:

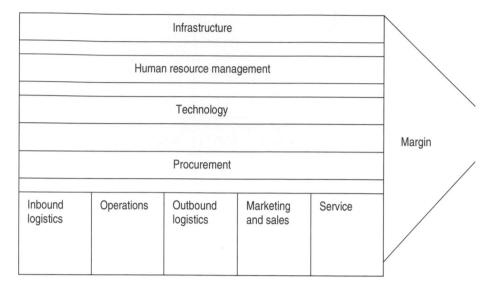

Figure 19.9   Value chain analysis.

- Infrastructure – structure of the organization, style of management, vision, culture and processes.
- Human resource management – training, support, rewards, recruitment and assistance.
- Technology – research and development, new product developments, information technology and process improvements.
- Procurement – policy, purchasing, suppliers, contractors, consultants, EU procurement, e-tendering, processing.

The advantages of value chain analysis are that it:

- focuses information analysis on specific areas of the business;
- aids matching capabilities to current and potential strategies;
- helps to maintain an overview of capabilities.

Its disadvantages are that it:

- only reviews the internal operation;
- is not related to external environmental influences, for example alliances.

Portfolio analysis

Portfolios use a wider range of information sources to produce quality information for analysis. They can be applied to almost any situation and provide greater flexibility to compare internal and external factors. The three most common portfolios are the Boston Consulting Group, General Electric and Shell directional matrices. Other auditing tools include gap analysis, the profit impact of marketing strategy (PIMS) and the Ansoff matrix.

The Boston Consulting Group (BCG) matrix

The BCG matrix compares relative market share with market growth. By splitting the matrix into four sections, it is possible by evaluating data to place a product or service into categories (Figure 19.10).

- *Stars* are products or brands that are experiencing above-average growth and good market share. Usually they bring in good income but require a lot of cash to support activities such as advertising.
- *Cash cows* are the stable parts of the business. They generate good income for relatively lower costs. Cash cows are usually operating in maturing markets, requiring a company decide what it will do to replace the present cash cow or to protect itself in the future. Surplus cash might be invested in *question marks*.
- *Question marks* are new products in development, new market opportunities or brands. Research and

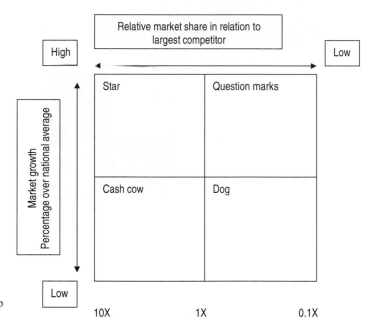

Figure 19.10   The Boston Consulting Group (BCG) matrix.

development costs are not covered by income. Here companies are nurturing ideas and developing markets.

- *Dogs* are products in decline. There is very little cash generation or cash used. Some companies keep one service going in order to achieve a strategic position.

On completing an exercise, strategies can be developed for each area. Nonetheless, the BCG matrix does have limitations.

The General Electric (GE) matrix
The GE matrix was developed to overcome some of the limitations of the BCG. Here the GE looks into a multiple number of factors in one matrix; using market/industry measures drawn from PEST, competitor and customer analysis, and comparing it with competitive/business strengths such as market share, distributors, suppliers and internal management and capabilities (Figure 19.11). Using a weighting system against selected criteria, it is possible to plot and identify possible strategies. Greater flexibility in the use of factors more relevant to the specific situation is one of the GE matrix benefits.

The Shell directional matrix
The only difference between GE and the Shell portfolios is the axis (Figure 19.12). The Shell directional axes are (1) prospects for profitability and (2) the enterprise's competitive capabilities.

Objectives

Objectives set the goals for a company or individual. They define the outcomes of the marketing strategy. It is important to set objectives for marketing and not the company as a whole. Marketing objectives could be market share size, growth rate, penetrating new markets and new product introduction. Essentially, objectives must be relevant to the department, and be precise and realistic. It is important that the objective is measurable to aid control further down the process.

When setting goals, it is important that these be objectives based on findings from the strategic analysis. An objective needs to be SMART:

- *Specific* – objectives need to be clearly definable, simple to understand and precise.
- *Measurable* – an objective needs a system to measure it by. If no measuring system is in place

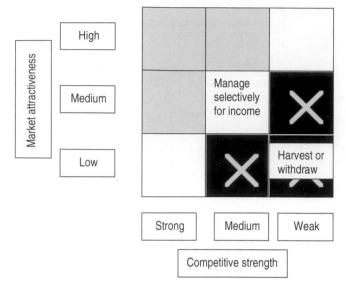

Figure 19.11   The General Electric matrix.

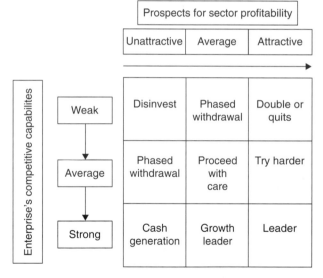

Figure 19.12   The Shell directional matrix.

then it is difficult to assess performance towards company goals.

- *Aspirational* – objectives should empower confidence in the company's ability to meet targets, whilst being challenging and realistic. They should be set at a level that is somewhat beyond the employees' own group ability.
- *Realistic* – objectives should be based on company ability to perform and implement plans. If objectives are too unrealistic they may have an adverse effect on staff.

- *Timely* – in line to meet targets and deadlines.

Objectives are normally expressed in time lines such as:

- Short term (6 months to 1 year)
- Medium (2 to 4 years)
- Long term in-line with vision; 5 to 7 years.

Objectives need to be flexible to the changing environment, acceptable to the individuals performing the task and comprehensive. Employees and managers must be able to understand and carry out tasks to meet targets.

A useful tool for reviewing and deciding on objectives is the balanced scorecard (BSC) approach (Figure 19.13). The BSC approach tries to link objectives and performance into an easily understood set of figures. Taking information from internal and external environments facilitates a balanced approach to objective setting. The four basic perspectives the method takes are:

(1) the customer's perspective;
(2) the internal perspective;
(3) the innovation and learning perspective;
(4) the financial perspective.

This approach helps the manager to appreciate wider objectives in the environment separately from financial objectives. This can help to form views based on quantitative and qualitative data.

Another good auditing tool for this exercise is gap analysis.

Strategies

At this stage most information has been gathered and managers are assessing which strategies they should adopt. By assessing past, present and future scenarios a marketing manager should form an impression of which markets are favourable and which are developing for later potential. All that is needed now is an appropriate strategy that meets marketing objectives. This stage considers the different options available, and prioritizes these to meet current objectives. The strategic section of the plan details the segments to be targeted and what position the company will occupy.

The direction a strategy takes depends on the objectives and therefore it is important to design a strategy based on all available information. The most important information includes that on competitive advantages, market factors and position.

Competitive advantage

A competitive advantage has to be relevant and defensible to the company. It can be asset or competence based. Value chain analysis is a useful tool to study potential sources of competitive advantage.

Porter (1988) identified three basic sources of competitive advantage (Figure 19.14):

(1) *Cost leadership*: strategic intent to pursue low margins in order to maximize returns and undercut the competition. Maintaining a cost leadership position requires all functional areas of an organization to reduce or keep costs to a minimum
(2) *Differentiation*: this tries to establish a difference, with the products or services offered focusing on specific characteristics and the promotion of these. For example, Volvo cars are not cheap, but they have carved out a competitive advantage through a reputation for safety and protection. Differentiation could be based on product performance, product perception (brand) and product augmentation (add-ons).
(3) *Focus*: focus activities and energies to a narrow audience. This can be defined as a geographic area or end-user focus or product line.

It is argued that these three strategies underpin a wide range of other strategies. The remaining deliberation is breadth of market. Does the company target a wide range of customers or operate in a niche?

Market factors

- Segments – size, identifiability, growth rate and profitability, as discussed earlier.
- Targets – definable groups.

On completing an understanding of the more favourable segments it is important to assess the potential target markets against a set of parameters, for example:

- sustainability of the target market;
- compatibility with corporate objectives;
- alignment with corporate values and beliefs;
- ability of the company to serve the target market;
- compatibility with internal operations;
- product/service compatibility.

With the use of auditing tools, decisions can be made regarding the most prosperous strategy. It is important to match the strategy with environmental concerns, company resources and value potential from the target market, now and into the future. Strategic examples include (1) push/pull for end user or distributors; (2)

	Strategic objectives	Strategic measurements
Financial	• Return on cash flow • Profitability • Cost of production • Ratio analysis	• Return on capital employed (ROCE) • Cash flow • Net margins • Profit forecast • Sales and future orders
Customer	• Perception by customers • Value for money • Satisfaction • Meet their needs	• Customer survey • Customer feedback • Mystery shopper • Buying habits • Usage survey
Internal	**Marketing** • Promotional plans • Customer acquitions • Market share **Manufacturing** • Production methods • New product developments • Lowering of manufacturing cost – lean manufacturing **Logistics** • Delivery costs • Time to market • Inventory management **Human resources** • Training • Staff moral **Quality** • Meeting quality management standards • Customer quality requirements	• Return on expenditure • Number of new customers • As percentage and comparing with previous figures • Efficiently: quicker, easier • Number of new products and success in the market place • Cost per item positive or negative • Unit cost of delivery • Cost of storage and management • Staff feedback • Staff retention • ISO9001 accreditation • Quality management standards
Innovation and learning	• Innovation • Research and development • Leadership skills	• Speed to market • Budgets, successes and breakthroughs • Number of new patents or licences

Figure 19.13   Balanced scorecard.

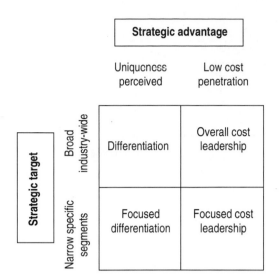

Figure 19.14   Three basic sources of competitive advantage (after Porter, 1998).

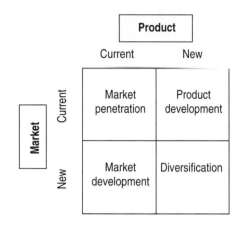

Figure 19.15   The Ansoff matrix (Ansoff, 1987).

defensive and offensive strategies; and (3) leader/follower/copier/low cost or niche market.

Models provide visual ideas or opportunities to pursue, and if company-wide skills meet the needs of a market or markets and the company has the resources to support the strategy then there is strategic fit.

In Figure 19.15, which depicts the Ansoff matrix, the boxes represent potential strategic choices. If the market and product are current then companies will need to increase market penetration or, alternatively, introduce new products into existing markets. Other possibilities include introducing current products into new markets, for example outside the current geographical area.

Other models used to produce information for analysis of market factors include the BCG, Shell and GE matrices.

Position

A company needs to decide on what basis it will compete. What position will it occupy in its customers' minds? To position a company it is vital to consider four points:

(1)   Credence – credibility in the customer's mind.
(2)   Competitiveness – Offers something others do not.
(3)   Consistency – of message from font to tome.
(4)   Clarity – specific message and image not foggy.

A useful framework is to build a perceptual map (Figure 19.16). This is a visual interpretation of the current market and provides an insight of where the competition sits and where new opportunities are, and helps to decide which strategies to pursue.

Positioning involves meeting and measuring customer perceptions. A company perception may differ from customer perception and it is a challenge of strategies to address these differences and align them correctly.

Tactics or the marketing mix

This section of the marketing strategy considers how the strategy will be implemented through key marketing platforms. These are known as the 7Ps: product, price, place, promotion, people, physical evidence and process. It is important to note that the marketing mix will mould the overall communications plan through the 4Ps – packaging, pricing, distribution (place) and promotion. Each element of the 4Ps requires objectives and strategies. Therefore, the final marketing strategy will include the manager's choice of strategy for the 4Ps.

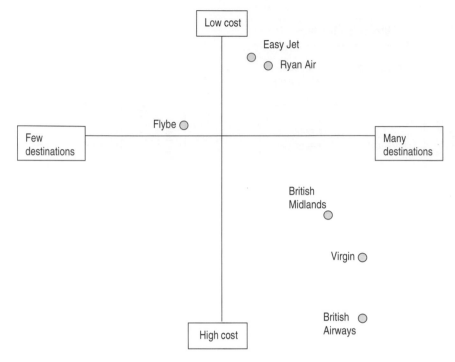

Figure 19.16   Product/service perceptual map.

Product/services

What products and services will be offered to meet customer needs? Size, variety, design, packaging and services are examples of the type of decisions that need to be made in order to successfully position a product at the right time in the right place for maximum exposure and purchase.

The difference between products and services are as follows:

- Products are tangible and available to be taken away. Products can be broken into core benefits to the customer, basic product, expected product, augmented product and potential product.
- Services are intangible, cannot be taken away from source and rely on heavy interaction and learning; they are inseparable and highly variable.

Price

Pricing strategies are complex but are required to pitch products correctly to the right customer, from distribu-

tion and warehousing to the end customer. For example, some companies enjoy knowing they can get big discounts, others prefer additional items included in the price, and yet others want flexibility in payment, all affecting the price.

Basic pricing objectives may include:

- Survival – faced with formidable competition, companies may set prices to cover variable costs and only some overhead costs. This is a short-term strategy because of the threat of losing too much money. It is usually applied to support a short-term promotion or to squeeze out competition or create a strategic advantage. For example, SKY decided to offer free set-top boxes in retaliation to ITV Digital.
- Maximum profit – maximum prices are set due to lack of competition or a premium brand and trying to match customer perceptions: for example, Rolex.
- Maximum revenue – prices are set according to variable demand at different times within the market. For example, plane seat prices are linked to certain times of the year. Going to New Zealand just

before Christmas may cost in excess of £1000, whilst in July the same plane ticket would cost £500.
- Maximum sales growth – low price set to penetrate a market: for example, Car Phone Warehouse free calls to other land-line users.
- Market skimming – used to launch new products.
- Product quality leadership – brand leader pricing strategies by charging more for premium services.

Pricing strategies (according to Kotler, 2003) may involve:

- prestige;
- psychology;
- penetration;
- marginality;
- skimming;
- differential;
- product line.

The pricing strategy must meet customers' perceptions, otherwise they will not purchase the product or service.

Place

Place is not just about where to buy. It includes everything from location, geography, logistics and support, right down to where a product can be found on supermarket shelving for maximum exposure.

Promotion

Promotion involves development of advertising, including e-media, which promotes use depending on the size of the company and target audience; different mediums and budgets are available.

A simple process of designing and planning a communication plan is SOSTAC. This stands for *situa*tional analysis, *o*bjectives, *s*trategy, *t*argets, *a*ction and *c*ontrol. Apart from the last two sections, this can be cribbed from strategic analysis. Action and control involve execution and controlling the execution through checks and balances, ensuring measurements are fed back through the system for further analysis.

Promotions cover all aspects of the communications plan and include brand strategies, which follow the same principles as the normal communication plan. The aim of the communications plan is to communicate with all customers, internal and external to the business,

ensuring the right customers know about the business, product or service. This can be accomplished through any number of channels, but it is important to be realistic. A small engineering firm would not embark on a television advertising campaign because most of its market will be clustered in either regional or diverse homogenous groups. Alternatively, a large multinational company selling consumer products to a national audience may decide it needs to use broadsheet advertisements to achieve a set of communication objectives.

Communications need to clearly deliver the correct message. This includes educating the public about good causes or informing customers that a service is still available when they need it, as with solicitors or action-orientated promotions that look at stimulating a positive reaction to 'do something now'.

People

This aspect of the marketing mix examines training, skill sets, knowledge building, attitude of staff and customers and quality of service. This includes the presentation of staff to customers and delivery of service or products.

Physical evidence

Evidence of physical presence, such as shops and offices, location of site and the image the physical assets present are other important marketing tactics.

Process

Process is an increasingly important aspect, particularly in the service sector. Traditionally 'buying' a holiday could take some time and may involve a range of providers (e.g. the hotel, air flight, excursions, car hire). Many businesses now attempt to cater for the many elements in a single, efficient process.

References

Ansoff, I. (1987) *Competitive Strategy*, revised edition. Penguin, Harmondsworth.

Bonoma, T.V. and Shapiro, B.P. (1983) *Organisational Models*. Prentice-Hall, Hemel Hempstead.

Dibb, S., Simpkin, L., Pride, W. and Ferrell, O.C. (1997) *Marketing: Concepts and Strategies, 4th edn*. Prentice-Hall, Hemel Hempstead.

Drummond, G. and Ensor, J. (2001) *Strategic Marketing: Planning and Control*, 2nd edn. Butterworth Heinemann, London.

Kotler, P. (2003) *Marketing Management*, 11th International edition. Prentice-Hall, Hemel Hempstead.

Porter, M.E. (1998) *Competitive Strategy; Techniques for Analysing Industries and Competitors*. Free Press, New York.

Further reading

Argyle, M. (1990) *The Psychology of Interpersonal Behaviour*. Penguin, Harmondsworth.

Buzzell, R.D. and Gale, B.T. (1987) *The PIMS Principle: Linking Strategy to Performance*. Free Press, New York.

Kotler, P., Armstrong, G., Saunders, J., and Wing, V. (2001) *Principles of Marketing*, 3rd European edition. Prentice-Hall, Hemel Hempstead.

Smith, C. (ed.) (2002) *Marketing: The Great British Brands, 1st August, p. 23*.

Useful websites

www.asa.org.uk – Advertising Standards Authority.
www.cim.co.uk – Chartered Institute of Marketing.
www.dma.org.uk – Direct Marketing Association.
www.ipr.org.uk – Institute of Public Relations.
www.mad.co.uk – marketing, media and advertising resource.
www.nrs.co.uk – National Readership Survey.
www.rab.org.uk – Radio Advertising Bureau.

20

Managing people

M.A.H. Stone

Introduction

Working in the countryside can often be messy and not just because of the mud! For many, there is no simple divide between employment in land-based industries and those that are just based in the countryside by choice or opportunity. Reasons for this include farmers taking second jobs or individuals running farms as a secondary business, so-called hobby farmers. Additionally, diversification of traditional rural industries into other ventures including tourism continues to accelerate. The countryside is the setting or context for organizations that range from the sole trader to the multinational or national public or voluntary body.

While it is important for all organizations to make the best use of their staff, effective management of people is essential for the large number of rural-based employers with small numbers of staff. This chapter is written with these organizations in mind. It is likely that owners and managers in these situations will not have the benefit of frequent practice of a wide range of people management skills. This includes managing people who may be widely dispersed, working from home or involved in more than one type of job. Therefore, this chapter aims to break down each process into stages, to define the terms used and to detail options.

There are three Ps that support any business – *people*, *product* and *profit*; all the three Ps need to be adequately addressed for the business to be successful. People are highly flexible but highly complex. Handling this resource will be dealt with under the following headings:

- Job analysis and job design.
- Recruitment and pay.
- Interviewing and discrimination.

- Control, guidance and negotiation.
- Training and development.
- Maintaining and increasing performance.
- Working patterns and environment.
- Glossary.
- References.
- Further reading.
- Useful websites.

Job analysis and job design

Systems and methods of work need to match both the goals of the business and the capacity of the workforce available. However, the use of labour is not based on rational economic decisions alone; there are many other influences. Social, cultural and legislative influences provide a frame of reference within which organizational decisions about labour use are made. The demand for labour is rarely constant, and some of the many causes of change in demand are:

- growth or decline in the business;
- seasonal fluctuations;
- cycles of production;
- change in demand;
- place in the product or service lifecycle;
- change in the product;
- change in the manpower itself;
- change in methods and equipment.

The purpose of a job is most commonly defined in functional terms:

- to do;
- to provide;
- to support;
- to be responsible for.

This is the first step, for until you have decided exactly what the job is designed to achieve, you will be unable to determine:

- how much labour you need;
- who will be best to do the job;
- how the job can be done best or even the best work pattern, e.g. full/part time.

This process is essential to ensure you are getting value for money from your staff. Each time the business requires additional labour or the workforce changes, the work to be done must be analysed and then the job/s designed or redesigned. This process helps to clearly determine what the organization wants the job holder to do.

The majority of job vacancies are those replacing leavers or for additional staff doing similar jobs. For this reason, the best source of information about what a job consists of is the person/s currently doing the job. The manager must carry out a *job study*, or in its more scientific form a *work study*, to determine working methods and required outputs.

Actively seeking the involvement of employees in the design, evolution and redesign of jobs is to be encouraged. The positive outcomes of worker involvement in job analysis and job design arise because the job holder:

- is close to the work;
- has evolved the process by learning from experience;
- can gain recognition for his/her work and skill

and because

- involvement is fostered and
- health and safety risks can be assessed.

However, there can be some problems because the job holder:

- may want to hide the ease of some of his/her work;
- may fear management motives regarding pay, grading or bonus payments.

The best results will be achieved in a climate of real dialogue between employee and employer.

Job information can also be obtained from leavers at an *exit interview*. The employer may discover frustrations, suggestions and details concerning employee

relations that will aid future job planning. This is best done after you have written the leaver's reference.

Job analysis and *work study* allow jobs to be broken down into the technical/non-technical functions required and for worker skills to be compared with the job. This allows managers to set standards, which can be built into a job description and person specification. A *job description* is a working document that details jobs, tasks, duties and responsibilities in a measurable way. A job description can be used:

- to supply information to candidates;
- as a reminder at the interview;
- as a guide for induction, training and appraisal.

A *person specification* can be thought of as a profile of the ideal candidate for the job defined in the job description. It should:

- detail the knowledge, skills and attitudes required;
- prioritize the criteria you will use to both select for interview and recruit.

Monitoring *labour turnover* allows the manager to check for trends, which might highlight problems. Some turnover cannot be avoided, and it is not desirable to try to avoid it, for new staff bring new ideas and approaches to a business. Turnover in jobs with little or no promotion or development prospects should be planned for. If labour turnover proves a problem, you will need to know if this is your business's problem alone or a local/industry problem. You can find answers or help with this from such places as:

- local employers/employer groups;
- Department for Work and Pensions/job centres;
- chambers of commerce;
- professional or industry groups.

Labour profiles help to highlight annual or seasonal changes in manpower requirements. They are also useful when considering a change in an aspect of the organization's workload or the introduction of a new one. A labour profile should also show casual labour requirements.

One element in *workforce planning* that can be overlooked is that managers or owners and their workforce grow old together. The problem for an heir or new manager is that many of the staff may have to be replaced in the same period, with the consequent loss of knowledge and skill.

Recruitment and pay

Recruitment

Employers need to make quick, accurate, relevant and legally defensible employment decisions. Managers must plan and organize to get and keep labour. Re-recruitment most commonly occurs because of poor preparation for or poor follow-up to recruitment.

Constant recruitment when the business is not growing can quickly have a negative impact on the credibility and efficiency of the organization. For example:

- Concentrating too much on recruitment can become a vicious circle by lowering the moral of experienced staff who may then leave.
- Output or quality can fall as initial training efforts reduce developmental training.
- Short-term coverage can become the norm, leading to tiredness, increased accidents, errors of skill and judgement, withdrawal of good will and strained working relationships.
- Management credibility falls and with it falls the ability to motivate and control.

We all frequently make initial judgements (first impressions) on the basis of little evidence and at great speed. To be objective in recruitment we must be aware of our bias and suspend judgement until we have sufficient relevant evidence to make a choice. The employer should concentrate on skills and behaviours because they are observable, describable and measurable. Past behaviour is the best guide to future behaviour that we have.

When measuring an applicant against a job description/person specification, attention should also be paid to the future. What skills and behaviours will aid the development of the organization? The relationship of behaviour to a job is illustrated in Figure 20.1.

Recruitment is a two-way selection process; potential employees also need information and the time to make a rational decision. The process can be viewed as attempting to deter the unsuitable as well as attracting the suitable. The advantage of supplying negative as well as positive information to potential applicants is that selection can be made on the basis of employee/employer choice without the sales pitch element of recruitment.

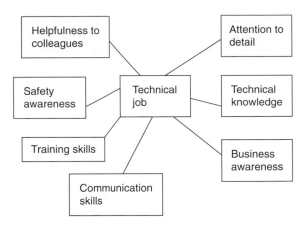

Figure 20.1   Aspects of a technical job.

As an employer, try to build into jobs those elements that will appeal to your target audience as well as serving the needs of the business; clear responsibilities in jobs for recent graduates, for example.

The owner's time is a commonly overlooked cost of managing an organization. If you do most jobs yourself, you can get very caught up duplicating the work best done by others. Use the skill and experience of job centres, newspapers and agencies, much of which is free: e.g. use job centre application forms rather than designing your own.

Advertising the job

The options include:

- job centres (national coverage);
- local papers;
- national papers;
- specialist journals (group or interest);
- agencies (recruitment and advertising);
- schools and colleges.

Payment and reward systems

The options include:

- payment for employees' time;
- payment for employees' output;
- payment for employees' service or loyalty.

The rules covering eligibility and the period or date of review for all reward packages should be fully detailed in both the offer of employment and the written terms and conditions of employment.

Time rates

This means payment for time worked, at a monetary rate per hour, including provision for extra or differing working hours. Some of the options for the length of the working period are:

- day, week, month;
- season;
- year;
- salaried positions;
- fixed hour contracts;
- flexi-time.

Production/performance rates

This refers to pay related wholly or in part to the achievement of production, sales, profit or other identified targets. For example:

- *piece work* is pay strictly limited to the quantity of output;
- *commission* is a method of payment that aims to give the employee motivation to sell, for each sale directly increases pay.

Payment based on performance or results relies on clear and effective measurement criteria. The manager must ensure that work is available in sufficient and measurable quantity to allow workers to earn a reasonable and consistent wage. The rules must be clear from the start and not require quick or frequent adjustment. The system must have sufficient flexibility to deal with changes in production and sales that may occur. Most organizations generate a mixture of productive and non-productive work. It can be hard to measure accurately and consistently the contribution to profit or success of all employees.

Service and loyalty rewards

Many businesses offer benefits in kind in addition to money, to reward or try to encourage loyalty and long service. For example:

- subsidized housing, child care or food;
- uniforms/clothing;
- loans;
- non-contributory pensions;
- training and development funding;
- cars/transport;
- professional membership fees.

The manager will need to measure the returns to the business of such rewards. These rewards are not of equal value to all employees or even the same employees over time; one option may be to offer your staff the choice from a range of benefits.

Establishing rates of pay and pay scales involves more than just establishing the going rate for a job; issues of power, personality, tradition and organizational culture are also involved. Pay levels are a way of establishing or reinforcing a hierarchy within an organization. As well as rewarding skill and experience, pay levels can signify levels of power and authority. Whatever pay system or method is chosen, compliance with the *national minimum* wage level and the *working time directive* must be ensured.

The opinion as to how fair a pay system is will depend upon the position of the individual within the system. For example, staff of differing ages offer differing things to an organization, but they may not fully understand the contribution of each other.

The effort reward bargain

The effort reward bargain is the personal perception of the exchange of effort, knowledge and time for the benefits of monetary and non-monetary reward. This perception will change over time and in comparison with the rewards of others.

Teams

Although many business goals are set and achieved through teams, managers must ensure that rewards to the individual have some relevance to individual performance.

Administration

Administration of pay and remuneration must be workable. Simplicity of measurement has meant that hourly pay is the most popular basis for reward. With this method, the control and measurement of output volume and quality will be a reflection of motivation and discipline. Before introducing a more complex form of payment, you must be sure the potential benefits of the new system outweigh the potential costs.

Interviewing

The interview is a setting in which issues of discrimination may be uppermost in the mind of the candidate. In preparing for any interview situation, the interviewer should consider potential discrimination issues.

Positive indicators to look for on application forms/CVs are that the applicant has:

- transferable skills;
- relevant experience – not just employment;
- learnt and developed;
- non-work interests;
- motivation to do this particular job;
- provided a full and clear account of his/her career.

The selection interview

The selection interview is a conversation with a purpose, the purpose being to gain evidence of each applicant's suitability for the job and to enable an objective assessment to be made against the person specification. Good interviews:

- allow the participants to get acquainted;
- help the organization make the best decision by collecting information in order to predict how successfully the individual will perform in the job;
- help the applicant make the right decision by providing full details of the job and organization;
- allow candidates to feel they have been given a fair hearing.

Preparing for the interview

Stress

A selection interview can put a lot of strain on the applicant; a potentially first-rate employee could be so nervous that you never discover how good he or she is. A good rule is that you should aim to stretch the candidates, but not to stress them.

Time

Give yourself enough time and tell interviewees how long you plan to keep them; before they come.

Interruptions

Try to minimize interruptions; they destroy concentration and do not allow for relaxation of interviewer or candidate.

The environment

If there is more than one applicant, make arrangements for their arrival, a place to wait, toilets and refreshments.

Documentation

You will need:

- person specification;
- job description;
- application form or CV;
- interview plan/prepared questions/notes;
- payment if travel expenses are to be paid;
- notes from any references taken up.

The WASP interview method and acronym

The acronym WASP stands for welcome, acquire, supply, plan and part.

Welcome

- Establish rapport by adopting a friendly approach and showing interest in the candidate.
- Greet the applicant, giving your name and position.
- Explain the purpose and structure of the interview.
- Tell the candidate you will be taking notes.

Acquire

- Establish applicant's experience, knowledge, skills and attitude.
- Ask questions using the 'funnel technique' (Fig. 20.3):
  - opening questions;
  - probing questions;
  - summarizing questions.

- Use a separate funnel of questions to deal with each new topic (see 'questions' below).

Supply

Tell applicant about:

- the job, using the job description;
- the conditions of employment;
- the organization;
- the benefits available;
- career prospects;
- the downside of the job as well as the highlights.

Plan and part

- Tell the applicant when a decision will be made.
- Tell applicants how they will be informed.
- Thank the applicant for attending the interview.
- Pass to second interview/tour of the enterprise/ arrange medical/testing.

Even if an applicant seems unsuitable, do not cut the interview abruptly short, but consider all applicants fully for the job. Likewise, even if you are impressed with an applicant, do not say anything that could be construed as a job offer. You may see a better applicant later, or a colleague may pick up an adverse point that you have missed. Employers should make it very clear to applicants at what stage a job offer is made, as acceptance of the offer forms part of a contract between applicant and employer.

Questions

Avoid long or multiple questions; encourage the applicant to speak by keeping your questions short and simple. You should always be encouraging in your manner, give the applicant's answers your full attention and make use of verbal and non-verbal prompts. Try not to react with your own opinions on specific subjects mentioned.

A behavioural approach to interviewing is based on gaining evidence of past performance behaviour as an indicator of future job performance. Investigating any of the areas shown in Figure 20.2 could provide evidence of relevant behaviours or skills.

Listen to the candidate and probe for more information where the body language signals do not seem consistent with the spoken message.

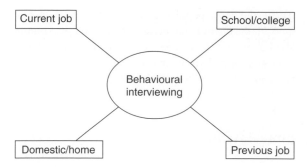

Figure 20.2   Key areas to probe with questions.

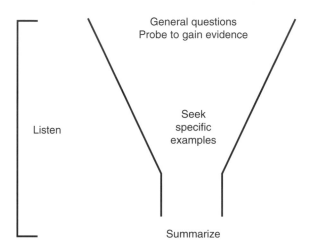

Figure 20.3   Question type funnel.

Figure 20.3 shows a question type funnel; this questioning process is equally applicable to other types of interview situations, e.g. counselling, disciplinary, grievance, appraisal and dismissal.

Examples of general opening questions

- What opportunities have you . . . ?
- Give me a recent example. . . .
- Describe a recent activity. . . .
- How did you go about . . . ?
- Tell me about the last occasion. . . .
- When have you been involved . . . ?
- What experience have you had . . . ?
- Describe a situation in which you. . . .
- To what extent . . . ?

Examples of follow-up probes and how to seek specific examples

- Describe specifically how. . . .
- What exactly did you do . . . ?
- Tell me more specifically. . . .
- How did you go about . . . ?
- What was your role . . . ?
- What were your reactions . . . ?
- Walk me through the process step by step. . . .
- How many times . . . ?
- Describe exactly what you did in detail. . . .
- How were you personally involved . . . ?
- Would you choose to do it this way again?
- What went wrong/what worked well?
- How did you assess your success?
- How would you approach a similar situation again?

Examples of how to summarize

- Am I right in saying . . . ?
- So you are saying . . . ?
- Is it fair to say . . . ?
- What you are saying is . . . ?

How to improve listening skills

Follow the acronym LISTEN:

- **L**ook interested.
- **I**nquire with questions.
- **S**tay on target.
- **T**est your understanding.
- **E**valuate the message.
- **N**eutralize your feelings; do not discriminate or allow personal opinions to influence your decision.

Listening is not talking; resist the temptation to launch into long discussions about your ideas or the company.

Areas of questioning you should not engage in during an interview

- Age.
- Race.
- Sexuality.
- Marital or relationship status.
- Dependants.
- Child care situation.

- Housing situation.
- Willingness to work different hours, unless this is a bona fide job requirement.

Minority/ethnic groups or females should not be asked questions not routinely put to white or male applicants.

Discrimination

In an employment context, discrimination is the application of unsound or irrelevant criteria to the process of selection, access and decision making. Equal opportunities form part of good management practice. It is a reality that half the workforce is female, that significant numbers of people are disabled and that Britain is a multi-cultural country. These people can and want to make a full contribution. Skills and behaviour qualities are not found only in specific geographical areas, among one sex or specific age or racial group; therefore exclusion on these grounds is therefore neither rational nor legal.

Everyone experiences discrimination at some time, but most experiences do not last for life; those of race, colour and sex, however, often do. These experiences will affect those discriminated against in their approach to getting a job. Discrimination in employment can persist even when members of these groups reach the interview stage due to the absence of cultural or social reference points; the lack of common experience between interviewer and interviewee.

Fair, legal and efficient recruitment processes are those that focus on the behaviour that will make an individual successful in your organization. This approach will enable you to deal positively with discrimination issues and assist the organization in making sound employment decisions about other groups, such as school leavers, the long-term unemployed and women returning to work.

Keep things simple; the more complex a recruitment system, the more chance that it will not be carried out in full or to the same standard every time.

Psychometric tests are devised by professionals and therefore involve costs. The values of the test must be compatible with those of the organization and its staff. Do not be tempted to use tests designed for another organization.

Remember that all applicants, their contacts (including their relatives and friends), along with current employees will form attitudes and expectations based upon the publicity and promises delivered in the recruitment process. Those you deal with will have an opinion as to how they have been treated and so act as either good or bad ambassadors for your organization.

Internal applicants should receive the same consideration and follow the same procedures as external candidates.

The Disability Discrimination Act

The Disability Discrimination Act establishes a framework of rights for disabled people covering employment and access to goods and services. The Act provides a statutory right of non-discrimination for disabled people. The Act applies to organizations employing 15 or more staff and covers both physical and mental impairments, which have substantial and adverse long-term effects on a person's ability to carry out normal day-to-day activities. This also applies to individuals suffering from HIV or AIDS.

The Act requires employers to make reasonable adjustments to the workplace to overcome the practical effects of disability. A new body, the National Disability Council, will monitor the effects of the legislation. Individuals who believe that they have been unlawfully discriminated against, within their jobs or when applying for a job, may take a claim to an employment tribunal (Crushway, 2003).

Control, guidance and negotiation

Identifiable responsibilities in conjunction with house rules and disciplinary/grievance procedures provide a clear framework for the management and guidance of staff behaviour throughout the employment relationship.

House rules

These are guidelines that should be written and displayed, that state the actions or behaviour demanded by the organization in set circumstances. Examples include:

* the use of telephones, equipment or vehicles for private use;

* arrangements for visitors;
* uniform and safety equipment policy.

House rules also signal the organization's attitude towards such things as:

* smoking;
* the protection of property and equipment;
* drinking, on or prior to duty, especially for those who work with machinery.

When organizations change or develop they can become vulnerable to dishonesty, and thus regular *audits* can help. Such an audit will scrutinize and review records and systems, in the same way that a financial audit checks the books. Periodic audits can reduce the temptation of staff, customers or visitors to break the rules through the knowledge that systems and records are checked. It is likely that as an organization grows, more financial and access controls will be required.

Disciplinary procedures

Discipline can seem very legalistic, but there is a clear distinction between what constitutes a breach of workplace discipline and what constitutes a breach of law. For example, to discipline a member of staff for removing equipment, the manager does not need to prove a worker intended to steal, but simply that equipment was taken off site without permission.

Employers should set out the responsibility they expect each of their staff to take. Some organizations will ask staff to sign these statements to signify their understanding and agreement. This approach is commonly found within larger established organizations. These organizations are not by nature less trusting, but have generally implemented these policies in the light of experience.

The purpose of a system of workplace discipline is not punishment but behaviour change. The legalistic way in which many of the disciplinary situations are described and the emotions that this generates often make the employee feel that punishment is at the root of the process.

Staff are not always willing to see the person disciplining them as someone reasonable, to whom they may have caused problems. Individuals who are the subject of workplace discipline commonly feel threatened and many react accordingly!

Establishing a system of discipline allows the employer to introduce order into what can be situations of conflict and high emotion. Procedures help to set the agenda and provide the manager with a tool to:

- calm;
- control;
- gain consensus;
- gain acceptance;
- reduce uncertainty.

Steps to ensure disciplinary procedures are seen as fair include:

- publicity – everyone needs to know the rules/system;
- detailing the procedures in the contract or terms and conditions of employment;
- agreeing procedures with workers or their representatives;
- following the 'best practice' advice of professional and advisory bodies such as ACAS;
- always following the agreed procedures;
- being seen to be consistent and impartial.

Remember: staff not involved in a disciplinary situation will be interested in the handling and outcome, due to curiosity and self interest.

Investigation

This must be your first action. You do not have a disciplinary situation on the basis of one person's statement. By starting the disciplinary process by the investigation of incidents or accusations you are immediately giving those concerned the chance to have their say (and let off steam). By not jumping to conclusions and taking the time to check the situation you are also allowing time for calming.

During an investigation you will need to:

- explain the process;
- talk to the relevant people in private;
- try to ignore pure opinion;
- make notes;
- be aware of bias and history;
- probe for the whole picture;
- *remember people do things for a reason.*

If the matter under consideration is serious and/or may take some time to investigate, you may wish to suspend the employee, on *full pay.*

Threats are often associated with disciplinary situations; they can come in the form of judgemental phrases, made on the spur of the moment. These do not help the orderly conduct of procedure, or the outcome. Do not endorse the threats or judgemental remarks of others and, if possible, follow these up and prevent repetition.

Do not commit yourself to anything at the outset except to investigate as an initial stage of the disciplinary process.

Ensure the person being disciplined understands why he or she is being disciplined. It is also worth stressing to those being disciplined that you will be spending time investigating the incident because you are interested in a positive outcome.

Time scales

Disciplinary issues should be resolved quickly. The uncertainty and stress of a long wait are unlikely to contribute positively to the organization. The only legitimate forms of delay are as follows:

- For the manager:
  - investigation;
  - to allow cooling off;
  - to allow time to consider your actions.
- For the subject of the disciplinary process:
  - to prepare evidence;
  - to find a supportive colleague to accompany him or her;
  - to consult with unions or other bodies.

Once a decision has been reached a line should be drawn under events. The only circumstances in which the event can be relevantly revisited are those of repetition and only then in private. As with criminal matters, once a sentence or fine has been paid the matter should be seen as closed.

Retrospective action

If a matter deserves investigation and/or implementation of disciplinary procedures, it must be fully dealt with at the time. If a blind eye is turned to a problem it is not legitimate to use the incident as evidence in a later disciplinary situation. If you ignore something, you can be perceived as condoning it. This will express to staff how seriously the issue is taken; ignored more than once and it becomes custom and practice.

Fundamental steps in handling disciplinary situations

- Investigation of events.
- Representation if wanted.
- Notice of meeting.
- Presentation of evidence.
- Fair hearing.
- Meeting, including note-taking.
- Pause for consideration.
- Consideration of any mitigation.
- Make decision.
- Clarify right of appeal.

Steps in a performance warning

- Explain what is being discussed.
- Re-state the standard required.
- Outline the shortfall.
- Discuss the target standard.
- Give a time span for improvement.
- Provide assistance.
- Impose warning or offer guidance.

Steps in a conduct warning

- Explain what is being discussed.
- Detail the incident of poor conduct.
- State the standard expected.
- Outline the consequences of repetition.
- Give a time span for improvement.
- Provide assistance.
- Impose warning or offer guidance.

Dismissal

This is the involuntary termination of the employment relationship, by the employer. Dismissal can be the culmination of a prolonged disciplinary process or immediate due to gross misconduct. The disciplinary process can start at any stage if the level of seriousness warrants this. The common disciplinary actions short of dismissal are:

- an initial verbal warning;
- a follow-up written warning;
- the final written warning.

Staff must be made aware of the actions that will result in disciplinary action and at what level. Failure to make this clear could result in complaints of constructive dismissal.

Employment tribunals

When the disciplinary system fails, the result may be a claim for unfair dismissal at an employment tribunal. The main concern of the tribunal is to establish:

- if there has been any breach of employment law;
- whether appropriate procedures were followed.

Some possible outcomes of tribunals are:

- out of tribunal settlements;
- compensation;
- re-engagement;
- re-instatement.

Ill health

Termination on health grounds can be a delicate issue, but failure to address the issue may prove costly, especially for small organizations. The common steps required prior to dismissal are:

- ask for medical checks;
- seek independent advice;
- keep the employee informed of the likely outcomes.

Retirement

It is sensible to plan for the changes in the level and type of input to the organization during an employee's working life; options close to retirement are:

- a stepped decline in effort/responsibility;
- gradual handover;
- consultancy.

The grievance interview

It is important to deal with grievances promptly to prevent them escalating. The objective is to:

- allow the employee to air the problem;
- discover the cause of the grievance;
- remove the problem/improve the situation.

When conducting the interview:

- let the employee state the case in his/her own way;
- distinguish between facts and feelings;
- listen, do not argue;
- get the employee to explore his/her own motives;
- ask the employee what he/she feels the solution is;
- do not commit yourself to anything;
- fix a date to come back to the employee.

Following up

Investigate the facts, obtain information and communicate by the agreed date. Check again later to see if the situation has improved.

Negotiation and bargaining

Industrial relations issues are closely tied up with power, control, conflict and personal interest. What helps to cloud industrial relations issues further is their linkage to politics and/or press coverage. The manager must stick closely to the issues for negotiation. Any help you can give your employees and their representative groups to do the same will be rewarded.

Once one party deviates from the substantive issues, both sides can quickly take refuge behind the posturing that has characterized many famous disputes. This also allows those who have no part in the negotiation, or knowledge of the issues, to become involved in looking for proof that one 'side' or the other is abusing their position or being irresponsible. Once you have lost sight of the substantive issues it is very hard to refocus the participants!

There are two types of negotiation and bargaining:

- Distributive bargaining: this is a fixed-sum game, which can be fine for simple issues. There are winners and losers and the situation can be characterized by suspicion and defensiveness.
- Integrative bargaining: this suits common problems that require a mutual solution. It is a positive-sum game; the outcome can be more for all, because co-operation can increase the size of the cake. It also takes longer and requires greater honesty.

Stages of negotiation and bargaining

Negotiation is rarely a one-off incident; there will always be some history to consider. Trust levels grow slowly and are best built up in low-stress situations.

Opening moves
These are commonly a statement of your position and a rejection of the other side's position. This is one extreme of your bargaining position, the other being the resistance point, beyond which you cannot go. For successful negotiation there needs to be some overlap in the bargaining range of each side.

Exploratory stage
Here you try to determine where your opponent stands. Frustration and time pressures are brought into play; your aim is to move your opponent's resistance point.

Consolidation
Here a change of concentration occurs, moving from divergent issues to convergent ones. Contentious issues are avoided and conflict is minimized.

Decision making
This starts when a compromise seems possible. Only then are concessions made to move from discussion to agreement. The goal changes from an ideal settlement to an acceptable one. It is sensible to help the other side accept the decision.

Negotiation has similarities to a game: it has rules, conventions, form, accepted moves and there is usually an audience who are passionate about the result. For example; offers once made are normally not withdrawn and moves are normally towards the other side's position.

High initial demands may help to:

- widen the bargaining range in your favour;
- change the other side's expectation of what is reasonable;
- lead your opponent to raise his or her estimate of your resistance point;
- give you scope to make big concessions.

However, you may lose credibility if your opponents refuse to negotiate or the final settlement is small compared to the demand.

If you play to win in the short run you may not achieve this in the long run.

Training and development

Spending time and money on the training and development of staff is easy. Managing the investment to get value for money is not. Before resources are allocated to the development of people you need to know the level from which you are starting; this is sometimes called a skills audit.

Skills audit

When looking for and measuring skill, the starting point should be goals and objectives to which the skills will be applied. In other words, what do you want your staff to achieve? The next step is breaking down the defined goals and activities of the organization, into measurable standards and objectives for each member of the team.

Once you have defined what the staff should be aiming for you can start to study the skills they will need. A skills audit could also look for unused or underused skills among the workforce; foreign language skills, for example.

Questions for a skills audit could be as follows:

- Are there defined standards of performance?
- Are the standards set at a correct level?
- Are these standards being met?
- Are competitors achieving the same results by more simple/cheaper methods?
- Are competitors achieving better results by the same/different means?

The skills audit will produce a comparison of the skills applied and the outcomes achieved. However, a shortfall in a standard of performance is not automatically a training need.

Training needs

A training need is an identifiable skill or body of knowledge, the learning or enhancement of which will generate positive, measurable benefits to the individual and/or the organization. The benefits of increased individual performance feed into company and personal career progression. Investing in people can be thought of as ensuring that the organization has a maintenance and development programme for its staff.

Some sources of information on training needs include:

- job analysis;
- appraisal;
- workplace assessment;
- changes in the nature of a job, equipment or markets;
- general communication.

As well as technical competence, a trainer needs to develop a working relationship with those to be trained and establish a climate in which learning can take place. Learning is the gradual extension of knowledge, skill and attitudes. Good trainers can be summed up by the acronym TRAINER. They are:

- **T**actful
- **R**espectful
- **A**ccurate
- **I**maginative
- **N**eat and tidy
- **E**nthusiastic
- **R**eliable

Employers should only pay for training that is achievable, relevant and when the need for the training is clear to the trainer and trainee. For example:

- to improve standards;
- to increase performance;
- to generate higher profit;
- in preparation for growth;
- to facilitate change;
- to motivate.

Standards of performance

A clear objective for any expenditure should have a quantifiable measure of success, in other words a standard. Training to a standard:

- helps to achieve continuity;
- lets staff know what is expected;
- allows trainers to plan and be effective;
- sets a measurable level of satisfactory performance;
- lets customers know what to expect, both internal and external customers;

- measures the standard so you know when to stop training;
- helps to focus competitive activity on the standard (not other staff).

Ideally, standards of performance should be written and made available to the people who need them, the staff as well as the manager.

Setting standards and the recording of training

For many organizations, standards of performance, training programmes and training records are all separate documents and systems. This need not be the case. The standards for the job can be both your training programme and record. A more detailed hard copy of standards may help trainers, but for the day-to-day monitoring and improvement of standards these are not required.

Establishing such systems provides a climate in which those who work for you can express their individual motivation to do a good job and improve their skills.

Workplace assessment

A true measure of job performance can often only be gauged on the job. The person best suited to assess job performance is someone who:

- knows what the standards are;
- knows what the obstacles to achievement are;
- knows, or can see, ways of improvement.

Workplace assessment is undertaken in order to:

- give feedback on job performance;
- help measure and build consistency;
- assess the impact of pressure and environment on standard delivery;
- identify true long-term strengths and weaknesses; of individuals, their equipment and their environment.

An environment of trust and a collaborative approach to organizational development is required for best results. If standards of performance are outlined from the start, the regular assessment of the performance of starters will form a natural part of the training and induction process. After establishing such an environment, workplace assessment can focus on the introduc-

tion or change of standards and the upkeep of existing standards.

Workplace assessment is one way in which a manager can show he or she is committed to getting it right. However, this approach will only gain acceptance if shortfalls in standards that are the responsibility of the organization or management are pursued with the same vigour as shortfalls by staff. For example:

- lack of equipment;
- lack of maintenance;
- lack of staff;
- poor planning;
- poor communication from above.

For workplace assessment to work best, it must be seen as a collaborative process. Remember, you are not there to conduct tests, but to facilitate staff achievement. During workplace assessment, new or amended duties can be agreed, progress can be monitored and praise given. Problems or barriers to progress can be approached jointly.

Training should be carried out prior to the need for the skill. The work method can then be assessed in action to determine the need for:

- additional training;
- further clarification of the standard;
- amendment of the standard.

This is a cyclical process.

Trainers also benefit from training because it:

- improves their handling of and confidence with people;
- makes the job easier;
- supports staff, while not doing their job for them;
- develops trust and respect;
- is part of managing instead of doing.

When offering training or development, be systematic; give trainees a structure to their learning, such as the model shown in Figure 20.4.

Many factors may potentially limit learning and the acquisition of skill, including:

- the situation and attitude of the person to be trained;
- their preferred learning style;
- the training techniques and environment;
- current skill limitations;
- lack of confidence;

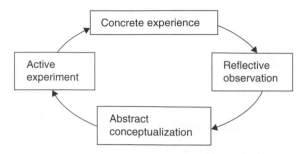

Figure 20.4 The Kolb (1984) cycle of learning and development.

- fear of the unknown;
- personal inhibitions;
- lack of practice opportunities;
- peer pressure;
- dynamics of the training group;
- loss of concentration (after 20 minutes);
- motivation; people only learn what they want to learn!

Aids to learning

- Use visual aids/props.
- Use more than one sense.
- Build from the known to the unknown.
- Show relevance to the work situation.
- Provide constructive feedback.
- Use clear objectives.
- Build on success.

Induction

Induction is a programme aiming to deliver the smooth and effective introduction of people into an established working environment. Break these programmes down according to what the employee needs to know by when. Use a variety of methods to allow a new employee to take on board potentially vast amounts of new information.

Consideration will need to be given to different groups, including:

- school leavers;
- long-term unemployed;
- graduates;
- women returning to work after a career break.

Maintaining and increasing performance

Appraisal

An appraisal is a meeting between an individual and his or her immediate manager to discuss that individual's job. The purpose is to maintain, guide and develop job performance. Past performance should not be looked at in a judgemental way, but reviewed in order to aid the process of work planning and goal setting for the period ahead. More sophisticated systems make provision for appraisees to carry out some self-assessment and provide comment and guidance to those that manage them (self and upward appraisal).

Appraisal benefits the manager by:

- allowing time to plan the work load and its allocation within the team for the period ahead;
- giving the manager feedback about his or her own performance;
- helping to foster a joint approach to work;
- providing a recorded plan.

Appraisal benefits staff because:

- career planning and development can be discussed;
- feedback about performance can be gained;
- future work and priorities can be clarified;
- it provides a recorded plan.

Job chats are an alternative or supplement to the appraisal. Shorter and less formal than an appraisal, they are still a forum for the discussion of job issues; past, present and future.

The major threats to positive appraisal are:

- lack of trust;
- lack of understanding/knowledge of the job;
- lack of time;
- being too closely linked to pay.

When appraisal is too closely linked to pay, individuals will concentrate on the assessment of past performance, negating the appraisal as a forward-looking collaborative exercise.

Activity plans

Typically, activity plans are written plans for activities covering a fixed period, including details of success criteria, support required and an interim review date. The plan should be reviewed at the conclusion of the period and a new one agreed. Plans run for months or weeks and cover normal areas of work and development as well as specific projects or objectives.

Management by objective

This management tool aims to break down the organization's goals into sub-goals and, in turn, individual goals. The starting point and concentration is the goal and not the method. This opens up the possibility of more individuality or creativity of method.

Feedback

Feedback is a powerful aid in learning about ourselves and the impact we have on others. To make this process constructive requires skill and practice on the part of the giver and recipient of feedback. The recipient needs something to build upon, positive or negative, and this information must be both understandable and usable. We often fail to give positive feedback because we forget, we are embarrassed or we are worried that someone may become too conceited.

Negative feedback is also often limited by the fear that the recipient may get upset, and we are not ready or willing to deal with this. We may be worried that it will affect our relationship in a lasting way; that the employee will not really understand what we are saying, or he/she will distort what we say. Finally, we may say nothing because we do not believe it will have any effect.

Although not always a good idea, giving feedback does provide an opportunity for change. The skills of giving feedback are as follows:

- Be clear about what you want to say in advance.
- Start with the positive.
- Be specific.
- Select priority areas.
- Focus on the behaviour not the person.
- Refer to behaviour that can be changed.
- Offer alternatives.
- Be descriptive rather than evaluative.
- Own the feedback (this is what I think).
- Leave the recipient with choices.
- Think what the feedback says about you.
- Give the feedback as soon as you can after the event.

When receiving feedback:

- listen to the feedback rather than immediately rejecting or arguing with it;
- be clear about what is being said;
- check with others rather than relying on only one source;
- ask for the feedback you want but do not get;
- decide what you will do as a result of the feedback.

Coaching

Coaching is a method of developing and/or correcting skills, attitudes, behaviours and techniques. With this approach the trainer, manager or mentor must have a basic agenda/outline of the subjects to be covered by the trainee.

Coaching uses situations that occur in the normal working day as triggers for training. One way to think of coaching is the use of mini case studies. In this way principles and applications can be looked at together and in a context that the trainee is more likely to understand.

It is important that the trainee understands the technique of coaching and is aware of when it is being practised, especially early in the coaching relationship, so as he or she can take advantage of the learning situations and not feel patronised.

Mentoring

Mentoring can work in the same way as coaching; however, the initiator of training or the coverage of issues is more likely to be the trainee. For best results the mentee must be involved in the choice of mentor. The mentor needs to be able to put him or herself in the place of the trainee. The biggest expert is not always the best trainer. The mentor must have an understanding of the mentee's personal goals and the need to make developmental activity personally relevant.

When looking at a scheme of mentoring you must decide what the basis of the relationship will be; some models are shown below.

Description	Application
Patron–Protégé	Commercial
Expert–Trainee	Training
Role model–Junior	Hierarchy
Confidant–Colleague	Partnership
Educator–Disciple	Developmental

Poor standards

Poor standards or the failure to fulfil potential may be the result of:

- poor initial training;
- forgotten guidance or instruction;
- cutting corners;
- faulty/inadequate equipment;
- lack of interest;
- personal problems;
- personality clashes.

An acronym of the steps to corrective action is OSCAR:

- **O**bserve.
- **S**tep in.
- **C**alm use of interpersonal skills.
- **A**sk questions.
- **R**epeat training.

Incentives

Considerations for the manager include:

- What are the aims of these incentives?
- Why are incentives needed over and above existing payment/remuneration policies?
- What happens when the scheme ends?
- What happens if the incentive does not work?

There are numerous circumstances that may cause incentive schemes to fail, many of them outside the control of an organization.

Steps for incentive introduction include:

- planning;
- consultation;
- explaining to staff why an incentive is needed and how the organization and the individual are going to benefit;
- trial;
- time to test the theory and assumptions;
- review.

The goals of an incentive need to be precise. Vague goals, such as to increase production or productivity, are the first step to failure. You need to know:

- how much;
- to what quality;
- for how long;
- for what reward.

Fairness

An incentive scheme is a deal between the organization and its staff. The required effort must deliver the promised benefits, for both individuals and the organization. It is rarely possible to apply the same scheme accurately to staff on different jobs. This is where life gets complicated. There are two main choices:

(1) development of additional criteria to allow more groups of staff to participate, while maintaining a benefit for the organization;
(2) exclusion of these staff from the scheme.

Teams

Incentives work best, administratively and practically, when applied to the individual because the motivation of a team is not the same as individual motivation. The evolution of team motivation is subject to group dynamics. The existing working relationships will need to be considered when designing incentives.

Quality and quantity

Incentives which concentrate on increasing output quantity require safeguards for quality. These can be contentious issues, especially if quality problems result in scheme changes that lower staff income.

The desire to be flexible and to take time out for training and peripheral work can be reduced by incentives. What gets measured and paid for gets done. Short-term income generation can become the opportunity for the worker, while longer-term concerns become problems for the manager. All rewards should be received as close to the effort that earned it as possible.

Working patterns and environment

Volunteers

The countryside is also the setting for significant levels of voluntary work as well as employment opportunities working for voluntary bodies. When recruiting and managing volunteers it is important to understand what their motivations are and what their level of commitment will be. It is likely that their motivation and commitment will be different to that of some employees. It is also important to recognize that while volunteers may have different roles within the organization to those of your employees, they are still likely to be working with employees and have many of the same rights as employees. Employees working with volunteers will benefit from training on issues specific to volunteers in your particular work context. Likewise, managers should seek advice on handling the interrelationship between staff and volunteer.

Working with family

Many rural organizations are also family concerns. Managing, working with or for members of your family can have many advantages; however, good people management practices and procedures are still needed. For example, it is essential that working conditions, pay and benefits packages, management decision making and ownership structures are both clear and documented. This will prevent future problems between family members and difficulties of equal treatment with non-family members employed by the organization.

Working from home

Many of those working in the countryside will be living on or close to the site of their work. This involves a range of people management issues including making sure that the standard of health and safety management and protection is equal to that which employees away from the home could expect. This includes ensuring that staff at all levels in the organization are able to maintain a healthy balance between their work and their private life, commonly known as the home/work balance.

Teleworking

The Department of Trade and Industry (DTI) has issued guidelines to businesses employing 'teleworkers'. Around two million workers are engaged in such work patterns, a large number of whom live in the countryside (DTI, 2003). This guidance is intended as a useful checklist of issues to consider when implementing teleworking. Management can use this guide to draw up company-specific policies on teleworking and to ensure that telework is introduced in such a way as to benefit both employers and employees.

The Data Protection Act

The Data Protection Act refers to data subjects – these are individuals about whom an organization holds or processes data. This includes everyone, not just employees. Although the Act deals mostly with data held electronically, it also allows for employee access to some manual systems maintained for a specific purpose, for example sickness records.

The 1998 Act extended the coverage of data protection to include any information stored about individuals in a 'relevant filing system'; a relevant filing system being one that is structured, either by reference to individuals or by reference to criteria relating to individuals, and thereby readily accessible. These include:

- databases;
- computer systems, email and CCTV;
- some manual record systems.

The data stored must be relevant, accurate and up to date, so out-of-date information should be identified and destroyed, unless there are other reasons for keeping it that can be justified. Justifiable reasons include the following:

- The data subject gives his/her consent.
- Processing the data is necessary to fulfil a contract to which the data subject is party, i.e. a contract of employment.
- Processing the data is necessary for legal obligations, e.g., taxation records.

If the employer uses data about employees to provide statistical information, employees should be told how the information held will be used. Employees' consent must be sought regarding sensitive personal data such as:

- racial or ethnic origins;
- political opinions;
- religious or other beliefs;
- membership of trade unions;

- physical or mental health or condition;
- sexual habits or orientation;
- actual or alleged offences and legal proceedings or sentences.

The onus is on the employer to tell employees what is kept, where, for what purpose and who may have access to it. It is good practice to:

- inform all employees what data are kept, the purposes for which the data are used and who may have access to the data;
- ask all employees to give consent to those data uses;
- help employees to keep their personal details up to date by providing them with a copy annually to check and amend.

Similar systems should be considered for external data subjects.

Individuals can access their data by making a request (usually in writing) and paying a fee if applicable of not more than £10. The business has 40 days to respond. Exceptions to the right of access include when the data are held/produced for management forecasting or planning and when release of the data would be likely to prejudice the conduct of the business. Examples may include manpower planning, and succession or career planning (Crushway, 2003).

Guidance on the provisions, exemptions and applicability of the Data Protection Act can be obtained from the Office of the Data Protection Commissioner, Wycliffe House, Water Lane, Wilmslow, Cheshire, SK9 5AF, or www.dataprotection.gov.uk. This should be the first step before setting up or amending a system.

The internet

It is increasingly common for organizations to have internet access within the workplace. Like any other business tool/practice it is sensible for the business to make it clear to staff how it expects staff to use this technology. Key issues include:

- the image of the business portrayed to outsiders;
- copyright and disclaimers;
- harassment and defamation;
- privacy, confidentiality and monitoring;
- private usage;
- the status of email compared to paper communication;

- data protection;
- viruses and system security/protection.

Health and safety at work

This vital area of managing people should be a central consideration of job design, work planning, recruitment, training and the development of staff. It is difficult to provide overall guidelines for the vast amounts of health and safety legislation. The interpretation and implications can be heavily dependent on the industrial sector concerned. However, there are excellent sources of advice for the manager, including local authorities, industrial bodies and the Health and Safety Executive (HSE). A well as providing advice and guidance, the HSE also produces excellent, clear and focused guides to specific legislation that makes the job of creating organizational policy and carrying out training a great deal easier.

When thinking about health and safety it is important to remember that the responsibility of the organization extends beyond your staff, to include your customers, clients, visitors to your premises and any contractors working under your instruction. The self-employed also have both rights and responsibilities for health and safety.

Fire regulations

Fire regulations require:

- an established means of raising the alarm;
- an evacuation procedure;
- safe and protected means of escape;
- an assembly area(s);
- records of training, equipment maintenance, fires and false alarms.

First aid

Legal and effective coverage for first aid requires more than just the correct ratio of first aiders to staff, supported by the correct level and maintenance of equipment. Employers need to consider the geographic distribution of staff and ensure that a first aider is available throughout the working day and week.

Glossary

Detailed below are working definitions for many of the terms used in the text; they are defined with reference to the context within which they are used in the text.

Activity plans: written plans for activities covering a fixed period, including details of success criteria, support required and review dates.

Appraisal: a meeting between an individual and his/her immediate manager to discuss the job, the purpose being to maintain, guide and develop job performance.

Behavioural interviewing: an approach to interviewing based on gaining evidence of past performance behaviour as an indicator of future job performance, regardless of background, the behaviours required being derived from the job description and person specification.

Coaching: a method the manager can use to develop and/or correct the skills, attitudes, behaviours and techniques of subordinates. The manager uses situations that occur in the normal working day as mini case studies in order to carry out training and development. Principles and applications are looked at together and in a context that the trainee is likely to understand.

Constructive dismissal: the situation when an employee leaves an organization, not because he or she wanted to, but because he/she felt it was the only course open to him/her; e.g. as a result of fear or frustration.

Discrimination: the application of unsound or irrelevant criteria to the process of selection, access and decision making.

Distributive bargaining: a negotiating situation in which the parties/sides seek to determine the allocation of a fixed quantity of resources, a fixed-sum game.

Equal opportunities: a situation in which there are no discriminatory barriers to prevent employees or applicants from making a full contribution to the organization.

Exit interviews: where the manager interviews an employee before he or she leaves so as to gain feedback on the performance of the job and organization and to ensure that positive suggestions based on experience are not lost to the organization.

Feedback: a method by which we learn about ourselves and the impact we have on others. Feedback provides individuals with an opportunity for change.

Flexitime and fixed hour contracts: systems designed to allow the manager to better balance the work to be done and the labour available; be that in the short term with flexitime or over the longer term with fixed hour contracts.

Induction: a programme to deliver the smooth and effective introduction of people into an established working environment.

Integrative bargaining: a negotiating situation in which both parties to the negotiation can gain through mutual co-operation; a positive-sum game.

Job analysis: where the manager assesses the work to be done, breaks this work down into the technical/non-technical duties required and sets standards which can be built into a job description and person specification.

Job chats: a discussion of issues arising from a job, past, present and future. Generally, job chats are shorter and less formal than an appraisal. They can be a useful alternative or supplement to the appraisal.

Job descriptions: a working document which details jobs tasks, duties and responsibilities in a measurable way. This will aid recruitment, training and planning.

Job design: the process by which systems and methods of work are matched to both the goals of the business and the capacity of the workforce available.

Labour profiles: a graphical representation of the labour required. They allow the manager to build up a picture of the labour needed in each time period by adding together the labour input needed for each task in that period.

Labour turnover: a comparison of the number of staff leaving over a set period with the total workforce, normally expressed as a percentage.

Management by objective: a management practice that works by breaking down organizational goals into team and, in turn, individual goals. The starting point and concentration is the goal and not the method.

Mentoring: a system in which the mentee or trainee can trigger training relevant to his or her work needs through access to a mentor (more experienced colleague). The mentor provides relevant development by being able to put him or herself in the place of the trainee and understand the mentee's personal goals.

Person specification: a written record of the ideal candidate and basis for any advertising, against which the manager can compare applicants.

Psychometric testing: a system by which job or promotion candidates are compared with a psychological profile of the ideal applicant, as defined by the organization and the test designer.

Teleworkers: staff that spend all or part of their time working from home and carry out this work using information technology.

Workplace assessment: the method by which an assessment of an employee's competence can be made in the workplace in relation to the standards defined for the job.

Work study: a formalized way of measuring the work that makes up a job; this can involve defining approved methods and allocating times for the completion of tasks.

References

Crushway, B. (exec. ed.) (2003) *Essential Facts, Employment*. Gee-Professional Publishing, London (CD-Rom).

Department of Trade and Industry (2003) *Teleworking*. Stationery Office, London.

Kolb, D. (1984) *Experiential Learning: Experience as the Source of Learning and Development*. Prentice Hall, New Jersey.

Further reading

Agere, S. and Joru, J. (eds) (2000) *Designing Performance Appraisals*. Commonwealth Secretariat, London.

Armstrong, M. and Barow, A. (1998) *Performance Management*. Institute of Personnel and Development, London.

Bell, A. and Smith, D. (1999) *Management Communication*. Wiley, Chichester.

Chapman, N. (2000) *Leadership: Essential Steps Every Manager Needs to Know*, 3rd edition. Prentice Hall, Hemel Hempstead.

Clarke, L. (1995) *Discrimination*, 2nd edition. Institute of Personnel and Development, London.

Cook, M. (1998) *Personnel Selection: Adding Value Through People*. Wiley, Chichester.

Denton, J. (1998) *Organisational Learning Effectiveness*. Routledge, London.

Dew, J. (1998) *Managing in a Team Environment*. Quorum, London.

Dickson, A. (2000) *Women at Work*. Kogan Page, London.

Fowler, A. (1996) *Negotiation Skills and Strategies*, 2nd edition. Institute of Personnel and Development, London.

Freemantle, D. (1995) *80 Things You Must Do to be a Great Boss*. McGraw Hill, Maidenhead.

Handy, C. (1990) *Understanding Voluntary Organisations*. Penguin, Harmondsworth.

Health and Safety Executive (1997) *Managing Health and Safety: An Open Learning Workbook*. HSE London.

Holbeche, L. (2001) *Aligning Human Resources and Business Strategy*. Butterworth Heinemann, Oxford.

Honey, P. (1992) *Problem People and How to Manage Them*. Institute of Personnel and Development, London.

Honey, P. (1997) *Improve Your People Skills*, 2nd edition. Institute of Personnel and Development, London.

Hyman, J. and Mason, B. (1995) *Managing Employee Involvement and Participation*. Sage, Sevenoaks.

Jackson, P. (1999) *Virtual Working: Social and Organisational Dynamics*. Routledge, London.

Klatt, B. (1998) *The Ultimate Training Workshop Handbook*. McGraw Hill, Maidenhead.

Kotter, J. (1996) *Leading Change*. McGraw Hill, Maidenhead.

Lock, D. (ed.) (1998) *The Gower Handbook of Management*, 4th edition. Gower, Aldershot.

Mill, C. (2000) *Managing for the First Time*. Institute of Personnel and Development, London.

Morris, S. (1993) *Discipline, Grievance and Dismissal: A Manager's Pocket Guide*. Industrial Society, London.

Osbourne, D. (1996) *Staff Training and Assessment*. Cassell, London.

Parker, S. (1998) *Job and Work Design*. Sage, Sevenoaks.

Parslow, E. (1999) *The Manager as a Coach and Mentor*, 2nd edition. Institute of Personnel and Development, London.

Pearce, J. (1993) *Volunteers – The Organisational Behaviour of Unpaid Workers*. Routledge, London.

Pettinger, R. (1997) *Managing the Flexible Workforce*. Cassell, London.

Salaman, M. (1998) *Industrial Relations: Theory and Practice*, 3rd edition. Prentice Hall, Hemel Hempstead.

Sargent, A. (1990) *Turning people on (the Motivation Challenge)*. Institute of Personnel Management, London.

Smith, D. (1998) *Developing People and Organisations*. Kogan Page, London.

Stredwick, J. (2002) *Managing People in a Small Business*. Kogan Page, London.

Tyson, S. and York, A. (2000) *Essentials of Human Resource Management.* Butterworth Heinemann, Oxford.

Vroom, V.H. and Deci, E.L. (1992) *Management and Motivation*, 2nd edition. Penguin, Harmondsworth.

Useful websites

www.acas.gov.uk – Arbitration Conciliation and Advisory Service (ACAS).

www.cipd.org.uk – Chartered Institute of Personnel Management (CIPD), Codes of Practice and People Management Journal.

www.dti.gov.uk – Department of Trade and Industry (DTI).

www.hse.gov.uk – Health and Safety Executive (HSE).

Index

Notes: The abbreviation CAP refers to the Common Agricultural Policy; page numbers in *italics* refer to tables; page numbers in **bold** refer to figures.